22nd Los Alamos Stellar Pulsation Conference Series Meeting 2016

Wide-Field Variability Surveys: A 21st Century Perspective

EPJ Web of Conferences Volume 152 (2017)

San Pedro de Atacama, Chile
28 November – 2 December 2016

Editors:

Marcio Catelan
Wolfgang Gieren

ISBN: 978-1-5108-4743-9

Printed from e-media with permission by:

Curran Associates, Inc.
57 Morehouse Lane
Red Hook, NY 12571

Printed by Curran Associates, Inc. (2017)

For additional information, please contact EDP Sciences – Web of Conferences
at the address below.

EDP Sciences – Web of Conferences
17, Avenue du Hoggar
Parc d'Activité de Courtabœuf
BP 112
F-91944 Les Ulis Cedex A
France

Phone: +33 (0) 1 69 18 75 75
Fax: +33 (0) 1 69 28 84 91

contact-edps@webofconferences.org

Additional copies of this publication are available from:

Curran Associates, Inc.
57 Morehouse Lane
Red Hook, NY 12571 USA
Phone: 845-758-0400
Fax: 845-758-2633
Email: curran@proceedings.com
Web: www.proceedings.com

TABLE OF CONTENTS

edp sciences

EPJ Web of Conferences

Statement of Peer review

In submitting conference proceedings to EPJ Web of Conferences, I certify to the Publisher that I adhere to the **Policy on Publishing Integrity** of the journal in order to safeguard good scientific practice in publishing.

1. All articles have been subjected to peer review administered by the proceedings editors.
2. Reviews have been conducted by expert referees, who have been requested to provide unbiased and constructive comments aimed, whenever possible, at improving the work.
3. Proceedings editors have taken all reasonable steps to ensure the quality of the materials they publish and their decision to accept or reject a paper for publication has been based only on the merits of the work and the relevance to the journal.

Title, date and place of the conference

Wide-field variability surveus: A 21st-century perspective
Nov. 28 - Dec. 2, 2016
San Pedro de Atacama, Chile

Proceedings editor(s):

Márcio Catelan and Wolfgang Gieren

Date and editor's signature

August 2, 2017

Preface

Márcio Catelan[1,2,*] and *Wolfgang* Gieren[2,3,**]

[1] *Pontificia Universidad Católica de Chile, Instituto de Astrofísica, Av. Vicuña Mackenna 4860, 782-0436 Macul, Santiago, Chile*
[2] *Millennium Institute of Astrophysics, Santiago, Chile*
[3] *Universidad de Concepción, Departamento de Astronomía, Casilla 160-C, Concepción, Chile*

Abstract. The 22nd edition of the Los Alamos Stellar Pulsation Conference Series was held in San Pedro de Atacama, northern Chile, between Nov. 28 and Dec. 2, 2016. Here is a short description of the event.

Between Nov. 28 and Dec. 2, 2016, about 140 astronomers from different parts of the world came to San Pedro de Atacama, northern Chile, to celebrate the 22nd edition of the Los Alamos Stellar Pulsation Conference Series. This was a very special occasion for the organizers, as it marked the very first time a meeting of the series took place in Latin America, and only the third time the event was held in the southern hemisphere.

On this occasion, we gathered together at Hotel Cumbres, in the outskirts of San Pedro, to discuss the subject of *Wide-field variability surveys: A 21st-century perspective*.

Astronomy is undergoing a major revolution in the way it approaches its core subject. With the increase in detector efficiency and size and the availability of ever more powerful computers, wide-field surveys, once few in number, have become increasingly common – and indeed, it is clear that the field will be largely dominated by massive wide-field surveys, photometric and spectroscopic alike, in the not-too-distant future. Many of these surveys are being, and will continue to be, conducted from Chile, not far from the chosen venue.

At this special transition moment between "classical" and survey-based astronomy, we are faced with important challenges on how to properly handle, process and analyze the increasingly large datasets that are continuously amassed by these surveys. Astronomers must increasingly work together with computer scientists and statisticians. Entirely new fields, such as the booming but fairly young field of Astrostatistics, are emerging in the process.

The challenge is particularly evident in the case of time-domain astronomy, where the databases needed can be larger by orders of magnitude, compared with the case of single-epoch surveys. Microlensing surveys such as MACHO, EROS, and OGLE have revealed to the community the enormous power of time-resolved surveys to study a vast number of different science topics, especially in the more general context of stellar variability. In this sense, pulsating variable stars, which were the main focus of this conference, come high at the top of the list.

Perhaps surprisingly, the last meeting that was aimed at discussing the impact of large-scale surveys on pulsating star research was held back in 1999, in Budapest, Hungary (IAU Colloq. 176,

*mcatelan@astro.puc.cl
**wgieren@astro-udec.cl

which was also the 14th edition of the Los Alamos Stellar Pulsation Conference Series). That meeting did help set the stage for the current state of affairs, but back in those days there were still few such massive surveys that were in operation – and those look rather small, in comparison with current and future surveys.

As a case in point, consider the OGLE project. Back in 1999, OGLE-II was taking data using a single 2k × 2k CCD. At present, OGLE-IV is scanning the southern skies from Las Campanas Observatory, again in northern Chile, using an array of thirty-two 2k × 4k CCDs, covering a much wider area of the sky, and with data flows that are higher by orders of magnitude than in the early phases of the OGLE project. In the near infrared, no time-resolved surveys were in operation back in 1999. Massive, wide-field variability surveys in the near-IR covering vast swathes of the sky are now being conducted for the first time in history. This includes the Vista Variables in the Vía Láctea (VVV) survey and the Vista survey of the Magellanic Cloud system (VMC), both of which are ESO Public Surveys running on ESO's Visible and Infrared Visible Telescope for Astronomy (VISTA), located in Cerro Paranal, northern Chile. Particularly revealing, in terms of the rise of the survey era, is the fact that ESO has built not just one, but actually two, state-of-the-art telescopes in Chile that are entirely dedicated to conducting surveys – the other one being the VLT Survey Telescope (VST). Wide-field spectroscopic survey facilities are also rising in prominence. The Large Sky Area Multi-Object Fiber Spectroscopic Telescope (LAMOST), located in Xinglong Station, Hebei Province, China, with its unique design, is a remarkable example. Among its achievements, LAMOST's key role in the spectroscopic follow-up of variable sources monitored in the *Kepler* space satellite fields is particularly noteworthy.

While LAMOST is now proving key in the follow-up of *Kepler* sources, back in 1999 space-based astronomy did not count on any instruments devoted to time-resolved astronomical observations even remotely approaching the power of CoRoT and *Kepler*, which have provided a major recent boost to the field of asteroseismology in particular. In fact, back in 1999, the field of extrasolar planet research (CoRoT and *Kepler*'s main *raison-d'être*, to be sure) barely existed, and so the variable star community could not profit much from the constant flow of information that is now coming from wide-field surveys whose main goal is to detect extrasolar planets through the transit method. In the late-1990's, data collected with the Hipparcos satellite provided a fundamental contribution to the field of variable stars, and Gaia was just a distant concept. By providing accurate parallaxes and proper motions to hundreds of thousands of variable stars across the Milky Way, Gaia will decidedly revolutionize our knowledge of the intrinsic properties of pulsating variable stars. Now Gaia has become a reality, and it was a privilege for us, as organizers, that some of its very first scientific results were presented precisely at this conference.

In like vein, many survey projects (often counting on dedicated telescopes) are now in operation, including several in Chile, whose main aim is to detect transient events, such as novae and (especially) supernovae. Examples include CRTS, PTF, ASAS-SN, HiTS, CHASE, Pan-STARRS, and many others. The most important (and largest) of all these projects, the Large Synoptic Survey Telescope (LSST), which is currently being built in Cerro Pachón, northern Chile, is slated to start operations in the early 2020's. The nightly LSST data flow will dwarf that from even the largest existing surveys. A data tsunami is in the horizon, and its first ripples have just started to hit the shore. A change in paradigm in the way variable star astronomy is pursued is clearly imminent. This is why we realized, a few years ago, that the time was ripe for a meeting specifically devoted to studying the impact of wide-field variability surveys upon pulsating star research.

At the San Pedro meeting, we made a concerted effort to bring together astronomers who have been involved in the planning and execution of wide-field variability surveys – past, current, and future – in order to share experiences, ponder what we have been able to learn about pulsating variables in this way, and discuss strategies to face the approaching data tsunami. We envisaged a meeting where leaders of some past variability surveys would highlight the lasting legacy of their work on the field of pulsating star research, and also where key players of ongoing and planned surveys – ground- and space-based alike – would present their projects and discuss the impact of their experiments on this type of science.

Importantly, we also saw our meeting as a perfect opportunity for players in the fields of transient and exoplanetary surveys to explore the synergies between their projects and pulsating star science. Last but not least, we also attempted to bring statisticians and computer scientists who make astronomy one of their main (if not the main) areas of research to San Pedro, so we could also hear their perspective on how to better "navigate the tsunami" and thus maximize the scientific return from the truly huge datasets that are now becoming a reality.

We hope that the many interesting contributions that are included in these proceedings will reflect our current view of this increasingly interdisciplinary field. We are especially grateful to our colleagues who have agreed to contribute review papers, several of which cover topics that, in our opinion, had heretofore been sorely missing from the the modern literature.

We would like to express our gratitude to the additional members of the Scientific Organizing Committee, who helped us put together an exciting program: L. Althaus, G. Clementini, L. Deng, A. J. Drake, F. Grundahl, N. Matsunaga, K. Pollard, A. Udalski, L. Walkowicz, and P. Whitelock. We are also tremendously grateful to the Local Organizing Committee, who helped ensure a nearly flawless meeting experience for all participants: J. Borissova, J. Alonso-García, M. Vučković, R. Contreras Ramos, A. Gallenne, and – most especially – A. Lagarini and S. Varela.

We gratefully thank the following organizations and projects, without whose generous support this meeting could not have taken place:

- European Organisation for Astronomical Research in the Southern Hemisphere (ESO);
- ALMA Observatory;
- Carnegie Observatories and Las Campanas Observatory;
- Association of Universities for Research in Astronomy (AURA);
- National Science Foundation (NSF);
- Giant Magellan Telescope (GMT);
- Tokyo Atacama Observatory (TAO) Project, University of Tokyo;
- Chile's Ministry for the Economy, Development, and Tourism's Millennium Science Initiative, through grant IC 120009, awarded to the Millennium Institute of Astrophysics (MAS);
- Centro de Astrofísica y Tecnologías Afines (CATA), Proyecto Basal/CONICYT PFB-06/2007;
- CONICYT/FONDECYT grant #1171273;
- CONICYT/PCI grant DPI20140066;
- Pontificia Universidad Católica de Chile;
- Universidad de Concepción.

Santiago, August 2, 2017

The impact of large-scale, long-term optical surveys on pulsating star research

Igor Soszyński[1,*]

[1] *Warsaw University Observatory, Al. Ujazdowskie 4, 00-478 Warszawa, Poland*

Abstract. The era of large-scale photometric variability surveys began a quarter of a century ago, when three microlensing projects – EROS, MACHO, and OGLE – started their operation. These surveys initiated a revolution in the field of variable stars and in the next years they inspired many new observational projects. Large-scale optical surveys multiplied the number of variable stars known in the Universe. The huge, homogeneous and complete catalogs of pulsating stars, such as Cepheids, RR Lyrae stars, or long-period variables, offer an unprecedented opportunity to calibrate and test the accuracy of various distance indicators, to trace the three-dimensional structure of the Milky Way and other galaxies, to discover exotic types of intrinsically variable stars, or to study previously unknown features and behaviors of pulsators. We present historical and recent findings on various types of pulsating stars obtained from the optical large-scale surveys, with particular emphasis on the OGLE project which currently offers the largest photometric database among surveys for stellar variability.

1 Introduction

Over the past quarter century, our knowledge of the pulsating stars has made a dramatic step forward and the lion's share of this progress came from large-scale optical surveys for stellar variability. The first such surveys were initiated in the early 1990s as a response to the suggestion of Bohdan Paczyński ([23]), who pointed out the possibility of using gravitational microlensing effects to detect massive compact objects in the Galactic halo. Paczyński predicted that photometric monitoring of a few million stars in the Magellanic Clouds over a time scale of two years should lead to a discovery of the first microlensing events (which in those days were hypothetical phenomena) and shed light on the nature of dark matter. This idea had a built-in "insurance policy" in case no gravitational microlensing events were detected. In the last sentence of his seminal paper Paczyński stated: "It is clear that the observational project is not simple, but one of its by-products, a systematic discovery of a large number of variable stars in a nearby galaxy, is attractive, even if no lensing events are discovered".

A few years later, three groups – EROS ([7]), MACHO ([1]), and OGLE ([59]) – started survey programs to search for microlensing events. The **EROS** (Expérience de Recherche d'Objets Sombres) experiment was a French collaboration between astronomers and particle physicists. This project was conducted from 1990 to 2003 and it was divided into two phases. In the first phase (EROS-1), it used two telescopes located at the ESO La Silla Observatory in Chile: a 1-meter Schmidt telescope and a

*soszynsk@astrouw.edu.pl

Table 1. Parts of the OGLE Collection of Variable Stars published so far.

Type of variable stars	Environments[1]	Number of stars	References
Classical Cepheids	LMC, SMC, GB, GD	9 587	[28], [48], [53]
Type II Cepheids	LMC, SMC, GB	603	[41], [45], [48]
Anomalous Cepheids	LMC, SMC	250	[52]
RR Lyrae stars	LMC, SMC, GB, GD	83 753	[28], [51], [54]
Long-Period Variables	LMC, SMC, GB	343 785	[43], [47], [49]
δ Scuti stars	LMC, GD	2 844	[28], [35]
Eclipsing and ellipsoidal binary systems	LMC, SMC, GB, GD	510 792	[24], [27], [55]
Total		951 614	

[1] GB = Galactic bulge; GD = Galactic disk.

0.4-meter telescope equipped with a CCD camera. In the second stage of the survey (EROS-2), the 1-meter Marly telescope at the same observatory was used. EROS monitored the Large (LMC) and Small (SMC) Magellanic Clouds, the center of the Milky Way, and four directions in the Galactic disk. The total number of sources in the EROS-2 database is about 87 million.

The **MACHO** (Massive Compact Halo Object) project was active for 8 years (1992-1999) and monitored 62 million stars in the Magellanic Clouds and the bulge of the Milky Way. This survey has had full-time use of the 1.27-meter telescope at Mount Stromlo Observatory, Australia, which was a rebuilt historical telescope known as the Great Melbourne Telescope from 1868. In the nineteenth century it was the second largest telescope in the world and the largest in the southern hemisphere. MACHO simultaneously observed in two colors, using eight 2048×2048 pixel CCDs.

The **OGLE** (Optical Gravitational Lensing Experiment) project is the only active one of these pioneering microlensing surveys which started in early nineties. Within a quarter of a century, OGLE has been evolving, gradually increasing its observing capabilities. Currently, in its fourth phase (OGLE-IV, [62]), the survey daily monitors over a billion objects in the densest stellar regions of the sky: central regions of the Milky Way, the Galactic disk, and the Magellanic Clouds. To date, the project has collected about one trillion (10^{12}) individual photometric measurements, making OGLE the world's largest survey aimed at searching for variability in the sky. The project uses a dedicated 1.3-meter Warsaw telescope at Las Campanas Observatory, Chile, equipped with a 32-chip mosaic CCD camera. The observations are conducted using the standard *V*- and *I*-band filters, which ensures direct use of the OGLE photometry in many astrophysical applications or comparison with existing older data.

During its long history the OGLE project contributed significantly to many fields of observational astrophysics: gravitational lensing and microlensing, extrasolar planets, cosmic distance scale, stellar clusters, interstellar extinction, Kuiper Belt objects, active galactic nuclei, etc. Variable stars occupy a special position among the most important achievements of the survey. OGLE is an unquestionable world leader in this field. The OGLE Collection of Variable Stars contains presently nearly one million objects of various types and this is the largest set of variable stars ever obtained by any astronomical project. Table 1 summarizes the number of variable stars published by OGLE so far.

In this paper, we present the most important results on pulsating stars obtained by the large-scale optical surveys. Our goal is to show the incredible progress that has been made in this field in the last 25 years. In the next sections we present respectively classical Cepheids, RR Lyrae stars, type II Cepheids, anomalous Cepheids, long-period variables, and other pulsating stars.

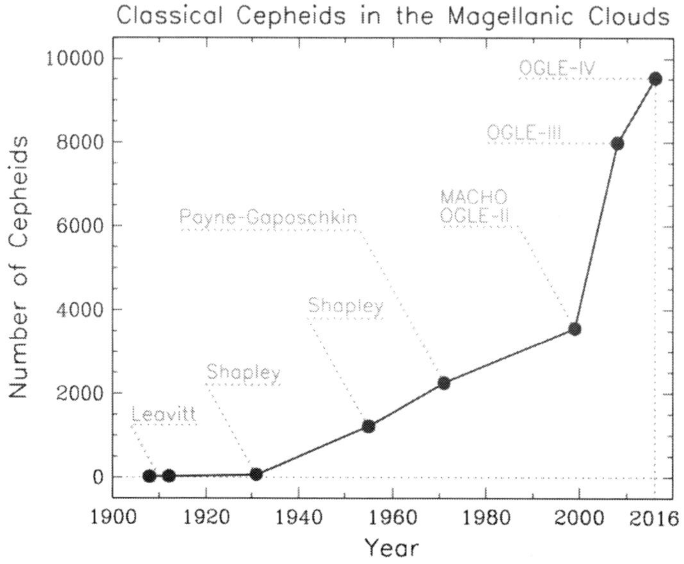

Figure 1. Number of known classical Cepheids in the Magellanic Clouds from the beginning of the twentieth century. References: [18], [19], [36], [37], [25], [26], [5], [60], [61], [40], [44], [53].

2 Classical Cepheids

Classical Cepheids are one of the most important distance indicators and tracers of the young stellar population. They are population I, relatively massive ($M > 4\,M_{\odot}$) giant stars obeying the famous period–luminosity (PL) relationship discovered in the Magellanic Clouds by Henrietta Leavitt in 1908 ([18]). Figure 1 shows how many classical Cepheids in the Magellanic Clouds have been known from the beginning of the twentieth century, starting from 16 Cepheids[1] discovered by Leavitt ([18]), through several review and catalog papers summarizing the number of Cepheids known in those days ([25, 26, 36, 37]), the first catalogs of Cepheids in the Magellanic Clouds released by the MACHO and OGLE-II projects in 1999 ([5, 60, 61]), ending with a real breakthrough that was made in recent years by OGLE-III ([40, 44]) and OGLE-IV projects ([53]). Currently, we almost reached the maximum value of this distribution. The total number of classical Cepheids in both Clouds is about 10 000.

2.1 Period–luminosity relations

The progress that has been made in this field is well illustrated in Figure 2. We present here four PL diagrams for classical Cepheids in the Magellanic Clouds. From left to right and from top to bottom we present here the historically first PL diagram plotted in 1912 by Leavitt ([19]) for 25 Cepheids in the SMC (two relationships result from the fact that each Cepheid is represented by two points corresponding to the minimum and maximum light), the period–Wesenheit index diagram obtained by Barry Madore in 1982 ([20]) containing all LMC Cepheids with high-quality photoelectric observations published in those days, the PL diagram with 1800 LMC Cepheids observed by the MACHO team ([5]), and the newest PL diagram with 4620 LMC classical Cepheids cataloged by the OGLE

[1] Actually, Leavitt ([18]) discovered a larger number of Cepheids, but in 1908 the classification was firm only for 16 of them for which periods were determined.

 DOI: 10.1051/epjconf/201715201001

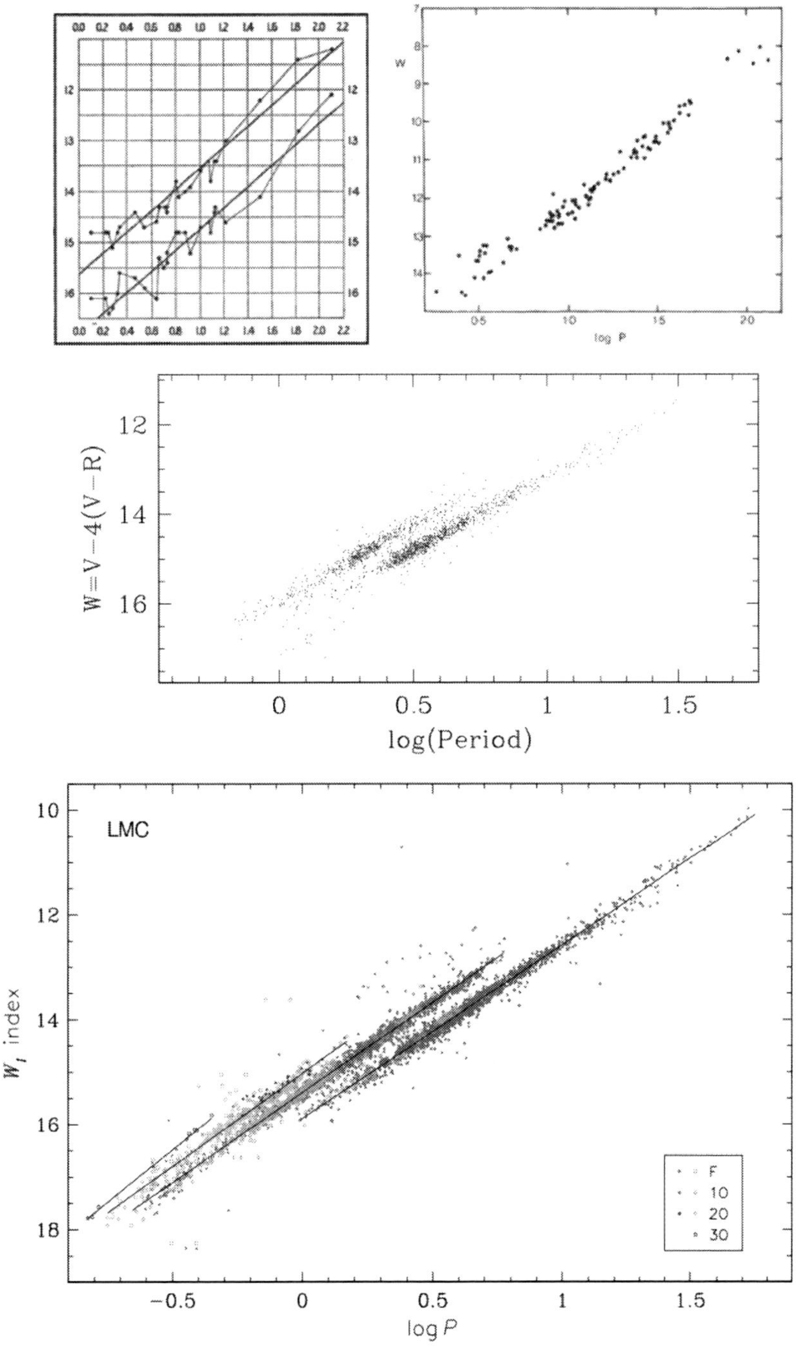

Figure 2. Period–luminosity diagrams for classical Cepheids in the Magellanic Clouds. From left to right and from top to bottom: the first PL diagram plotted in 1912 by Henrietta Leavitt for 25 Cepheids ([19]), the period–Wesenheit index plot obtained in 1982 by Barry Madore for 96 Cepheids with high-quality photoelectric observations ([20]), the PL diagram with 1800 Cepheids identified by the MACHO survey in 1999 ([5]), and the most recent PL diagram containing 4620 classical Cepheids in the LMC detected by the OGLE project ([53]).

survey ([53]). Note that only the microlensing surveys showed the two PL sequences of Cepheids – for those pulsating in the fundamental mode and the first overtone mode. In the most recent OGLE diagram, we can distinguish as many as four PL relations populated by the fundamental, first overtone, second overtone and third overtone Cepheids.

The Magellanic Clouds played in the past and still play a crucial role in the studies and calibrations of the PL relations of Cepheids and other pulsating stars. There is still an open question for the linearity and universality of the PL relations and its dependence on the chemical composition of the stars. The extraordinary large number of Cepheids discovered by the microlensing surveys allowed many authors to test the linearity of the PL relations with unprecedented precision. The first non-linearity around the periods of 2.5 days in the SMC Cepheids were noticed on the basis of the data published by the EROS survey ([8]). Then, the studies of the PL and period–color relations for Cepheids have been presented in countless papers. Cepheids in the LMC and SMC pulsating in the fundamental and first-overtone modes, using various pass-band from visual to infrared, have been thoroughly tested.

These studies are essential for using Cepheids as distance indicators, but they are also used to study the three-dimensional structure of the Magellanic System. Among many papers on this topic, we would like to draw attention to a recent work by Jacyszyn-Dobrzeniecka et al. ([17]), who used the most complete collection of classical Cepheids from the OGLE-IV project to describe the structures of both Clouds. They also studied the young stellar population in the Magellanic Bridge, since several classical Cepheids were recently discovered in this region ([53]).

2.2 Multi-mode Cepheids

Multi-mode Cepheids and other classical pulsators deserve special attention, since they show well the progress that has been made thanks to the large-scale surveys. In the pre-microlensing era the picture was simple – only about ten double-mode Cepheids had been known. Most of them pulsated simultaneously in the fundamental- and first-overtone modes (F/1O) and only one classical Cepheid – CO Aur – had the first and second overtones (1O/2O) excited, although the excitation of the second overtone in this star was disputed.

The revolution in this field began in 1995 when the MACHO team announced the discovery of 45 double-mode Cepheids in the LMC ([2]) – four times more than had been known before. The observations from the microlensing survey gave the first unambiguous evidence that the second overtone can be excited in Cepheids. The most recent collection of Cepheids published by OGLE ([53]) contains more than 700 double-mode Cepheids. The Petersen diagram (period ratio versus the longer period) for the Magellanic Clouds Cepheids is shown in Figure 3. Here, we also plot two double-mode classical Cepheids belonging to previously unknown classes: oscillating simultaneously in the first and third overtones (1O/3O) and second and third overtones (2O/3O). Triple-mode Cepheids are marked with triangles in Figure 3. About ten such unique pulsators are currently known and all of them have been discovered in the databases of the large-scale surveys. Finally, the non-radial modes in classical Cepheids manifest themselves in the form of three sequences in the Petersen diagram around the period ratios between 0.60 to 0.65 (empty circles in Fig. 3). Recently, Dziembowski ([9]) proposed an interesting theoretical interpretation of these modes. More information about the "revolution" in the Petersen diagram can be found in the papers by Radosław Smolec and Henryka Netzel in these proceedings.

2.3 Cepheids in eclipsing binary systems

Classical Cepheids which are members of eclipsing binary systems are very rare objects. No such star is currently known in the Milky Way. All of the known objects of this type belong to the Magellanic

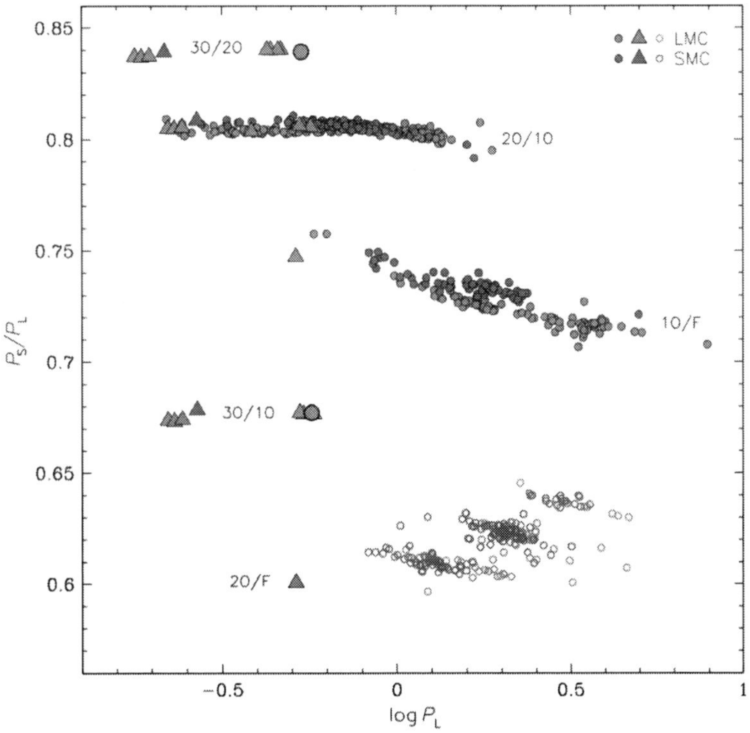

Figure 3. Petersen diagram for multi-mode classical Cepheids in the Magellanic Clouds. Solid dots show double-mode variables, triangles mark triple-mode Cepheids (three points per star) and empty circles represent Cepheids with non-radial modes excited. Blue and red symbols indicate stars in the LMC and SMC, respectively.

Clouds and all of them were discovered by the OGLE survey ([63]). Eclipsing binary systems hosting Cepheids have several interesting astrophysical applications. First, photometric and spectroscopic analyses of such objects may provide masses and radii of pulsating stars with great accuracy. It is particularly important in the context of the longstanding mass discrepancy problem for classical Cepheids. Second, detached double-lined eclipsing binary systems containing Cepheids are promising tools to calibrate the Cepheid PL relation, since a careful analysis of such systems may provide their accurate distances. Third, it is possible to precisely determine the so called *p*-factor – the crucial parameter in the Baade-Wesselink method of distance determinations to Cepheids. The Araucaria project ([11]) specializes in the in-depth analysis of pulsating stars in eclipsing binary systems (e.g., [12, 31, 34]).

2.4 Cepheids in the Milky Way

Besides the Magellanic Clouds, a number of classical Cepheids were identified by the OGLE team in the central regions of our Galaxy ([48]). Using this sample, Feast et al. ([10]) found an unexpected young stellar component of the flared outer disk located behind the Galactic center. This discovery clearly showed that the structure of the Galaxy is far from being fully understood. The cause behind the flaring of the disk remains uncertain, but one possible explanation is the presence of dark matter.

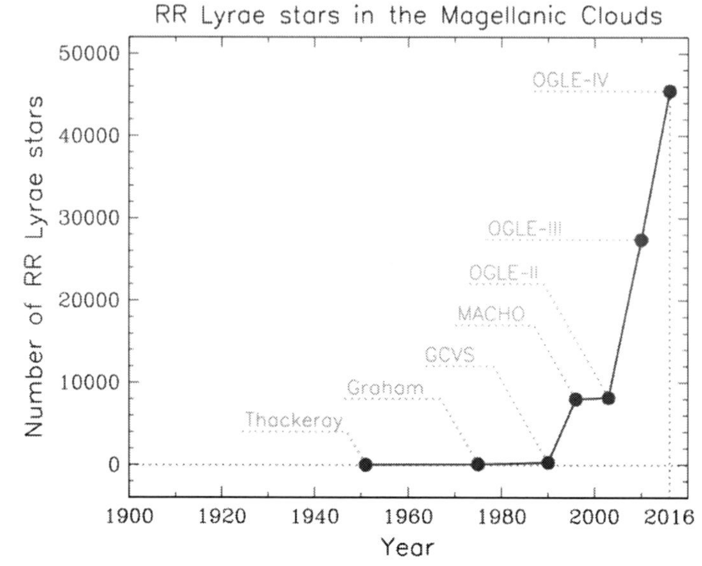

Figure 4. Number of known RR Lyrae stars in the Magellanic Clouds from the beginning of the twentieth century. References: [57], [14], [6], [3], [42], [46], [54].

Also the collection of known classical Cepheids in the Milky Way will be soon significantly extended thanks to a new subproject of the OGLE project, the OGLE Galaxy Variability Survey – a long-term photometric survey for stellar variability in the Galactic disk and in the outer regions of the Galactic bulge. This project covers nearly 2/3 of the Galactic plane with Galactic longitudes from $-170°$ to $+60°$ and latitudes from $-3°$ to $+3°$ in the disk and from $-15°$ to $+15°$ in the bulge region. The total observed area is about 2000 square degrees covered by about 1200 individual pointings. For the preliminary results of the search for Cepheids and RR Lyrae stars see the paper by Andrzej Udalski in these proceedings.

3 RR Lyrae stars

RR Lyrae stars are indispensable tracers of, and distance indicators for, the old stellar population in our own and nearby galaxies. These radially pulsating horizontal branch stars populate galactic halos, thick disks and globular clusters. RR Lyr variables are numerous, present in various stellar environments, and are relatively easy to detect, when precise, well covered light curves are available.

Figure 4 illustrates how the number of known RR Lyr stars in the Magellanic System has increased over the years. The first RR Lyr variables in the Magellanic Clouds were identified by A. David Thackeray in 1951 ([57]), but 40 years later only about 300 RR Lyr stars were known in both galaxies ([6]), mostly in the fields around globular clusters. The breakthrough came with the large-scale surveys, in particular the OGLE project, which increased the number of identified RR Lyr stars in the Magellanic Clouds to over 45 000 ([54]). Huge samples of RR Lyr stars were also found in our Galaxy: over 38 000 objects were discovered in the bulge by the OGLE team ([51]), about 22 000 stars in the Milky Way halo were identified by the Catalina survey ([58]), and an impressive number of 150 000 RR Lyr candidates were selected in the database of the Pan-STARRS survey ([16]). However, the purity and completeness of the latter sample is significantly reduced due to a limited number of observing points collected to date by the Pan-STARRS project.

These numerous sets of RR Lyrae stars have proven to be a powerful probe of the Galactic structure. Pietrukowicz et al. ([30]) used the OGLE RR Lyr variables to show that the old stellar population traces the barred structure of the Galactic bulge. In the Galactic halo several overdensities have been discovered ([16, 58]) with the most pronounced being the Sagittarius Stream, which consists of tidally stripped stars from the Sagittarius Dwarf Galaxy. It is worth mentioning that the central regions of the Sagittarius Dwarf Galaxy were recently explored by Hamanowicz et al. ([15]) based on RR Lyr stars published by OGLE.

Moreover, the collection consisting of tens of thousands of RR Lyr variables with long-term, well covered light curves is an inexhaustible source of unique pulsators exhibiting exotic behaviors. Let us mention the discovery of a new class of double-mode RR Lyr variables (anomalous RRd stars, [56]), the detection of non-radial modes in RR Lyr stars (e.g., [22]), the first observations of the Blazhko effect in RRd stars ([38]), RR Lyr stars switching their pulsation modes ([50, 51]), and a candidate for an RR Lyr star in an eclipsing binary system (OGLE-BLG-RRLYR-02792), which turned out to be the first known representative of a new type of pulsating stars mimicking classical RR Lyr stars ([32]).

4 Type II Cepheids

Type II Cepheids belong to the same population as RR Lyr stars, but are much less numerous. There are several hundreds of type II Cepheids known in the Galaxy and Magellanic Clouds. This class of variable stars is traditionally divided into three groups, each one at a different evolutionary stage: BL Her stars with periods between 1 and 4 days, W Vir stars with periods in the range 4–20 days and RV Tau stars with periods above 20 days and characteristic alternating deep and shallow minima.

Observations obtained by the large-scale programs, in particular MACHO and OGLE, led to very important discoveries about type II Cepheids. Alcock et al. ([4]) found the first certain RV Tau stars in the LMC and derived their absolute magnitudes for the first time confirming that they are luminous, post-AGB stars. This discovery showed that RV Tau stars obey a PL relation which is an extension of the PL relation delineated by other type II Cepheids – BL Her and W Vir stars.

One of the most important achievements of the OGLE project was the discovery of the fourth subclass of type II Cepheids, which was called peculiar W Vir stars ([41]). These pulsating variables exhibit distinct features of their light curves and are on average bluer and brighter than regular W Vir stars. Moreover, a large fraction of peculiar W Vir stars show eclipsing or ellipsoidal modulation superimposed on the pulsation light curves, which suggests that all these stars are members of binary systems.

5 Anomalous Cepheids

Until the discovery of the first bona fide anomalous Cepheids in the Magellanic Clouds by the OGLE team ([41, 52]), the vast majority of this kind of pulsators was known in dwarf galaxies. The sample of 250 anomalous Cepheids detected by OGLE in the LMC and SMC multiplied the total number of such stars known in the Universe.

The origin of anomalous Cepheids is still a subject of debate. There is a consensus that they are metal-deficient core-helium-burning stars with masses 1–2 $M_{\odot}$. They pulsate in the fundamental mode or in the first-overtone mode and their PL relations are located between the relations for classical and type II Cepheids. There are two leading hypotheses for the origin of anomalous Cepheids: that they are intermediate-age stars (1–6 Gyr old) with exceptionally low metallicity or that they are coalesced old (> 10 Gyr) binary stars.

An analysis of the spatial distribution of anomalous Cepheids in the Magellanic Clouds gives a hint for their origin ([52]). Their distribution is much more similar to the old stellar population represented by RR Lyr stars than to the young classical Cepheids. However, some anomalous Cepheids in the LMC seem to trace spiral arms, which makes them similar to the younger stars. The conclusion is that anomalous Cepheids are probably a mixture of old and intermediate-age stars.

The large and homogeneous sample of anomalous Cepheids discovered by OGLE in the Magellanic Clouds was also used to demonstrate that the coefficients derived from the Fourier light curve decomposition are unambiguous discriminators between classical and anomalous Cepheids. This finding enabled Soszyński et al. ([52]) to discover four Galactic anomalous Cepheids in the foreground of the Magellanic Clouds – the first fundamental-mode stars of that type in the field (outside clusters) of the Milky Way. This discovery showed that anomalous Cepheids must be quite numerous in the Galactic halo.

6 Long-period variables

Long-period variables (LPVs), or in other words pulsating red giants, constitute a very numerous group of variable stars. They are bright, some of them (Miras) exhibit very large amplitudes, but they belong to the least studied types of pulsating stars. The reason for this lies in their long periods, from weeks to years, and usually complex nature of their light curves. The study of LPVs benefited the most from the long-term monitoring by the microlensing surveys.

Figure 5 illustrates the history of discovering the PL relations for LPVs. The first such relation was noticed in 1981 by Glass and Lloyd-Evans ([13]) for Miras. Then, the second PL sequence was found in 1996 by Wood and Sebo ([64]) for semiregular variables. Other sequences in the PL plane were discovered on the basis of the data collected by the large-scale surveys. Wood ([65]) used the MACHO database to find five PL relations obeyed by LPVs, while the OGLE data allowed us to distinguish more than ten sequences ([39]). Most of these relations are already linked to particular radial and non-radial modes of pulsation ([21]); however, sequence D, populated by the so called Long Secondary Periods, still does not have a satisfactory explanation.

7 Blue pulsating stars

The variability surveys have strongly contributed to our knowledge about pulsating stars on the blue side of the HR diagram. An achievement of the ASAS project (All Sky Automated Survey, one of the first whole sky surveys conducted with fully automated small telescopes) may be given as example. Pigulski & Pojmański ([33]) reported the discovery of nearly 300 new β Cep stars in the ASAS survey – three times more than were known before. Such a large sample opened the possibility to perform a series of statistical analyses of β Cep stars and to compare the results with pulsation theory.

Another very fresh discovery comes from the OGLE project. OGLE-GD-DSCT-0058 is a star in the Galactic disk which was initially classified as a δ Sct star ([28]). Its period is only 28 minutes, it has a large amplitude, and its light curve morphology is typical for fundamental-mode pulsating stars. Surprisingly, the spectrum obtained by Pietrukowicz et al. ([29]) showed that the effective temperature of this star is 30 000 K, which corresponds to spectral type O9 – so this object cannot be a δ Sct star. The surface gravity places this star between main-sequence stars and subdwarf OB stars. OGLE-GD-DSCT-0058 is probably the first representative of a new type of pulsating stars, tentatively called Blue Large-Amplitude Pulsators (BLAPs). A careful search in the OGLE photometric databases resulted in the discovery of more than a dozen additional BLAPs in the Galactic bulge.

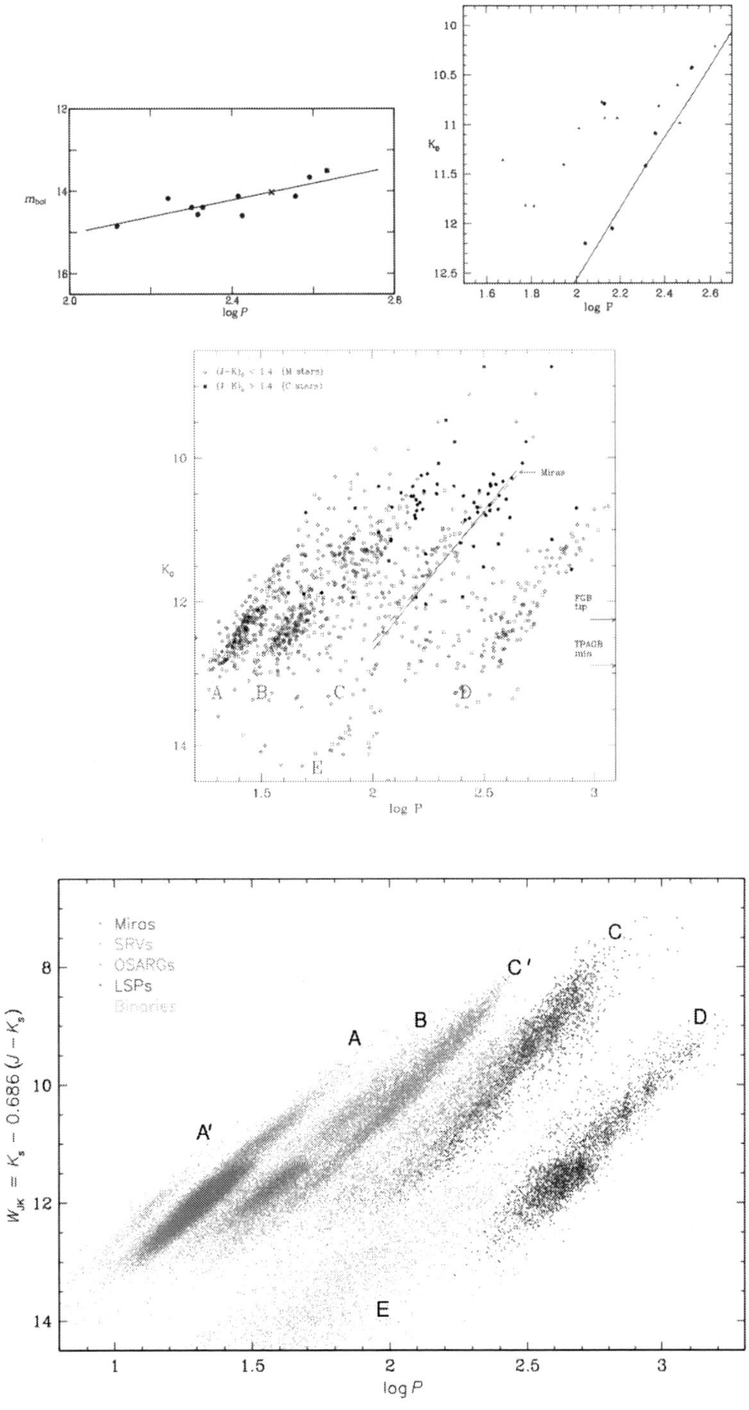

Figure 5. Period–luminosity diagrams for long-period variables in the LMC. From left to right and from top to bottom: the near-infrared PL diagram obtained by Glass and Lloyd Evans in 1981 for 11 Miras ([13]), two PL relations (for Miras and semiregular variables) discovered by Wood & Sebo in 1996 ([64]), the PL diagram plotted by Peter Wood in 1999 on the basis of the MACHO observations ([65]), and the modern PL diagram for LPVs from the OGLE Collection of Variable Stars ([43]).

8 Conclusions

The large-scale optical observational projects have radically changed the way we understand pulsating stars. Hundreds of thousands of new variable objects have been identified and cataloged on the basis of the photometry collected in the course of these projects. Studies of such huge samples of variable stars have at least three advantages. First, it is possible to examine in detail the statistical properties of pulsators (e.g., PL relations). Second, in the large sets of variable stars of a given type it is easier to find objects of very rare subtypes, or even those previously unknown. Third, complete and homogeneous samples of pulsators can be used to study the structure and evolution of our and other galaxies.

Large-scale photometric surveys practically demonstrated that it is feasible to monitor billions of stars per night, making up to 10^{11} photometric measurements per year. Long-term, well covered light curves are ideal to probe the photometric behavior of pulsating stars for timescales from minutes to years. The technology is still relatively cheap. The question arises, what is the future of large-scale sky surveys? The existing projects cover the densest stellar regions of the sky: the Galactic bulge and disk and the Magellanic Clouds. Outside these regions the number of stars is much smaller, so further expanding the sky coverage in existing projects will not significantly increase the number of monitored stars. It seems that further development will go toward deeper photometry (LSST), higher time resolution, and broader spectral coverage.

Acknowledgments: This work has been supported by the Polish National Science Centre grant MAESTRO no. 2014/14/A/ST9/00121.

References

[1] Alcock, C., Akerlof, C. W., Allsman, R. A., et al., Nature, **365**, 621 (1993)
[2] Alcock, C., Allsman, R. A., Axelrod, T. S., et al., AJ, **109**, 1654 (1995)
[3] Alcock, C., Allsman, R. A., Axelrod, T. S., et al., AJ, **111**, 1146 (1996)
[4] Alcock, C., Allsman, R. A., Alves, D. R., et al., AJ, **115**, 1921 (1998)
[5] Alcock, C., Allsman, R. A., Alves, D. R., et al., AJ, **117**, 920 (1999)
[6] Artyukhina, N. M., Durlevich, O. V., Frolov, M. S., et al., General Catalogue of Variable Stars, 4rd ed., vol. V. Extragalactic Variable Stars, "Kosmosinform", Moscow (1995)
[7] Aubourg, E., Bareyre, P., Bréhin, S., et al., Nature, **365**, 623 (1993)
[8] Bauer, F., Afonso, C., Albert, J. N., et al., A&A, **348**, 175 (1999)
[9] Dziembowski, W. A., Communications of the Konkoly Observatory Hungary, **105**, 23 (2016)
[10] Feast, M. W., Menzies, J. W., Matsunaga, N., & Whitelock, P. A., Nature, **509**, 342 (2014)
[11] Gieren, W., Pietrzyński, G., Bresolin, F., et al., The Messenger, **121**, 23 (2005)
[12] Gieren, W., Pilecki, B., Pietrzyński, G., et al., ApJ, **815**, 28 (2015)
[13] Glass, I. S., & Evans, T. L., Nature, **291**, 303 (1981)
[14] Graham, J. A., PASP, **87**, 641 (1975)
[15] Hamanowicz, A., Pietrukowicz, P., Udalski, A., et al., Acta Astron., **66**, 197 (2016)
[16] Hernitschek, N., Schlafly, E. F., Sesar, B., et al., ApJ, **817**, 73 (2016)
[17] Jacyszyn-Dobrzeniecka, A. M., Skowron, D. M., Mróz, P., et al., Acta Astron., **66**, 149 (2016)
[18] Leavitt, H. S., Annals of Harvard College Observatory, **60**, 87 (1908)
[19] Leavitt, H. S., & Pickering, E. C., Harvard College Observatory Circular, **173**, 1 (1912)
[20] Madore, B. F., ApJ, **253**, 575 (1982)
[21] Mosser, B., Dziembowski, W. A., Belkacem, K., et al., A&A, **559**, A137 (2013)

[22] Netzel, H., Smolec, R., & Moskalik, P., MNRAS, **447**, 1173 (2015)
[23] Paczyński, B., ApJ, **304**, 1 (1986)
[24] Pawlak, M., Soszyński, I., Udalski, A., et al., Acta Astron., **66**, 421 (2016)
[25] Payne-Gaposchkin, C., & Gaposchkin, S., Smithsonian Contrib. Astrophys., **9**, 1 (1966)
[26] Payne-Gaposchkin, C. H., Smithsonian Contrib. Astrophys., **13** (1971)
[27] Pietrukowicz, P., Mróz, P., Soszyński, I., et al., Acta Astron., **63**, 115 (2013a)
[28] Pietrukowicz, P., Dziembowski, W. A., Mróz, P., et al., Acta Astron., **63**, 379 (2013b)
[29] Pietrukowicz, P., Latour, M., Angeloni, R., et al., Acta Astron., **65**, 63 (2015a)
[30] Pietrukowicz, P., Kozłowski, S., Skowron, J., et al., ApJ, **811**, 113 (2015b)
[31] Pietrzyński, G., Thompson, I. B., Gieren, W., et al., Nature, **468**, 542 (2010)
[32] Pietrzyński, G., Thompson, I. B., Gieren, W., et al., Nature, **484**, 75 (2012)
[33] Pigulski, A., & Pojmański, G., AIPC, **1170**, 351 (2009)
[34] Pilecki, B., Graczyk, D., Gieren, W., et al., ApJ, **806**, 29 (2015)
[35] Poleski, R., Soszyński, I., Udalski, A., et al., Acta Astron., **60**, 1 (2010)
[36] Shapley, H., Harvard College Observatory Bulletin, **883**, 16 (1931)
[37] Shapley, H., & McKibben Nail, V., Proc. Nat. Acad. Sci., **41**, 829 (1955)
[38] Smolec, R., Soszyński, I., Udalski, A., et al., MNRAS, **447**, 3756 (2015)
[39] Soszyński, I., Dziembowski, W. A., Udalski, A., et al., Acta Astron., **57**, 201 (2007)
[40] Soszyński, I., Poleski, R., Udalski, A., et al., Acta Astron., **58**, 163 (2008a)
[41] Soszyński, I., Udalski, A., Szymański, M. K., et al., Acta Astron., **58**, 293 (2008b)
[42] Soszyński, I., Udalski, A., Szymański, M. K., et al., Acta Astron., **59**, 1 (2009a)
[43] Soszyński, I., Udalski, A., Szymański, M. K., et al., Acta Astron., **59**, 239 (2009b)
[44] Soszyński, I., Poleski, R., Udalski, A., et al., Acta Astron., **60**, 17 (2010a)
[45] Soszyński, I., Udalski, A., Szymański, M. K., et al., Acta Astron., **60**, 91 (2010b)
[46] Soszyński, I., Udalski, A., Szymański, M. K., et al., Acta Astron., **60**, 165 (2010c)
[47] Soszyński, I., Udalski, A., Szymański, M. K., et al., Acta Astron., **61**, 217 (2011a)
[48] Soszyński, I., Udalski, A., Pietrukowicz, P., et al., Acta Astron., **61**, 285 (2011b)
[49] Soszyński, I., Udalski, A., Szymański, M. K., et al., Acta Astron., **63**, 21 (2013)
[50] Soszyński, I., Dziembowski, W. A., Udalski, A., et al., Acta Astron., **64**, 1 (2014a)
[51] Soszyński, I., Udalski, A., Szymański, M. K., et al., Acta Astron., **64**, 177 (2014b)
[52] Soszyński, I., Udalski, A., Szymański, M. K., et al., Acta Astron., **65**, 233 (2015a)
[53] Soszyński, I., Udalski, A., Szymański, M. K., et al., Acta Astron., **65**, 297 (2015b)
[54] Soszyński, I., Udalski, A., Szymański, M. K., et al., Acta Astron., **66**, 131 (2016a)
[55] Soszyński, I., Pawlak, M., Pietrukowicz, P., et al., Acta Astron., **66**, 405 (2016b)
[56] Soszyński, I., Smolec, R., Dziembowski, W. A., et al., MNRAS, **463**, 1332 (2016c)
[57] Thackeray, A. D., & Wesselink, A. J., Nature, **171**, 693 (1953)
[58] Torrealba, G., Catelan, M., Drake, A. J., et al., MNRAS, **446**, 2251 (2015)
[59] Udalski, A., Szymański, M., Kałużny, J., Kubiak, M., & Mateo, M., Acta Astron., **42**, 253 (1992)
[60] Udalski, A., Soszynski, I., Szymanski, M., et al., Acta Astron., **49**, 223 (1999a)
[61] Udalski, A., Soszynski, I., Szymanski, M., et al., Acta Astron., **49**, 437 (1999b)
[62] Udalski, A., Szymański, M. K., & Szymański, G., Acta Astron., **65**, 1 (2015a)
[63] Udalski, A., Soszyński, I., Szymański, M. K., et al., Acta Astron., **65**, 341 (2015b)
[64] Wood, P. R., & Sebo, K. M., MNRAS, **282**, 958 (1996)
[65] Wood, P. R., PASA, **17**, 18 (2000)

OGLE Cepheids and RR Lyrae stars in the Milky Way

Andrzej Udalski[1,*] on behalf of the OGLE survey team

[1] *Warsaw University Observatory, Al. Ujazdowskie 4, 00-478 Warszawa, Poland*

Abstract. We present new large samples of Galactic Cepheids and RR Lyrae stars from the OGLE Galaxy Variability Survey.

1 Introduction

The Optical Gravitational Lensing Experiment (OGLE) is one of the largest sky variability surveys. In its fourth phase it regularly monitors over a billion stars. The main OGLE observing targets are the densest stellar regions of the sky: the Galactic center and Magellanic Clouds ([10]).

One of the most important outcome of the project is the *OGLE Collection of Variable Stars* (OCVS). It currently consists of almost a million well characterized periodic variable stars of many different types ([9]). The OCVS contains a full variety of pulsating stars from the main astrophysical "laboratories" like the Magellanic System and the Galactic bulge. For example, the current sample of RR Lyrae stars counts over 38000 objects in the Galactic bulge and over 45000 stars in the Magellanic System (Magellanic Clouds and Magellanic Bridge, [6, 8]). The number of Cepheids in the Magellanic System reaches 10000 pulsators ([7]). The OGLE samples are very pure and complete. They were used for many interesting projects related, among others, to investigations of the structure of the Galactic center ([4]) and Magellanic System ([1, 2]).

In 2013 a new OGLE long-term survey – OGLE Galaxy Variability Survey (OGLE GVS) – was initiated in the frame of the OGLE-IV phase. An area of about 2000 square degrees in the sky is regularly monitored. The survey covers a strip of ±3 degree width along the Galactic plane for all Galactic longitudes accessible from the OGLE observing site – Las Campanas Observatory, Chile. Additionally, a large area of the external Galactic bulge is also covered.

After collecting a large sample of photometric data (over 100 epochs), spanning three observing seasons, the region of the Galactic disk with $190 < l < 345$ degrees is ready for exploration. Here we present the preliminary report on the first quick search for Cepheids and RR Lyrae stars in these fields of the OGLE GVS.

2 Observations

The OGLE-IV survey has been conducted using the 1.3-m Warsaw telescope with a 32 CCD detector wide field mosaic camera located at the Las Campanas Observatory, Chile. The field of view of the OGLE-IV camera is 1.4 square degrees with the pixel scale of 0.26 arsec/pixel. The OGLE GVS has been started in February 2013 and it is still conducted albeit with different cadence and priority of fields compared to previous seasons. The main variability survey is a shallow one with the exposure

*udalski@astrouw.edu.pl

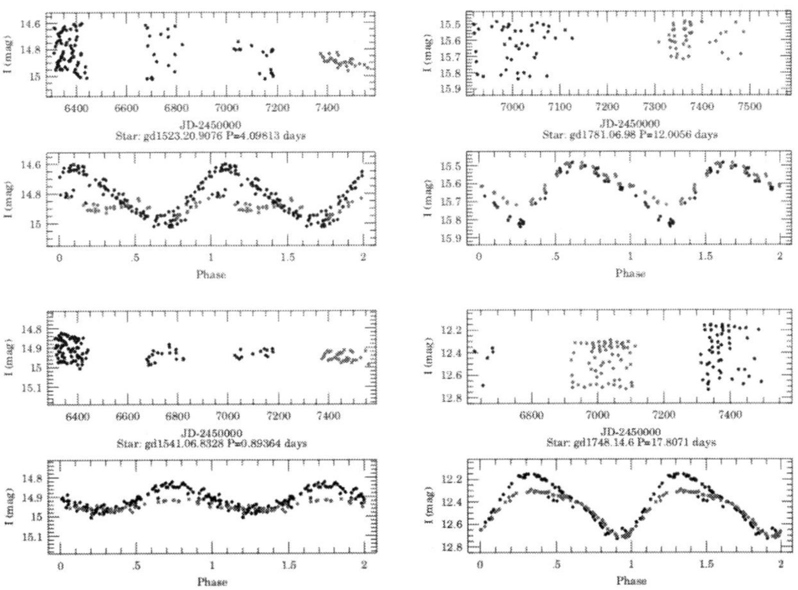

Figure 1. Example light curves of variable stars mimicking pulsating-like variability behavior. Red points mark observing season with considerably different shape of the light curve.

time of 25 seconds. Most observations are collected in the I-band filter for variability search. A total of 120–150 epochs in each of the fields presented here have been collected so far. Also, about ten or more observations have been secured in the V-filter for color information.

The collected images have been reduced with the OGLE standard photometric pipeline which is based on the image subtraction technique – DIA ([10]). The faint limit for variability search in the OGLE GVS is about $I \approx 18.5$ mag.

3 Cepheids and RR Lyrae stars from the OGLE GVS

We carried out the pilot search for pulsating variables of Cepheid and RR Lyrae types among about 560 million stellar sources detected on the OGLE GVS images of the Galactic plane. This "fast search" is different than the typical global OGLE variable stars searches. It has been conducted on pre-selected subsample of stars – only those which displayed clear variability signatures on the difference images. Minimum 15 detections of variable flux in the difference images were required to include a star to this sample.

The pre-selected variable star candidates were then subject to period search. Two techniques were used: analysis of variance, AoV ([5]), and Fourier analysis implemented as FNPEAKS program ([3]). Light curves of objects with significant periodic signals were further investigated. First, their shape was fitted with Fourier series and the Fourier parameters R_{21} and ϕ_{21} were calculated. In the next step of pre-selection, only candidates with Fourier parameters in a wide range covering typical regions occupied by Cepheids and RR Lyrae stars on the R_{21} vs. $\log P$ and ϕ_{21} vs. $\log P$ diagrams were left. Finally, the candidates were visually inspected and objects with light curves of non-pulsating origin (e.g., eclipsing binaries) were removed from the ultimate list of Cepheid/RR Lyrae candidates.

DOI: 10.1051/epjconf/201715201002

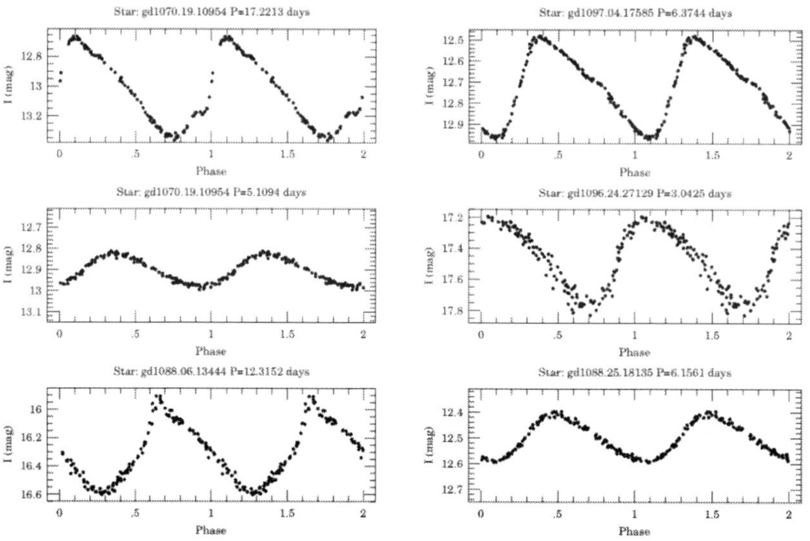

Figure 2. Example light curves of newly discovered Galactic disk Cepheids.

Surprisingly, the list of objects at this stage was still severely contaminated. The main source of fake identifications were spotted stars which can mimic very well the light curve of pulsating star. On the other hand the shape of the variability in such stars evolves with time and in the case of long-term monitoring, as during OGLE GVS, spotted stars can be extracted and separated from genuine pulsators. Figure 1 presents light curves of a few examples of spotted stars mimicking in some observing seasons pulsating stars.

After visual inspection of the OGLE Cepheid/RR Lyrae candidates, 1930 and 2170 objects were left as very likely Galactic Cepheids and RR Lyrae type stars, respectively. At this stage of analysis the boundary between these two groups was set artificially at $P = 0.7$ day. Figure 2 presents a small gallery of newly discovered Galactic Cepheids. The list of Cepheids contains classical Cepheids as well as type II objects.

Preliminary lists of the OGLE Cepheid and RR Lyrae candidates will additionally be refined when the color data are available. Nevertheless, the purity of the sample is already high. The completeness of these samples is different. While the Galactic Cepheids are bright enough to be detected on the OGLE subtracted images, fainter RR Lyrae stars located farther in the halo are certainly missed in our preselected sample of analyzed stars. We expect that many additional fainter RR Lyrae stars will be detected during the final exensive search for variables in the OGLE GVS fields.

Figures 3 and 4 show the distribution of new OGLE Galactic Cepheids and RR Lyrae stars on the map of the sky, respectively. Figures 5 and 6 present distribution of these groups of stars in the real sky. Even without deeper analysis one can easily notice differences in distributions. For example, while RR Lyrae stars are generally located symmetrically around the Galactic plane, Cepheids lie mostly below the plane over a long range of observed Galactic longitudes.

The new OGLE samples of Galactic Cepheids and RR Lyrae stars will be released to the community in the next months *via* the OCVS. They will constitute the next huge OGLE datasets of these important pulsating stars.

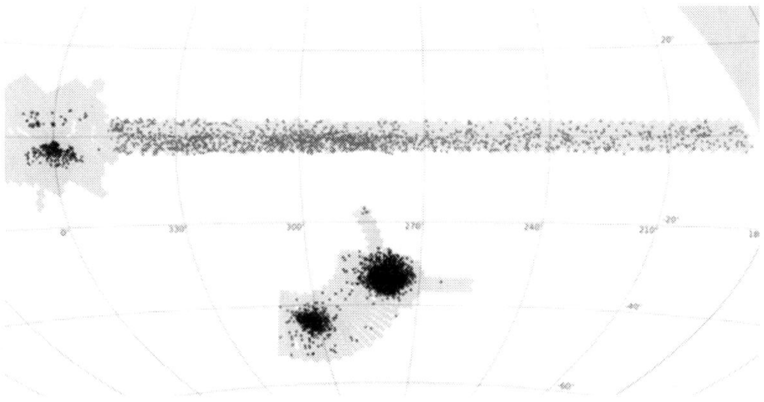

Figure 3. OGLE Cepheids in the sky. The sky map is in the Galactic coordinates (l, b). Light gray shaded areas show the regions regularly monitored by the OGLE-IV survey. Dark gray shaded region in the upper right corner indicates part of the sky not accessible from Las Campanas Observatory. Red dots present Cepheid star candidates found in the OGLE GVS fields. Black dots mark positions of the OGLE Cepheids in other OGLE-IV targets.

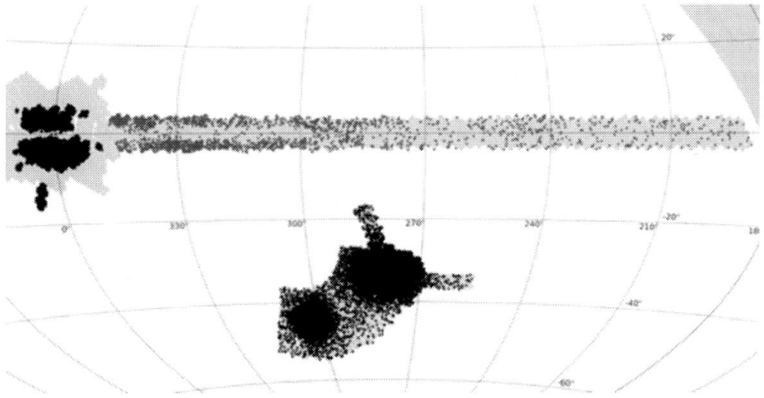

Figure 4. OGLE RR Lyrae stars in the sky. The sky map is in the Galactic coordinates (l, b). Light gray shaded areas show the regions regularly monitored by the OGLE-IV survey. Dark gray shaded region in the upper right corner indicates part of the sky not accessible from Las Campanas Observatory. Red dots present RR Lyrae star candidates found in the OGLE GVS fields. Black dots mark positions of the OGLE RR Lyrae stars in other OGLE targets.

Acknowledgments: The OGLE project has received funding from the National Science Centre, Poland, grant MAESTRO 2014/14/A/ST9/00121.

References

[1] Jacyszyn-Dobrzeniecka, A.M., Skowron, D.M., Mróz, P., et al., Acta Astron., **66**, 149 (2016)
[2] Jacyszyn-Dobrzeniecka, A.M., Skowron, D.M., Mróz, P., et al., Acta Aston., **67**, 1 (2017)
[3] Kołaczkowski, Z., `http://helas.astro.uni.wroc.pl/deliverables.php?lang=en&active=fnpeaks` (2016)

Figure 5. Details of the distribution of OGLE Cepheids in the Galactic plane. Green dots indicate positions of the OGLE-IV Cepheid candidates from the OGLE GVS. Yellow dots are OGLE Cepheids from the Galactic bulge fields.

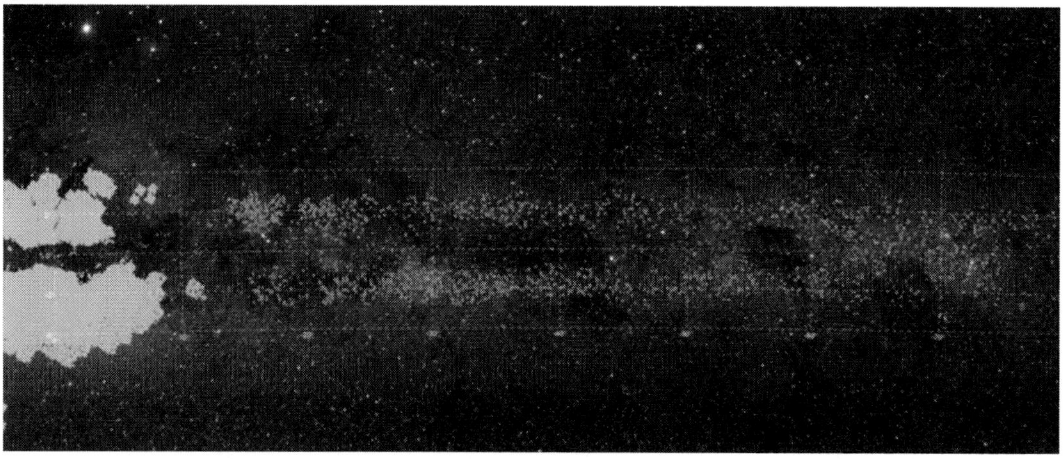

Figure 6. Details of the distribution of OGLE RR Lyrae stars in the Galactic plane. Green dots indicate positions of the OGLE-IV RR Lyrae star candidates from the OGLE GVS. Yellow dots are OGLE RR Lyrae stars from the Galactic bulge fields.

[4] Pietrukowicz, P., Kozłowski, S., Skowron, J., et al., ApJ, **811**, 113 (2015)
[5] Schwarzenberg-Czerny, A., MNRAS, **241**, 153 (1989)
[6] Soszyński, I., Udalski, A., Szymański, M.K., et al., Acta Astron., **64**, 177 (2014)
[7] Soszyński, I., Udalski, A., Szymański, M.K., et al., Acta Astron., **65**, 297 (2015)
[8] Soszyński, I., Udalski, A., Szymański, M.K., et al., Acta Astron., **66**, 131 (2016)
[9] Soszyński, I., Pawlak, M., Pietrukowicz, P., et al., Acta Astron., **66**, 405 (2016)
[10] Udalski, A., Szymański, M.K., & Szymański, G., Acta Astron., **65**, 1 (2015)

Blazhko modulation in first-overtone RR Lyrae stars from the OGLE Collection

Henryka Netzel[1,2]*, *Radosław* Smolec[2]**, *Igor* Soszyński[1], and *OGLE* Team

[1]*Warsaw University, Astronomical Observatory, Al. Ujazdowskie 4, 00-478 Warsaw, Poland*
[2]*Nicolaus Copernicus Astronomical Centre, Polish Academy of Sciences, Bartycka 18, 00-716 Warsaw, Poland*

Abstract. We present results of the first comprehensive study of the Blazhko effect in first-overtone RR Lyrae stars from the Galactic bulge based on the Optical Gravitational Lensing Experiment data.

1 Introduction

RR Lyrae stars are classically pulsating stars. Most of them are simple single-periodic radial pulsators, which pulsate either in the radial fundamental mode (RRab type) or in the radial first overtone (RRc type). We also know double-mode RR Lyrae stars pulsating simultaneously in the radial fundamental mode and the first overtone (RRd type) or in the fundamental mode and the second overtone, but the latter are scarce.

Despite their astrophysical importance and relative simplicity there are still unsolved problems concerning RR Lyrae stars. One of the most stubborn problems is the Blazhko effect, which is a quasi-periodic modulation of pulsation amplitude and/or phase. It was discovered by Sergey Blazhko in 1907 ([3]), who observed changes in moments of maximum light of RW Dra. Since its discovery, the Blazhko effect has lacked a satisfactory explanation.

In these proceedings we present a first comprehensive study of the Blazhko effect in first-overtone RR Lyrae stars from the Galactic bulge. As a data source, we used data from the fourth phase of the Optical Gravitational Lensing Experiment (OGLE; [15]). OGLE regularly monitors the brightness of more than a billion sources. Excellent data quality together with numerous observations and long time base are perfect for a search of Blazhko effect in RR Lyrae stars.

2 Data and method

In the OGLE collection of variable stars there are more than 38 000 RR Lyrae stars from the Galactic bulge. 10 826 among them are of RRc type ([13]). Because of the numerous input sample, we developed an automatic algorithm for analysis and selection of Blazhko RRc stars. The algorithm uses Fourier analysis and consecutive prewhitening method. In principle, the Blazhko effect manifests itself in the power spectrum as equally spaced multiplets centered on the main frequency and its

*henia@netzel.pl
**smolec@camk.edu.pl

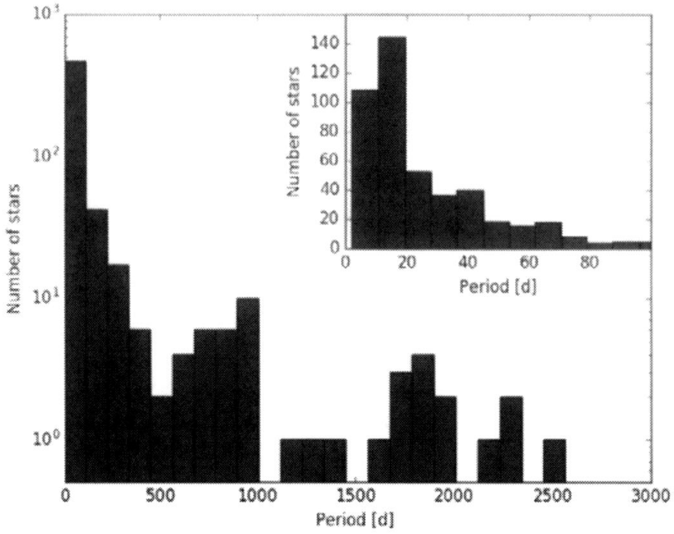

Figure 1. Histogram of Blazhko periods for RRc stars. In the top right corner of the plot we include a zoom-in for the short-period range. Note that the number of stars in the zoom-in is plotted in a linear scale, whereas the whole distribution is plotted in a logarithmic scale.

harmonics. Separation between the multiplet components corresponds to the modulation frequency. In ground-based photometry we usually detect triplets or doublets. The algorithm searched for these characteristic features in the power spectrum. If the algorithm detected at least two sidepeaks with the same separation, it considered a star to be a Blazhko candidate. The algorithm selected 3066 candidates, which constitutes 28 per cent of the input sample. However, many RRc stars exhibit period changes, which manifest themselves in power spectra as a remnant power in the position of the first overtone after the prewhihtening. This remnant power together with its annual aliases can lead to misclassification of a star as a long-period Blazhko variable. To verify the candidates we visually inspected the power spectra of all 3066 candidates. Stars which passed the visual inspection were all analysed manually.

3 Results

The analysis resulted in 573 Blazhko RRc stars, which constitutes 5.3 per cent of the input sample. In 384 stars we detected triplets (BL2), which constitute 67 per cent of the Blazhko sample. Doublets were detected in 189 stars (BL1), which constitute 33 per cent of the Blazhko sample. For a full table of Blazhko stars together with their properties, see Netzel et. al. (in prep.). The distribution of Blazhko periods in stars from our sample is presented in Figure 1. In the top right corner of this figure we present a zoom-in for periods shorter than 100 days. Note that the full distribution is plotted in a logarithmic scale, whereas the zoom-in is plotted in a linear scale. Stars with short periods of modulation are the most numerous. The shortest detected Blazhko period is 2.23 days for a BL1 star. Among BL2 stars the shortest period is 4.9 days. The longest period of the Blazhko effect depends on the time base of observations. OGLE-IV data allow on average to detect modulation with periods up to 1000 days. This is the reason for a sudden drop in the number of stars in Figure 1. However,

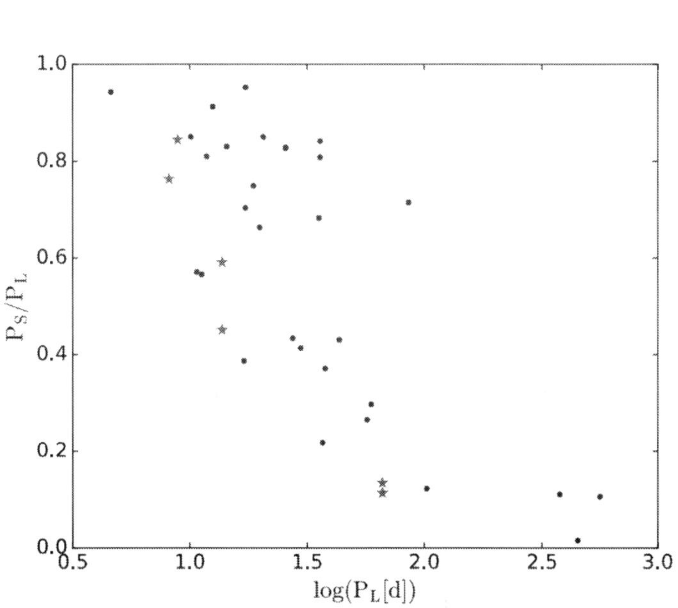

Figure 2. Relation between the ratio of shorter to longer Blazhko period versus longer Blazhko period for stars with multi-periodic Blazhko effect. Stars with two Blazhko periods are marked with blue dots. Stars with three Blazhko periods are marked with red asterisks.

combining OGLE data from the previous phases of the experiment allowed us to detect a few stars with modulation periods longer than this limit. The longest modulation period detected is 2477 days.

In 32 stars from our sample we detected two Blazhko periods. Three Blazhko periods were detected in 2 stars from our sample. Blazhko periods of these stars are presented in Figure 2 in the form of a "Petersen diagram", i.e. we plot the ratio of the shorter to longer modulation period versus the longer period. The majority of these stars have short periods of modulations (below 50 days). Only in three stars do we observe a significantly longer period. Observed period ratios cover a wide range, and no particular grouping that could correspond e.g. to a ratio of small integer numbers is observed.

Besides the main pulsation frequency, its harmonics and Blazhko sidepeaks, additional significant signals were detected in dozens of stars. In Figure 3 we present the Petersen diagram for already known double-mode RR Lyrae stars. On top of these stars, we plot Blazhko stars in which we detected additional signals (orange symbols). Many of these stars fall into the known groups: several stars fall into 0.61 and 0.68 groups, four stars are close to the RRd sequence and might be anomalous RRd stars ([11, 12]; see also Smolec et al. in these proceedings). There is another numerous group on the Petersen diagram with high period ratios (above 0.9). In these stars we observe additional signal with frequency very close to the first-overtone frequency. It can be either a non-radial mode with period close to the main pulsation period or a manifestation of another Blazhko modulation (but since we did not detect other sidepeaks, the star was not classified as a Blazhko variable).

One of the most intriguing phenomena discovered in Blazhko RRab stars is period doubling, which manifests as alternating minima and maxima in the light curve. In the power spectrum, it manifests as signals close to half-integer frequencies. This phenomenon was detected for the first time in RR Lyr from *Kepler* data ([5]), and later studies showed that it affects many modulated RRab stars (see, e.g., [2, 14]). It motivated a new model explaining the Blazhko effect ([4]). Period doubling

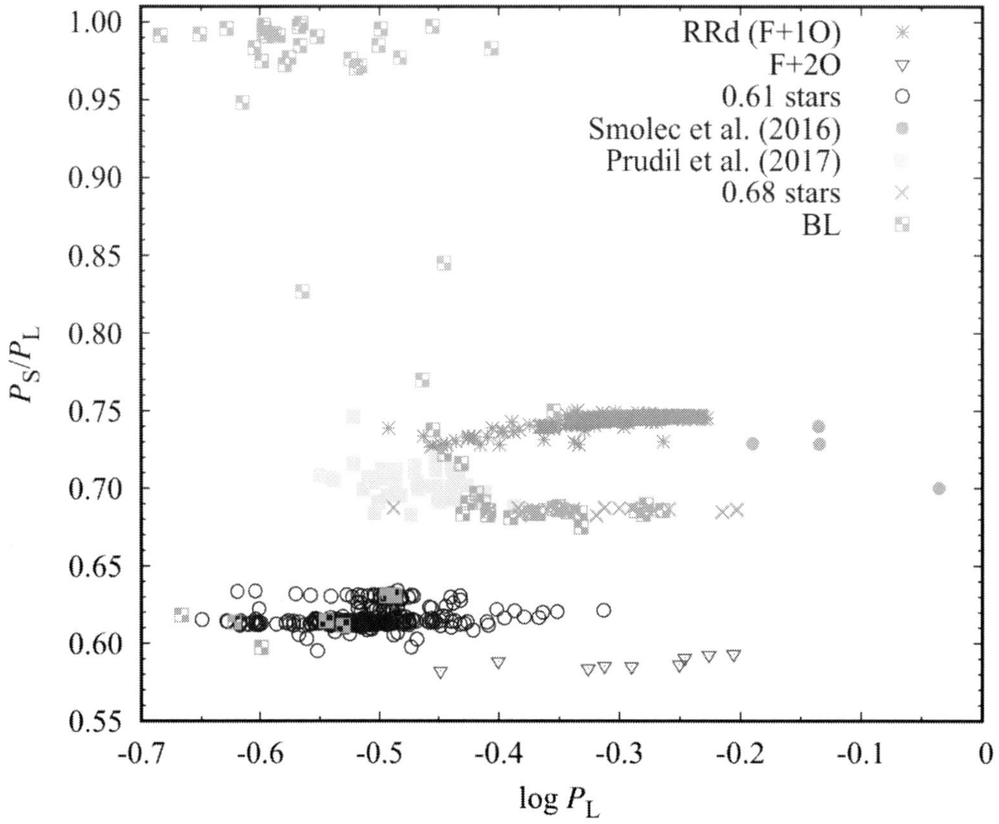

Figure 3. Petersen diagram for the known groups of double-periodic RR Lyrae stars. RRd stars are marked with green asterisks, fundamental mode and second overtone stars are marked with red triangles, first overtone ones with a shorter-period non-radial mode stars forming a period ratio 0.61 are marked with blue circles ([8]), first overtone and long-periodic variability forming a period ratio 0.68 are marked with pink crosses ([7]), RR Lyrae stars with short-period signal are marked with yellow squares ([9]) and RRab stars with another short-period signal are marked with blue filled circles ([10]). Blazhko stars with additional signals from our sample are marked with orange symbols.

was not detected in Blazhko RRc stars. During manual analysis of our Blazhko candidates we did not find any significant signal close to half-integer frequencies. In order to detect this effect we followed a procedure outlined in [1], and averaged the power spectra of all Blazhko stars in our sample, after prewhitening with pulsation and modulation components. All spectra were scaled, so the main pulsation frequency of each star corresponds to 1. The average spectrum is presented in Figure 4. In the case of stars in which we observed a remnant signal at the position of the main pulsation frequency due to e.g. a period change, we removed the effect using a time-dependent prewhitening method ([6]). Use of this method resulted in a sharp drop visible in Figure 4 at $f/f_{1O} = 1$. If Blazhko RRc stars exhibit the period doubling effect, in Figure 4 we should see signals around values 0.5 and 1.5, which is not the case. We note however that detection of the period doubling effect in the ground-based data is challenging and our results do not eliminate the possibility that it plays a role also in Blazhko RRc stars.

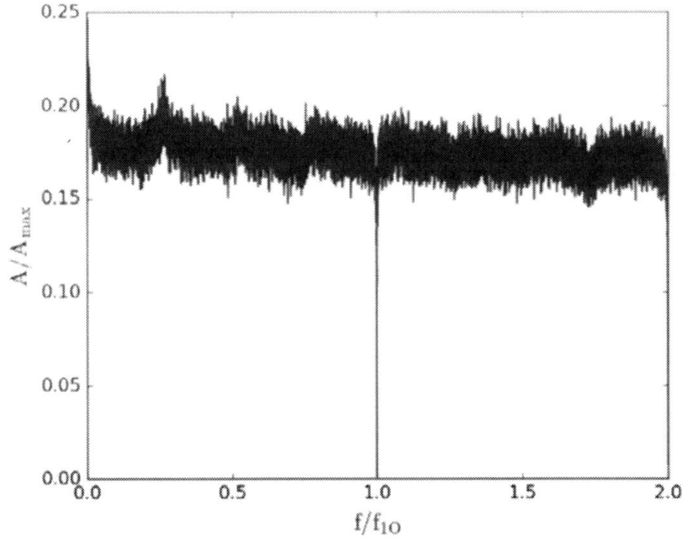

Figure 4. Averge spectrum of all Blazhko stars from our sample. We removed the period change present in some stars using time-dependent prewhitening. Spectra are scaled, so 1 on the horizontal axis corresponds to the frequency of the first overtone.

Full results of this study will be published elsewhere (Netzel et al., in preparation).

Acknowledgments: This research is supported by the Ministry of Science and Higher Education through grant no. 0192/DIA/2016/45 and by the Polish National Science Centre through grant no. DEC-2015/17/B/ST9/03421. HN is grateful to the organizers for their financial support.

References

[1] Alcock, C., Alves, D. R., Becker, A., et al., ApJ, **598**, 597 (2003)
[2] Benkő, P. E., Szabó, R., Molnár, L., & Kolláth, Z., ApJS, **213**, 13 (2014)
[3] Blazko, S., AN, **175**, 325 (1907)
[4] Buchler, J. R., & Kolláth, Z., ApJ, **731**, 24 (2011)
[5] Kolenberg, K., Szabó, R., Kurtz, D. W., et al., ApJ, **713**, L198 (2010)
[6] Moskalik, P., Smolec, R., Kolenberg, K., et al., MNRAS, **447**, 2348 (2015)
[7] Netzel, H., Smolec, R., & Dziembowski, W., MNRAS, **451**, L25 (2015)
[8] Netzel, H., Smolec, R., & Moskalik, P., MNRAS, **453**, 2022 (2015)
[9] Prudil, Z., Smolec, R., Skarka, M., & Netzel, H., MNRAS, **465**, 4074 (2017)
[10] Smolec, R., Prudil, Z., Skarka, M., & Bakowska, K., MNRAS, **461**, 2934 (2016)
[11] Smolec, R., Soszyński, I., & Udalski, A., MNRAS, **447**, 3753 (2015)
[12] Soszyński, I., Smolec, R., & Dziembowski, W. A., MNRAS, **463**, 1332 (2016)
[13] Soszyński, I., Udalski, A., Szymański, M.K., et al., Acta Astron., **64**, 177 (2014)
[14] Szabó, R., Benkő, J. M., Paparó M., et al., A&A, **570**, A100 (2014)
[15] Udalski, A., Szymański, M.K., & Szymański, G., Acta Astron., **65**, 1 (2015)

Pulsating star products from the Palomar Transient Factory: Ultra-long period Cepheids in M31 and RR Lyrae in *Kepler* field

Chow-Choong Ngeow[1,*]

[1] *Graduate Institute of Astronomy, National Central University, Jhongli 32001, Taiwan*

Abstract.
The Palomar Transient Factory (PTF) and its successor, the intermediate PTF (iPTF), are wide-field synoptic sky surveys aimed to detect transients. Even though the main science goal for PTF/iPTF is to detect various types of transients, the synoptic nature of the surveys can also be used for the study of variable stars. In this proceedings contribution, I will first give a brief introduction to PTF/iPTF, followed by the two pulsating stars studies using the PTF/iPTF data: the Ultra-Long Period Cepheids (ULPC) in M31 and the RR Lyrae in the *Kepler* field. For the formal study, we searched the M31's ULPC using PTF imaging data, and follow up the candidates with other telescopes. Our finding revealed that there are only two ULPC in M31. I will give a brief implication of our finding in distance scale studies. For the latter study, I will present our work on the derivation of metallicity-light curve relation in native PTF/iPTF R-band using the RRab stars in the *Kepler* field.

1 Introduction

The 48-inch Samuel Oschin Telescope, known as P48, located at the Palomar Observatory, possesses a distinct history in modern observational astronomy. For examples, the P48 Telescope has been used in various surveys in the past, including the Palomar Observatory Sky Survey (POSS), the Near-Earth Asteroid Tracking (NEAT) program, and the Quasar Equatorial Survey Team (QUEST). Today, the fully robotic P48 is used for dedicated wide-field transients surveys, known as the Palomar Transient Factory (PTF, 2009-2012, [1, 2]) and intermediate-PTF (iPTF, 2013-2017). Both PTF and iPTF use the same wide-field mosaic camera, which consists of eleven 2K $\times$ 4K CCDs, equipped on the P48 to produce a field-of-view (FOV) of $\sim$ 7.3 degree-squared with a pixel scale of $1.01''$/pixel. Nominal time series observations of PTF/iPTF were carried out mainly in Mould R filter; additional but less observed data was taken in the Sloan g filter. With an exposure time of 60 seconds, the PTF/iPTF data can reach to a depth of $R \sim 20.5$ mag. The PTF/iPTF surveys employed a mixed cadence for their time series observations, ranging from few minutes to few days, that depends on various experiments. Besides the studies of transients – the main science goals of PTF/iPTF –, the time series PTF/iPTF data has also been used in the variable stars research with noticeable examples presented in [3–12]. In this proceedings contribution, I present another two examples of using the PTF/iPTF data in pulsating variable stars works: the study of ultra-long period Cepheids (ULPC) and RR Lyrae.

*cngeow@astro.ncu.edu.tw

2 Examples of variable stars studies with PTF/iPTF data

2.1 The search for ULPC in M31

It has been known that Cepheids with very long period appear to be outliers in the period-luminosity (P-L) relation (for example, see [13]), and subsequently they were ignored in distance scale applications. In 2009, [14] (hereafter B09) proposed that Cepheids with periods longer than 80 days follow a different P-L relation and period-Wesenheit (P-W) relation than their shorter period counterparts, hence called them the ULPC. Later, [15] (hereafter F12) compiled a list of 37 ULPC based on literature search, nearby doubling the sample size of ULPC as originally presented in B09. These 37 ULPC are located in 10 nearby galaxies, ranging from dwarf galaxies, such as the Large and Small Magellanic Clouds, to large spiral galaxies like M81. However, the nearest spiral galaxy – the Andromeda Galaxy (M31) – is not included in F12's list. Since M31 holds the promise to be an anchoring galaxy in local distance scale ladder to calibrate a number of secondary distance indicators, it is of great interest to search and identify ULPC in M31. Furthermore, ULPC found in M31 can also be served as test case for applicability of ULPC in distance scale work, because a large number of distance measurements to M31 with various techniques is available in the literature (for a review, see [16]). Therefore, we initiated a program to search for ULPC in M31 using the PTF data, because the FOV of PTF image is large enough to cover the entire M31 (which subtended $\sim$ 3 degrees in the sky) and the multi-year time series PTF data is ideal to search for variable stars with period longer than 80 days.

To search for ULPC in M31, we downloaded the reduced R-band PTF images for M31 that spanned from January 2010 to January 2012, for a total of 172 frames. Cadence of this dataset ranges from 1 day to few days. We applied an image subtraction technique to search for variable sources, and employed various selection criteria to look for the ULPC candidates, including the variable sources have to be periodic with period longer than 80 days. We identified 8 ULPC candidates out of $\sim 10^5$ variable sources. A more detailed description of our selection criteria and results can be found in [17]. To verify the ULPC nature of these candidates, V- and I-band light curves data are needed so that their mean VI-band magnitudes and $(V - I)$ colors can be compared to other known ULPC in the plane of color-magnitude diagram (CMD) and P-W relation as presented in F12, by adopting a recommended distance modulus of M31 ([16]). We used the 60-inch robotic telescope (P60), located at the Palomar Observatory, and the Lulin One-meter Telescope (LOT), located at the Lulin Observatory (in central Taiwan), to perform the VI bands follow-up observations ([18]). Based on their locations in CMD and P-W planes with respect to the known ULPC and other types of long period variables, only two out of the eight candidates were identified as true ULPC.

We then derived the distance modulus of M31 with the two confirmed ULPC in M31. The resulting distance modulus broadly agreed with the recent values based on Cepheids given in the literature (as collected in [16]), however the associated error of the distance modulus is a few times larger than other measurements. The main reason for this large error is the small number statistics that, in the case of M31, relies on only two ULPC. Since ULPC are young and high mass supergiants, we do not expect many of them to be found in a single galaxy, and hence the small number statistics will always be the dominant source of error for the derived distance modulus. Therefore, ULPC may not be a good standard candle to derive the distance to individual galaxies. Other reasons that prevent the application of ULPC in distance scale work include the large dispersion of the ULPC P-W relation (as compared to the shorter period Cepheids) and the fact that the two available P-W relations in B09 and F12 are in discrepancy with each other.

2.2 Metallicity-light curve relation for ab-type RR Lyrae in the *Kepler* field

Besides Cepheids, RR Lyrae are also a type of standard candle that has been widely used in distance scale studies. We have initiated a program to search and characterize RR Lyrae in the PTF/iPTF data. The first step in our program is to derive a metallicity-light curve relation in the native PTF/iPTF R-band for ab-type RR Lyrae (i.e. the fundamental mode RR Lyrae; hereafter RRab stars). This is because the absolute V-band magnitude, M_V, for RRab stars is correlated with metallicity [Fe/H]. The actual form of the M_V-[Fe/H] relation might be a linear, quadratic, or segmented function, nevertheless a measurement of [Fe/H] is needed. Obtaining spectroscopic [Fe/H] could be expensive in terms of observing time, hence an alternate approach is to estimate the [Fe/H] for RRab stars with a metallicity-light curve relation ([19]). Faint RRab stars found in the PTF/iPTF data could belong to the Galactic halo, at which they are good tracer to map out the sub-structure of Galactic halo (for example, see [20, 21]), however they need to be spectroscopically confirmed. Estimation of photometric [Fe/H] via the metallicity-light curve relation for these faint RRab stars can help in reducing the targets that are needed for spectroscopic follow-up, because their [Fe/H] should be similar to metallicity expected in Galactic halo.

We use a sample of RRab stars in *Kepler* field to derive the PTF R-band metallicity-light curve relation, because these RRab stars possess accurate [Fe/H] based on high resolution spectroscopic observations ([22]). Another advantage of using this sample of RRab stars is the fact that their periods have been determined with very high accuracy from the almost continuous *Kepler* light curves. The PTF/iPTF R-band light curves are available for the majority of the RRab stars in this sample, some of which have been re-observed with shorter exposure time to ensure the photometry is not saturated. We then derived the following metallicity-light curve relation: $[Fe/H]_{PTF} = -4.089 - 7.346\,P + 1.280\,\phi_{31}$, where P is the pulsation period in days, and ϕ_{31} is one of the Fourier parameters that describe the light curve shape. We tested our derived relation to RRab stars in the K2-E2 sample ([23]) and a sample of halo RRab stars presented in [21], as both samples contain independent [Fe/H] measurements. The mean difference of our estimated [Fe/H] and the published values are −0.09 dex and −0.15 dex for the K2-E2 sample and the halo sample, respectively, validating our derived PTF R-band metallicity-light curve relation.

3 Conclusion and future prospect

The time series data from the PTF/iPTF surveys provides a great opportunity to the variable stars research in addition to transients studies. In this proceedings contribution I provided two examples of using the PTF/iPTF data in the study of pulsating variable stars:

- We searched and identified two ULPC in M31, and found that ULPC in general are not a suitable standard candle. This work has been published in [24].
- We derived the metallicity-light curve relation in native PTF R-band with a sample of RRab stars in the *Kepler* field, which can be used to estimate the metallicity for newly found RRab stars in PTF/iPTF. This work has been published in [25].

Part of the PTF/iPTF data has been released to public, details on accessing the data can be found in the following URL: `http://www.ptf.caltech.edu/page/data_access`. The iPTF will cease its operation in early 2017 in preparing and commissioning for the next generation wide-field synoptic survey – the Zwicky Transient Facility (ZTF, late 2017 – ~2020). ZTF will be using the same P48 Telescope with a new ultra wide-field mosaic CCD camera. This camera consists of sixteen 6 K × 6 K e2v CCDs that provide a FOV of ~ 47 degree-squared while still maintaining a pixel scale of 1″/pixel. The filters used in ZTF will be the same as in PTF/iPTF (i.e., the gR-band filters). Combining the very large FOV and a typical exposure time of 30 seconds, ZTF can observe $\sim$ 3750 degree-squared of the sky in an hour, and the 3π sky can be surveyed in 8 hours. The ZTF time series data will definitely be a gold mine for variable stars research before the Large Synoptic Survey Telescope (LSST) comes on-line in 2020 or after.

Acknowledgments: The author thanks the funding from the Ministry of Science and Technology (Taiwan) under the contract 104-2112-M-008-012-MY3.

References

[1] Law, N. M., Kulkarni, S. R., Dekany, R. G., et al., PASP, **121**, 1395 (2009)
[2] Rau, A., Kulkarni, S. R., Law, N. M., et al., PASP, **121**, 1334 (2009)
[3] Miller, A. A., Hillenbrand, L. A., Covey, K. R., et al., ApJ, **730**, 80 (2011)
[4] Levitan, D., Fulton, B. J., Groot, P. J., et al., ApJ, **739**, 68 (2011)
[5] Agüeros, M. A., Covey, K. R., Lemonias, J. J., et al., ApJ, **740**, 110 (2011)
[6] van Eyken, J. C., Ciardi, D. R., Rebull, L. M., et al., AJ, **142**, 60 (2011)
[7] Law, N. M., Kraus, A. L., Street, R., et al., ApJ, **757**, 133 (2012)
[8] Tang, S., Cao, Y., Bildsten, L., et al., ApJL, **767**, L23 (2013)
[9] Levitan, D., Kupfer, T., Groot, P. J., et al., ApJ, **785**, 114 (2014)
[10] Schindewolf, M., Levitan, D., Heber, U., et al., A&A, **580**, A117 (2015)
[11] Kao, W., Kaplan, D. L., Prince, T. A., et al., MNRAS, **461**, 2747 (2016)
[12] Marsh, F. M., Prince, T. A., Mahabal, A. A., et al., MNRAS, **465**, 4678 (2017)
[13] Freedman, W. L., ApJ, **326**, 691 (1988)
[14] Bird, J. C., Stanek, K. Z., & Prieto, J. L., ApJ, **695**, 874 (2009)
[15] Fiorentino, G., Clementini, G., Marconi, M., et al., Ap&SS, **341**, 143 (2012)
[16] de Grijs, R., & Bono, G., AJ, **148**, 17 (2014)
[17] Lee, C.-H., Ngeow, C.-C., Yang, T.-C., et al., in *2013 International Conference on Space Science and Communication (IconSpace2013)*, Proc. 2013 IEEE Int. Conf. on Space Science and Communication, ed. M. Abdullah et al. (IEEE Xplore), p. 1 (2013)
[18] Lee, C.-H., Ngeow, C.-C., & PTF Collaboration, in *The Tenth Pacific Rim Conference on Stellar Astrophysics*, ed. H.-W. Lee, Y Woon Kang & K.-C. Leung, ASP Conference Series, **482**, 55 (2014)
[19] Jurcsik, J., & Kovács, G., A&A, **312**, 111 (1996)
[20] Sesar, B., Cohen, J. G., Levitan, D., et al., ApJ, **755**, 134 (2012)
[21] Sesar, B., Grillmair, C. J., Cohen, J. G., et al., ApJ, **776**, 26 (2013)
[22] Nemec, J. M., Cohen, J. G., Ripepi, V., et al., ApJ, **773**, 181 (2013)
[23] Molnár, L., Szabó, R., Moskalik, P. A., et al., MNRAS, **452**, 4283 (2015)
[24] Ngeow, C.-C., Lee, C.-H., Ting-Chang Yang, M., et al., AJ, **149**, 66 (2015)
[25] Ngeow, C.-C., Yu, P.-C., Bellm, E., et al., ApJS, **227**, 30 (2016)

Synergies between exoplanet surveys and variable star research

Geza Kovacs[1,*]

[1]*Konkoly Observatory of the Hungarian Academy of Sciences, H-1121, Budapest, Konkoly Thege M. ut 15-17, Hungary*

Abstract. With the discovery of the first transiting extrasolar planetary system back in 1999, a great number of projects started to hunt for other similar systems. Because the incidence rate of such systems was unknown and the length of the shallow transit events is only a few percent of the orbital period, the goal was to monitor continuously as many stars as possible for at least a period of a few months. Small aperture, large field of view automated telescope systems have been installed with a parallel development of new data reduction and analysis methods, leading to better than 1% per data point precision for thousands of stars. With the successful launch of the photometric satellites CoRoT and *Kepler*, the precision increased further by one-two orders of magnitude. Millions of stars have been analyzed and searched for transits. In the history of variable star astronomy this is the biggest undertaking so far, resulting in photometric time series inventories immensely valuable for the whole field. In this review we briefly discuss the methods of data analysis that were inspired by the main science driver of these surveys and highlight some of the most interesting variable star results that impact the field of variable star astronomy.

1 Introduction – some historical background

Up until 1999, telephoto lenses were not acknowledged as a source of major astronomical discoveries. This has changed dramatically when, after the discovery of the first extrasolar planet 51 Peg ([61, 62]), many research teams started concentrated efforts to observe the photometric transit signals of the already discovered planets, and find new ones. Because some of the systems discovered by radial velocity methods had short periods of a few days and the estimated masses were in the range of Jupiter masses, these extrasolar planets (dubbed as Hot Jupiters, or HJs for short) with their 5 – 20% transit probabilities and ~ 1% transit depths were excellent targets for photometric detection, even for telescopes with apertures of ~ 5 – 10 cm. Indeed, five years after the discovery of 51 Peg, the photometric monitoring of HD 209458 resulted in the multiple observations of the transit events of the companion ([23, 42]). This important discovery immediately fixed the ambiguity of the planet mass due to orbital inclination, proved that the source of radial velocity variation is an orbiting companion and not a stellar surface phenomenon, and allowed for the first time to derive the two basic parameters (mass and radius) of an extrasolar planet.

After this ground-breaking discovery, over twenty different projects started to monitor large (~ 100 square degrees) chunks of the sky and look for the same kind of events in millions of stars,

*kovacs@konkoly.hu

without a priori radial velocity monitoring. Interestingly, the first photometric discovery came not from these small-telescope projects but from the OGLE project, operating a 1.3 m telescope and originally devoted to microlensing surveys. Unlike HD 209458, OGLE-TR-56 ([92]) is a faint, $V = 16.6$ mag star. This posed rather strong constraints on the possible spectrographs used for the validation of the system. Nevertheless, in spite of its faintness, the system was successfully verified (see [49]). It was soon realized that small telescopes are vital for the more massive discovery of additional systems that are considerably brighter.

In 2004, the first transiting extrasolar planet (TEP) was discovered by using small aperture wide-field automated telescopes, namely TrES-1 ([5]). The Trans-Atlantic Exoplanet Survey (TrES) network hosted the STARE telescope, the one that was instrumental in the discovery of the first transiting exoplanet, HD 209458. Although the discovery of TrES-1 was a good sign that small telescope projects are, indeed, capable of making significant discoveries, it was not until the fall of 2006 when these projects started to yield the rate of discovery hoped for. The relatively long delay between conceiving the idea and the first science output is attributed to several factors. This includes the time needed for the technical implementation (please note that these telescopes are autonomous with sophisticated software making decisions on telescope operation, data acquisition and basic image reduction); development of new methods for massive signal search and, especially, to solve the problem of filtering out colored noise; and accessibility of proper instrumentation for precise follow-up observations with the proper amount of telescope time to get accurate light curve and radial velocity solutions. Last, but not least, these objects are rather rare, with a $\sim 0.4\%$ (or lower) true occurrence rate (see, e.g., [37]) and an even lower (by a factor of $10 - 30$) observability rate.

With the discovery of HAT-P-1 and WASP-1 ([10, 25]), the speed of the discoveries substantially increased, yielding by now over 200 well-characterized HJ systems, covering a wide variety of stellar and planet parameters. Meanwhile, the long-planned space mission CoRoT was launched in 2006 and started to deliver high-quality, uninterrupted photometric time series on thousands of stars, leading to the discovery of the first super-Earth object, CoRoT-7b. Three years later, after many years of planning, the *Kepler* satellite was launched and a new era began with startling discoveries, such as Kepler-9, the first system with multiple transiting planets ([45]), or Kepler-186f, the first of the several systems hosting planets in the habitable zone ([74]), some of which might even be similar to Earth, such as Kepler-22b ([17]). The *Kepler* mission still continues, from 2014 under the name K2, showing the ingenuity of engineering and community support for the brilliant solution for the loss of the two reaction wheels by the time the main mission was completed. The K2 mission has proven to be quite successful. By visiting ecliptic fields, a broad spectrum of astrophysically interesting objects are visited, driven by researchers across the various disciplines of astronomy.

The main science (discover and characterize planets outside the solar system) that drives the above ground- and space-based projects resulted in an increase in the quality and amount of photometric data never seen before. It is obvious that this fundamental change has a serious impact also on variable star research. In this review we focus on the ground-based wide-field surveys as the results of the variable star works on the space mission data are more widely known due to the large number of researchers working on these data. On the other hand, works on the variable star aspects of the ground-based surveys are less organized and sporadic. Nevertheless, as we will see, these surveys are both complementary and in many respects competitive to the space surveys. Therefore, it is very important to be aware of the results obtained so far and also of what these data could be used for in future works.

2 Wide-field photometric surveys – the overall importance

Unlike variable searches, targeting specific stars in earlier surveys (e.g., globular cluster studies, works on RR Lyrae stars in microlensing survey data [3]), current photometric projects, while looking for periodic transit signals, scan through the whole list of objects passing some minimum precision condition. Photometric surveys targeting extrasolar planets have to satisfy high standards in data quality (continuity and precision) and in volume (due to the rare occurrence and short duration of the event, even though it is periodic). This results in time series that surpass the quality of most of the photometric data gathered earlier and yields valuable sources for a large diversity of studies, not directly related to extrasolar planetary science. These science "by-products" may often be as important as those targeted by the original idea that has led to the initialization of the project.

Many projects started after the first discovery of the transit of HD 209458 in 2000. Some of them were not realized (e.g., Permanent All Sky Survey, or PASS[1], [28]), some of them were abandoned (e.g., Wise observatory Hungarian-made Automated Telescope, or WHAT[2]), and some of them survived, expanded and became important/major contributors to the exoplanet inventory (e.g., Wide Angle Search for Planets, or WASP, [73]).

In the design of the instrumentation of these surveys, one might consider the optimization of the system (including hardware and observational strategy) for the most effective survey of large numbers of stars. One of the issues is the size of the optics used. In an interesting paper by [71], the authors come up with a solution that employing a 2 inch telescope, with a 4K × 4K CCD camera, would be optimum. Indeed, this setting is very close to the ones used by most of the "classical" surveys that range from 4 cm (KELT) to ~ 20 cm (e.g., HATSouth), with 10 cm as the most often used, and also a combination of these under the same closure (Qatar Exoplanet Survey, or QES[3], [6]).

Figure 1. The KELT-North building at the Winer Observatory (Arizona, USA), and the instrument hosted. The instrument is equipped with a 42 mm diameter Mamiya f/1.9 telephoto lens, a 4 K × 4 K Apogee AP16E CCD camera put on a Paramount mount. The observations are fully robotic without the need for onsite supervision. The data are periodically transferred to the Ohio State University.

For illustration, the somewhat unique setting for KELT-North is shown in Figure 1. The instrument has a field of view of 26 × 26 square degrees, almost seven times as large as that of the instrument aboard the *Kepler* satellite. The southern station is hosted by the South African Astronomical Observatory and consists also of a single instrument, a replica of the one at the northern station. Both instruments are fully robotic. They covered so far ~ 70% of the sky, and supplied 4.5×10^6 light curves. As of today, they discovered 15 TEPs[4] and made possible visits to several important "side topics" (see Sect. 3).

[1] `http://www.iac.es/proyecto/pass/`
[2] `http://wise-obs.tau.ac.il/~what/index.html`
[3] `http://www.qatarexoplanet.org/`
[4] `http://exoplanet.eu`

Figure 2 summarizes the major photometric surveys that have contributed (or will contribute) to the discovery of TEPs. For a broader view, we included also related space missions[5]. It is impossible to give even a brief description of all these projects. Therefore, we focus on HAT-N/S and WASP, the two main surveys contributing most of the TEPs discovered from the ground, and describe some of the main parameters of these projects. However, it is important to note that several other projects also made very significant contributions. For example, MEarth[6] (M dwarf stars in search of new Earth-like exoplanets, see [47]) has provided an important contribution to the variability survey of M dwarfs ([67]), and made the first discoveries of a transiting Super-Earth/Mini-Neptune ([24]) and an Earth-sized planet from the ground ([15]). Yet another, highly significant discovery comes from the TRAPPIST[7] (TRAnsiting Planets and PlanetesImals Small Telescope) project. The system consists of two 60 cm robotic telescopes that are located at La Silla Observatory (Chile) and at the Oukaïmden Observatory (Morocco). In 2016 TRAPPIST made the first ground-based discovery ([34]) of a multi-planetary system (consisting of three Earth-sized planets) around a nearby M8 dwarf. The proximity of the system makes it a prime candidate for a deep study of the long-chased class of multiplanetary transiting planets around M dwarfs. And indeed, just before completing this review, four additional planets were announced for this system, discovered by the *Spitzer* Space Telescope ([35]).

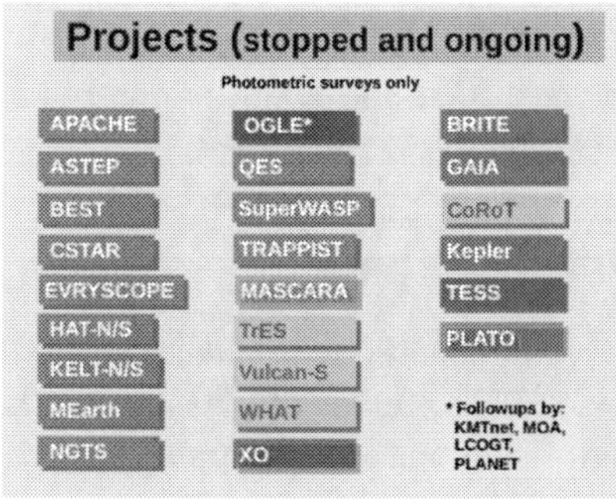

Figure 2. Summary of the photometric survey programs with past/current/future contributions to exoplanet science. Different coloring/shading refer to the current status of these projects. The first two columns are for the ground-based projects, whereas the third column is for the space missions. Green and blue are for ongoing, gray for past/no longer running, red/orange for immediate/near future projects. The MASCARA project is at an early stage of operation ([88]). The XO project has recently been revitalized ([27]). The microlensing planet search conducted by the OGLE project tightly relies on the follow-up collaborative efforts with the participating projects/instruments shown in the lower right.

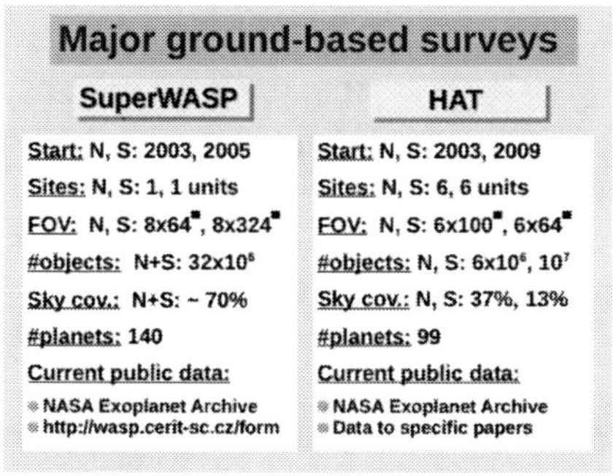

Figure 3. Brief list of the main characteristics of the two major ground-based transiting extrasolar planet surveys. SuperWASP has recently switched to smaller diameter lenses at the southern site. Earlier field of view was the same as that at the northern site. Current sky coverages are based on private communications (SuperWASP) and on recently published papers (HAT-N/S, [16, 48]). Planet statistics are based on `http://exoplanet.eu` and contain a few commonly discovered systems.

Some of the basic characteristics of the two major survey programs are listed in Figure 3. SuperWASP[8] consists of two observing units. One (SuperWASP-North) is in La Palma (Spain), while

[5]Gaia is not specified for planetary system search but will obviously yield a great contribution to it – mostly via the precise position measurements, but even through its low cadence photometry, once it is combined with dedicated ground-based follow-up observations (see [32]).

[6]`https://www.cfa.harvard.edu/MEarth/Welcome.html`

[7]`http://www.ati.ulg.ac.be/TRAPPIST/Trappist_main/Home.html`

[8]`http://www.superwasp.org`

the other (SuperWASP-South) is at the South African Astronomical Observatory. Each unit consists 8 telephoto lenses (of 11 cm and 7 cm diameter) attached to 2K × 2K CCD cameras enabling to take images on a very large chunk of the sky. The more detailed description of SuperWASP can be found in [73] and an update in [84].

The HAT project went through two major steps in the development. The survey on the northern hemisphere, HATNet[9], started in 2004, and consists of two sites (one at Mauna Kea Observatory/Hawaii and the other at the Fred Lawrence Whipple Observatory/AZ) with altogether 6 observing units, each equipped with a 11 cm diameter telephoto optics and a 4K × 4K format CCD camera. The southern counterpart (HATSouth[10]) is a bigger undertaking, with observing units different in design from those of HATNet. Established in 2009, HATSouth consists of three sites: Las Campanas Observatory (Chile), High Energy Stereoscopic System (HESS, Namibia) and Siding Spring Observatory (Australia). At each site there are two units with four 4K × 4K CCD cameras attached to 18 cm diameter astrographs. The detailed description of HATNet and HATSouth can be found in [9] and [12], respectively.

Both projects made (and are making) important discoveries in the field of extrasolar planets. With the spread in longitude, careful data acquisition, in-depth treatment of the data and longevity of the projects, these surveys are capable of detecting short period Hot Neptunes (e.g., HAT-P-11b [11] and HAT-P-26b [39]), Warm Jupiters with periods longer than 10 days (e.g., HAT-P-15 [52], WASP-130 [41], HATS-17 [20]) and HJs with shallow transits around evolved stars (e.g., WASP-73 [31], HAT-P-50 [40]). Both projects make the data used in their published papers available for the public. In addition, SuperWASP deposited all light curves gathered between 2004 and 2008 in a searchable format at the NASA/IPAC exoplanet site[11].

3 Impact on data analysis – curing the red noise syndrome

As emphasized in Sect. 2, wide-field surveys have to satisfy certain conditions in order to meet the original science goal. Because in the transit surveys all "photometrically sound" objects are searched for signals, any minuscule detail counts. Therefore, it is not surprising that a considerable effort has been made to achieve the highest data quality possible for a given instrumentation. In this section we briefly summarize the various methods that are intended to reach this goal.

As follows from their very names, wide-field surveys cover a large part of the sky even if we consider only the single exposure images. Therefore, already the basic method of reduction (ensemble photometry, e.g., [46]) should be modified ([9]), considering vignetting, differential reddening and refraction, change in the size and shape of the point spread function (PSF). Furthermore, even after taking all these into account, most of the stars will still contain some residual scatter due to uncured (or improperly considered) observational and environmental effects. Although a considerable filtering of these effects is possible with the aid of Differential Image Analysis (DIA, [4], or OIS, [2]), even this, more in-depth treatment cannot cure all remaining systematics ([29]). Therefore, *post-processing* is an inevitable step in modern photometric surveys.

Differing in the details in the implementation, all these methods are based on the general idea that systematics are characterized by some combination of: a) the common perturbations in many objects in a given field, b) perturbations depending on external parameters, associated with each star, allowing peculiar (non-common) types of variability. Examples of type a) systematics are the nightly variation of the transparency, the trace of the improperly corrected track of an airplane or that of a cosmic ray.

[9] `http://hatnet.org/`

[10] `http://hatsouth.org/`

[11] Similar (albeit less extensive) depository exists for the KELT project and fractional data releases for TrES and XO – see `http://exoplanetarchive.ipac.caltech.edu`.

Table 1. Major red noise filtering methods developed during the past 12 years.

Acronym	Expanded name	Brief description*	Ref.
TFA	Trend Filtering Algorithm	Cotrending by the stars in the field using simple least squares	[53]
SysRem	Systematics Removal	Iterative step-by-step cotrending by the stars in the field, using PCA-related method	[89]
SARS	Simultaneous Additive and Relative Sysrem	**SysRem** extended to additive and multiplicative red noise	[68]
EPD	External Parameter Decorrelation	Using parameters (e.g., object pixel coordinates) external to the light curve	[11]
PDC	Presearch Data Conditioning	Cotrending by using Bayesian approach with PCA-selected template light curves	[87] [85]
MarPLE	Marginalized Probability of a Lone Eclipse	Using Bayesian method to find single transit events and the period to combine them	[14]
TERRA	Transiting Exoearth Robust Reduction Algorithm	Cotrending by employing least squares fit, PCA and template outlier selection	[72]
ARC	Astrophysically Robust Correction	Cotrending by using Bayesian approach with entropy criterion and PCA to select the essential contribution from the templates	[75]
SFF	Self Flat Fielding	Systematics, due to the roll correction of the *Kepler* (K2) spacecraft, are filtered out by using pixel positions as external parameters	[93]
CPM	Casual Pixel Model	Using pixel-level autoregressive model to predict systematics during transit	[95]
GPM	Gaussian Process Model**	Signal search by using Gaussian models for a simultaneous fit of the systematics and the underlying signal	[1]
FFF	Full-Fledged-Fit**	Periodic transit search by simultaneous fit for systematics and the signal	[33]
SIP	Systematics Insensitive Periodogram	FFF extended to sinusoidal signals	[7]
DOHA	capital of Qatar	Cotrending by best-correlating templates on full and nightly time bases	[64]

* The description given may not fully reflect the full depth of the method.
** Our acronym attached to the method.

Type b) systematics include flux variability due to changes in the size and shape of the PSF, or to the spatial-dependent pixel sensitivity. The idea of systematics correction can be represented perhaps in the simplest way within the framework of TFA/EPD (see Table 1). We approximate the observed flux variation $F(t)$ of our target of interest by the following expression:

$$\hat{F}(t) \quad = \quad F_0(t) + N(t) + \sum_{j=1}^{M} a_j X_j(t) + \sum_{i=1}^{K} b_i E_i(t) \quad , \tag{1}$$

where $F_0(t)$ is the "true" signal (free from systematics and random – uncorrelated – noise); $N(t)$ is the instrumental/environmental white noise component; and $\{X_j(t) : j = 1, 2, ..., M\}$, $\{E_i(t) : i = 1, 2, ..., K\}$ are, respectively, the M template/cotrending time series of a representative group of objects in the field and the K external parameters. The regression coefficients $\{a_j\}$ and $\{b_i\}$ are determined by some multiple regression method, in the simplest case by standard least squares. Since $F_0(t)$ is not known a priory, the standard approach in the case of signal search is to assume that systematics dominate the observed signal and set $F_0(t)$ equal to zero. The period search is performed on the time series obtained after the subtraction of the so-determined contribution of systematics. Once the period is found, the signal reconstruction phase follows, in which the complete signal model, represented by Eq. (1), is solved. For transit signals this step is iterative, whereas for multiperiodic signals, representable by Fourier series, the solution can be obtained in a single step ([51]).

Please note that the philosophy of all systematics filtering methods is fundamentally different from the standard ensemble approach. In the latter we *do not fit* the target time series but simply use the time-dependent total (ensemble) flux to correct for the effects of atmospheric variations by dividing the instantaneous target fluxes by the corresponding ensemble fluxes. Under ideal circumstances, this method filters out only the atmospheric component of the variation, but leaves the intrinsic variation of the object intact. On the other hand, all systematics correction methods listed in Table 1 *distort* the signal to some degree, and the main purpose of many methods is to minimize this effect without jeopardizing the filter efficiency too much. This can be attempted by using only the "essential" cotrending time series (estimated by Principal Component Analysis, e.g., TERRA) and formulate the problem in the Bayesian framework to allow proper weightings of the different contributions to the systematics (e.g., PDC, ARC). One may also try to employ the full model (see eq. 1) already in the period search phase, by assuming a certain type of signal with limited number of parameters. Surprisingly, this highly time consuming method can be implemented by a proper grid-search algorithm that runs with an acceptable speed ([7, 33]). Yet another way of running "full-fledged" period searches is to employ a Gaussian Process model that allows to weigh the different contributions to the signal by using properly chosen correlation kernels ([1]). MarPLE, the method developed for transit search on the sparse data of MEarth ([14]), is aimed also at conserving the signal shape, by using single nights to find single transits and combine these results within the Bayesian framework on the full dataset to find the period.

Unfortunately, these works lack a thorough comparison with the more traditional approach, in which the data are filtered first for systematics, and then, the signal search is performed on these filtered data (containing considerably less systematics and the signal – albeit the latter with various degrees of distortion). In a subsequent study by [54] this comparison was performed on the subsets of the HATNet and K2 databases. They found that full-fledged methods may have limited applicability due to problems related to the proper disentangling of the signal and systematics in the observed signal. It seems that the traditional approach (filter first, then analyze) works better.

4 Selected highlights – the variable star aspects

In addition to the highly competitive nature of the field of extrasolar planets, ground-based wide-field surveys have been making a significant impact also on various fields of variable star research. So far, altogether over a hundred papers have been published by these projects on objects (directly) unrelated to exoplanets. Several of them have received a considerable attention from the particular field (e.g., stellar rotations among Galactic field K–M stars [38] and in the open clusters Come Berenices [26] and in the Hyades and Praesepe [30]). The fields covered by these "side" studies range from rotating asteroids ([70]) through classical pulsating stars ([81]) to supernovae ([80]). It is impossible to summarize all these topics in the depth they deserve. Therefore, here we go through the main fields

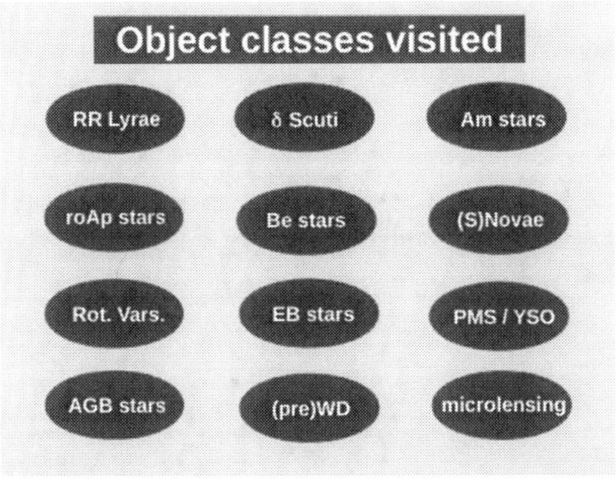

Figure 4. Classes of variable stars studied in various works using the data acquired by ground-based wide-field surveys.

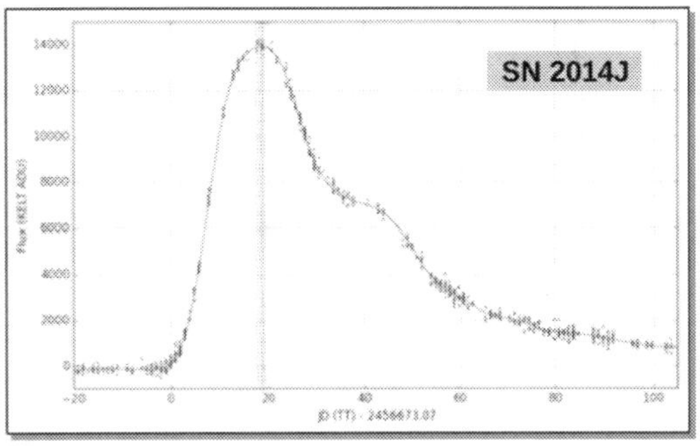

Figure 5. Detection of the pre- and post-outburst states of SN 2014J by using the survey data of the KELT project ([80]). (Reproduced with the authors' permission from `https://arxiv.org/`.)

of study only briefly. Nevertheless, a few topics (reflecting mainly the prejudice of the author of this review) will be discussed somewhat deeper.

Except for some special topics (e.g., Solar System bodies [70]), the main types of objects visited by the survey-related investigations are shown in Figure 4. For curiosity, we also list "microlensing", albeit this topic is limited to a single project only (for testing the ability of HATNet to detect microlensing events toward the Galactic Bulge [66]).

- *SN 2014J and eruptive stars:* By the very nature of the supernova explosions, they are usually discovered in the brightest phase, although some of them are caught already during rising light (e.g., [79]). On the other hand, the observation of the *pre-supernova* phase of SN 2014J in M82 ([80]) was absolutely unique until three similar events, back to 2011 and 2012, were disclosed in [69], by using the archive of the *Kepler* satellite. The full light curve, as observed by the KELT project, is shown in Figure 5. Analysis of the pre-burst state is very important for making a distinction between the various progenitor scenarios leading to type Ia supernovae explosions. In the particular case of SN 2014J, other (e.g., HST, X-ray and radio) observations exclude the single degenerate+large companion model, but white dwarf mergers are allowed both by the above data and by the pre-burst light curve ([36]).

  Similarly exciting are the discovery of the precursor of Nova Sco 2008 ([91]) and the quite recent announcement of the nova candidate KIC 9832227 for an outburst in 2022 ([65]). In both cases the precursors are W UMa binaries, and the prediction in the latter case relies on the observation of a similar steep decrease in the orbital period as for V1309 Sco, the binary progenitor of Nova Sco 2008. Both works utilize survey data, including those coming from wide-field projects (NSVS, ASAS, SuperWASP).

  For other eruptive phenomena, we draw attention to several studies performed by utilizing the SuperWASP survey. A handful of novae (recurrent and transient) were investigated by [63]. Other works on eruptive variables can be found in [22] and [90].

- *Pulsating subdwarf B (sdB) stars:* SuperWASP pioneered the extension of variability search in the frequency regime close to or above the sampling rate used in the standard exoplanet survey mode.

 DOI: 10.1051/epjconf/201715201005

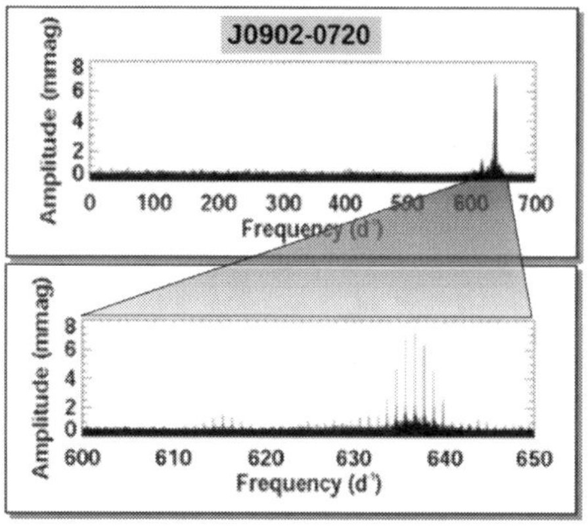

Figure 6. Frequency spectrum of the sdB variable J0902 discovered by [44] using the survey data of the SuperWASP project. *Upper panel:* full spectrum. *Lower panel:* spectrum zoomed in around the high-power regime. Please note the absence of low-frequency alias due to the lack of strong periodicity in the data sampling. (Adapted with the authors' permission from `https://arxiv.org/`.)

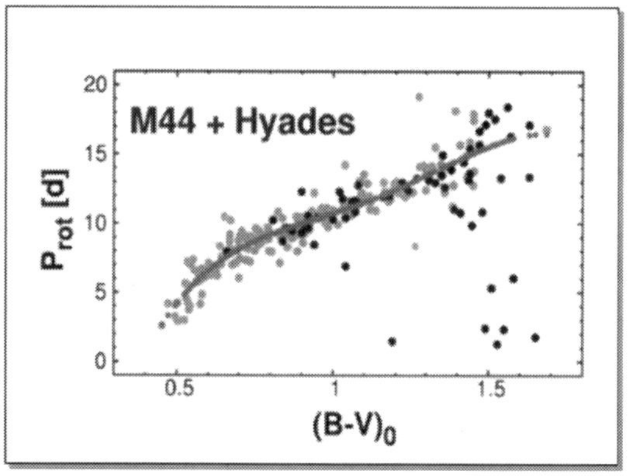

Figure 7. Color-period diagram for the rotational variables of the open clusters Praesepe and Hyades ([50]). Red dots show the fourth-order polynomial fit to the combined data of Praesepe (gray points, HATNet) and the Hyades (black points, SuperWASP [30]). The two clusters have similar ages, leading to similar ridges for the rotationally settled stars. (© ESO; reproduced with permission.)

This has led to various exciting discoveries, including rapidly oscillating Ap stars (see later) and pulsating sdB stars (see Fig. 6).

- *Rotating stars in open clusters:* Stellar rotation can be exhibited through the photometric variation induced by the varying spot coverage of the visible hemisphere of the star. Prior to the wide-field surveys, specific targets in the field and open clusters were the subjects of individual projects, often facing with the difficulties detecting low-amplitude non-stationary signals on the time scale of few days to months. On the other hand, this kind of variability is in the comfort zone of the wide-field surveys. As already noted at the beginning of this section, several pioneering works have been made on open clusters and on the field by SuperWASP, KELT and HATNet.

  Figure 7 shows the combined diagnostic diagram of stellar rotation for two famous clusters of the same age. The method of gyrochronological age determination of field stars is based on these kinds of diagrams ([13]; see, however, the current controversy on the applicability of this method to field stars: [21, 50, 94].)

- *δ Scuti stars and variability of A-type stars:* Stars on and close to the main sequence with spectral types A to F are known to exhibit low-amplitude, short period pulsations, often identified as p- or g-mode pulsations or some mixture of these. One of the unanswered questions concerning these variables is the dependence of the occurrence of the pulsations on the physical parameters. In spite of the theoretical expectations, only a fraction of the stars pulsate in the instability strip. Low sensitivity of the variability surveys is no longer an issue. Indeed, even with the exquisite accuracy of the *Kepler* space telescope, we find that only 12% of the stars in the instability strip are δ Scuti or γ Dor (or mixed) variables (44% are spotted or eclipsing variables, and the remaining 44% are constant [19] – see also [18] for a more extended survey). Most of the variables have amplitudes above 1 mmag, i.e., within the reach of ground-based wide-field surveys. Indeed, many papers have been published on these stars, most of them utilizing the SuperWASP data.

One of the most exciting findings was the discovery of the large occurrence of pulsators among Am stars (although rare occurrence of Am star pulsations was known from sparse earlier works – e.g., [56]). From the examination of 1600 Am stars, [82] found 200 variables exhibiting pulsations similar to those of "normal" δ Scuti or γ Dor stars. The excitation mechanism of these stars could result from some delicate mixture of gravitational depletion of metals, rotation and turbulence (see the spectroscopic survey of Am stars by [83]). Further examination of 1.5 million stars in the F–B spectral range also in the SuperWASP database by [43] has led to the discovery of additional Am stars and 10 rapidly oscillating Ap stars (a class of chemically peculiar stars, discovered unexpectedly by [55] in 1982).

- *Other diversities:* Unfortunately, we have space only to name a few of the many exciting results: dimming episodes in young stellar objects ([76, 77]); active main sequence B stars ([57]); RR Lyrae stars with Blazhko effects ([81, 86]); AGB stars ([8]); eclipsing variables: quadruple (quintuple?) system ([59]), the longest eclipse ever measured ([78]; see also [58]), eclipsing variable close to the short period limit ([60]), etc.

5 Conclusions

We cannot conclude without emphasizing the great importance of the Northern Sky Variability Survey, NSVS (based on the observations of the Robotic Optical Transient Search Experiment, ROTSE-I, PI: *Carl Akerlof*) and the All Sky Automated Survey, ASAS (PI: *Grzegorz Pojmanski*) that proved the concept, the significance and the competitive scientific role of small, fully robotic instruments. We should also recall the indefeasible role of *Bohdan Paczynski*, who was a very strong supporter of small telescope variability surveys.

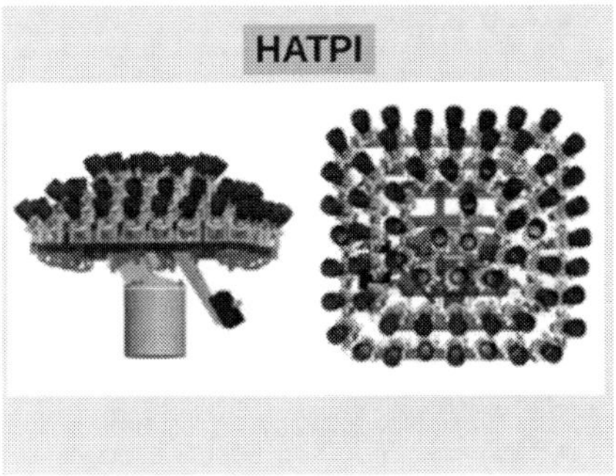

Figure 8. Computer drawing of the HATPI platform, hosting 63 instrument-holder units (with optics, CCD and fine-pointer mechanics). The cameras will take images in every 30 seconds virtually on the full visible sky at the Las Campanas Observatory. The project is in the active phase of development. When operational, it will be the biggest undertaking relative to other similar projects, such as EVRYSCOPE (`http://evryscope.astro.unc.edu/`), FLY'S EYE (`https://flyseye.net/`) and MASCARA (`http://mascara1.strw.leidenuniv.nl/`).

Today's wide field surveys grew up from the exciting and highly challenging idea of discovering extrasolar planetary systems by using the rare events of tiny dimmings of light when the planet moves across the line of sight between the star and the observer. These projects have reached the level of efficiency when the discovery of this kind of events became standard. New methods of time series analysis have been developed, fertilizing other fields of research and helping in the development of the proper filtering of the data acquired by the various space missions. The data gathered by these projects are immense in size and we are only at the beginning of utilizing the millions of light curves observed over the past 15 years. The examples shown in this summary suggest the wide range of applicability of these databases.

Continuing and further developing ground-based wide field surveys will remain a significant goal in observational astronomy, in spite of highly competitive ground- and space-based projects (e.g., LSST, TESS). Full utilization of these data (including merging the different databases and their combination with other data from the ground and space) is still ahead. New projects are in progress (e.g., see Fig. 8), aiming at more continuous sky monitoring with the goal of covering a great variety of astrophysical phenomena from transiting planets to supernovae. We are sure that these assets will constitute an integral part of future variable star studies.

Acknowledgments: Many thanks are due to the organizers of this conference to keep alive the reputation of the Los Alamos stellar pulsation meetings and for the cordial spirit throughout the conference. We gratefully acknowledge the information received on current project statuses from Don Pollacco, Richard West and Andrew Collier Cameron (SuperWASP), from Gáspár Bakos (HAT) and from Joshua Pepper (KELT). Prompt responses to my questions from Keivan Stassun and Daniel Holdsworth are appreciated. Thanks are due for the financial support received from the organizers and from the Directory Fund of the Konkoly Observatory.

References

[1] Aigrain, S., Hodgkin, S. T., Irwin, M. J., et al., MNRAS, **447**, 2880 (2015)
[2] Alard, C., & Lupton, R. H. ApJ, **503**, 325 (1998)
[3] Alcock, C., Allsman, R. A., Axelrod, T. S., et al., AJ, **111**, 1146 (1996)
[4] Alcock, C., Allsman, R. A., Alves, D., et al., ApJ, **521**, 602 (1999)
[5] Alonso, R., Brown, T. M., Torres, G., et al., ApJ. Letters, **613**, 153 (2004)
[6] Alsubai, K. A., Mislis, D., Tsvetanov, Z. I., et al., AJ, **153**, 200 (2017)
[7] Angus, R., Foreman-Mackey, D., Johnson, J. A., ApJ, **818**, 109 (2016)
[8] Arnold, R. A., Pepper, J., Rodriguez, J. E., et al., AAS, meeting#225, id.449.16 (2015)
[9] Bakos, G., Noyes, R. W., Kovacs, G., et al., PASP, **116**, 266 (2004)
[10] Bakos, G. Á., Noyes, R. W., Kovacs, G., et al., ApJ, **656**, 552 (2007)
[11] Bakos, G. Á., Torres, G., Pál, A., et al., ApJ, **710**, 1724 (2010)
[12] Bakos, G. Á., Csubry, Z., Penev, K., et al., PASP, **125**, 154 (2013)
[13] Barnes, S. A., ApJ, **586**, 464 (2003)
[14] Berta, Z. K., Irwin, J., Charbonneau, D., et al., AJ, **144**, 145 (2012)
[15] Berta-Thompson, Z. K., Irwin J., Charbonneau, D., Nature, 527, 204 (2015)
[16] Bhatti, W., Bakos, G. Á, Hartman, J. D., et al., preprint (arXiv:1607.00322) (2016)
[17] Borucki, W. J., Koch, D. G., Batalha, N., et al., ApJ, **745**, 120 (2012)
[18] Bowman, Dominic M., Kurtz, Donald W., Breger, Michel, et al., MNRAS, **460**, 1970 (2016)
[19] Bradley, P. A., Guzik, J. A., Miles, L. F., AJ, **149**, 68 (2015)
[20] Brahm, R., Jordán, A., Bakos, G. Á., et al., AJ, **151**, 89 (2016)
[21] Buzasi, D., Lezcano, A., & Preston, H. L., J. Space Weather Space Clim., **6**, A38 (2016)
[22] Byckling, K., Osborne, J. P., & Wheatley, P. J., MNRAS, **399**, 1576 (2009)
[23] Charbonneau., D., Brown, T., Latham D., & Mayor, M., ApJL, **529**, 45 (2000)
[24] Charbonneau, D., Berta, Z. K., & Irwin, J., Nature, **462** 891 (2009)
[25] Collier Cameron, A., Bouchy, F. Hébrard, G., et al., MNRAS, **375**, 951 (2007)
[26] Collier Cameron, A., Davidson, V. A., Hebb, L., et al., MNRAS, **400**, 451 (2009)
[27] Crouzet, N., McCullough, P. R., Long, D., et al., AJ, **153**, 94 (2017)
[28] Deeg, H. J., Alonso, R., Belmonte, J. A. et al., PASP, **116**, 985 (2004)
[29] Dekany, I. & Kovacs, G., A&A, **507**, 803 (2009)
[30] Delorme, P., Collier Cameron, A., Hebb, L., et al. MNRAS, **413**, 2218 (2011)
[31] Delrez, L., Van Grootel, V., Anderson, D. R., et al., A&A, **563**, A143 (2014)
[32] Dzigan, Y., & Zucker, S., MNRAS, **428**, 3641 (2013)

[33] Foreman-Mackey, D., Montet, B. T., Hogg, D. W. et al., ApJ, **806**, 215 (2015)
[34] Gillon, M., Jehin, E., Lederer, S. M., et al., Nature, **533**, 221 (2016)
[35] Gillon, M., Triaud, A. H. M. J., Demory, B.-O., et al., Nature, **542**, 456 (2017)
[36] Goobar, A., Kromer, M., Siverd, R., et al., ApJ, **799**, 106 (2015)
[37] Guo, X., Johnson, J. A., Mann, A. W., et al., ApJ, 838, 25 (2017)
[38] Hartman, J. D., Bakos, G. Á., Noyes, R. W., et al., AJ, **141**, 166 (2011)
[39] Hartman, J. D., Bakos, G. Á., Kipping, D. M., et al., ApJ, **728**, 138 (2011)
[40] Hartman, J. D. W. Bhatti, W., Bakos, G. Á., et al., AJ, **150**, 168 (2015)
[41] Hellier, C., Anderson, D. R., Cameron, A. C., Delrez, L., et al., MNRAS, **465**, 3693 (2017)
[42] Henry, G. W., Marcy, G. W., Butler, R. P., Vogt, S. S., ApJL, **529**, 41 (2000)
[43] Holdsworth, D. L., Smalley, B., Gillon, M., et al., MNRAS, **439** 2078 (2014)
[44] Holdsworth, D. L., Østensen, R. H., Smalley, B., et al., MNRAS, **466**, 5020 (2017)
[45] Holman, M. J., Fabrycky, D. C., Ragozzine, D., et al., Science **330**, 51 (2010)
[46] Honeycutt, R. K., PASP, **104**, 435 (1992)
[47] Irwin, J. M., Berta-Thompson, Z. K., Charbonneau, D., et al., in *18th Cambridge Workshop on Cool Stars, Stellar Systems, and the Sun*, ed. G. van Belle & H.C. Harris, p. 767 (2015)
[48] Juncher, D.; Buchhave, L. A., Hartman, J. D., et al., PASP, **127**, 851 (2015)
[49] Konacki, M., Torres, G., Jha, S., & Sasselov, D. D., Nature, **421**, 507 (2003)
[50] Kovacs, G., A&A, **581**, A2 (2015)
[51] Kovacs, G., & Bakos, G. A., CoAst, **157**, 82 (2008)
[52] Kovacs, G., Bakos, G., Hartman, J. D., et al., ApJ, **724**, 866 (2010)
[53] Kovacs, G., Bakos, G., Noyes, R. W., MNRAS, **356**, 557 (2005)
[54] Kovacs, G., Hartman, J. D., Bakos, G. Á., A&A, **585**, 57 (2016)
[55] Kurtz, D. W., MNRAS, **200**, 807 (1982)
[56] Kurtz, D. W., Garrison, R. F., Koen, C., MNRAS, **276**, 199 (1995)
[57] Labadie-Bartz, J., Pepper, J., McSwain, M. V., et al., AJ, **153**, 252 (2017)
[58] Lipunov, V., Gorbovskoy, E., Afanasiev, V. et al., A&A, **588**, A90 (2016)
[59] Lohr, M. E., Norton, A. J., Kolb, U. C., et al., A&A, **549**, A86 (2013)
[60] Lohr, M. E., Hodgkin, S. T., Norton, A. J., et al., A&A, **563**, A34 (2014)
[61] Marcy, G. W., & Butler, R. P., AAS meeting#187, id.70.04 (1995)
[62] Mayor, M. & Queloz, D., Nature, **378**, 355 (1995)
[63] McQuillin, R., Evans, A., Wilson, D., et al., MNRAS, **419**, 330 (2012)
[64] Mislis, D., Pyrzas, S., Alsubai, K. A., et al., MNRAS, **465**, 3759 (2017)
[65] Molnar, L. A., Van Noord, D., Kinemuchi, K., et al., AAS Meeting #229, id.417.04 (2017)
[66] Nataf, D. M., Stanek, K. Z., Bakos, G. A., AcA, **59**, 255 (2009)
[67] Newton, E. R., Irwin, J., Charbonneau, D., et al., ApJ, **821**, 93 (2016)
[68] Ofir, A., Alonso, R., Bonomo, A. S., et al., MNRAS, **404**, L99 (2010)
[69] Olling, R. P., Mushotzky, R., Shaya, E. J., et al., Nature, **521**, 332 (2015)
[70] Parley, N. R., Green, S. F., McBride, N., et al., LPI Contr. #1405, 8049 (2008)
[71] Pepper, J., Gould, A., & Depoy, D. L., AcA, **53**, 213 (2003)
[72] Petigura, E. A., & Marcy, G. W., PASP, **124**, 1073 (2012)
[73] Pollacco, D. L., Skillen, I., Collier Cameron, A., et al., PASP, **118**, 1407 (2006)
[74] Quintana, E. V., Barclay, T., Raymond, S. N., et al., Science, **344**, 277 (2014)
[75] Roberts, S., McQuillan, A., Reece, S., Aigrain, S., MNRAS, **435**, 3639 (2013)

[76] Rodriguez, J. E., Pepper, J., Stassun, K. G., et al., AJ, **146**, 112 (2013)
[77] Rodriguez, J. E., Stassun, K. G., Cargile, P., et al., ApJ, **831**, 74 (2016)
[78] Rodriguez, J. E., Stassun, K. G., Lund, M. B., et al., AJ, **151**, 123 (2016)
[79] Shappee, B. J., Piro, A. L., Holoien, T. W.-S., ApJ, **826**,144 (2016)
[80] Siverd, R. J., Goobar, A., Stassun, K. G., Pepper, J., ApJ, **799**, 105 (2015)
[81] Skarka, M., A&A, **562**, A90 (2014)
[82] Smalley, B., Kurtz, D. W., Smith, A. M. S., et al., A&A, **535**, A3 (2011)
[83] Smalley, B., Antoci, V., Holdsworth, D. L., et al., MNRAS, **465**, 2662 (2017)
[84] Smith, A. M. S., & The WASP consortium, Contrib. Astr. Obs. Sklanate Pleso, **43**, 500 (2014)
[85] Smith, J. C., Stumpe, M. C., Van Cleve, J. E., et al., PASP, **124**, 1000, (2012)
[86] Sódor, Á., Jurcsik, J., Molnár, L., et al., ASPC, **462**, 228
[87] Stumpe, M. C., Smith, J. C., Van Cleve, J. E., et al., PASP, **124**, 985, (2012)
[88] Talens, G. J. J., Spronck, J. F. P., Lesage, A.-L., et al., A&A, **601**, A11 (2017)
[89] Tamuz, O., Mazeh, T., Zucker, S., MNRAS, **356**, 1466 (2005)
[90] Thomas, N. L., Norton, A. J., Pollacco, D., et al., A&A, **514**, A30 (2010)
[91] Tylenda, R., Hajduk, M., Kamiński, T., et al., 2011, A&A, **528**, 114 (2011)
[92] Udalski, A., Zebrun, K., Szymanski, M., et al., AcA, **52**, 115 (2002)
[93] Vanderburg, A., & Johnson, J. A., PASP, **126**, 948 (2014)
[94] van Saders, J. L., Ceillier, T., Metcalfe, T. S., Nature, **529**, 181 (2016)
[95] Wang, D., Hogg, D. W., Foreman-Mackey, D., & Schölkopf, B., PASP, **128**, 4503 (2016)

Pulsating stars in SuperWASP

Daniel L. Holdsworth[1,*]

[1]*Jeremiah Horrocks Institute, University of Central Lancashire, Preston PR1 2HE, UK*

Abstract. SuperWASP is one of the largest ground-based surveys for transiting exoplanets. To date, it has observed over 31 million stars. Such an extensive database of time resolved photometry holds the potential for extensive searches of stellar variability, and provide solid candidates for the upcoming TESS mission. Previous work by e.g. [15], [5], [12] has shown that the WASP archive provides a wealth of pulsationally variable stars. In this talk I will provide an overview of the SuperWASP project, present some of the published results from the survey, and some of the on-going work to identify key targets for the TESS mission.

1 Introduction

The WASP project is a wide-field survey for transiting exoplanets. The project is a two-site campaign: the first instrument is located at the Observatorio del Roque de los Muchachos on La Palma and achieved first light in 2003 November; the second is located at the Sutherland Station of the SAAO and achieved first light in 2005 December. Each instrument consists of eight 200-mm, f/1.8 Canon telephoto lenses backed by Andor CCDs of 2048 × 2048 pixels observing ∼61 deg^2 each through broad-band filters covering a wavelength range of 4000 − 7000 Å (Figure 1; [13]). This set-up enables simultaneous observations of up to eight fields with a pixel size of 13.7 arcsec. The instruments capture two consecutive 30-s integrations at a given pointing, and then move to the next observable field. Typically, fields are revisited every 10 min.

The images collected are passed through the reduction pipeline, where the data are corrected for primary and secondary extinctions, the instrumental colour response and the system zero-point. The atmospheric extinction correction uses a network of stars with a known $(B-V)$ colour to determine the extinction terms, which are then applied to all extracted stars using an assumed colour of G-type stars. This process results in a "WASP V" magnitude which is comparable to the Tycho-2 V_t passband. The data are also corrected for systematic errors using the SysRem algorithm of [18].

Aperture photometry is performed at stellar positions provided by the USNO-B1.0 input catalogue ([10]). Stars brighter than ∼15$^{\rm th}$ magnitude are extracted. Data are stored in FITS format with labels of the observed field, camera and date of observation. Such a configuration and extended time-base allow the extraction of multiple light curves for each object based on either date, field or camera. The extracted photometry achieves a precision of better than 1% for stars brighter than 11$^{\rm th}$ magnitude, and 0.5% for stars brighter than magnitude 9.4.

*dlholdsworth@uclan.ac.uk

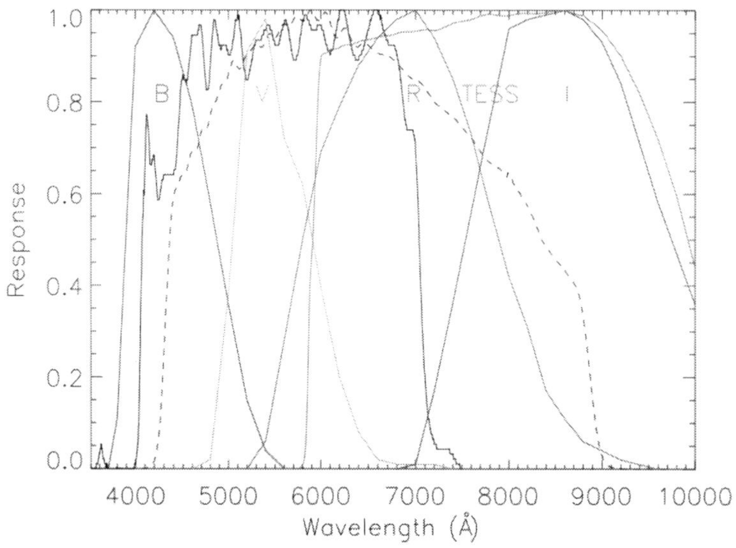

Figure 1. Comparison of the WASP filter (solid black line) with that used by *Kepler* (dashed line), *TESS* (pink line) and Johnson filters (coloured lines as labelled).

To date, there are over 428 billion data points in the archive covering over 31 million unique objects. With such a large data base of objects it is possible to search for a wide variety of stellar variability.

2 WASP capabilities

Before any analysis can be conducted, a thorough investigation into the capabilities of the data is required. An initial characterisation was conducted by [15] when investigating pulsations in Am stars, but a more detailed investigation was conducted by [7] as part of a global search for pulsations in hot stars.

By using the results of the *Kepler* mission as a benchmark, the latter study showed that for stars of magnitude 11 and brighter, it is possible to detect variability to an amplitude of 0.5 mmag (Fig. 2). This is, however, also dependant on the length of time-series available, blending of sources in the photometric aperture, and inherent noise in the data (seeing conditions, moon phase, etc.).

With knowledge of the detection limits of the WASP data, we are able to rule out variable stars which we know have amplitudes below that of our limit. We are, therefore, able to restrict variability studies to hot stars which show "high" amplitude pulsations, such as δ Scuti stars, subdwarf stars, RR Lyrae stars, and Cepheids.

3 Low-frequency variables

[7] identified over 80 000 stars with variability in the frequency range $5 - 50\,\mathrm{d}^{-1}$. However, studies of identified targets require detailed and time consuming analysis, on a star by star basis. This frequency range is plagued by the daily aliases suffered by single-site ground-based observations. Alongside this, harmonics of binary stars serve to add many false-positive detections of pulsational variability.

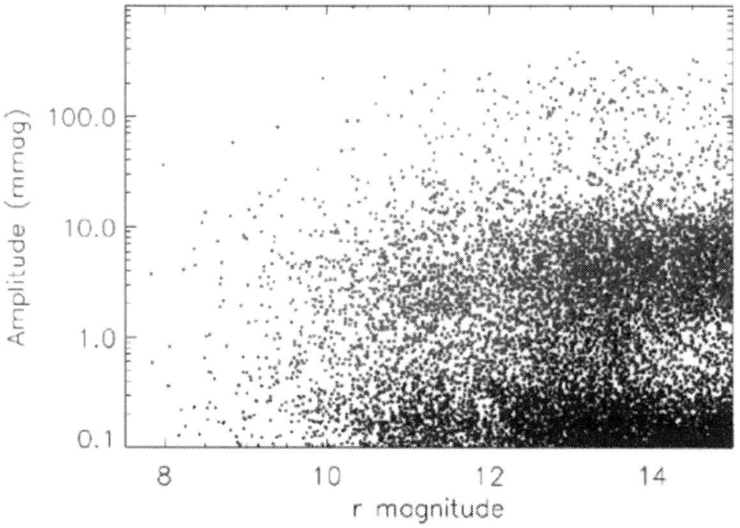

Figure 2. The detected peaks (red) and the undetected peaks (blue) of *Kepler* pulsations as analysed with WASP photometry. The detection rate is a function of stellar brightness, with WASP able to detect variability to 0.5 mmag for stars brighter than 11th magnitude.

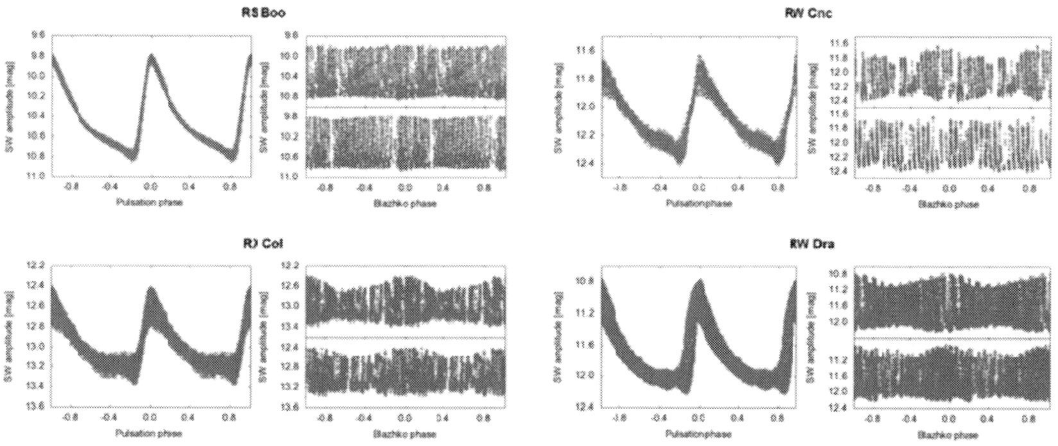

Figure 3. A selection of RR Lyrae stars identified in WASP photometry. The amplitude variability as a result of the Blazhko effect can be clearly seen. Figure from [14]. Reproduced with permission.

However, it is not all bad news in this frequency range. Studies by [15] and [17] detected γ Doradus and δ Scuti pulsations in Am stars, and [2] successfully identified rotational modulation signals in chemically peculiar stars.

Furthermore, stars with high amplitude variability are much easier to detect above the low-frequency noise and artefacts. [14] studied RRab stars in the SuperWASP archive and found nine new Blazhko stars in their sample (see Figure 3). More recently, Greer (`http://www.star.uclan.ac.uk/stars2016/slides/Greer.pdf`) is working on almost 5 000 RR Lyrae stars (of the ab type) in the SuperWASP archive – the most extensive survey of these stars in the SuperWASP archive to date – and is identifying hundreds of new Blazhko candidates.

4 High-frequency variables

Moving into the high-frequency range, [5] published an extensive list of variable stars with frequencies above 50 d^{-1}. Within the results were δ Sct pulsators, rapidly oscillating Ap (roAp) stars, and subdwarf B (sdB) variables.

The most significant results are the discovery of 11 new roAp stars. These are rare stars, with only 61 known (see [16] for a catalogue), which pulsate in high-overtone, low-degree modes. Follow up observations of some of these discoveries are revealing that these stars are more peculiar than previously thought, and that current excitation mechanisms might not be able to explain all the observed frequencies (e.g., [6, 8]). The use of SuperWASP removed the sample biases which have been present in previous surveys of these stars, thus providing new and interesting stars to study in detail.

The sdB stars are also rather rare, with a little over 100 known in their class. [9] published three newly discovered stars of this type, after searching the SuperWASP archive. This demonstrates the power of ground-based surveys to probe to very high frequencies in the search for rapid variability.

Pre-main-sequence (PMS) stars are also present in the archive. HD 34282 was discovered to be a pulsating PMS star by [1] in a search for pulsations in Herbig Ae/Be stars to constrain the PMS instability strip. An investigation into the WASP archive for these objects also identified HD 34282 as a pulsating star (see Fig. 4). Although the light curve is very noisy at low frequency, the presence of the pulsations is clearly evident in the periodogram. There is undoubtably more of this type of star in the archive to be exploited.

5 Combining data

Large survey archives provide the possibility of extending data sets by, in some cases, decades or more. In the case of SuperWASP, this was well demonstrated by [11] and [4]. In the case of [4], the authors were able to combine WASP data with *Kepler* data on the star KIC 7106205. This star shows dramatic amplitude modulation of just one of its many modes ([3]).

By including the WASP data for this star, it was shown that the amplitude of the one mode was much greater before the start of *Kepler* observations, making the loss of pulsation energy much more dramatic. Figure 5 shows the amplitude of the variable mode over the period of both the WASP (red) and *Kepler* (blue) observations. This is just one example of such a study, with more being conducted.

6 Summary

We have shown that the SuperWASP archive is an excellent source of data for the study of a whole host of variable stars. With its short cadence and pseudorandom sampling, it is possible to probe to very high frequencies in the search for variable stars.

As well as for discovery purposes, SuperWASP is very useful for extending the data of other projects or missions before their first light. Although the WASP data do not have the precision of dedicated observations, or space-based observations, to be able to extend the time-base of any observations can often be invaluable in the long-term study of stellar variability.

With the launch of the TESS mission in 2018, and the ongoing K2 mission, the WASP archive is an excellent source of complimentary data.

Acknowledgments: The WASP project is funded and maintained by Queen's University Belfast, the Universities of Keele, St. Andrews and Leicester, the Open University, the Isaac Newton Group, the Instituto de Astrofisica Canarias, the South African Astronomical Observatory and by the STFC.

 DOI: 10.1051/epjconf/201715201006

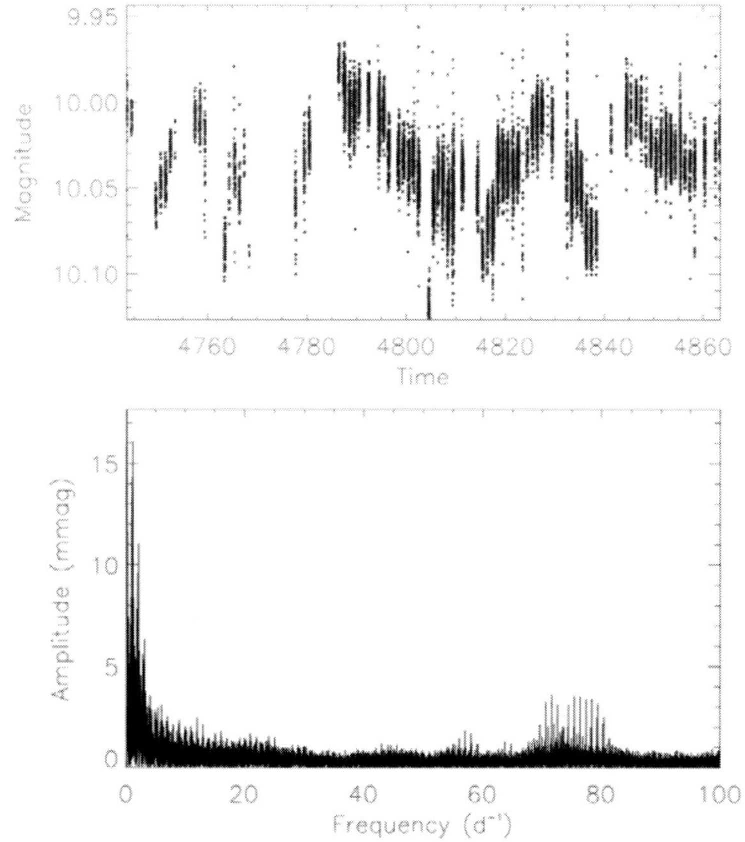

Figure 4. The variability of the known pulsating PMS star HD 34282 in the WASP data. The top panel shows the noisy light curve, with the periodogram shown in the bottom panel. The pulsations are clearly visible in the lower plot.

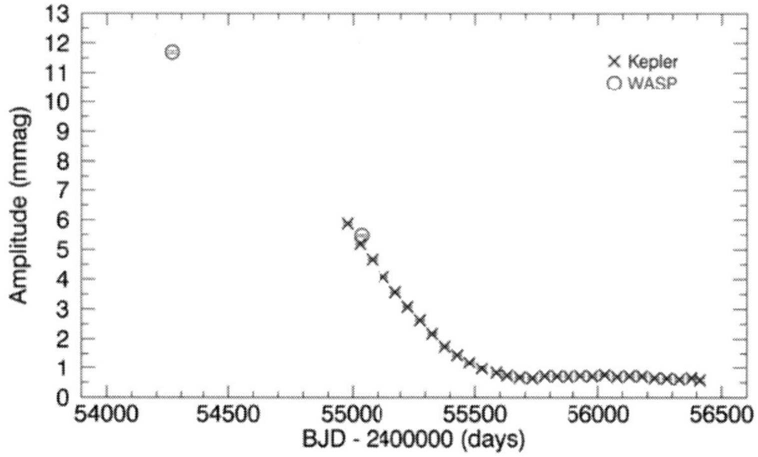

Figure 5. The amplitude variation of one mode in KIC 7106205. The original detection was made with *Kepler* data (blue), but has been extended with WASP data (red). Figure adapted from [4].

References

[1] Amado, P. J., Moya, A., Suárez, J. C., Martín-Ruiz, S., Garrido, R., Rodríguez, E., Catala, C., & Goupil, M. J., MNRAS, **352**, 11 (2004)

[2] Bernhard, K., Hümmerich, S., & Paunzen, E., AN, **336**, 981 (2015)

[3] Bowman, D. M., & Kurtz, D. W., MNRAS, **444**, 1909 (2014)

[4] Bowman, D. M., Holdsworth, D. L., & Kurtz, D. W., MNRAS, **449**, 1004 (2015)

[5] Holdsworth, D. L., Smalley, B., Gillon, M., et al., MNRAS, **439**, 2078 (2014)

[6] Holdsworth, D. L., Smalley, B., Kurtz, D. W., Southworth, J., Cunha, M. S., & Clubb, K. I., MNRAS, **443**, 2049 (2014)

[7] Holdsworth, D. L., Ph.D. Thesis (Keele University, 2015)

[8] Holdsworth, D. L., Kurtz, D. W., Smalley, B., Saio, H., Handler, G., Murphy, S. J., & Lehmann, H., MNRAS, **462**, 876 (2016)

[9] Holdsworth, D. L., Østensen, R. H., Smalley, B., & Telting, J. H., MNRAS, **466**, 5020 (2017)

[10] Monet, D. G., Levine, S. E., Canzian, B., et al., AJ, **125**, 984 (2003)

[11] Murphy, S. J., Pigulski, A., Kurtz, D. W., et al., MNRAS, **432**, 2284 (2013)

[12] Paunzen, E., Skarka, M., Walczak, P., Holdsworth, D. L., Smalley, B., West, R. G., & Janík, J., MNRAS, **453**, 1241 (2015)

[13] Pollacco, D. L., Skillen, I., Collier Cameron, A., et al., PASP, **118**, 1407 (2006)

[14] Skarka, M., A&A, **562**, A90 (2014)

[15] Smalley, B., Kurtz, D. W., Smith, A. M. S., et al., A&A, **535**, A3 (2011)

[16] Smalley, B., Niemczura, E., Murphy, S. J., et al., MNRAS, **452**, 3334 (2015)

[17] Smalley, B., Antoci, V., Holdsworth, D. L., et al., MNRAS, **465**, 2662 (2017)

[18] Tamuz, O., Mazeh, T., & Zucker, S., MNRAS, **356**, 1466 (2005)

Time-series surveys and pulsating stars: The near-infrared perspective

Noriyuki Matsunaga[1,*]

[1] *Department of Astronomy, The University of Tokyo, 7-3-1 Hongo, Bunkyo-ku, Tokyo 113-0033, Japan*

Abstract. The purpose of this review is to discuss the advantages and problems of near-infrared surveys in observing pulsating stars in the Milky Way. One of the advantages of near-infrared surveys, when compared to optical counterparts, is that the interstellar extinction is significantly smaller. As we see in this review, a significant volume of the Galactic disk can be reached by infrared surveys but not by optical ones. Towards highly obscured regions in the Galactic mid-plane, however, the interstellar extinction causes serious problems even with near-infrared data in understanding the observational results. After a review on previous and current near-infrared surveys, we discuss the effects of the interstellar extinction in optical (including *Gaia*) to near-infrared broad bands based on a simple calculation using synthetic spectral energy distribution. We then review the recent results on classical Cepheids towards the Galactic center and the bulge, as a case study, to see the impact of the uncertainty in the extinction law. The extinction law, i.e. the wavelength dependency of the extinction, is not fully characterized, and its uncertainty makes it hard to make the correction. Its characterization is an urgent task in order to exploit the outcomes of ongoing large-scale surveys of pulsating stars, e.g. for drawing a map of pulsating stars across the Galactic disk.

1 Introduction

What we call "near-infrared" (hereafter near-IR) here is the wavelength range covered by the photometric bands of JHK. A broad and general review on the photometric bands is found in [3]. Note that the K band, at around 2 μm, has variations of filter transmission in different systems, and we mainly consider the K_s band which lacks the longest-λ part ($\lambda > 2.3$ μm) of the broader K in the following discussions. The wavelength around 3.5 μm will be called "mid-IR", partly because most important datasets in this range tend to be collected with space facilities rather than ground-based telescopes today, but we also discuss some results obtained in this mid-IR range. A review on studies in the shorter-λ (I-band and shorter) range is given by Soszyński in this volume ([93]).

There are mainly three advantages of IR data for observations of pulsating stars. First, they are less affected by interstellar extinction. At around 2 μm, for example, the extinction is around 10% of that at the optical V band ([10, 80]) or even smaller (~1/16, [72]). Secondly, some objects are enshrouded in circumstellar dust, so that they are only visible or bright in the IR ([95, 96]). IR data are crucial to study properties of such dust. Thirdly, we know pulsating stars tend to show simpler characteristics in

*matsunaga@astron.s.u-tokyo.ac.jp

the IR. In particular, the period-luminosity relation of Cepheids is known to be tight and less affected by metallicity. There has been a wealth of literature on its advantage (e.g., [5, 8, 35, 42, 89]). This is of great value for high precision cosmology ([81]) and for other applications.

The following five facilities have been actively used for observing pulsating stars in the IR: Infrared Survey Facility (IRSF) in South Africa, VISTA in Chile, *Hubble Space Telescope*, *Spitzer Space Telescope*, and CTIO 1.5-m telescope with CPAPIR camera. Catalogs from all-sky surveys like 2MASS[1] and WISE[2] are very conveniently used and naturally making a giant impact on studies of pulsating stars, but they are not discussed in this review.

- The 1.4-m telescope IRSF is working with SIRIUS near-IR camera in the SAAO Sutherland Observatory, South Africa ([69, 70]). Even with the moderate field-of-view, $7.7' \times 7.7'$, the efficiency of the three detectors taking simultaneous images in three bands (JHK_s) has been producing important catalogs of variable stars with color information for many stellar systems and regions: the Galactic center and bulge ([46, 47, 49, 51]), globular clusters ([45, 91]), the Magellanic Clouds ([36, 37]), and nearby dwarf galaxies ([21, 54, 56–59, 100, 101]). Some of these results will be discussed in more detail below.
- VISTA is a 4.1-m telescope in the ESO Cerro Paranal Observatory, Chile, and is working with a wide-field near-IR camera VIRCAM. The sixteen detectors, 67 million pixels effectively covering ~0.6 deg^2, together with the relatively large telescope aperture have a very high survey ability ([94]). Among the six public surveys, two surveys are targeted at surveying pulsating stars: VISTA Variables in the Via Lactea (VVV, [60]) and VISTA survey of the Magallanic Clouds (VMC, [12]). See also Minniti et al. ([65]) and Cioni et al. ([15]) in these proceedings. While the comprehensive photometric catalog provided by the VVV has been used in many investigations about the Galactic bulge and disk ([9, 29, 30, 61, 87, 98]), its time-series data have led to new discoveries and insights into various pulsating stars like classical Cepheids ([11, 18, 19]) and RR Lyrs ([1, 17, 31, 32, 62–64]). VMC has been also providing very useful near-IR datasets for pulsating stars, classical Cepheids ([44, 68, 82, 84]), type II Cepheids ([83]), and RR Lyrs ([67]) as well as other topics such as star formation history in the Magellanic Clouds ([39, 85, 86]). In addition to time variations in brightness, both of the VVV and VMC surveys are providing proper motions ([13, 14, 33, 40]) which will be very useful to discuss the nature of variable stars and other populations. Besides the two large surveys, there is an important contribution (and will be more) from VISTA: McDonald et al. ([52, 53]) studied variable stars in the Sagittarius dwarf spheroidal galaxy.
- *Spitzer Space Telescope* is actively used to obtain important time-series datasets in the mid IR through some conspicuous projects. The Carnegie Chicago Hubble Program has been making a considerable effort on establishing the period-luminosity relation of Cepheids in the mid IR (see e.g. [25, 26, 88, 89]). The *Spitzer* Legacy Program "Surveying the Agents of a Galaxy's Evolution" (SAGE, [55]) and its related programs (SAGE-Var, in particular; [79]) collected a comprehensive dataset in the mid IR for pulsating stars and other objects in the Magellanic Clouds ([77–79]). In this proceedings book, Whitelock et al. ([102]) discuss mid-IR characteristics of large amplitude variables in the LMC and IC 1613 by combining the SAGE-Var data and other datasets.
- *Hubble Space Telescope (HST)* has been also producing important results on pulsating stars mainly for the purpose of cosmology and determining the H_0 constant in particular. While the *HST* Key Project to measure H_0 in the 1990s used optical data for detecting Cepheids in distant galaxies as much as possible ([24], and references therein), Riess and collaborators have been using near-IR data exclusively collected with an identical instrument to reduce the systematic uncertainties caused

[1]`http://irsa.ipac.caltech.edu/Missions/2mass.html`
[2]`http://irsa.ipac.caltech.edu/Missions/wise.html`

by the intrinsic characteristics of Cepheids like the metallicity effect on the period-luminosity relation and by the cross-instrument errors ([81]).

- The CPAPIR near-IR camera attached to the CTIO 1.5-m telescope was used to carry out the LMC Near-IR Synoptic Survey ([41]). Its good collection of time-series data for a wide area of the LMC has been producing important results on classical Cepheids and type II Cepheids ([4–7]).

It is clear that these surveys are providing us with useful data on pulsating stars. The IR surveys are particularly useful for exploring the large space of the Milky Way including the disk obscured by interstellar extinction, which is the main focus of the following discussions. When interesting and reddened objects are found in obscured regions, follow-up spectroscopy in the IR will be important. The demands for IR spectrographs will grow rapidly considering that a large number of pulsating stars and other objects are being found in IR surveys. APOGEE and its successor APOGEE-2 ([43]) together with other modern near-IR high-resolution spectrographs like GIANO ([74]) and WINERED ([34, 75]) will play important roles in collecting detailed information such as radial velocities and chemical abundances of pulsating stars, but this is beyond the scope of this review.

2 Expected limits of optical and near-IR surveys

In this section, we discuss limits of optical and near-IR surveys by calculating broad-band photometry with different extinction laws. We consider classical Cepheids and examine how far surveys can reach. Previous surveys of Cepheids are far from complete (see, e.g., Fig. 1 in [48]), which is mainly due to the interstellar extinction. Windmark et al. ([103]) predicted the total number and the distribution of Cepheids which can be detected in the *Gaia* survey. They obtained ~20000 based on a simple exponential-disk model[3] and the local density. They also predicted about half of them will be detected by *Gaia* with the limiting magnitude of $G = 20$ mag. While *Gaia* will detect a large number of new Cepheids, such an optical survey is affected by the interstellar extinction and is limited up to several kpc along the Galactic plane. The accurate limit, however, depends on the extinction law as we see below.

In order to see the effects of the extinction on the broad-band photometry, we present results of integrating the spectral energy distribution after the extinction applied. This calculation is summarized as

$$m_\lambda = \int F_\lambda A_\lambda T_\lambda d\lambda, \tag{1}$$

where F_λ, A_λ, and T_λ indicate flux density, wavelength-dependent extinction function, and filter transmission, respectively. For F_λ, we adopt the synthetic spectrum[4] of the Sun whose effective temperature is within the range of Cepheids' temperatures. For comparing the effects of different extinction laws, we here consider a power law, $A_\lambda \propto \lambda^{-\alpha}$, with $\alpha = 2$, and the law of Cardelli ([10]). The former is close to the extinction law in Nishiyama et al. (2006, [71]), which we call the N06 law hereinafter, while the Cardelli law (henceforth C89) corresponds to a power law with $\alpha = 1.61$, significantly shallower than the N06 law. We consider the filters, V, G, J, H and K_s for T_λ. The transmission curves are taken from [3] for V, [38] for G, and the technical information on the SIRIUS camera for JHK_s. Figure 1 plots the SED with varying amounts of extinction, i.e. $F_\lambda A_\lambda$, from $A_{K_s} = 0$ to 10 mag, in addition to the filter transmission, T_λ. Note that in our calculation the integration of the SED is done after the λ-dependent extinction is applied.

[3]Such a simple exponential disk, see their prediction in Figure 5 of [103], is not consistent with the distribution of Cepheids we found as described in Section 3.

[4] The solar spectrum obtained by using the ATLAS9 code with the solar abundance in [2] was taken from `http://wwwuser.oats.inaf.it/castelli/sun.html`.

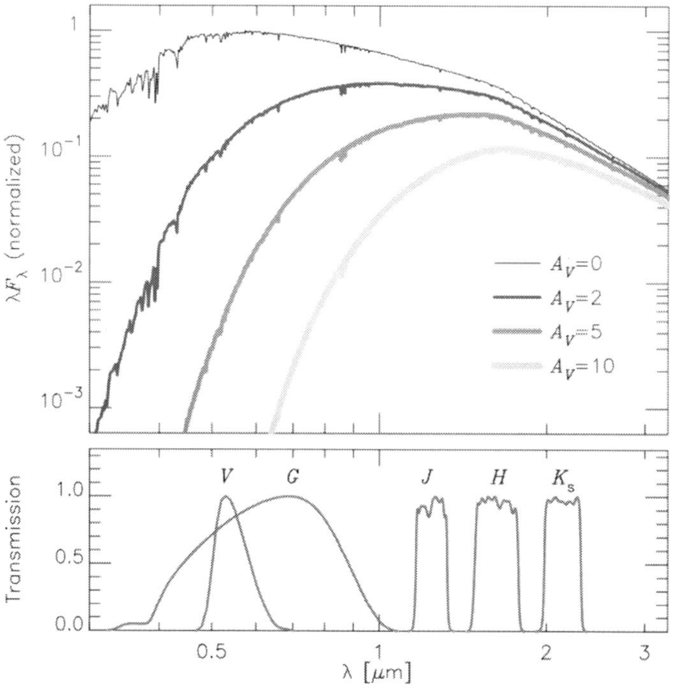

Figure 1. (Top) The synthetic spectrum of the Sun based on ATLAS9 by Kurucz and Castelli and those affected by the power law extinction, ~ λ^{-2}. (Bottom) Filter transmission curves, normalized to the peak of each bandpass.

Figure 2 plots the extinctions in V and *Gaia*'s G bands against that in K_s. For example, A_V/A_{K_s} is around 14 with the N06 law (almost consistent with the result in [72], ~16) and around 8 with the C89 law. The extinction in G is clearly smaller than that in V. *Gaia* can reach relatively deeper than previous V-band surveys in obscured regions of the Milky Way. Furthermore, the curves for A_G/A_{K_s} in Figure 2 get shallower towards the larger A_{K_s}. This is because the G band is broad and the effective wavelength changes with increasing the amount of extinction. The central wavelengths of G and V bands are 0.673 and 0.551 μm respectively ([38]). The SED affected by a large extinction, however, keeps more photons in the longer wavelength part of the G band, as illustrated in Figure 1, so that the sensitivity to the extinction gets closer to that of a photometric band at for a longer-λ range. This leads to an even larger difference between A_V/A_{K_s} and A_G/A_{K_s} at a larger A_{K_s}; at around A_{K_s} = 2.5 mag, A_G gets nearly half of the corresponding A_V.

2.1 How deep will *Gaia* and optical surveys be able to reach?

Figure 3 plots the maximum extinction that can be reached with a given magnitude in each photometric band as a function of distance. The limiting magnitudes of G = 20, V = 20, H = 15, and K_s = 14 mag are adopted as an example. The interstellar extinction varies from one line of sight to another, and we here consider two cases: the three-dimensional extinction map towards the bulge taken from [90] and a moderate increase of extinction at the rate of 0.1 mag/kpc. With the N06 law adopted (indicated by the gray filled curve), the *Gaia*'s limiting magnitude ([99]) allows one to detect Cepheids with A_{K_s} ~ 1.2 mag at ~6.5 kpc towards the bulge or at over 10 kpc in the latter case of the moderate

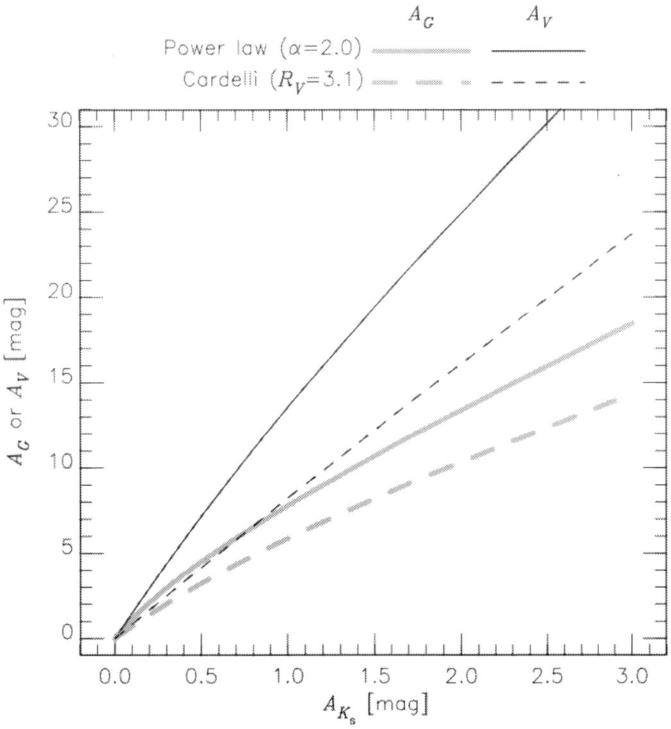

Figure 2. The optical extinctions, A_G (gray) and A_V (black), are plotted against A_{K_s}. The filled curves are obtained for the power law with $\alpha = 2$ while the dashed curves for the Cardelli law with $R_V = 3.1$ as indicated in the legend.

extinction. These limits are significantly deeper than a V-band survey. If the extinction law were closer to the C89 law, the limits would be even deeper, but at least towards the bulge the N06 law is considered to be more likely ([72]). Towards less obscured regions, *Gaia* and some other optical surveys can see to larger distances. The sensitivity of *Gaia* and more importantly its all-sky coverage will extend the frontier of the variability survey. Other current and future optical surveys will also make important contributions to finding new variables and exploring their population in the wide range of the Milky Way. For example, Feast et al. ([22]) identified classical Cepheids, among OGLE-III Cepheids ([92]), which are located in the flared part of the Galactic disk beyond the bulge. These objects are slightly off the Galactic mid-plane ($2° < |b| < 5°$). The extinctions of these objects, 0.17–0.57 mag, at the distances of 22–30 kpc correspond to roughly 0.01 mag/kpc but with a scatter of 50%. The OGLE-IV survey for the Galactic disk has actually revealed a rich population of Cepheids in the southern hemisphere as presented by Udalski et al. in this volume ([97]). In their maps, however, newly found Cepheids and RR Lyrs have patchy distributions which clearly shows the limit of optical surveys and the needs of IR surveys.

2.2 How deep can near-IR surveys reach?

Even with the moderate limiting magnitudes, H- and K_s-band surveys can reach significantly further than 10 kpc except through thick molecular clouds. The extinctions of objects towards the Galactic

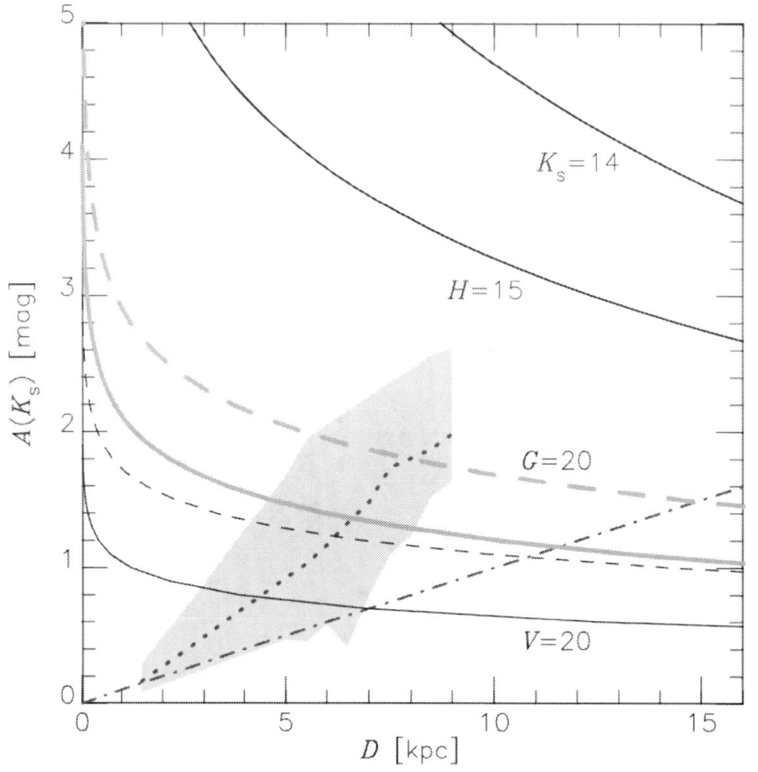

Figure 3. The maximum extinction, A_{K_s}, with which a Cepheid with P = 10 days can be detected with a given limiting magnitude. The limiting magnitudes are given within the figure, while a Cepheid with P = 10 days has absolute magnitudes of −4, −4, −5.6, and −5.7 mag in $VGHK_s$ bands, respectively. For V and G, the same styles are used as in Figure 2: filled curves when the power law with α = 2 is used and dashed curves with the Cardelli law used. Only the power law is used for the H and K_s bands. The red dotted curve and the shaded area illustrate the three-dimensional extinction map by [90] towards the bulge (see also Fig. 3 of [51]), while the dashed-dotted line indicates the extinction increasing at the rate of 0.1 mag/kpc.

center are ~2.5 mag in A_{K_s}, which makes it impossible to detect them in the optical ranges. The *Gaia* all-sky map, accompanying its first data release, clearly shows the dark lane caused by the disk extinction (Fig. 2 in [27]). The Galactic plane within 1 deg tends to have the extinctions around 1 mag in A_{K_s} or stronger ([20, 90]) and thus near-IR surveys are more effective than optical surveys at around 5 kpc and further. As mentioned in the Introduction, some results from near-IR surveys carried out with IRSF and VISTA have already demonstrated that many obscured variables can be found in a large space of the Galactic mid-plane. Such surveys seeing through the Galactic plane is necessary, in particular, to study the thin disk component; the Galactic latitude of 1 deg corresponds to the distance of 140 pc from the plane at the distance of 8 kpc. It is difficult to predict how much we can ultimately see through the entire range of the disk. Towards the Galactic center and bulge, for example, if the same amount of extinction needs to be overcome on the opposite side of the disk, we would expect an extinction of A_{K_s} ~ 5 mag. It is possible to observe such reddened objects if they are intrinsically bright, although thick molecular clouds may well interrupt the lines of sight to prevent us from reaching to the opposite end even in the IR.

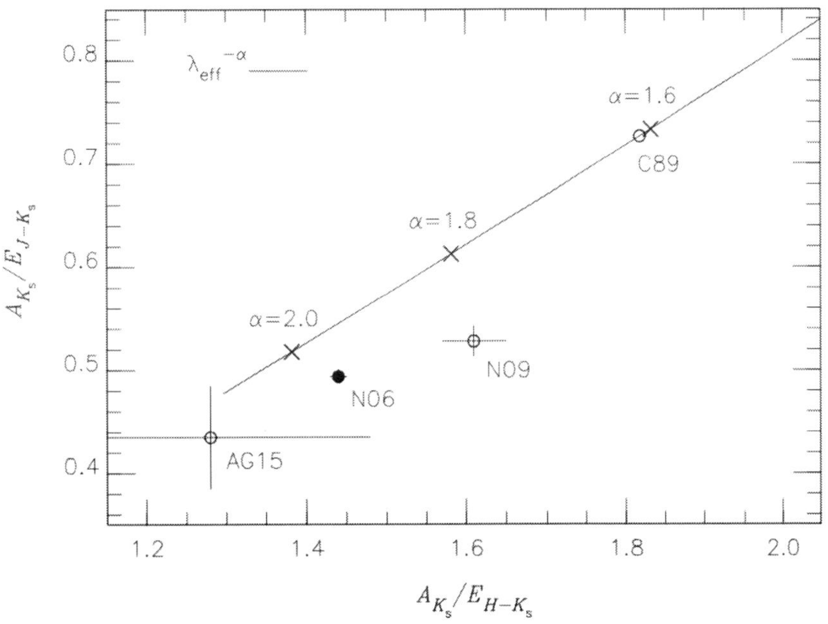

Figure 4. Comparison of the extinction coefficients, A_{K_s}/E_{J-K_s} and A_{K_s}/E_{H-K_s}. The straight line gives the relation between the two coefficients with varying power α as indicated by crosses. The points with the labels of AG15, C89, N06, and N09 indicate the values obtained by [1], [10], [71], and [73], respectively. The C89 law corresponds to the power law with $\alpha = 1.61$.

3 Impact of the uncertainty in the extinction law

The near-IR surveys will open our path to the obscured regions of the Milky Way as we discussed in the previous section, but the interstellar extinction still poses a serious problem with interpreting observational data we'll obtain. For many applications, we need to make corrections of the extinction and reddening, and this requires using a proper law of extinction. When we combine H- and K_s-band data with the period-luminosity relation of Cepheids to estimate the distances and foreground extinctions, for example, we need to derive the amount of extinction, A_{K_s}, considering an extinction coefficient, A_{K_s}/E_{H-K_s}, and the reddening, $E_{H-K_s} = (H - K_s) - (M_H - M_{K_s})$, where M_H and M_{K_s} are predicted by the period-luminosity relation[5]. Figure 4 plots some extinction coefficients in the near IR in the literature and shows a large scatter among them. Figure 5 illustrates how the different extinction laws (N06 and C89) give different reddening vectors on the color-magnitude diagram (left) and the color-color diagram (right). For example, a reddening of $E_{H-K_s} = 1.5$ mag would give $A_{K_s} = 2.16$ and 2.73 mag with the N06 and C89 laws, respectively. The differences in the extinction laws thus introduce large uncertainties in estimating distances of such reddened objects (the reddenings of stars at around the Galactic center are around 1.5–2 mag in E_{H-K_s}). On the other hand, Figure 5 also indicates that the reddening vector on the color-color diagram is relatively insensitive to the difference in the extinction law; the N06 and C89 laws would predict different $(H - K_s, J - H)$ by only ~0.2 mag for the stars at around the Galactic center.

[5]When one uses the relation of a Wesenheit index, e.g. $W_{K_s} = K_s - \alpha(H - K_s)$, which is often considered as a reddening-free index, it is still necessary to decide the coefficient α properly in order to make this index reddening free.

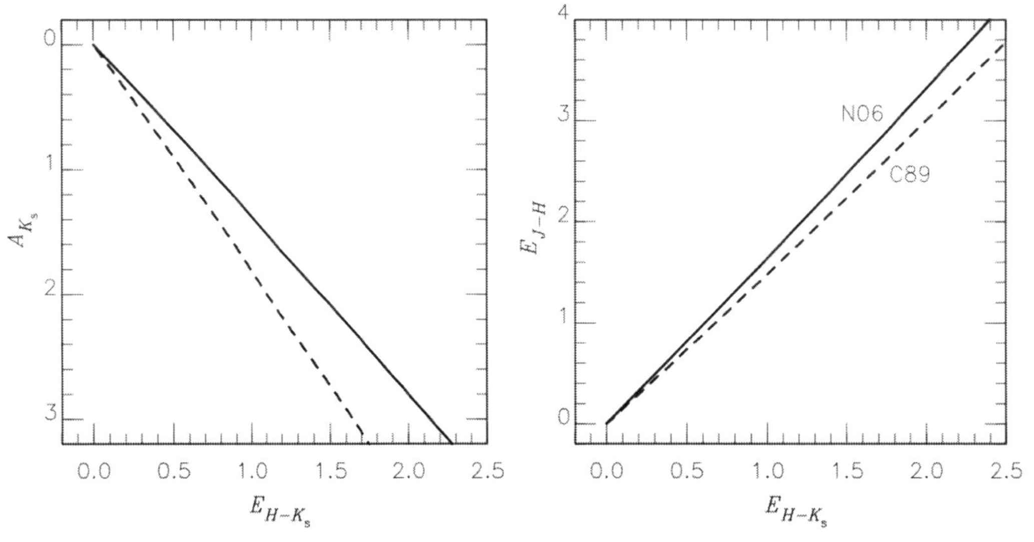

Figure 5. Reddening vectors according to the extinction laws of C89 ([10], dashed) and N06 ([71], solid) on the color-magnitude diagram (left) and the color-color diagram (right).

A good example of the impact of the extinction law is found and discussed in [51]. Dékány et al. ([19]) reported a few dozens of classical Cepheids towards the Galactic bulge and concluded that they form a thin disk surrounding the Galactic center. Matsunaga et al. ([51]), however, reported their discoveries of classical Cepheids which are located behind the bulge and concluded that the region around the Galactic center lacks classical Cepheids. Some of the Cepheids are common in the two works, but they obtained significantly different distances for the Cepheids, $\Delta\mu_0$ = 0.4–0.5 mag. Matsunaga et al. ([51]) found that such a large difference was caused because two different extinction laws were used: the N06 law in [51] and the N09 law from Nishiyama et al. (2009, [73]) in [19]. The A_{K_s}/E_{H-K_s} in the N06 and N09 laws are 1.44 and 1.61 (Fig. 4), respectively, and these led to offsets of ~0.3 mag and the significantly different conclusions on the distribution of the Cepheids.

Readers are referred to [51] for details, but we here outline how the extinction law towards the Galactic bulge can be constrained together with adding a new plot which supports our conclusion. An important group of Cepheids are four classical Cepheids which belong to the nuclear stellar disk (NSD, henceforth). The NSD is a disk-like system around the Galactic center, with a radius of about 200 pc. This region is also known as the Central Molecular Zone (CMZ). Unlike the more extended bulge, this region is known to host young stars like the famous massive clusters Arches and Quintuplet, and those young stars are rotating within this disk. We can conclude that the four Cepheids belong to the NSD for a few reasons: the projected distances, similarity in the physical parameters of Cepheids, the fact that their distances must be similar independent of the adopted extinction law, and that their radial velocities are consistent with the rotation of the NSD. These strongly indicate that their distances should agree with that of the Galactic center ($R_0 = 8.0 \pm 0.5$ kpc, [16, 28]) within the size of the NSD (~0.2 kpc). Note that some estimates of R_0, e.g. the orbit modeling of the S stars around the central black hole ([28]), are independent of the extinction law. As illustrated in the left panel of Figure 6, A_{K_s}/E_{H-K_s} should be around 1.44, rather than 1.61, in order to keep the four Cepheids at around 8 kpc. Note that, with the period-luminosity relations and observed magnitudes given, the location of each Cepheid on this diagram doesn't depend on the adopted extinction law or its distances. The four

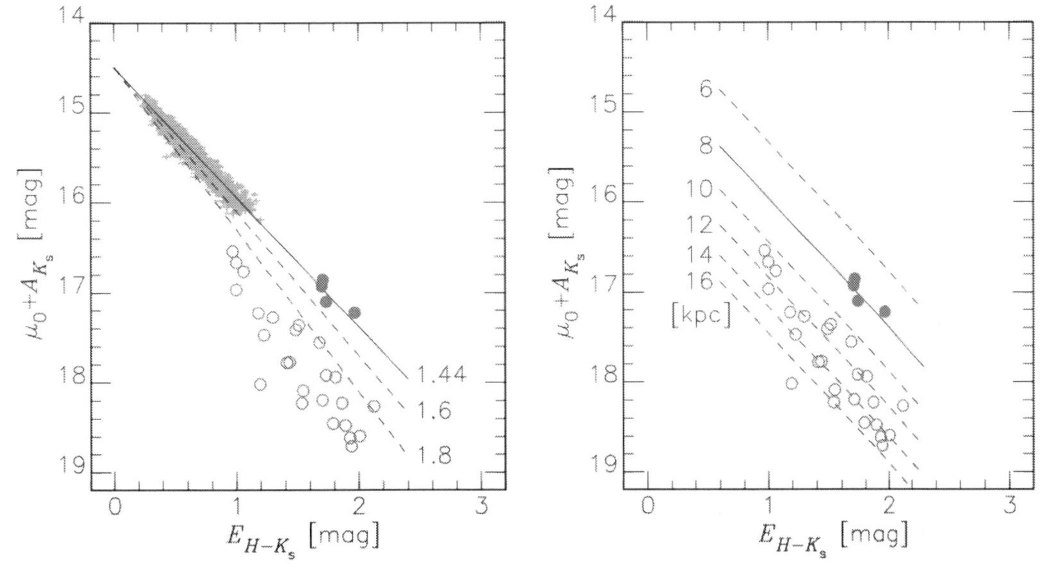

Figure 6. (Left) Reddening vectors on the color-magnitude diagram with different coefficients of A_{K_s}/E_{H-K_s} starting from the distance of the Galactic center, 8 kpc, at the zero color excess, E_{H-K_s} = 0 mag. The sequence of cyan points near the top-left corner, taken from N06 [71], indicates the red clump peaks affected by various amounts of reddening. Filled circles indicate the four classical Cepheids in the nuclear stellar disk ([47, 50]), and open circles indicate the other classical Cepheids reported in [51]. (Right) Same as the left panel, but lines have the slope of A_{K_s}/E_{H-K_s} = 1.44 and correspond to different distances from 6 to 16 kpc.

NSD Cepheids are precisely on the extension of Nishiyama's fit to the red clumps. Assuming that the Cepheids and the red clumps are at the same distance, this strongly supports the extinction value of 1.44. Once the coefficient of 1.44 is adopted, the distances of other Cepheids can be determined as one can see on the right panel of Figure 6. The vertical offset in this panel corresponds to the offset in distance. Other Cepheids are located significantly further than the Galactic Center, showing the gap between the NSD (marked down by the four Cepheids) and the inner edge of the distribution of disk Cepheids. Also, we found no Cepheids on the nearer side, which supports the absence of Cepheids around the Galactic Center; the saturation limit of our survey could have allowed us to detect Cepheids between ~5 and 8 kpc from the Sun if the density were high enough. These support our conclusion on the lack of Cepheids within 2.5 kpc around the Galactic center except the NSD. More details are given in the original paper ([51]) as mentioned above, and some remarks are also found in the conference summary for this conference ([23]).

Acknowledgments: The author is grateful to Michael Feast for reading the draft and for his useful comments. We appreciate financial support from the Japan Society for the Promotion of Science (JSPS) through the Grant-in-Aid, No. 26287028.

References

[1] Alonso-García, J., Dékány, I., Catelan, M., Contreras Ramos, R., Gran, F., Amigo, P., Leyton, P., & Minniti, D., AJ, **149**, 99 (2015)

[2] Asplund, M., Grevesse, N., & Sauval, A. J., ASPC, **336**, 25 (2005)

[3] Bessell, M. S., ARA&A, **43**, 293 (2005)
[4] Bhardwaj, A., Kanbur, S. M., Singh, H. P., Macri, L. M., & Ngeow, C.-C., AJ, **447**, 3342 (2015)
[5] Bhardwaj, A., Kanbur, S. M., Macri, L. M., Singh, H. P., Ngeow, C.-C., Wagner-Kaiser, R., & Sarajedini, A., AJ, **151**, 88 (2016a)
[6] Bhardwaj, A., Kanbur, S. M., Macri, L. M., Singh, H. P., Ngeow, C.-C., & Ishida, E. E. O., MNRAS, **457**, 1644 (2016b)
[7] Bhardwaj, A., Macri, L. M., Rejkuba, M., Kanbur, S. M., Ngeow, C.-C., & Singh, H. P., AJ, **153**, 154 (2017)
[8] Bono, G., Caputo, F., Marconi, M., & Musella, I., ApJ, **715**, 277 (2010)
[9] Borissova, J., Chené, A.-N., Ramírez Alegría, S., et al., A&A, **569**, A24 (2014)
[10] Cardelli, J. A., Clayton, G. C., & Mathis, J. S., ApJ, **345**, 245 (1989)
[11] Chen, X., de Grijs, R., & Deng, L., MNRAS, **464**, 1119 (2017)
[12] Cioni, M.-R. L., Clementini, G., Girardi, L., et al., A&A, **527**, A116 (2011)
[13] Cioni, M.-R. L., Girardi, L., Moretti, M. I., et al., A&A, **562**, A32 (2014)
[14] Cioni, M.-R. L., Bekki, K., Girardi, L., et al., A&A, **586**, A77 (2016)
[15] Cioni, M. R.,L., these proceedings (`arXiv:1703.06769`) (2017)
[16] de Grijs, R., & Bono, G., ApJ, **227**, 5 (2016)
[17] Dékány, I., Minniti, D., Catelan, M., Zoccali, M., Saito, R. K., Hempel, M., & Gonzalez, O. A., ApJL, **776**, L19 (2013)
[18] Dékány, I., Minniti, D., Majaess, D., et al., ApJL, **812**, L29 (2015a)
[19] Dékány, I., Minniti, D., Hajdu, G., et al., ApJL, **799**, L11 (2015b)
[20] Dutra, C. M., Santiago, B. X., Bica, E. L. D., & Barbuy, B., MNRAS, **338**, 253 (2003)
[21] Feast, M. W., Whitelock, P. A., Menzies, J. W., & Matsunaga, N., MNRAS, **421**, 2998 (2012)
[22] Feast, M. W., Menzies, J. W., Matsunaga, N., & Whitelock, P. A., Nature, **509**, 342 (2014)
[23] Feast, M. W., these proceedings (`arXiv:1702.07147`) (2017)
[24] Freedman, W. L., Madore, B. F., Gibson, B. K., et al., ApJ, **553**, 47 (2001)
[25] Freedman, W. L., Madore, B. F., Scowcroft, V., et al., AJ, **142**, 192 (2011)
[26] Freedman, W. L., Madore, B. F., Scowcroft, V., Burns, C., Monson, A., Persson, S. E., Seibert, M., & Rigby, J., ApJ, **758**, 24 (2012)
[27] Gaia Collaboration, Brown, A. G. A., Vallenari, A., et al., A&A, **595**, A2 (2016)
[28] Gillessen, S., Eisenhauer, F., Fritz, T. K., Pfuhl, O., Ott, T., & Genzel, R., IAUS, **289**, 29 (2013)
[29] Gonzalez, O. A., Rejkuba, M., Zoccali, M., Valenti, E., Minniti, D., & Tobar, R., A&A, **552**, A110 (2012)
[30] González-Fernández, C., Asensio Ramos, A., Garzón, F., Cabrera-Lavers, A., & Hammersley, P. L., ApJ, **782**, 86 (2012)
[31] Gran, F., Minniti, D., Saito, R. K., Navarrete, C., Dékány, I., McDonald, I., Contreras Ramos, R., & Catelen, M., A&A, **575**, A114 (2015)
[32] Gran, F., Minniti, D., Saito, R. K., et al., A&A, **591**, A145 (2016)
[33] Gromadzki, M., Kurtev, R., Geamín, J. C., et al., AcA, **66**, 293 (2016)
[34] Ikeda, Y., Kobayashi, N., Kondo, S., et al., Proc. SPIE, **9908**, 5 (2016)
[35] Inno, L., Matsunaga, N., Bono, G., et al., ApJ, **764**, 84 (2013)
[36] Ita, Y., Tanabé, T., Matsunaga, N., et al., MNRAS, **347**, 720 (2004a)
[37] Ita, Y., Tanabé, T., Matsunaga, N., et al., MNRAS, **353**, 705 (2004b)
[38] Jordi, C., Gebran, M., Carrasco, J. M., et al., A&A, **523**, A48 (2010)
[39] Kerber, L. O., Girardi, L., Rubele, S., & Cioni, M. R. L., A&A, **499**, 697 (2009)

[40] Kurtev, R., Gromadzki, M., Beamín, J. C., et al., MNRAS, **464**, 1247 (2017)
[41] Macri, L. M., Ngeow, C.-C., Kanbur, S. M., Mahzooni, S., & Smitka, M. T., AJ, **149**, 117 (2015)
[42] Madore, B. F., & Freedman, W. L., PASP, **103**, 933 (1991)
[43] Majewski, S. R., APOGEE Team, & APOGEE-2 Team, AN, **337**, 863 (2016)
[44] Marconi, M., Molinaro, R., Ripepi, V., et al., MNRAS, **466**, 3206 (2017)
[45] Matsunaga, N., Fukushi, H., Nakada, Y., et al., MNRAS, **370**, 1979 (2006)
[46] Matsunaga, N., Kawadu, T., Nishiyama, S., Nagayama, T., Hatano, H., Tamura, M., Glass, I. S., & Nagata, T., MNRAS, **399**, 1709 (2009)
[47] Matsunaga, N., Kawadu, T., Nishiyama, S., et al., Nature, **188**, 477 (2011)
[48] Matsunaga, N., Journal of Physics: Conference Series, **372**, 12026 (2012)
[49] Matsunaga, N., Feast, M. W., Kawadu, T., et al., MNRAS, **429**, 385 (2013)
[50] Matsunaga, N., Fukue, K., Yamamoto, R., et al., ApJ, **799**, 46 (2015)
[51] Matsunaga, N., Feast, M. W., Bono, G., et al., MNRAS, **462**, 414 (2016)
[52] McDonald, I., Zijlstra, A. A., Sloan, G. C., et al., MNRAS, **436**, 413 (2013)
[53] McDonald, I., Zijlstra, A. A., Sloan, G. C., Kerins, E., Lagadec, E., & Minniti, D., MNRAS, **439**, 2618 (2014)
[55] Meixner, M., Gordon, K. D., Indebetouw, R., et al., AJ, **132**, 2268 (2006)
[54] Menzies, J. W., Feast, M. W., Tanabé, T., Whitelock, P. A., & Nakada, Y., MNRAS, **335**, 923 (2002)
[56] Menzies, J. W., Feast, M. W., Whitelock, P. A., Olivier, E., Matsunaga, N., & Da Costa, G., MNRAS, **385**, 1045 (2008)
[57] Menzies, J. W., Whitelock, P. A., Feast, M. W., & Matsunaga, N., MNRAS, **406**, 86 (2010)
[58] Menzies, J. W., Feast, M. W., Whitelock, P. A., & Matsunaga, N., MNRAS, **414**, 3492 (2011)
[59] Menzies, J. W., Whitelock, P. A., & Feast, M. W., MNRAS, **452**, 910 (2015)
[60] Minniti, D., Lucas, P. W., Emerson, J. P., et al., New Astronomy, **15**, 433 (2010)
[61] Minniti, D., Saito, R. K., Alonso-García, J., Lucas, P. W., & Hempel, M., ApJL, **733**, L43 (2011)
[62] Minniti, D., Contreras Ramos, R., Zoccali, M., Rejkuba, M., Gonzalez, O. A., Valenti, E., & Gran, F., ApJL, **830**, L14 (2016)
[63] Minniti, D., Palma, T., Dékány, I., et al., ApJL, **838**, L14 (2017a)
[64] Minniti, D., Dékány, I., Majaess, D., et al., AJ, **153**, 179 (2017b)
[65] Minniti, D., these proceedings (2017)
[66] Monson, A. J., Freedman, W. L., Madore, B. F., Persson, S. E., Scowcroft, V., Seibert, M., & Rigby, J. R., ApJ, **759**, 146 (2012)
[67] Moretti, M. I., Clementini, G., Muraveva, T., et al., MNRAS, **437**, 2702 (2014)
[68] Moretti, M. I., Clementini, G., Ripepi, V., et al., MNRAS, **459**, 1687 (2016)
[69] Nagashima, C., Nagayama, T., Nakajima, Y., et al., Proc. "Star Formation 1999", 397 (1999)
[70] Nagayama, T., Nagashima, C., Nakajima, Y., et al., Proc. SPIE, **4841**, 459 (2003)
[71] Nishiyama, S., Nagata, T., Kusakabe, N., et al., ApJ, **638**, 839 (2006)
[72] Nishiyama, S., Nagata, T., Tamura, M., Kandori, R., Hatano, H., Sato, S., & Sugitani, K., ApJ, **680**, 1174 (2008)
[73] Nishiyama, S., Tamura, M., Hatano, H., Kato, D., Tanabé, T., Sugitani, K., & Nagata, T., ApJ, **696**, 1407 (2009)
[74] Origlia, L, Oliva, E., Baffa, C., et al., Proc. SPIE, **9147** , 1 (2014)
[75] Otsubo, S., Ikeda, Y., Kobayashi, N., et al., Proc. SPIE, **9908**, 79 (2016)

[76] Palma, T., Minniti, D., Dékány, I., Clariá, J. J., Alonso-García, J., Gramajo, L. V., Ramírez Alegría, S., & Bonatto, C., New Astronomy, **424**, 1807 (2012)

[77] Polsdofer, E., Seale, J., Sewilo, M., Vijh, U. P., Meixner, M., Marengo, M., & Terrazas, M., AJ, **149**, 78 (2015)

[78] Riebel, D., Meixner, M., Fraser, O., Srinivasan, S., Cook, K., & Vijh, U., ApJ, **723**, 1195 (2010)

[79] Riebel, D., Boyer, M. L., Srinivasan, S., et al., 2015, ApJ, **807**, 1 (2015)

[80] Rieke, G. H., & Lebofsky, M. J., ApJ, **288**, 618 (1985)

[81] Riess, A. G., Macri, L. M., Hoffmann, S. L., et al., ApJ, **826**, 56 (2016)

[82] Ripepi, V., Moretti, M. I., Marconi, M., et al., MNRAS, **424**, 1807 (2012)

[83] Ripepi, V., Moretti, M. I., Marconi, M., et al., MNRAS, **446**, 3034 (2015)

[84] Ripepi, V., Marconi, M., Moretti, M. I., et al., ApJS, **224**, 21 (2016)

[85] Rubele, S., Kerber, L., Girardi, L., et al., A&A, **537**, A106 (2012)

[86] Rubele, S., Girardi, L., Kerber, L., et al., MNRAS, **449**, 639 (2015)

[87] Saito, R. K., Zoccali, M., McWilliam, A., Minniti, D., Gonzalez, O. A., & Hill, V., AJ, **142**, 76 (2011)

[88] Scowcroft, V., Freedman, W. L., Madore, B. F., Monson, A. J., Persson, S. E., Seibert, M., Rigby, J. R., & Melbourne, J., ApJ, **773**, 106 (2013)

[89] Scowcroft, V., Seibert, M., Freedman, W. L., Beaton, R. L., Madore, B. F., Monson, A. J., Rich, J. A., & Rigby, J. R., MNRAS, **459**, 1170 (2016)

[90] Schultheis, M., Chen, B. Q., Jiang, B. W., et al., A&A, **566**, A120 (2014)

[91] Sloan, G. C., Matsunaga, N., Matsuura, M., et al., ApJ, **719**, 1274 (2010)

[92] Soszyński, I., Udalski, A., Pietrukowicz, P., et al., AcA, **61**, 285 (2011)

[93] Soszyński, I., these proceedings (2017)

[94] Sutherland, W., Emerson, J., Dalton, G., et al., A&A, **575**, A25 (2015)

[95] Tanabé, T., Nishida, S., Matsunaga, S., et al., Nature, **385**, 509 (1997)

[96] Tanabé, T., Kučinskas, A., Nakada, Y., Onaka, T., & Sauvage, M., ApJS, **155**, 401 (2004)

[97] Udalski, A., these proceedings (`arXiv:1703.02980`) (2017)

[98] Valenti, E., Zoccali, M., Gonzalez, O. A., et al., A&A, **587**, L6 (2016)

[99] van Leeuwen, F., Evans, D. W., De Angeli, F., et al. A&A, **599**, A32 (2017)

[100] Whitelock, P. A., Menzies, J. W., Feast, M. W., Matsunaga, N., Tanabé, T., & Ita, Y., MNRAS, **394**, 795 (2009)

[101] Whitelock, P. A., Menzies, J. W., Feast, M. W., Nsengiyumva, F., & Matsunaga, N., MNRAS, **428**, 2216 (2013)

[102] Whitelock, P. A., Kasiwal, M., & Boyer, M., these proceedings (`arXiv:1702.06797`) (2017)

[103] Windmark, F., Lindegren, L., & Hobbs, D., A&A, **530**, A76 (2011)

Pulsating stars in the VMC survey*

Maria-Rosa L. Cioni[1,2,**], *Vincenzo* Ripepi[3], *Gisella* Clementini[4], *Martin A.T.* Groenewegen[5], *Maria I.* Moretti[3], *Tatiana* Muraveva[4], and *Smitha* Subramanian[6]

[1] *Leibniz-Institüt für Astrophysik Potsdam, An der Sternwarte 16, 14482 Potsdam, Germany*
[2] *University of Hertfordshire, PAM, College Lane, Hatfield AL10 9AB, United Kingdom*
[3] *INAF-Osservatorio Astronomico di Capodimonte, Salita Moiariello 16, 80131 Napoli, Italy*
[4] *INAF-Osservatorio Astronomico di Bologna, via Gobetti 93/3, 40129 Bologna, Italy*
[5] *Koninklijke Sterrenwacht van België, Ringlaan 3, 1180 Brussels, Belgium*
[6] *Kavli Institute for Astronomy and Astrophysics, Peking University, Yi He Yuan Lu 5, Hai Dian District, Beijing 100871, China*

Abstract. The VISTA survey of the Magellanic Clouds system (VMC) began observations in 2009 and since then, it has collected multi-epoch data at K_s and in addition multi-band data in Y and J for a wide range of stellar populations across the Magellanic system. Among them are pulsating variable stars: Cepheids, RR Lyrae, and asymptotic giant branch stars that represent useful tracers of the host system geometry.

1 Introduction

The Magellanic Clouds are prominent sky markers in the southern hemisphere since the time of the first circumnavigation of the globe by Fernão de Magalhães in 1519-1522 ([15]). They are the most luminous and largest dwarf satellite galaxies of the Milky Way. They are metal poor, nearby, and are interacting with each other and with the Milky Way. Signatures of these interactions are associated to a Bridge and Stream; both features are predominately traced by neutral hydrogen that was stripped from the galaxies as a result of dynamical interactions. Most likely they were accreted only recently (about 2 Gyr ago) and arrived under the influence of the Milky Way potential together with their own satellites ([10]).

2 The VMC survey

The near-infrared YJK_s VISTA ESO public survey of the Magellanic Clouds system (VMC) began in 2009 and is expected to be completed in 2018. This is a photometric survey in three filters, Y, J, and K_s performed at the VISTA telescope using the VIRCAM camera. The latter provides a spatial resolution of 0.34″ per pixel and a field-of-view of 1.65 deg^2 sampled by 16 Raytheon detectors. Details about the telescope and instrument are given in [7]. The VMC survey covers an area of approximately 170 deg^2 of the Magellanic system and includes stars as faint as $\sim$ 22 mag (5σ Vega) at each filter. Stars

*Based on observations made with VISTA at ESO under programme ID 179.B-2003.
**mcioni@aip.de

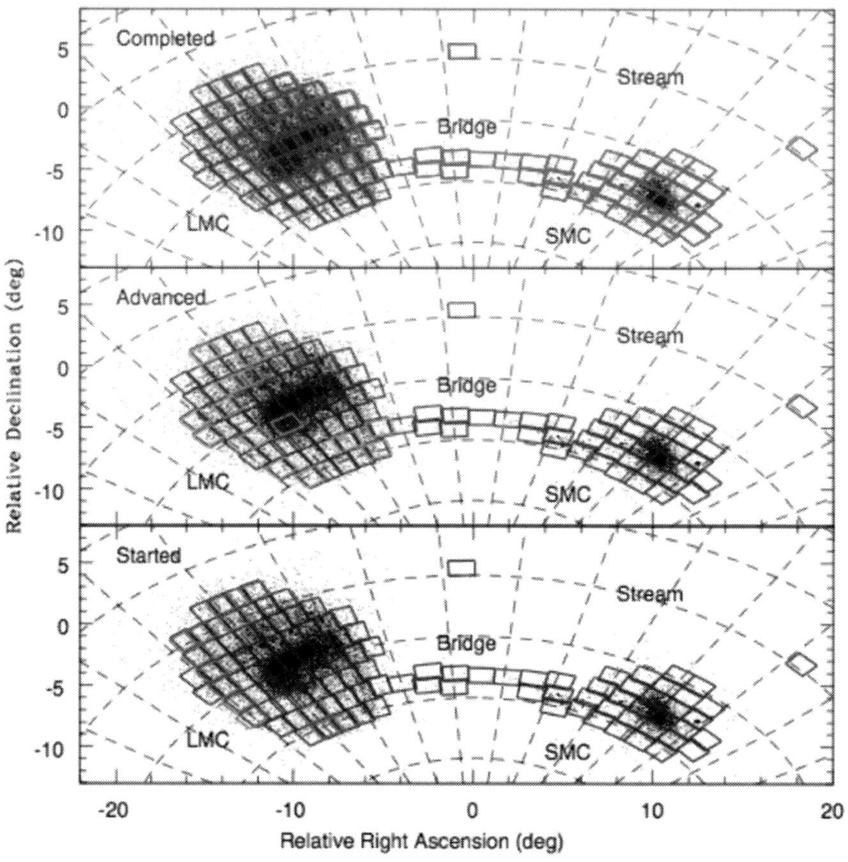

Figure 1. Progress of VMC observations. Rectangles identify individual VMC tiles as follows: all 110 survey tiles (blue), tiles fully observed at all three filters and epochs (red), tiles that are in an advanced stage of observation, namely tiles for which most of the YJ observations and at least 6 epochs at K_s have been obtained (green), and tiles that have only a few observations acquired (magenta). The image is centred at (α = 03:24:00, δ = −69:00:00) and dashed lines indicate steps of 40 min and 5 deg in RA and DEC, respectively.

brighter than ~ 10 mag at K_s are usually saturated. The survey provides three independent epochs at the Y and J filters and at least 12 independent epochs at the K_s filter. Details about the survey strategy are given in [3].

The main scientific goals of the VMC survey are to derive the spatially resolved star formation history across the Magellanic system and to measure its three-dimensional geometry. The first goal defines the depth of the survey because it is necessary to observe stars as faint as below the old main-sequence turnoff to derive age and metallicity from the interpretation of colour-magnitude diagrams using theoretical models. The second goal defines the monitoring strategy of the survey in order to derive distances from variable stars (Cepheids and RR Lyrae stars); red clump giant stars are also used to measure distances. The VMC survey also addresses a wide range of legacy science: from the study

 DOI: 10.1051/epjconf/201715201008

of individual objects in the Magellanic Clouds and in the Milky Way, star formation, stellar clusters, the dynamics of the Magellanic system, to specific investigations of background galaxies and quasars.

The effective area covered in one observation (a pawprint) is not contiguous because of large gaps among the detectors. To cover homogeneously the field-of-view (a tile) it is necessary to fill these gaps with a six-point mosaic. In this way pixels in the central region of 1.5 deg^2 in size are observed 2 – 6 times while in two external regions (tile wings) they are observed only once, within a given tile. The VMC survey covers the Magellanic system with 110 overlapping tiles (Fig. 1). The spacing between the K_s epochs follows a cadence with a minimum spacing of 0, 1, 3, 7, and 17 days for each subsequent epoch, while there is no requirement on the spacing of the Y and J band epochs; these can all be obtained during the same night. To observe one VMC tile, in a given filter, it takes about 1 hour. The overall time to complete the VMC survey is of $\sim$ 2000 hours or $\sim$ 250 nights. Observations for the VMC survey are obtained in service mode which guarantees homogeneity in sky quality (clouds, airmass, and seeing).

The VMC data are first processed at the Cambridge Astronomy Survey Unit using the VISTA Data Flow System pipeline (VDFS; [9]). Pawprint and tile images are produced and the corresponding source catalogues per filter and epoch are created. Next, deep images resulting from stacking individual epoch observations, with their associated catalogues are created at the Wide-Field Astronomy Unit. Individual epochs are also linked with each other. The VMC data are ingested in the VISTA science archive (VSA; [5]) which is a point of access for both the VMC team and the astronomical community. At yearly intervals, some VMC data are made publicly available and they can be accessed at the VSA or at the ESO archive. The VMC team creates additional data products, such as catalogues with point spread function (PSF) photometry and catalogues of variable stars, which are also released to the community via the two points of entry.

The VMC public web page is at `http://star.herts.ac.uk/~mcioni/vmc`, where it is possible to view the deep three-colour images with a user friendly zoom-in facility that allows us to appreciate the richness and quality of the data. As of 10th March 2017 the VMC observations are 88% complete. In particular, 83 tiles are fully observed in each filter and epoch, the SMC and the Bridge regions are also fully observed, and the LMC is $\sim$ 80% complete (Fig. 1). The fourth public data release of the VMC data comprises 19 tiles (10 in the LMC, 5 in the SMC, 2 in the Bridge, and 2 in the Stream). For these tiles deep images in each filter with their associated pipeline and PSF catalogues, single tile images per filter with associated source lists, and individual pawprint images also with their associated catalogues are made available, as well as catalogues of some variable stars.

3 VMC results on pulsating stars

3.1 Cepheids

Cepheids are hydrogen-shell burning stars crossing the instability strip. They are often classified as: classical or type I Cepheids that refer to 4 – 20 $M_\odot$ stars a few 100 Myr old, anomalous Cepheids that refer to 1.3 – 2.1 $M_\odot$ stars a few Gyr old and metal poor, and type II Cepheids that refer to 0.5 $M_\odot$ stars 9 – 10 Gyr old. Cepheids are important standard candles via period-luminosity, period-luminosity-colour, and Wesenheit relations. In the near-infrared chemical effects (from He and Fe) on these relations are reduced, reddening is low, light-curves are sinusoidal, and the relations are linear with a small dispersion. These properties represent clear advantages in using these stars and near-infrared pass-bands to measure distances.

In the ($J - K_s$, K_s) diagram (Fig. 2), the instability strip for the combined modes of pulsation (fundamental, first and second overtone) spans about 0.5 mag in colour and 8 mag in K_s band. Using VMC multi-epoch data up to eight different light-curve templates were built and subsequently used

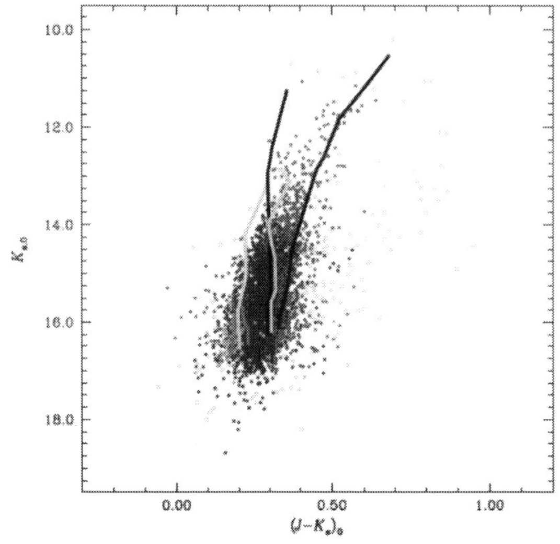

Figure 2. Color-magnitude diagram of classical Cepheids in the SMC. Theoretical instability strips for fundamental (black lines; red points), first overtone (grey lines; blue points), and second overtone (magenta lines; cyan points) pulsators are indicated. See Ripepi et al. ([17]) for details.

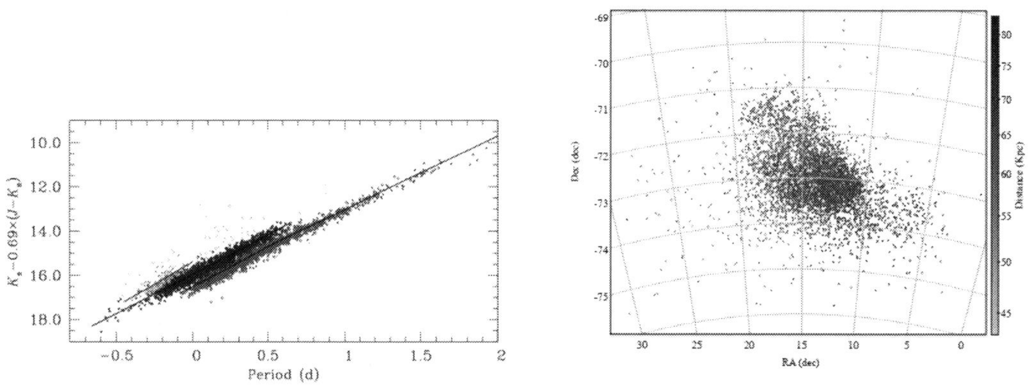

Figure 3. (Left) Period-Wesenheit relation for classical Cepheids pulsating in the fundamental (red), first overtone (blue), and second overtone (cyan) modes. The scatter represents the intrinsic depth of the SMC. (Right) Spatial distribution of classical Cepheids across the SMC; points are colour coded by distance as indicated in the legend. Details are given in Ripepi et al. (2017, submitted).

to measure accurate mean magnitudes for the Cepheids ([17]). Combining these magnitudes with the period obtained from the Optical Gravitational Lensing Experiment (OGLE) project we constructed period-luminosity relations and alike with a small scatter (Fig. 3, left). Considering that the depth of the Magellanic Clouds is about 0.2 mag for the LMC and 0.7 mag for the SMC, we are able to use these relations to quantify distance effects throughout the galaxies. Preliminary maps of the spatial distribution of classical Cepheids in the different combination of Cartesian coordinates indicate that the de-projected SMC resembles the LMC galaxy with an elongated bar and a region above it that can be associated to a spiral arm. We confirm a complex structure for the galaxy and a large depth along the line-of-sight ($\sim$ 20 kpc) where stars to the northern east region are closer to us, and to the LMC, than stars in the southern west region (Fig. 3, right); see also [11] and [16].

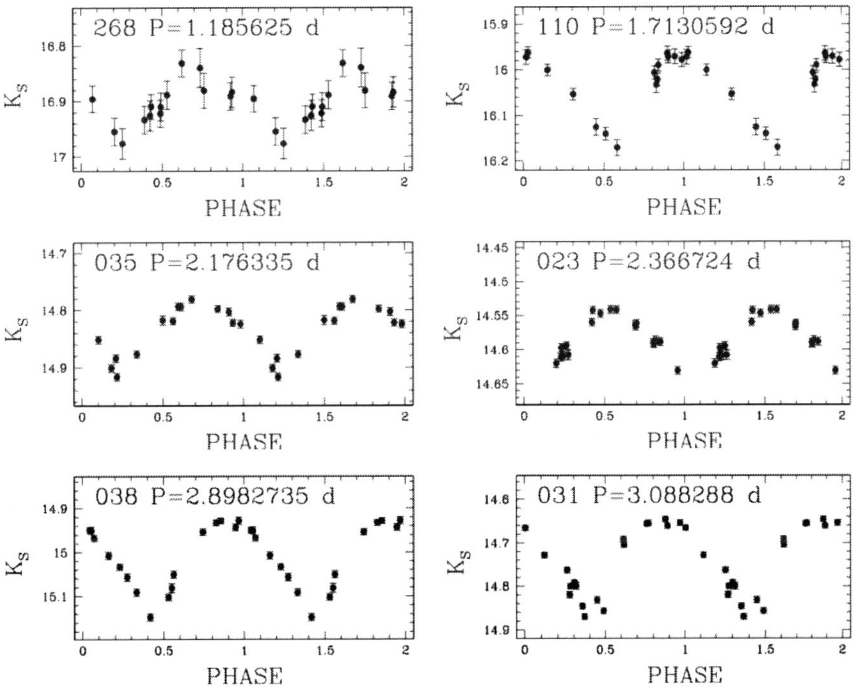

Figure 4. K_s-band curves for six candidate classical Cepheids newly discovered by the VMC survey. Reference numbers and the periods of pulsation are indicated in each panel. Further details are given in [13].

Based on the distribution of known classical Cepheids in the VMC colour-magnitude and colour-colour diagrams, on the period-luminosity relation, and on the shape of their K_S-band light-curve, we have identified about 290 new candidate classical Cepheids in the external regions of the SMC, using only the VMC near-infrared photometry ([13]) . A few light-curves are shown in Figure 4. The number of SMC Cepheids we have detected is consistent with the predictions of Rubele et al. ([16]) for the star formation history recovery in these regions of the galaxy. About 13% of these Cepheid candidates were subsequently confirmed by the OGLE IV project.

VMC JK_s photometry of nine classical Cepheids in the SMC was used together with optical photometry and radial velocity data to find best fit models that provide distances, masses, luminosities, and radii of the stars (Marconi et al. [12]; see also these proceedings).

3.2 RR Lyrae stars

RR Lyrae are old (> 10 Gyr), low mass (0.6 – 0.8 $M_\odot$) stars burning helium in their core. They populate the horizontal branch. The amplitude of their light variation in the K_s band is significantly smaller than in the visual and they are fainter than Cepheids. This makes them more difficult to detect and study. Nevertheless, we have analysed a sample of 70 RR Lyrae stars in the LMC for which K_s-band magnitudes from the VMC survey, spectroscopically determined metallicities ([8]) and periods from the OGLE III survey are available ([14]). We computed a period-luminosity-metallicity (PK_sZ) relation for these stars. Figure 5 shows the projections of the derived PK_sZ relation on the $\log(P) - K_s$ and K_s – [Fe/H] planes.

 DOI: 10.1051/epjconf/201715201008

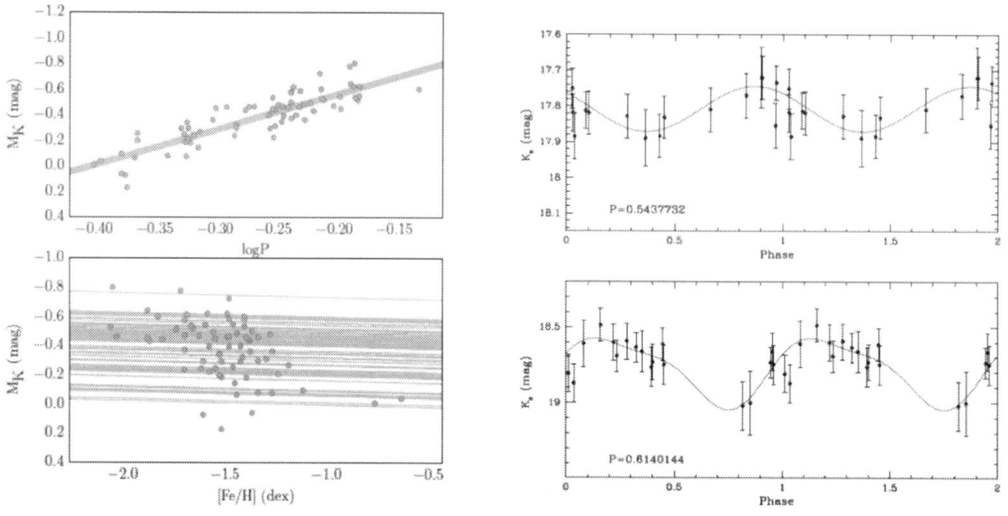

Figure 5. (Left) Projections of the PK_sZ relation of RR Lyrae stars in the LMC on the log(P) – K_s (top) and K_s – [Fe/H] (bottom) planes. Grey lines are lines of equal metallicity (top) or equal period (bottom). (Right) Phased K_s-band light curves of SMC RR Lyrae stars observed by the VMC survey. Periods are from the OGLE III catalogue and measured in days. Red lines show the best fits.

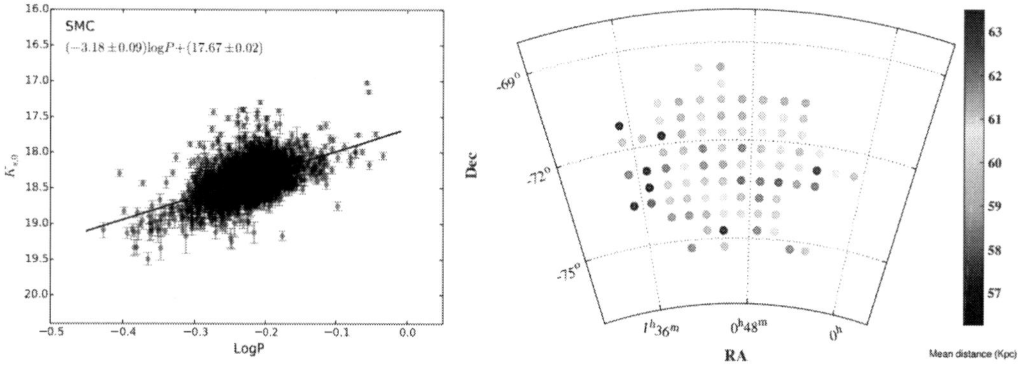

Figure 6. (Left) $PK_{s,0}$ relation of 2662 RR Lyrae stars in the SMC. (Right) Two dimensional distribution of mean distance measured in different sub-regions of the SMC. Each point corresponds to an area of 0.6×0.5 deg^2. Details are given in Muraveva et al. (2017, submitted).

The derived PK_sZ relation was used to determine individual distances to 2662 RR Lyrae stars pulsating in the fundamental mode and distributed across the SMC and, thus, study the structure of this galaxy. Examples of the light-curves of RR Lyrae stars in the SMC are shown in Figure 6, where the red line represents the best fit. We have derived mean magnitudes and constructed a PK_s relation for these stars (Fig. 6). The large scatter (~ 0.20 mag) of the relation reflects the extended structure of the SMC along the line-of-sight. Preliminary results show that RR Lyrae stars in the south-eastern region of the SMC are at closer distances (Fig. 6) . This is suggestive of tidal interactions. The spatial

DOI: 10.1051/epjconf/201715201008

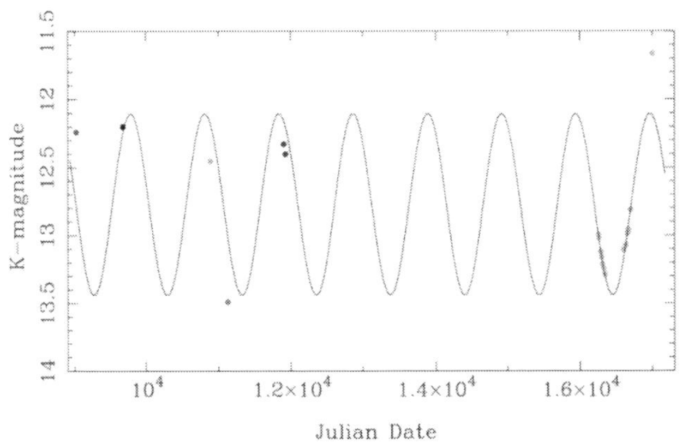

Figure 7. K_s-band curve for the AGB star IRAS05506 combining VMC data (cluster of green points) with literature data from other sources (other coloured points). The best fit periodic variation is indicated with a thin grey line. Small shifts are probably due to the different filter curves associated to each survey.

distribution is almost uniform and the three-dimensional structure is ellipsoidal. The line-of-sight depth varies from 500 pc to 10 kpc.

3.3 Asymptotic giant branch stars

Stars with initial masses up to 8 $M_\odot$ evolve to the asymptotic giant branch (AGB) phase. In particular, intermediate-age evolved giant stars are classified into M-type or C-type depending on their atmospheric chemical composition. They are also characterised by pulsation with long periods (a few hundred days) and large amplitudes (exceeding one magnitude in some bands). These stars trace the extended structure of the Magellanic Clouds (e.g. [4]) and have recently been found by Gaia in tidal tails ([1]). The VMC survey provides a relatively short sampling of their K_s-band curve, but combined with literature observations from 2MASS and DENIS they allow us to derive the amplitude and period of variation with greater accuracy than by using only fewer and sparser measurements. Ultimately, an accurate measurement of the mean K_s magnitude of AGB stars will provide better period-magnitude relations that, similarly to those obtained for Cepheids and RR Lyrae stars, will allow us to estimate distances, and trace the geometrical distribution of the Magellanic Clouds at intermediate ages.

4 Conclusions

The VMC observations are at least 80% complete. These unique data produce accurate mean K_s-band magnitudes, for different types of stars, from at least 12 independent epochs. Within the next years we would like to use these data to go a step further into our understanding of the Magellanic Clouds and of specific stars. The analysis of the structure of the SMC using Cepheids and RR Lyrae stars will soon be finalised and the results will be compared with the recent studies based on the OGLE data. The analysis of the structures of the LMC and of the Bridge will follow and AGB stars will provide a complementary view of the system.

The study of the distribution and properties of stellar populations of different ages will allow us to set constraints on the orbital history of the Magellanic system. Recent simulations of the two body system, Milky Way and LMC, support the scenario that the LMC is either on a first passage or on a long period orbit (5 Gyr) around the Milky Way ([2]). A comparison with analogous systems in the Illustris simulations supports a first approach. In this case the mass of the Milky Way would be limited to 1.5×10^{12} $M_\odot$ and the LMC would have retained most of its cosmological mass.

To further characterise, dynamically and chemically, the Magellanic system we need to obtain a large set of spectra that in radial velocity match the proper motions obtained at present with VMC and soon with Gaia, that sample substructures and all stellar populations. In addition, iron abundances will permit us to estimate better ages from the colour-magnitude diagrams and to assess period-luminosity dependencies that may influence the accuracy via which distances are measured. These are the goals of a new large scale project to be undertaken at VISTA with the 4MOST instrument ([6]). The One Thousands and One Magellanic Fields survey (1001MC; PI Cioni) will obtain both low (R = 5000) and high (R = 20000) resolution spectra across an area of about 1000 deg^2 for about half a million stars with r < 20 mag. These data will allow us to measure on average radial velocities with an accuracy < 2 km/s and iron abundance with an accuracy of < 0.2 dex. This survey will begin in 2022 and will last for approximately 5 years. For details see `https://www.4most.eu/cms/surveys/galactic/`.

Acknowledgments: This project has received funding from the European Research Council (ERC) under European Union's Horizon 2020 research and innovation programme (grant agreement No 682115). We thank the Cambridge Astronomy Survey Unit (CASU) and the Wide Field Astronomy Unit (WFAU) in Edinburgh for providing calibrated data products under the support of the Science and Technology Facility Council (STFC) in the UK. M.-R.L. Cioni acknowledges support from the STFC [grant number ST/M001008/1].

References

[1] Belokurov, V., Erkal, D., Deason, A. J., et al., MNRAS, **466**, 4711 (2017)
[2] Besla, G., Martínez-Delgado, D., van der Marel, R. P., Beletsky, Y., Seibert, M., Schlafly, E. F., Grebel, E. K., & Neyer, F., ApJ, **825**, 20 (2016)
[3] Cioni, M.-R. L., Clementini, G., Girardi, L., et al., A&A, **527**, A116 (2011)
[4] Cioni, M.-R. L., Habing, H. J., & Israel, F. P., A&A, **358**, L9 (2000)
[5] Cross, N. J. G., Collins, R. S., Mann, R. G., et al., A&A, **548**, A119 (2012)
[6] de Jong, R. S., Barden, S. C., Bellido-Tirado, O., et al., *4MOST:the 4-metre Multi-Object Spectroscpic Telescope project at preliminary design review*, Proceeding of the SPIE, **9908**, 99081O (2016)
[7] Emerson, J., & Sutherland, W., The Messenger, **139**, 2 (2010)
[8] Gratton, R. G., Bragaglia, A., Clementini, G., Carretta, E., Di Fabrizion, L., Maio, M., & Taribello, E., A&A, **421**, 937 (2004)
[9] Irwin, M., Lewis, J., Hodgkin, S., et al., *VISTA data flow system: pipeline processing for WFCAM and VISTA*, Proceeding of the SPIE, **5493**, 411 (2004)
[10] Koposov, S. E., Belokurov, V., Torrealba, G., & Evans, N. W., ApJ, **805**, 130 (2015)
[11] Jacyszyn-Dobrzeniecka, A. M., Skowron, D. M., Mróz, P., et al., AcA, **66**, 149 (2016)
[12] Marconi, M., Molinaro, R., Ripepi, V., et al., MNRAS, **466**, 3206 (2017)
[13] Moretti, M. I., Clementini, G., Ripepi, V., et al., MNRAS, **459**, 1687 (2016)
[14] Muraveva, T., Palmer, M., Clementini, G., et al., ApJ, **807**, 127 (2015)
[15] Pigafetta, A., *Magellan's Voyage A Narrative Account of the First Circumnavigation* (Dover Publications, 1994)
[16] Rubele, S., Girardi, L., Kerber, L., et al., MNRAS, **449**, 639 (2015)
[17] Ripepi, V., Marconi, M., Moretti, M. I., et al., ApJS, **224**, 21 (2016)

Spitzer observations of large amplitude variables in the LMC and IC 1613

Patricia A. Whitelock[1,⋆], *Mansi* Kasliwal[2], and *Martha* Boyer[3,4]

[1] *South African Astronomical Observatory & Astronomy Department, University of Cape Town, South Africa*
[2] *Division of Physics, Mathematics and Astronomy, CIT, Pasadena, CA91125, USA*
[3] *NASA Goddard Space Flight Center, MC 665, 8800 Greenbelt Road, Greenbelt, MD 20771, USA*
[4] *Department of Astronomy, University of Maryland, College Park, MD 20742, USA*

Abstract. The 3.6 and 4.5 μm characteristics of asymptotic giant branch variables in the LMC and IC 1613 are discussed. For C-rich Mira variables there is a very clear period-luminosity-colour relation, where the [3.6] – [4.5] colour is associated with the amount of circumstellar material and correlated with the pulsation amplitude. The [4.5] period-luminosity relation for dusty stars is approximately one mag brighter than for their naked counterparts with comparable periods.

1 Introduction

The material discussed here involves observations made with the *Spitzer* spacecraft at 3.6 and 4.5 μm, and comes from the SAGE Var Collaboration [10] for the LMC and from the SPIRITS ([7]) and DUSTiNGS ([2]) collaborations for IC 1613. We focus on the large amplitude asymptotic giant branch (AGB) stars, which were found by SPIRITS as a byproduct of their search for infrared (IR) transients and by DUSTiNGS because of their strong dust emission.

These stars have pulsation periods that are mostly in the range 100 to 1000 days, and peak-to-peak amplitudes over 0.3 mag at [3.6] and/or [4.5]. They are very luminous bolometrically and in the IR. They are near the very top of the AGB, so they represent the brightest evolutionary phase for stars of a particular mass/metallicity. They are bright in the IR because they are cool and because they become enshrouded in dust during their AGB evolution, so they are particularly bright at the wavelengths of interest to JWST and the next generation of extremely large ground-based telescopes. The variables with the largest amplitudes are traditionally known as Miras, although those with very thick dust shells are often identified as "extreme-AGB" (x-AGB) stars (often, but not necessarily, C-rich), or as OH/IR stars (if they are O-rich).

The Miras are potentially useful distance indicators (rivalling Cepheids; [4]) as well as for studies of resolved old and intermediate mass populations in a range of galaxies. They are astrophysically important because they are responsible for some of the processed material returned to the interstellar medium, including important elements such as carbon, lithium and elements produced by the s-process. The details of this are still rather sketchy because the process of mass loss is not yet well understood.

⋆paw@saao.ac.za

DOI: 10.1051/epjconf/201715201009

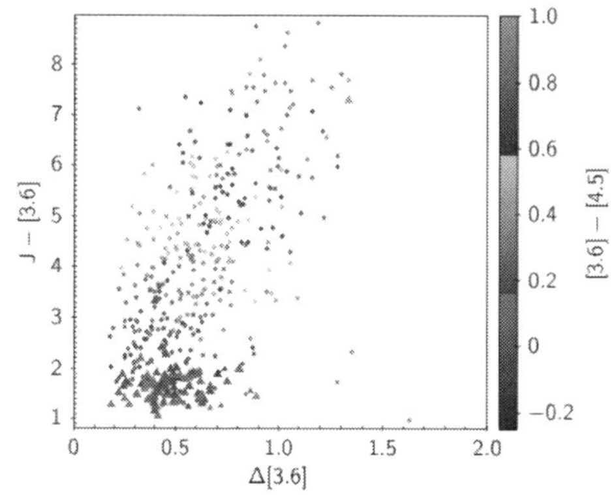

Figure 1. Colour (J − [3.6]) against amplitude at 3.6 μm (Δ[3.6]). C- and O-rich stars (as determined from SED fits; [10]) are shown as circles and triangles, respectively. Note that the redder C-stars tend to have larger amplitudes.

AGB variables in the Magellanic Clouds fall into three classes ([11]): Miras, semi-regular variables (SRVs) and OGLE small amplitude red variables (OSARGS), of which only the Miras and SRVs have sufficiently large amplitudes to be of interest here. Wood et al. ([14]) noted that AGB stars occupied various period-luminosity (PL) relations. Miras pulsate in the fundamental mode[1] while SRVs pulsate in the fundamental and/or in overtones. The Miras generally show only a single mode while SRVs often show two or more modes. Most Miras have larger amplitudes than most SRVs, but the two types are often difficult to distinguish without a large number of observations ([11]).

2 Large Magellanic Cloud (LMC)

Riebel et al. ([10]) obtained observations of the LMC bar at 3.6 and 4.5μm at four epochs, which they combined with the two earlier epochs of the SAGE legacy survey to give a total of six epochs. There are good periods for these variables from the OGLE III survey ([11]) at V and/or I, and single epoch JHK_S photometry from the InfraRed Survey Facility (IRSF) in South Africa ([8]).

Figure 1 shows a plot of J − [3.6] against Δ[3.6]. For the O-rich stars, there is no clear correlation of amplitude with colour, while the C-stars show a relation with redder stars having larger amplitudes. Of course with just six epochs Δ[3.6] is only a lower limit and the J measurement comes from a single epoch, hence the large scatter. The difference between the O- and C-rich stars may be related to the relative transparency of the O-rich grains ([1]). Riebel et al. analysed the *Spitzer* data, together with periods from MACHO, to establish PL relations for Mira variables with J − [3.6] < 3.1 only, i.e. excluding the redder x-AGB stars.

Figure 2 shows the [3.6] and [4.5] PL relations for all of the variables that the OGLE group ([11]) identify as Miras. OGLE does not distinguish variability class on the basis of colour and this therefore includes the redder x-AGB stars as well as those Riebel et al. ([10]) considered. Riebel et al. identified only 10 AGB candidates in their *Spitzer* selected variables, that were not previously known variables. These are probably Miras or SRVs that were either not observed or not classified by OGLE, possibly because they were too red. We can therefore be confident that the sample of Miras considered here is very nearly complete.

[1]Possibly hot bottom burning (HBB) Miras pulsate in the first overtone ([5]).

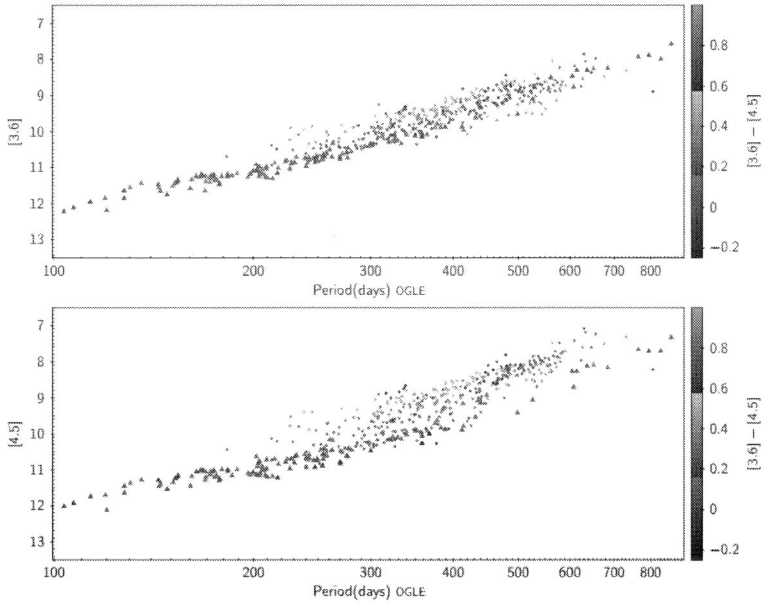

Figure 2. The [3.6] and [4.5] PL relations for *Spitzer* variables in the LMC that are classified as Miras by OGLE. O-rich and C-rich stars are identified as triangles and circles respectively, while the colour coding is according to [3.6] – [4.5] as shown on the right-hand side.

Riebel et al. discussed the PL relations for O- and C- rich stars and noted the surprisingly large scatter, particularly in the [4.5] relation; by including the x-AGB stars we see even more spread. Some of this may be due to variations in the CO fundamental vibration-rotation feature which falls within the [4.5] band, but it is clear from the colour dependence that most of it is due to dust emission. The spread at a given period is over one magnitude and the brighter sources have larger [3.6] – [4.5] indices. This is very clear for the C-rich Miras in Figure 3, where the [4.5] relation is contrasted with the K_S relation. At [4.5] the reddest stars are the brightest because of dust emission, while at K_S they are the faintest because of dust absorption. It is also notable in this figure that the spread in K_S of the reddest stars is around three mag whereas that in [4.5] is less than one mag. This is partially because the K_S observations are single epoch, but the asymmetric character of the dust shells around C-stars must also contribute to this scatter. There is a good deal of evidence that C-rich Miras eject dust in random directions ([3, 15]), in the same way as R Coronae Borealis stars, possibly because the dust forms over convection cells ([6]). The dust will be optically thin at [3.6] and [4.5] so the entire shell will be measured in those bands.

Riebel et al. colour-selected Miras for their analysis so that most of their 3 to 4 μm flux originated from the star rather than from the shell. The spread we see in the [4.5] PL relation (Fig. 3, left panel) is due to the increasing contribution from dust as the star gets redder, i.e. there is a period-luminosity-colour relation for these stars. It is particularly interesting to see that once the dust shell dominates the flux the PL relation reestablishes itself, almost parallel to the original relation, but about 1.2 mag brighter than that seen for the naked stars. The effect is much smaller at [3.6], because the contribution from dust emission is weaker. It will be interesting to examine the bolometric luminosities of these variables and to compare with models to see if those with thicker shells can be understood as more evolved than those without. There is evidence that Miras do not change their periods dramatically

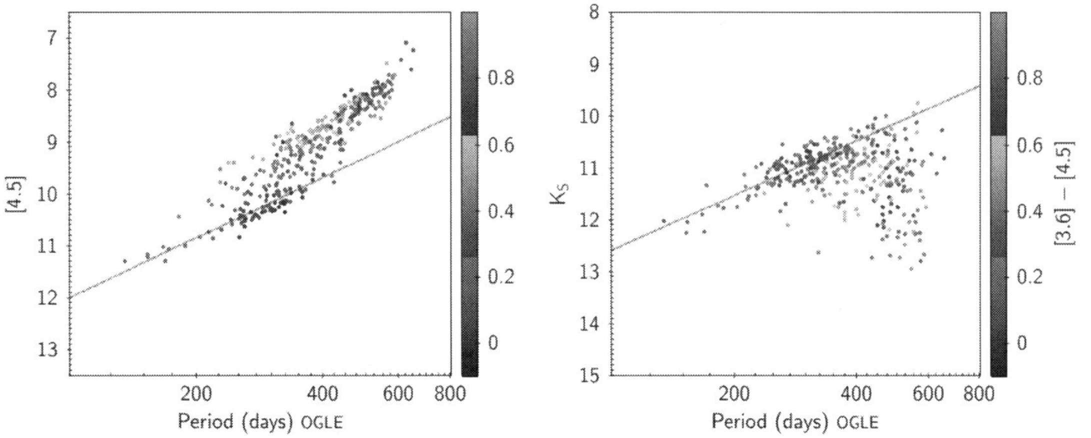

Figure 3. PL relationships for the same C-rich stars at [4.5] and at K_S. The straight lines are the Mira PL relationships from Riebel et al. ([10], derived from the bluer group of this sample of stars) and Whitelock et al. ([16], for a completely different sample of stars).

once they start pulsating in the fundamental mode (e.g., [4]), but some change in period may occur as the star evolves through the instability strip ([13]). Only stars with a limited range of initial masses will become C-rich as HBB will keep stars over 3 or 4 $M_{\odot}$ O-rich.

3 IC1613

Menzies et al. ([9]) discussed the variable AGB stars in IC 1613, identifying several O- and C-rich Miras with periods ranging from 260 to 880 days from JHK photometry. Of particular interest were four O-rich Miras, with periods ranging from 460 to 580 days and luminosities above the K_S and m_{bol} PL relations. The detection of lithium in one of these demonstrated that it is undergoing HBB, and we can presume that this is why these stars depart from the PL relation (which is linked to the core-mass luminosity relation at the tip of the AGB). This survey was sensitivity limited and would not have detected the most dusty variables, which are faint at J.

The SPIRITS and DUSTiNGS surveys do not cover exactly the same region of IC 1613, but they recover those variables found by [9] in the overlap region, and detect several more that are particularly dusty. Although we can determine periods for many of these, there are insufficient observations to distinguish SRVs from Miras. Therefore, in Figure 4, we compare them with OGLE and *Spitzer* data from the LMC for both SR and Mira variables. We distinguish large (Δ[3.6] and Δ[4.5] > 0.5 mag) and small pulsation amplitudes, noting that those with large amplitudes are almost certainly Miras. Within the LMC most of the large amplitude variables have periods over 300 days and are more luminous than their low amplitude counterparts at a given period, although there are a few large amplitude stars at shorter periods and lower luminosities. The variables in IC 1613, which have periods from 180 to 580 days, follow very much the same trends. There are a few of the lower amplitude stars that appear to be overtone pulsators. The four HBB stars cluster closely together as they do at shorter wavelengths; they are amongst the brightest variables at 3.6 μm, but are are not as red as the most luminous C-stars, which have comparable periods.

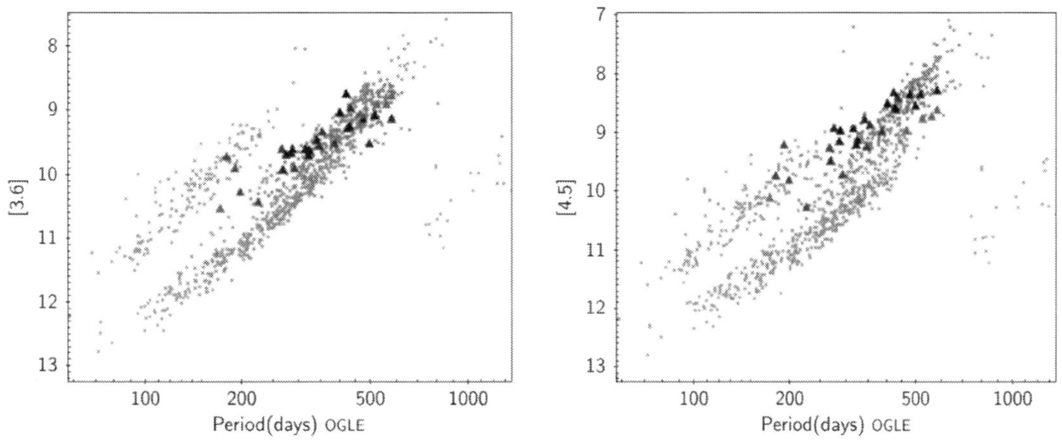

Figure 4. [3.6] and [4.5] PL relations for the SR and Mira variables in the LMC (small circles: cyan, except for stars with $\Delta[3.6]$ and $\Delta[4.5] > 0.5$ mag, which are magenta), and in IC 1613 (triangles: O-rich stars in red and presumed C-rich stars in blue or if $\Delta[3.6]$ and $\Delta[4.5] > 0.5$ mag, in black; the single yellow symbol is a Cepheid). The magnitudes of the IC 1613 variables have been corrected by 5.83 mag, for the relative distances of the two galaxies. The two cyan sequences that stand out show the fundamental and first overtone pulsators.

Acknowledgments: This work makes use of VizieR, the CDS catalogue service, and the TOPCAT ([12]) data processing package. PAW acknowledges the receipt of a research grant from the South African National Research Foundation (NRF).

References

[1] Bladh, S. Höfner, S., Nowotny, W., Aringer, B., & Eriksson, K., A&A, **553**, A20 (2013)

[2] Boyer, M. L., McQuinn, K. B. W., Barmby, P., et al., ApJS, **216**, 10, (2015)

[3] Feast, M. W., Whitelock, P. A., & Marang, F., MNRAS, **346**, 878 (2003)

[4] Feast, M. W., & Whitelock, P. A., in *Setting the scene for Gaia and LAMOST*, ed. S. Feltzing, G. Zhao, N. A. Walton, & P. A. Whitelock, IAUS, **298**, 2013 (2014)

[5] Feast, M. W., in *AGB Stars and Related Phenomena*, ed. T. Ueta, N. Matsunaga, & Y. Ita, p. 48 (2009)

[6] Feast, M. W., these proceedings (2017)

[7] Kasliwal, M. M., Bally, J., Masci, F., et al., ApJ, **839**, 88 (2017)

[8] Kato, D., Nagashima, C., Nagayama, T., et al., PASJ, **59**, 615 (2007)

[9] Menzies, J. W., Whitelock, P. A., & Feast, M. W., MNRAS, **459**, 910 (2015)

[10] Riebel, D., Boyer, M. L., Srinivasan, S., et al., ApJ, **807**, 1 (2015)

[11] Soszyński, I., Udalski, A., Szymańński, M. K., et al., Acta Astron., **59**, 239 (2009)

[12] Taylor, M. B., ASPC, **347**, 29 (2005)

[13] Wood, P. R., MNRAS, **448**, 3829 (2015)

[14] Wood, P. R., Alcock, C., Allsman, R. A., et al., in *Asymptotic Giant Branch Stars*, ed. T. Le Bertre, A. Lebre, & C. Waelkens, IAUS, **191**, 151 (1999)

[15] Whitelock, P. A., Feast, M. W., Marang, F., & Groenewegen, M. A. T., MNRAS, **369**, 751 (2006)

[16] Whitelock, P. A., Feast, M. W., & van Leeuwen, F., MNRAS, **386**, 313 (2008)

Multiwavelength light curve parameters of Cepheid variables

Anupam Bhardwaj[1,2,*], *Shashi M.* Kanbur[3], *Marcella* Marconi[4], *Marina* Rejkuba[2], *Harinder P.* Singh[1], and *Chow-Choong* Ngeow[5]

[1]*Department of Physics & Astrophysics, University of Delhi, Delhi 110007, India*
[2]*European Southern Observatory, Karl-Schwarzschild-Straße 2, 85748, Garching, Germany*
[3]*State University of New York, Oswego, NY 13126, USA*
[4]*INAF-Osservatorio astronomico di Capodimonte, Via Moiariello 16, 80131 Napoli, Italy*
[5]*Graduate Institute of Astronomy, National Central University, Jhongli 32001, Taiwan*

Abstract. We present a comparative analysis of theoretical and observed light curves of Cepheid variables using Fourier decomposition. The theoretical light curves at multiple wavelengths are generated using stellar pulsation models for chemical compositions representative of Cepheids in the Galaxy and Magellanic Clouds. The observed light curves at optical (*VI*), near-infrared (JHK_s) and mid-infrared (3.6 & 4.5-μm) bands are compiled from the literature. We discuss the variation of light curve parameters as a function of period, wavelength and metallicity. Theoretical and observed Fourier amplitude parameters decrease with increase in wavelength while the phase parameters increase with wavelength. We find that theoretical amplitude parameters obtained using canonical mass-luminosity levels exhibit a greater offset with respect to observations when compared to non-canonical relations. We also discuss the impact of variation in convective efficiency on the light curve structure of Cepheid variables. The increase in mixing length parameter results in a zero-point offset in bolometric mean magnitudes and reduces the systematic large difference in theoretical amplitudes with respect to observations.

1 Introduction

Classical Cepheids are radially pulsating variables that display periodic light curves and follow a well-defined period-luminosity relation (PLR, [1]). Cepheids play a vital role in extragalactic distance scale to determine an accurate and precise value of the Hubble constant ([2]). In addition, a comparison of observed pulsation properties of these variables with theoretical pulsation models provides fundamental test of the theory of stellar pulsation and evolution.

First quantitative study of light curves of Cepheid variables based on Fourier decomposition method was presented for a sample of 57 Cepheids ([3]). It was suggested that the lower order Fourier amplitude and phase coefficients contain the most characteristic features of the light curve structure ([4, 5]). The Cepheid light curves display bump features as a function of period known as Hertzsprung progressions (hereafter HP, [6]). The variation of HP for Classical Cepheids was investigated using pulsation models and it was found that the central period of the progression occurs around 11 days for LMC Cepheids ([7]). The comparison of pulsation properties of Cepheids from theoretical models

*anupam.bhardwajj@gmail.com

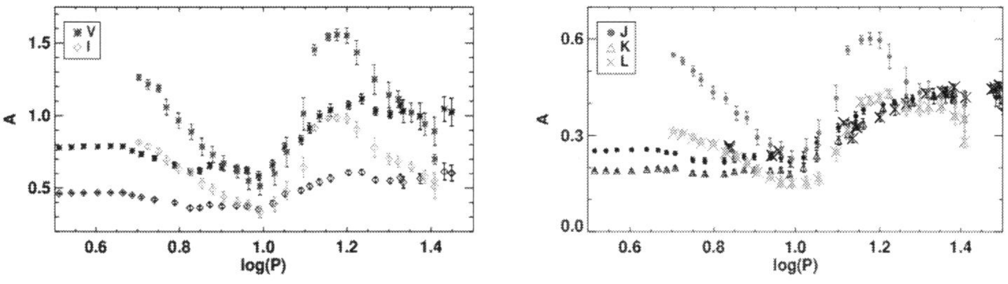

Figure 1. A comparison of observed (black symbols) and theoretical (colored symbols) mean amplitudes for fundamental-mode Cepheids in the LMC.

with observations in terms of PLR has been a subject of many studies in the past decade ([8, 9]) but no rigorous comparison of light curve structure was carried out at multiple wavelengths for different compositions.

The observed light curve structure of Cepheid variables in the Galaxy and LMC was studied in detail and the variation of light curve parameters as a function of period and wavelength was presented in [10]. This work was extended to theoretical models of Cepheid variables and a comparison was presented at multiple wavelengths ([11]). We summarize the results from these analyses in the following sections.

2 Analysis and results

The theoretical light curves are obtained from full amplitude, nonlinear, convective hydrodynamical models and follow the same assumptions as discussed in [12]. For a fixed chemical composition and mass, the luminosity levels are adopted from stellar evolutionary calculations, i.e., canonical mass-luminosity (M-L) relations. In addition, models with a brighter luminosity level by 0.25 dex (non-canonical M-L relations) are also computed to account for possible overshooting and mass loss efficiency. The topology of the instability strip is explored for each combination of M-L levels and the bolometric light curves are obtained. These light curves are transformed to optical and near-infrared filters by empirical relations from static atmospheric models. The observed optical and infrared band light curves are compiled from several sources in literature and the details can be found in [10].

The theoretical and observed light curves were fitted with a Fourier series in the following form:

$$m = m_0 + \sum_{k=1}^{N} A_k \sin(2\pi k x + \phi_k), \tag{1}$$

where x represents the phase along the pulsation cycle. The optimum order of fit (N) is obtained using the minimization of the successive residuals and the Fourier coefficients are used to formulate Fourier amplitude ratios and phase differences ($R_{k1} = A_k/A_1$ & $\phi_{k1} = \phi_k - k\phi_1$, for $k > 1$).

The amplitude of the light curve is defined as the difference of maximum and minimum magnitudes obtained from the Fourier fit. Figure 1 displays the variation of mean amplitudes for the Cepheids in the LMC. The average values are obtained by sliding mean calculations in a bin-size of $\log(P) = 0.1$ dex and in steps of 0.03 dex. The theoretical mean amplitudes are systematically larger than observed amplitudes at VIJ wavelengths, as a function of period except close to the period of 10 days. At longer wavelengths (KL) the amplitudes are consistent between observations and theory.

 DOI: 10.1051/epjconf/201715201010

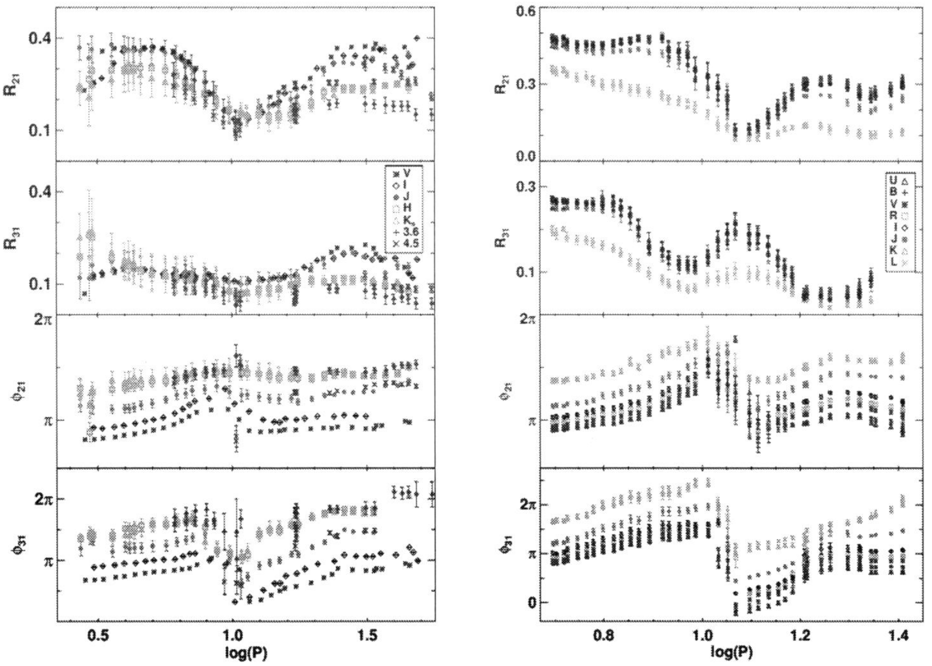

Figure 2. Multiband mean Fourier parameters for observed Galactic Cepheid light curves (left panel) and corresponding theoretical light curves (right panel) with chemical composition $Y = 0.28$, $Z = 0.02$.

2.1 Fourier parameters as a function of wavelength and metallicity

We discuss the variation of mean Fourier parameters as a function of period and wavelength for different metallicities. Figure 2 displays the variation of the mean Fourier parameters for Galactic Cepheids (left panels) and for theoretical models with metallicity $Z = 0.02$ (right panels) at multiple wavelengths. We find a decrease in Fourier amplitude parameters with increase in wavelength while the phase parameters increase with wavelength for both theoretical and observed light curves. In the long period range, $\log(P) > 1$, R_{21} increases sharply to a peak value around $\log(P) = 1.3$, at optical wavelengths. This variation is visible for both theoretical and observed Fourier parameters and separates optical wavelengths to redder bands. Similar features in Fourier parameters are also seen for Magellanic Cloud Cepheids ([11]). The mean phase parameters display a systematic offset at all wavelengths for the Galaxy and Magellanic Cloud Cepheid models.

Figure 3 displays variation in theoretical I-band mean Fourier parameters as a function of metallicity. The amplitude parameters in short period range [$\log(P) < 1$] increase with decrease in metallicity. The phase parameters also provide evidence of a decrease with metallicity but the progression is not distinctly clear. The central minimum in case of R_{21} shifts to longer periods with decrease in metallicities. Similar variation in HP was also seen for observed light curves of Cepheids in [10].

2.2 A comparison of observed and theoretical Fourier parameters

Figure 4 compares the Fourier parameters of the theoretical and observed I-band light curves for Cepheid variables in the LMC. The theoretical models are plotted for two different sets of mixing

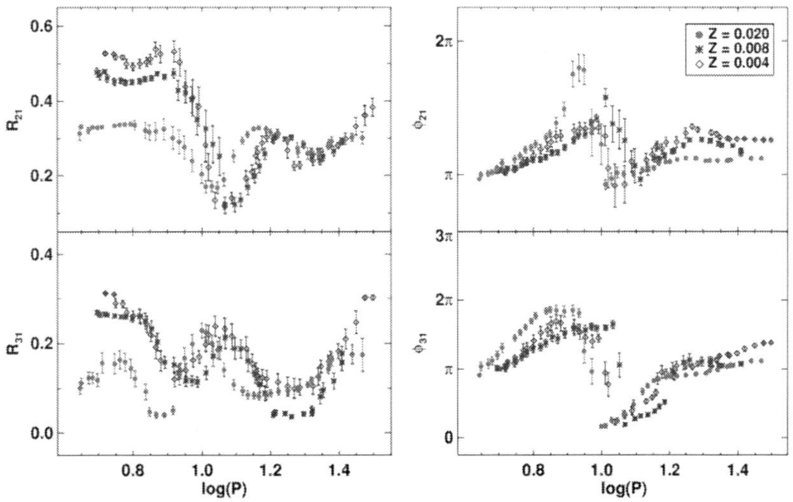

Figure 3. Mean Fourier parameters in *I*-band for theoretical models of Cepheids with different metal abundances.

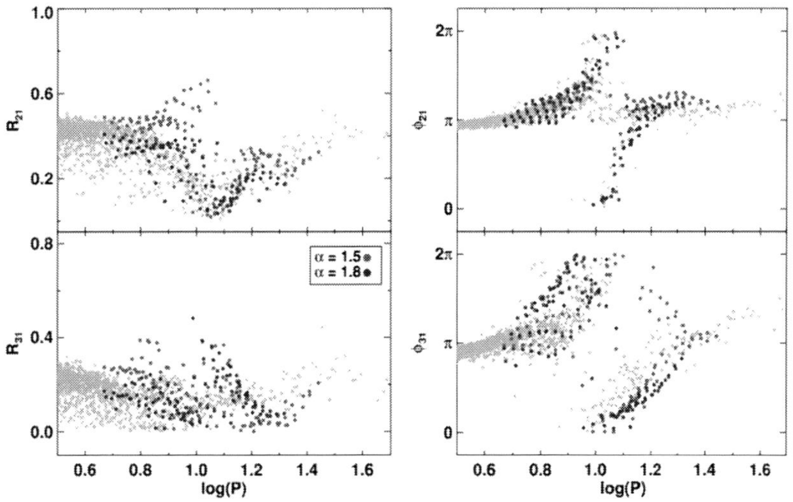

Figure 4. Fourier parameters for theoretical and observed (grey) *I*-band light curves for Cepheids in the LMC. The theoretical models correspond to two mixing lengths, standard 1.5 (red) and enhanced 1.8 (blue).

length parameters. For standard mixing length (α = 1.5), there is a large discrepancy in theoretical amplitude parameters at short period range [log(P) < 1], when compared to observations. We find that canonical set of models displays larger offset in the overlapping period range. In contrast, the non-canonical Cepheid models are more consistent with observations. Increase in mixing length (α = 1.8) enhances the convective efficiency and damps down pulsation. In turn, it affects the width of the instability strip as the red edge becomes bluer. The increase in convection very likely leads to decrease in theoretical amplitudes. This is evident from Figure 4, where increased mixing length (blue circles) helps in minimizing the discrepancy in theoretical amplitudes with respect to the observations. However, this also leads to zero-point offset in bolometric mean magnitudes. This may cause biases in distance estimates based on theoretical P-L relations with different mixing length parameters ([13]).

3 Summary

We discuss the variation of the light curve parameters as a function of period, wavelength and metallicity for the theoretical and observed light curves of Cepheid variables. We present a quantitative comparison between theoretical and observed light curve structure posing strong constraints for stellar pulsation codes that incorporate stellar atmosphere models to produce wavelength-dependent theoretical Cepheid light curves. We find a large discrepancy in theoretical amplitude parameters with respect to the observations and suggest that this difference can be reduced by adopting a combination of non-canonical set of models and increased mixing length. We note that the Fourier amplitude parameters can also be used to discriminate between canonical and non-canonical set of models in the overlapping period range. However, a more detailed analysis is required to further constrain the mass-luminosity relations obeyed by Cepheid variables and to explore the impact of different input parameters on the light curve structure. The mass-luminosity relations for Cepheids are based on stellar evolutionary calculations and therefore a thorough analysis with additional model computations on a smoother grid can potentially help constrain theories of stellar evolution and pulsation.

Acknowledgments: AB acknowledges the Senior Research Fellowship grant 09/045(1296)/2013-EMR-I from the Council of Scientific and Industrial Research (CSIR), India.

References

[1] H.S. Leavitt, E.C. Pickering, Harvard College Observatory Circular **173**, 1 (1912)
[2] A.G. Riess, L.M. Macri, S.L. Hoffmann, D. Scolnic, S. Casertano, A.V. Filippenko, B.E. Tucker, M.J. Reid, D.O. Jones, J.M. Silverman et al., ApJ **826**, 56 (2016)
[3] N.R. Simon, A.S. Lee, ApJ **248**, 291 (1981)
[4] N.R. Simon, T.J. Teays, ApJ **261**, 586 (1982)
[5] G. Kovacs, I. Shlosman, J.R. Buchler, ApJ **307**, 593 (1986)
[6] E. Hertzsprung, BAIN **3**, 115 (1926)
[7] G. Bono, M. Marconi, R.F. Stellingwerf, A&A **360**, 245 (2000)
[8] G. Bono, M. Marconi, R.F. Stellingwerf, ApJs **122**, 167 (1999)
[9] F. Caputo, M. Marconi, I. Musella, A&A **354**, 610 (2000)
[10] A. Bhardwaj, S.M. Kanbur, H.P. Singh, L.M. Macri, C.C. Ngeow, MNRAS **447**, 3342 (2015)
[11] A. Bhardwaj, S.M. Kanbur, M. Marconi, M. Rejkuba, H.P. Singh, C.C. Ngeow, MNRAS **466**, 2805 (2017)
[12] M. Marconi, R. Molinaro, G. Bono, G. Pietrzyński, W. Gieren, B. Pilecki, R.F. Stellingwerf, D. Graczyk, R. Smolec, P. Konorski et al., ApJ **768**, L6 (2013)
[13] G. Fiorentino, M. Marconi, I. Musella, F. Caputo, A&A **476**, 863 (2007)

Pulsating white dwarfs

S. O. Kepler[1,*] and *Alejandra D.* Romero[1,**]

[1] *Departamento de Astronomia, Instituto de Física, Universidade Federal do Rio Grande do Sul, 91501-970 Porto Alegre, RS, Brazil*

Abstract. The Sloan Digital Sky Survey has allowed us to increase the number of known white dwarfs by a factor of five and consequently the number of known pulsating white dwarfs also by a factor of five. It has also led to the discovery of new types of variable white dwarfs, as the variable hot DQs, and the pulsating Extremely Low Mass white dwarfs. With the *Kepler* Mission, it has been possible to discover new phenomena, the outbursts present in a few pulsating white dwarfs.

1 Introduction

White dwarf stars are the final evolutionary state of stars with initial masses up to 8.5 – 10.6 $M_\odot$ ([1]), corresponding to 95–97 % of all stars. The fraction depends on the stellar metallicity, which affects both the Initial-Mass-Function and the Initial-to-Final-Mass Relation. For single stars, the minimum mass of a present day white dwarf is around 0.30 – 0.45 $M_\odot$ ([2]), because progenitors that would become lower mass white dwarfs have main sequence evolution time larger than the age of the Universe. Such masses correspond, considering the mass-radius relation of white dwarfs, to a minimal $\log g \simeq 6.5$. Evolutionary models e.g. by [3] indicate that the maximum surface gravity for main sequence A stars, which have similar optical spectra to DA white dwarfs, corresponds to $\log g \leq 4.75$, including very low metallicity. There is therefore a gap between low mass white dwarfs and main sequence stars, $4.75 \leq \log g \leq 6.5$.

Most white dwarfs do not generate energy from nuclear fusion, but radiate due to residual gravitational contraction. Because of the degenerate equation of state, contraction is accompanied by a loss of thermal energy instead of increase as in the case of ideal gases; the evolution of white dwarfs is therefore often simply described as cooling. The radius of an average white dwarf star is of the same order of the Earth's radius, which implies that they have small surface area, resulting in very large cooling times; it takes approximately 10^{10} years for the effective temperature of a $\sim 0.6\,M_\odot$ white dwarf to decrease from 100 000 K to near 5 000 K. Consequently, the cool $\sim 0.6\,M_\odot$ ones are still visible and among the oldest objects in the Galaxy ([4]). Therefore, studying white dwarfs is extremely important to comprehend the processes of stellar formation and evolution in the Milky Way ([5, 6]).

The progenitors of white dwarfs lose most of their envelope in the giant phases, where mass loss depends on metallicity. If the remainder H mass were above $\simeq 10^{-4}M*$, or the He mass above $\simeq 10^{-2}M*$, there would be observable nuclear burning in the white dwarf phase. The limits depend on

*kepler@if.ufrgs.br
**alejandra.romero@ufrgs.br

the mass of the white dwarf. Most white dwarfs have atmospheres dominated by H, and the remainder by He. All other elements are only small traces, much less abundant than in the Sun, due to separation in the strong gravitational field ([7]). The lightest elements float to the surface once the white dwarf cools. The He-core white dwarf stars in the mass range 0.2 – 0.45 $M_\odot$, referred to as low-mass white dwarfs, are usually found in close binaries, often double degenerate systems ([8]), being most likely a product of interacting binary star evolution. More than 70% of those studied by [9] with masses below 0.45 $M_\odot$ and all but a few with masses below 0.3 $M_\odot$ show radial velocity variations ([10, 11]). [2] suggests single low-mass white dwarfs result from the evolution of old metal-rich stars that truncate evolution before the helium flash due to severe mass loss. They also conclude all white dwarfs with masses below $\simeq$ 0.3 $M_\odot$ must be a product of binary star evolution involving interaction between the components, otherwise the lifetime of the progenitor on the main sequence would be larger than the age of the Universe.

In Figure 1 we show the results of our effective temperature and surface gravity determinations for all candidates from SDSS. We calculated the single star evolutionary models shown in the figure with the MESA ([12]) evolutionary code, including diffusion. In Figure 2 the evolutionary models are those with rotation and diffusion of [13].

Even though the low resolution hydrogen lines observed in SDSS spectra are poor surface gravity indicators below $T_{\rm eff} \simeq 10\,000$ K, considering the SDSS spectra are concentrated mainly outside the Galaxy disk, we were surprised that several thousand stars were classified by the SDSS pipeline as A and B stars. Considering their lifetimes on the main sequence smaller than 1 Gyr, and their distance modulus $(m - M) \geq 14.5$ at the SDSS bright saturation, if these stars were main sequence stars, there would be a considerable population of young stars very far from the Galactic disk. Using their measured radial velocities, and proper motions if available, [14] estimated their U, V, W velocities and show there would be a large number of hypervelocity A stars, not detected up to date. If these stars are in fact low mass counterparts of interacting binary evolution, similar to the models of [13, 15], they are mainly concentrated in the Galactic disk. Considering we do not know their metallicities, and that low ionization potential metals contribute significantly to the electron pressure, we estimated their surface gravities with two sets of models, a pure hydrogen model and a solar composition model. The estimated surface gravities with solar metallicity models were on average $\Delta \log g \simeq 0.5$ dex smaller, but not systematically. Our plotted values are the solar metallicity ones.

1.1 Interacting binaries

[17] found an RR Lyrae with 0.26 $M_\odot$, and [16] found a 0.23 $M_\odot$ pulsating subdwarf (sdBV). Interacting binary with mass exchange models by [13] show that during a hydrogen shell burning pulse, an extremely low mass white dwarf can cross the main sequence, horizontal branch and even giant instability strip. [18] estimate that up to 5% of stars that cross the RR Lyrae and Cepheid instability strip are binaries.

DA white dwarf stars with masses $M \leq 0.45\,M_\odot$ and $T_{\rm eff} < 20\,000$ K are Low Mass and Extremely Low Mass (ELM) as found by [9–11, 19–22]. [23]–[27] found pulsations in eight of these ELMs, similar to the pulsations seen in DAVs (ZZ Ceti stars), as described in [29]. [30] found 17 pre-ELMs, i.e., helium–core white dwarf precursors, and [31, 32] report pulsations in six of them. Pulsations are an important tool to study the stellar interior, and [33] – [37] report on theoretical models and pulsations of ELMs. [38] and [39] show there are thousands of stars, photometrically classified as blue horizontal branch stars by [41–43], that have spectroscopic estimated surface gravities much higher than main sequence stars ($\log g \geq 4.75$) and therefore must have radii smaller than the Sun, classifying them as sdAs, in line with the hot subdwarfs reviewed by [44]. [14] discuss they are possibly ELM white dwarf stars. [38, 45] show that photometrically selected white dwarfs have a

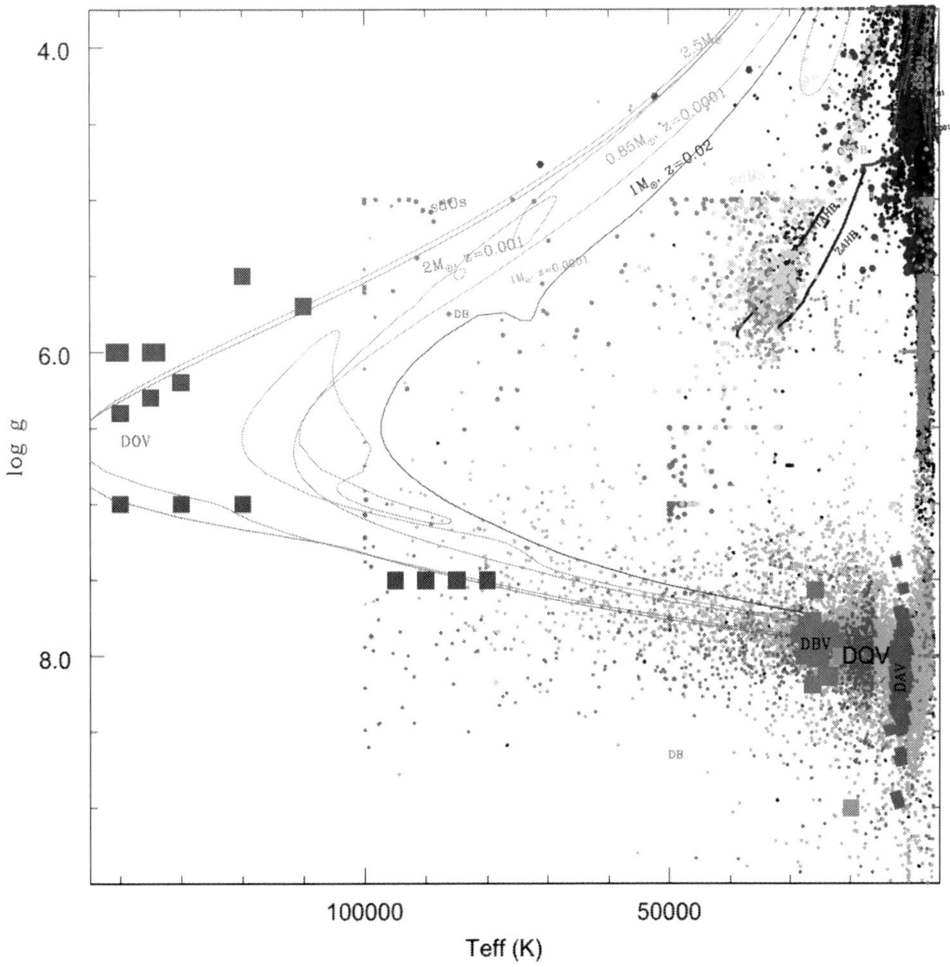

Figure 1. Estimated effective temperature and superficial gravity for 79 440 blue stars in the Sloan Digital Sky Survey up to Data Release 13, and single star evolutionary models for different metallicities, showing there is a region in the middle of the diagram that no single star evolutionary models cover. Data of different colors represent distinct spectral classification and catalogs, DAs in light green, DBs in red, sdBs and sdOs in yellow, sdAs in dark green and black and the pulsating stars. The Zero Age Horizontal Branch (ZAHB) and Terminal Age Horizontal Branch (TAHB) plotted were calculated with solar composition models.

contamination around 40%. Even the ones selected also from proper motion by [46] show significant contamination by non-white dwarf objects, when spectra are available.

Most stars that produced white dwarfs are born in binaries or multiple systems. [47] demonstrates that while around 70% of stars more massive than the Sun are in binaries, two-thirds of the most common stars, M type dwarf stars, are single. More than 10% of the spectroscopically identified white dwarfs in SDSS have red companions ([38, 48]). [49, 50] show that nearly 25% of all main sequence binaries are close enough that mass transfer interactions occur when the more massive star becomes a red giant or an asymptotic giant star. If mass transfer exceeds the Eddington limit, the secondary star is not able to accrete the transferred material and the system evolves through a common envelope

DOI: 10.1051/epjconf/201715201011

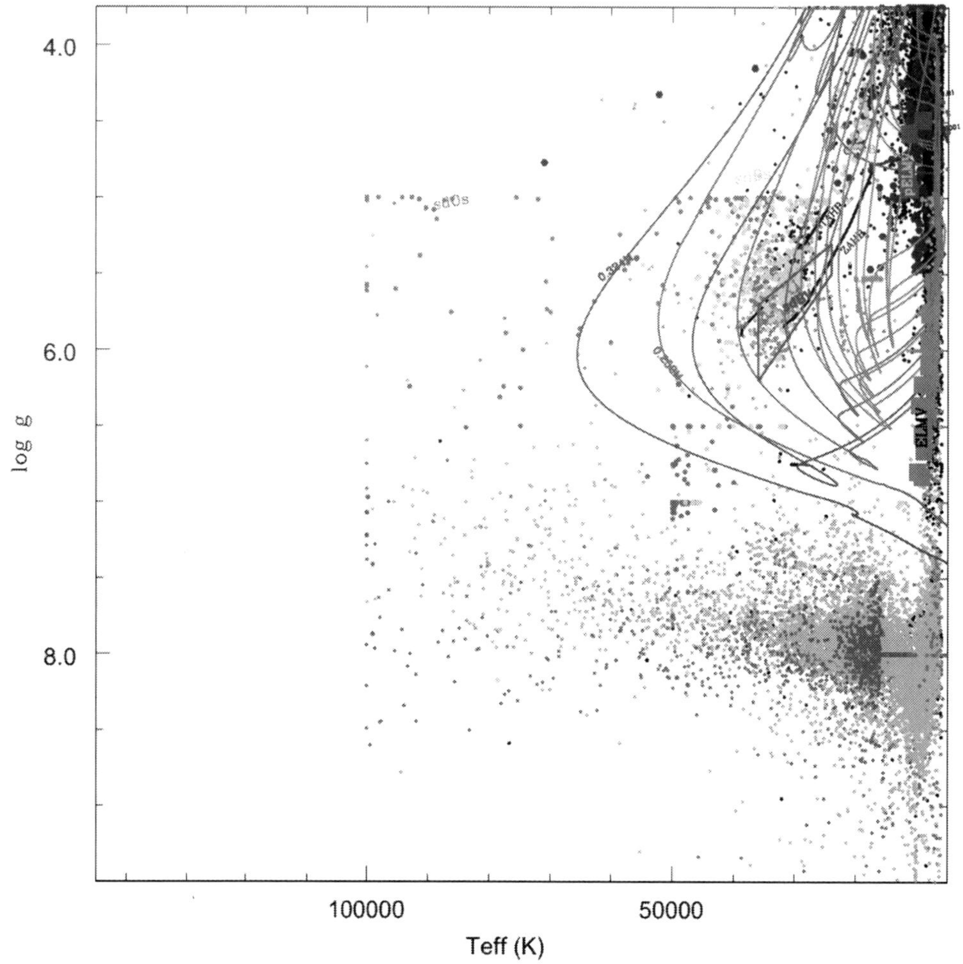

Figure 2. Estimated effective temperature and superficial gravity for 79 440 blue stars in SDSS Data Release 13, and interacting binary star evolutionary models ([13]), showing the region in the middle of the diagram is covered by interacting binary models. sdBs, SX Phe, ELMs and pre-ELMs are in these regions.

phase, i.e., the core of the giant and the main sequence companion orbit within the outer layers of the giant star, leading to the shrinkage of the orbit and the release of orbital energy. The orbital energy deposited into the envelope eventually ejects it. Therefore a close binary is formed by the core of the giant star and a main sequence companion, later a close white dwarf-main sequence binary. An ELM will be formed if the envelope is ejected before the helium-flash, which would happen if the star has a low initial mass, i.e., $M \lesssim 2\,M_{\odot}$, to reach conditions to fuse helium in the core before it becomes degenerate.

2 Mass distribution

We estimated the masses of all DA white dwarfs found by [51], [52] and [38]. There were no new optical stellar spectra in SDSS Data Release 13. For the DA mass distribution, we only consider

Table 1. Number of known pulsating white dwarfs.

Class	Number
DAVs	181
DOVs[1]	22
DBVs	23
ELMVs	11
pre-ELMVs	5
DQVs	3

[1] Pulsating PG 1159 stars.

spectra with S/N ≥ 15 to have reliable mass determinations. From $T_{\rm eff}$ and $\log g$ values obtained from our fits, after correcting to 3D convection following [55], we use the mass-radius relations of [53], [54] and [3] to calculate the stellar mass.

Considering that white dwarfs with larger mass have smaller radius, and therefore can only be seen to smaller distances in a magnitude limited survey as SDSS, we calculated the density by correcting the visible volume with the $1/V_{\rm max}$ method of [56], up to a maximum $g = 19$ magnitude. For DAs with $T_{\rm eff} \geq 10\,000$ K, $N = 4054$, we estimate $\langle M \rangle = 0.647 \pm 0.002\ M_\odot$, $T_{\rm eff} \geq 13\,000$ K, $N = 3637$, $\langle M \rangle = 0.646 \pm 0.002\ M_\odot$, $T_{\rm eff} \geq 16\,000$ K, $N = 3012$, $\langle M \rangle = 0.641 \pm 0.002\ M_\odot$, $T_{\rm eff} \geq 25\,000$ K, $N = 1121$, $\langle M \rangle = 0.613 \pm 0.003\ M_\odot$. The analysis shows that the DA and DB distributions have very different shapes. The DA's has a tail to larger masses, while the DB's is extended to lower masses. This is probably reflecting some limitation in the progenitors that can undergo very-late thermal pulses and become DBs.

3 Pulsations

During the cooling of the white dwarf star, partial ionization zones of C, O, He and H develop at subsequently lower $T_{\rm eff}$. Such partial ionization zones increase the opacity and cause pulsations. For C/O, the stars are called pulsating PG 1159 stars, DOVs or GW Vir stars, and occur at 140 000 K $\lesssim T_{\rm eff} \lesssim$ 75 000 K. For He, 32 000 K $\lesssim T_{\rm eff} \lesssim$ 22 000 K and are called DBVs. For H, 13 000 K $\lesssim T_{\rm eff} \lesssim$ 10 500 K and are called DAVs or ZZ Ceti stars. Recent additions are the DQVs, with 22 000 K $\lesssim T_{\rm eff} \lesssim$ 18 000 K, and with 10 000 K $\lesssim T_{\rm eff} \lesssim$ 8 000 K, the ELMVs and the pre-ELMVs or EL CVn stars.

4 Magnetic fields

[57] presents a review on magnetic fields in white dwarf stars. When examining each candidate SDSS spectrum by eye, [38, 51, 52] found 822 stars with Zeeman splittings indicating magnetic fields above 2 MG — the limit where the line splitting becomes too small to be identified at the SDSS spectral resolution ([58]). The mean fields, estimated following [59], range from 2 MG to 700 MG. We caution that stars with large fields are difficult to identify because fields above around 30 MG intermix subcomponents between different hydrogen series components so much that, depending on effective temperature and signal-to-noise, it becomes difficult to identify the star as containing hydrogen at all, and affecting even the colors significantly. Both the low field limit and the high field limit are totally dominated by systematic effects, not the real limits. The effect of the magnetic field on pulsations has been estimated by, e.g., [80, 81].

Table 2. Rotation periods of white dwarfs as determined via asteroseismology.

Star	P_{rot} [h]	T_{eff} [K][1]	Type	M [$M_\odot$]
RX J2117.1+3412	28	170000	GW Vir	0.72
PG 1159-035	33	140000	GW Vir	0.54
NGC 1501	28	134000	[WCE]	0.56
PG 2131+066	5	95000	GW Vir	0.55
PG 1707+427	16	85000	GW Vir	0.53
PG 0122+200	37	80000	GW Vir	0.53
PG 0112+104	10.17	31040	DBV	0.58
KIC 8626021	43	29700	DBV	0.56
EC 20058-5234	2	25500	DBV	0.65
GD 358	29	23740	DBV	0.54
G 226-29	9	12510	ZZ Ceti	0.83
G 185-32	15	12470	ZZ Ceti	0.67
SDSS J113655.17+040952.6	2.6	12330	ZZ Ceti	0.55
SDSS J161218.08+083028.1	0.93	12330	ZZ Ceti	0.79
Ross 548	37	12300	ZZ Ceti	0.63
GD 165	50	12220	ZZ Ceti	0.68
LP 133-144	41.8	12150	ZZ Ceti	0.59
KIC 11911480	86.4	12160	ZZ Ceti	0.58
L 19-2	13	12070	ZZ Ceti	0.69
HS 0507+0435	41	12010	ZZ Ceti	0.73
EC 14012-1446	14.4	12020	ZZ Ceti	0.72
KUV 11370+4222	5.56	11940	ZZ Ceti	0.72
G 29-38	32	11910	ZZ Ceti	0.72
HL Tau 76	53	11470	ZZ Ceti	0.55
SDSS J171113.01+654158.3	16.4	11130	ZZ Ceti	0.90
GD 154	50.4	11120	ZZ Ceti	0.65
KIC 4552982	15.0	10860	ZZ Ceti	0.71
SDSS J094000.27+005207.1	11.8	10590	ZZ Ceti	0.82

[1] The effective temperatures and masses are corrected to 3D convection ([40]).

5 Rotation

In general the measured rotation period for single white dwarfs ranges from 1 h to 18 d, with a median around 1 d ([60]). The fastest single white dwarf rotator from asteroseismological measurements (Table 2) is the 0.79 $M_\odot$ DAV SDSS J161218.08+083028.1 discovered by [61], assuming the two observed periods at 115.0 s and 117.0 s are two components of a rotation triplet.

Differential rotation in white dwarfs was studied in [62]–[65], using the change in rotation splitting of non-radial pulsations.

6 Axions and dark mass

Axions are the best candidates for dark mass ([68]). [69–74] show white dwarf pulsations and luminosity function are consistent with extra cooling caused by axions of masses around 17 ± 4 meV.

7 Fitting models

The pulsation spectra exhibited by ZZ Ceti stars strongly depend on their inner chemical profile. There are several processes affecting the chemical profiles that are still not accurately determined. See [75] for a study of the impact of the current uncertainties in stellar evolution on the expected pulsation properties of ZZ Ceti stars.

Each single period is determined by the integral of the pulsation kernel (or work function) over the whole star. It cannot distinguish among different distributions, as demonstrated for example by [79]. If different modes are not independent, i.e., they sample the same regions of the distributions, they carry less information than independent modes. [76] propose one could determine the whole chemical distribution profile from pulsations. [77] performed a test using ten periods, namely the modes $l = 1$, $k = 2, 3, 4, 5, 6$ and $l = 2$, $k = 3, 4, 5, 6, 7$. Note that in these tests, a sequence of consecutive modes $k = 2, 3, 4, 5, 6$ is needed to sample the structure, a sequence that is usually not seen in real stars. ZZ Ceti stars show different pulsation spectra: hot stars show few short modes that sample the inner parts of the star while the cool stars show many long modes, which sample the outer parts of the stars. These characteristics must be taken into account when asteroseismology is applied to white dwarfs. A good example of a hot ZZ Ceti pulsator is G 117-B15A, with three modes. Were it not for the convection description problem, that introduces an uncertainty of $\Delta T_{\rm eff} \simeq 500$ K, and the problem with line broadening that gives different $\log g$ from different spectral lines, we could use the 3 modes of G 117-B15A to get 3 structural parameters, plus dP/dt to estimate the core mean molecular weight.

8 *Kepler* satellite

Observations of pulsating white dwarfs with the *Kepler* satellite are limited to the brightest objects due to the relatively small size of the telescope, but its long observations have allowed not only an exquisite precision in the pulsation spectra but also the discovery of outbursts lasting hours ([82–84]). These outbursts resemble the *forte* episode observed in 1996 for the DBV GD 358 by [85, 86].

Acknowledgments: S.O.K. and A.D.R. are financed by Conselho Nacional de Desenvolvimento Científico e Tecnológico, Brasil. This research has made use of NASA's Astrophysics Data System and of the cross-match service provided by CDS, Strasbourg. Funding for the Sloan Digital Sky Survey has been provided by the Alfred P. Sloan Foundation, the U.S. Department of Energy Office of Science, and the Participating Institutions. The SDSS web site is `www.sdss.org`. Part of the work has been based on observations obtained at the Southern Astrophysical Research (SOAR) telescope, which is a joint project of the Ministério da Ciência, Tecnologia, Inovações e Comunicações (MCTIC) da República Federativa do Brasil, the U.S. National Optical Astronomy Observatory (NOAO), the University of North Carolina at Chapel Hill (UNC), and Michigan State University (MSU).

References

[1] Woosley, S. E., & Heger, A., ApJ, **810**, 34 (2015)
[2] Kilic, M., Stanek, K. Z., & Pinsonneault, M. H., ApJ, **671**, 761 (2007)
[3] Romero, A. D., Campos, F., & Kepler, S. O., MNRAS, **450**, 3708 (2015)
[4] García-Berro, E., & Oswalt, T. D., New Ast. Rev., **72**, 1 (2016)
[5] Winget, D. E., Hansen, C. J., Liebert, J., et al., ApJL, **315**, L77 (1987)
[6] Campos, F., Bergeron, P., Romero, A. D., et al., MNRAS, **456**, 3729 (2016)
[7] Schatzman, E., Nature, **161**, 61 (1948)
[8] Marsh, T. R., Dhillon, V. S., & Duck, S. R., MNRAS, **275**, 828 (1995)
[9] Kilic, M., Brown, W. R., Allende Prieto, C., et al., ApJ, **727**, 3 (2011)

[10] Brown W. R., Kilic M., Allende Prieto C., Gianninas A., Kenyon S. J., ApJ, **769**, 66 (2013)
[11] Gianninas, A., Hermes, J. J., Brown, W. R., et al., ApJ, **781**, 104 (2014)
[12] Paxton, B., Marchant, P., Schwab, J., et al., ApJS, **220**, 15 (2015)
[13] Istrate, A. G., Marchant, P., Tauris, T. M., et al., A&A, **595**, A35 (2016)
[14] Pelisoli, I., Kepler, S. O., Koester, D., & Romero, A. D., ASPC 20th European White Dwarf Workshop, **509**, 447 (2017)
[15] Althaus, L. G., Miller Bertolami, M. M., & Córsico, A. H., A&A, **557**, A19 (2013)
[16] Latour, M., Heber, U., Irrgang, A., et al., A&A, 585, A115 (2016)
[17] Pietrzyński, G., Thompson, I. B., Gieren, W., et al., Nature, **484**, 75 (2012)
[18] Karczmarek, P., Wiktorowicz, G., Iłkiewicz, K., et al., MNRAS, **466**, 2842 (2017)
[19] Brown, W. R., Kilic, M., Allende Prieto, C., & Kenyon, S. J., ApJ, **723**, 1072 (2010)
[20] Brown W. R., Kilic M., Allende Prieto C., Kenyon S. J., ApJ, **744**, 142 (2012)
[21] Gianninas, A., Kilic, M., Brown, W. R., Canton, P., & Kenyon, S. J., ApJ, **812**, 167 (2015)
[22] Brown, W. R., Gianninas, A., Kilic, M., Kenyon, S. J., & Allende Prieto, C., ApJ, **818**, 155 (2016)
[23] Hermes, J. J., Montgomery, M. H., Winget, D. E., et al., ApJL, **750**, L28 (2012)
[24] Hermes, J. J., Montgomery, M. H., Winget, D. E., et al., ApJL, **765**, 102 (2013)
[25] Hermes, J. J., Montgomery, M. H., Gianninas, A., et al., MNRAS, **436**, 3573 (2013)
[26] Bell, K. J., Kepler, S. O., Montgomery, M. H., et al., 19th European Workshop on White Dwarfs, **493**, 217 (2015)
[27] Bell, K. J., Gianninas, A., Hermes, J. J., et al., ApJ, **835**, 180 (2017)
[28] Hermes, J. J., Charpinet, S., Barclay, T., et al., ApJ, **789**, 85 (2014)
[29] Van Grootel, V., Fontaine, G., Brassard, P., & Dupret, M.-A., ApJ, **762**, 57 (2013)
[30] Maxted, P. F. L., Bloemen, S., Heber, U., et al., MNRAS, **437**, 1681 (2014)
[31] Maxted, P. F. L., Serenelli, A. M., Marsh, T. R., et al., MNRAS, **444**, 208 (2014)
[32] Gianninas, A., Curd, B., Fontaine, G., Brown, W. R., & Kilic, M., ApJL, **822**, L27 (2016)
[33] Córsico, A. H., & Althaus, L. G., A&A, **569**, A106 (2014)
[34] Córsico, A. H., & Althaus, L. G., ApJL, **793**, L17 (2014)
[35] Istrate, A. G., Tauris, T. M., & Langer, N., A&A, **571**, A45 (2014)
[36] Istrate, A. G., Tauris, T. M., Langer, N., & Antoniadis, J., A&A, **571**, L3 (2014)
[37] Istrate, A. G., Fontaine, G., Gianninas, A., et al., A&A, **595**, L12 (2016)
[38] Kepler, S. O., Pelisoli, I., Koester, D., et al., MNRAS, **455**, 3413 (2016)
[39] Kepler, S. O., Koester, D., Romero, A. D., Ourique, G., & Pelisoli, I., ASPC 20th European White Dwarf Workshop, **509**, 421 (2017)
[40] Tremblay, P.-E., Ludwig, H.-G., Steffen, M., & Freytag, B., A&A, **552**, A13 (2013)
[41] Xue, X. X., Rix, H. W., Zhao, G., et al., ApJ, **684**, 1143 (2008)
[42] Xue, X.-X., Rix, H.-W., Yanny, B., et al., ApJ, **738**, 79 (2011)
[43] Carollo, D., Beers, T. C., Placco, V. M., et al., Nature Phys., **12**, 1170 (2016)
[44] Heber, U., PASP, **128**, 082001 (2016)
[45] Gentile Fusillo N. P., Gänsicke B. T., Greiss S., MNRAS, **448**, 2260 (2015)
[46] Munn, J. A., Harris, H. C., von Hippel, T., et al., AJ, **153**, 10 (2017)
[47] Lada, C. J., ApJL, **640**, L63 (2006)
[48] Rebassa-Mansergas, A., Ren, J. J., Parsons, S. G., et al., MNRAS, **458**, 3808 (2016)
[49] Farihi, J., Hoard, D. W., & Wachter, S., ApJS, **190**, 275 (2010)

DOI: 10.1051/epjconf/201715201011

[50] Nebot Gómez-Morán, A., Gänsicke, B. T., Schreiber, M. R., et al., A&A, **536**, A43 (2011)
[51] Kleinman, S. J., Kepler, S. O., Koester, D., et al., ApJS, **204**, 5 (2013)
[52] Kepler, S. O., Pelisoli, I., Koester, D., et al., MNRAS, **446**, 4078 (2015)
[53] Althaus, L. G., García-Berro, E., Isern, J., & Córsico, A. H., A&A, **441**, 689 (2005)
[54] Renedo, I., Althaus, L. G., Miller Bertolami, M. M., et al., ApJ, **717**, 183 (2010)
[55] Tremblay P.-E., Ludwig H.-G., Steffen M., Freytag B., A&A, **559**, A104 (2013)
[56] Schmidt, M., ApJ, **151**, 393 (1968)
[57] García-Berro, E., Kilic, M., & Kepler, S. O., Intl. Journal Modern Phys. D, **25**, 1630005 (2016)
[58] Kepler S. O., et al., MNRAS, **429**, 2934 (2013)
[59] Külebi B., Jordan S., Euchner F., Gänsicke B. T., Hirsch H., A&A, **506**, 1341 (2009)
[60] Kawaler, S. D., 19th European Workshop on White Dwarfs, ASPC, **493**, 65 (2015)
[61] Castanheira, B. G., Kepler, S. O., Kleinman, S. J., Nitta, A., & Fraga, L., MNRAS, **430**, 50 (2013)
[62] Charpinet, S., Fontaine, G., & Brassard, P., Nature, **461**, 501 (2009)
[63] Córsico, A. H., Althaus, L. G., Kawaler, S. D., et al., MNRAS, **418**, 2519 (2011)
[64] Fontaine, G., Brassard, P., & Charpinet, S., 18th European White Dwarf Workshop, **469**, 115 (2013)
[65] Hermes, J. J., Kawaler, S. D., Bischoff-Kim, A., et al., ApJL, **841**, L2 (2017)
[66] Cantiello, M., Mankovich, C., Bildsten, L., Christensen-Dalsgaard, J., & Paxton, B., ApJ, **788**, 93 (2014)
[67] Fuller, J., Lecoanet, D., Cantiello, M., & Brown, B., ApJ, **796**, 17 (2014)
[68] Ringwald, A., preprint (`arXiv:1612.08933`) (2016)
[69] Isern, J., & García-Berro, E., Nuclear Phys. B Proc. Suppl., **114**, 107 (2003)
[70] Isern, J., García-Berro, E., Althaus, L. G., & Córsico, A. H., A&A, **512**, A86 (2010)
[71] Córsico, A. H., Althaus, L. G., Miller Bertolami, M. M., et al., MNRAS, **424**, 2792 (2012)
[72] Córsico, A. H., Althaus, L. G., Romero, A. D., et al., Journal of Cosm. Astrop. Phys., **12**, 010 (2012)
[73] Córsico, A. H., Romero, A. D., Althaus, L. G., et al., Journal of Cosm. Astrop. Phys., **7**, 036 (2016)
[74] Battich, T., Córsico, A. H., Althaus, L. G., & Miller Bertolami, M. M., Journal of Cosm. Astrop. Phys., **8**, 062 (2016)
[75] De Gerónimo, F. C., Althaus, L. G., Córsico, A. H., Romero, A. D., & Kepler, S. O., A&A, **599**, A21 (2017)
[76] Giammichele, N., Fontaine, G., Brassard, P., & Charpinet, S., ApJS, **223**, 10 (2016)
[77] Giammichele, N., Charpinet, S., Fontaine, G., & Brassard, P., ApJ, **834**, 136 (2017)
[78] Giammichele, N., Charpinet, S., Brassard, P., & Fontaine, G., A&A, **598**, A109 (2017)
[79] Montgomery, M. H., Metcalfe, T. S., & Winget, D. E., MNRAS, **344**, 657 (2003)
[80] Jones, P.W., Pesnell, W.D., Hansen, C.J., and Kawaler, S.D., ApJ, **336**, 403 (1989)
[81] Tremblay, P.-E., Fontaine, G., Freytag, B., et al., ApJ, **812**, 19 (2015)
[82] Bell, K. J., Hermes, J. J., Bischoff-Kim, A., et al., ApJ, **809**, 14 (2015)
[83] Hermes, J. J., Montgomery, M. H., Bell, K. J., et al., ApJ, **810**, L5 (2015)
[84] Bell, K. J., Hermes, J. J., Montgomery, M. H., et al., ApJ, **829**, 82 (2016)
[85] Nitta, A., Kepler, S. O., Winget, D. E., et al., Baltic Astr., **7**, 203 (1998)
[86] Kepler, S. O., Nather, R. E., Winget, D. E., et al., A&A, **401**, 639 (2003)

A survey of pulsating DA and DB white dwarfs

Observations with the Whole Earth Telescope

J. L. Provencal[1,2,⋆], *M.* Montgomery[3,⋆⋆], and *H.* Shipman[2]

[1] *University of Delaware, Dept. of Physics and Astronomy, Newark DE 19716, USA*
[2] *Mt. Cuba Astronomical Observatory, Wilmington DE, USA*
[3] *University of Texas, Dept. of Astronomy, Austin TX, USA*

Abstract.
White dwarfs represent the end point of stellar evolution for the majority of stars. As such, they are excellent astrophysical laboratories. They are structurally simple, with electron degenerate cores surrounded by thin surface layers of helium and/or hydrogen. The g-mode pulsations provide a window into their internal structure. The Whole Earth Telescope has been conducting a long-term survey of pulsating white dwarfs with the goal of providing an empirical map of convection parameters across the DA and DB instability strips. We present an overview of white dwarf asteroseismology, and discuss the current status of our survey.

1 White dwarf asteroseismology and diagnostics

Asteroseismology provides us with a window through which to peer into stellar interiors. Through asteroseismology, we can study fundamental problems in stellar evolution such as energy transport, thermodynamics, and magnetism. White dwarfs (WDs) are particularly important targets for asteroseismology. WDs are the most numerous inhabitants of the stellar graveyard. WDs are structurally simple: an electron degenerate carbon/oxygen core surrounded by thin non-degenerate layers of hydrogen and helium. In broad brush strokes, WDs are divided into two classes by their atmospheric compositions. DA white dwarfs have nearly pure hydrogen layers on top of a layer of helium. DAs represent ≈80% of the population ([2]). DB white dwarfs are characterized by a layer of nearly pure helium overlying their carbon/oxygen cores. Lacking substantial nuclear reactions, white dwarfs simply cool as they age, passing through specific temperature ranges (the DBV and DAV instability strips) within which they pulsate. These pulsators are otherwise normal objects, so what we learn about their structure can be applied to the WD entire population to further our understanding of stellar evolution.

WDs are g-mode pulsators, and we assume that each pulsation mode can be described by a spherical harmonic of degree ℓ, radial overtone k, and azimuthal number m, where m takes integer values between $-\ell$ and ℓ. A key diagnostic for g-mode pulsators is the mean period spacing ΔP between modes of the same (ℓ, m) but consecutive radial overtone k (Fig. 1). ΔP depends primarily on stellar mass. Deviations of individual spacings from this mean value provide information on both the

⋆`jlp@udel.edu`
⋆⋆`mikemon@astro.as.utexas.edu`

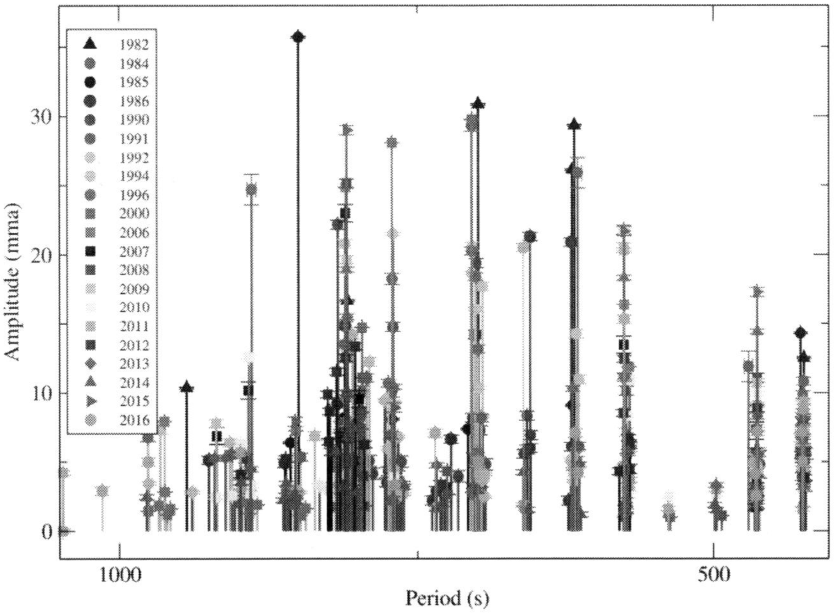

Figure 1. A schematic representation of the distribution of pulsation frequencies for the prototype DBV GD358. The systematic distribution of frequencies into a series of modes is evident.

structure of the surface layers and the degenerate core ([4]). A second diagnostic is the presence of multiplets. The multiplet components have the same (ℓ, k) but different values of the azimuthal index m. In the limit of slow rotation, the frequency difference between multiplet components is an indicator of rotation as sampled by a given mode. Multiplet structure is also a strong indication of a mode's ℓ value. We expect a triplet for $\ell = 1$, a quintuplet for $\ell = 2$, and so on. Deviations from equal frequency splitting within a single multiplet and changes in splittings from one multiplet to the next reveal information about differential rotation and magnetic field strength. Additional details can be found in [6].

2 Convection in white dwarfs

Convection is one of the largest sources of uncertainties in stellar models. The Whole Earth Telescope (WET, [5]) is engaged in a long term survey to provide an empirical map of WD convective parameters across the DA and DB instability strip. The basic premise is that the photospheric flux is delayed and/or attenuated relative to flux at the bottom of the convection zone (CZ) by an amount that depends on the thickness of the CZ. As an additional complication, the depth of the CZ varies during a pulsation cycle. As a pulsation propagates through the star, it interacts with the surface CZ. The CZ acts as a nonlinear mixer, and the result is an observed distortion of the shape of the light curve. Convective light curve fitting utilizes the nonsinusoidal shapes of white dwarf light curves to recover the thermal response timescale at the base of the CZ. Full details can be found in [3] and [6].

Figure 2 shows the current status of the survey for DA WDs. We find two preliminary conclusions. First, the measured parameter τ_0 is the thermal response timescale at the base of the CZ, and is related to the CZ's depth and mass. Theory predicts that a WD convection zone will deepen and increase in mass as the star cools. We should find an increase in τ_0 as we move to lower temperatures, and this is

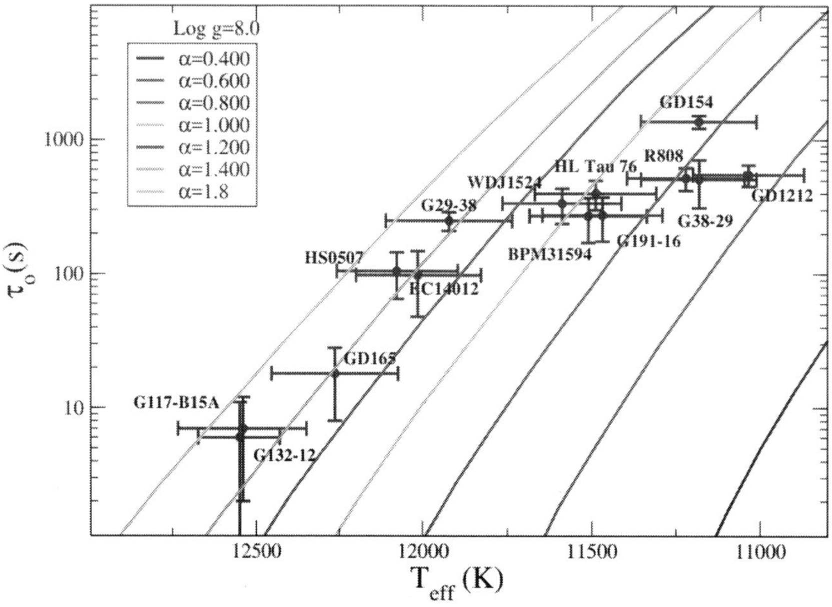

Figure 2. Current empirical map of the thermal response at the base of the convection zone τ_0 as a function of $T_{\rm eff}$ for DA white dwarfs. Predicted values of τ_0 for different values of the MLT parameter α are given.

indeed what we see for WDs with $T_{\rm eff} > 12{,}000$ K. For WDs with $T_{\rm eff} < 12{,}000$ K, τ_0 increases from $\approx$150 s to only $\approx$550 s at $T_{\rm eff} = 11{,}000$ K. τ_0 values for stars below $T_{\rm eff} < 12{,}000$ K do not follow the predictions of any single mixing-length theory (MLT) α parameterization. The behavior of τ_0 is also in rough agreement with the 3D convective simulations of [7]. If one must use MLT in modelling, the proper value of α will depend on the WD temperature. We mention the caveat that τ_0 is the thermal timescale at the base of the CZ. There is no reason to require that the value of α matching the observed τ_0 in this region is also appropriate for representing α at the photosphere.

We find further evidence that the CZ in DAs does not behave as expected when we consider the predicted maximum pulsation periods and the observed longest period pulsations for stars in our sample (Fig. 3). [1] demonstrates that g-mode pulsations should occur when convective driving exceeds radiative damping. This condition is met when the pulsation period $P < P_{\rm max}$, where $P_{\rm max} = 2\pi\tau_0$. In Figure 3, we compare the observed longest period pulsations (points) with the predicted $P_{\rm max}$ (solid line). We find fair agreement for our hottest stars ($T_{\rm eff} > 12{,}000$ K, with low values of τ_0), but diverge rapidly for cooler stars. Some mechanism currently not understood is functioning to both limit τ_0 and prevent observed pulsation periods from increasing as rapidly as predicted in cooler pulsators.

Finally, Figure 4 shows the current status of the survey for DB WDs. Interpreting the behavior of τ_0 for the helium instability strip is hampered by the large uncertainties associated with the optical spectrographic temperature determinations for PG1456, PG1351, KUV05134 and CBS114. The remaining objects have UV temperature determinations with much smaller uncertainties. However, on first look, it does not seem as if τ_0 for helium convection behaves as we see for hydrogen (Fig. 2). We find a general increase in τ_0 with increasing $T_{\rm eff}$.

Acknowledgments: JLP acknowledges the support of the Mt. Cuba Astronomical Observatory.

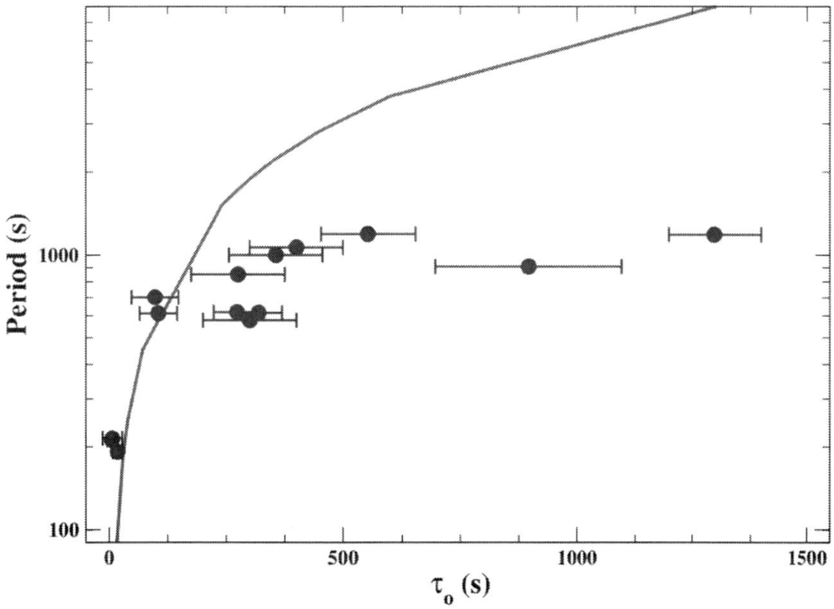

Figure 3. Comparison of the dominant large amplitude observed pulsation period for the stars in Figure 2 (points) with the theoretical value of $P_{\rm max}$ (solid line).

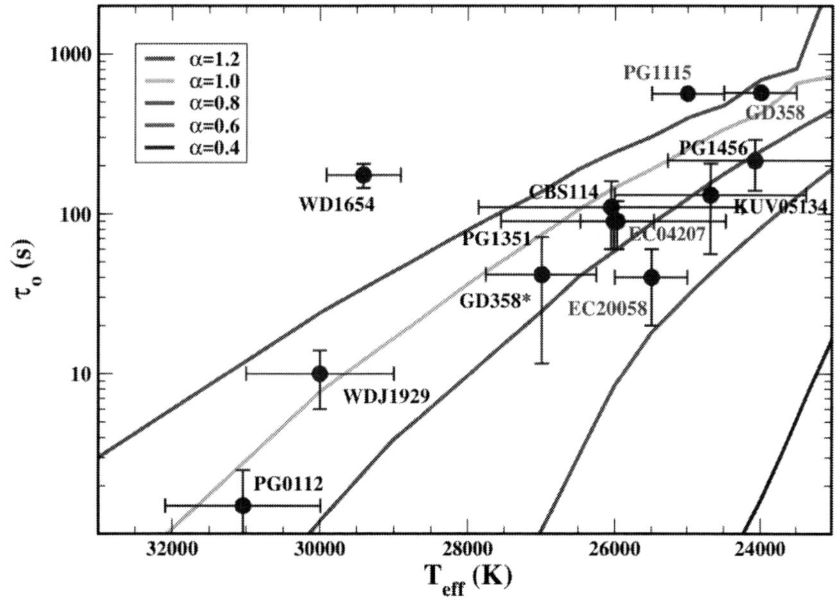

Figure 4. Current empirical map of the thermal response at the base of the convection zone τ_0 as a function of $T_{\rm eff}$ for DB white dwarfs. Predicted values of τ_0 for different values of the MLT parameter α are given.

References

[1] Brickhill, A. J., MNRAS, **252**, 334 (1991)
[2] Eisenstein, D. J., et al., ApJS, **167**, 40 (2006)
[3] Montgomery, M., ApJ, **633**, 91 (2005)
[4] Montgomery, M., et al., ApJ, **716**, 84 (2010)
[5] Nather, R. E., et al., ApJ, **361**, 309, (1990)
[6] Provencal, J. L., et al., ApJ, **751**, 91 (2012)
[7] Tremblay, P.-E., Gianninas, A., Kilic, M., et al., 2015, ApJ, **809**, 148 (2015)

Look behind you! The past can contribute uniquely to the future

Elizabeth Griffin[1,*]

[1] *Dominion Astrophysical Observatory, 5071 West Saanich Road, Victoria, BC, V9E 2E7, Canada*

Abstract. The surveys which are the chief foci of this meeting are almost exclusively concerned with recent and upcoming electronic surveys. But the practice of surveying the night sky is not new; some of the first coordinated observatory activities involved surveys, of various hues and sizes, and for many decades the IAU paid tribute to that activity by supporting a special Commission for it. The wealth of past surveys is described briefly. Though now regarded as limited in quality, early surveys carry date-stamps which enable their contents to add usefully, sometimes critically, to modern science.

1 The Future

Science progresses forwards and upwards. New information creates new pathways and ladders. New knowledge adds another layer on the summit of the pyramid already established by generations of effort and achievement. Like the arrow of time, evolution in our science is inexorably onwards to the Future, and to new things. The Future is the target; the Past was but a facilitator, its parameters superseded, its restrictions overcome, its horizons expanded. Virtually all the contributions reported in these *Proceedings* demonstrate fascinating individual or collective ability to put ingenuity into practice. Doing so at multiple wavelengths simultaneously can reveal new types of object that had escaped notice through the exercise of sometimes inflexible conventional methods. Moreover, what is freshly learned is not forgotten; the *data* (the observations) can be mined, the *information* in whole data-sets can be federated, and new evidence of variability, better-quantified characteristics, and more complete samples, can be analysed, communicated and published. Generations of researchers yet to graduate will work still further on the *knowledge* which the present ones leave behind them today – a never ending search for the *wisdom* towards which all the effort is striving. Yet despite so many cogent reasons to push onwards and upwards, there will always be some situations in which the Past can also contribute uniquely to modern-day studies, contributions that can be critical when resolving an ambiguity, or (as illustrated in Fig. 1) extending a base-line significantly.

2 The Past

Astronomical surveys are certainly not new in concept. As with any pursuit in which a vast uncharted terrain presents itself, an obvious way to begin is to map it – to get a working appraisal of its population, range and scope, and to record that for re-use and for posterity. So with astronomy. The first surveys recorded on physical media (i.e., could be shared) commenced as soon as the resources –

*elizabeth.griffin@nrc-cnrc.gc.ca

manpower, equipment and detectors – were assembled. Scarcely had the process of photography been commercialized than the *Astrographic Catalogue* and its deeper-reaching cousin, the *Carte du Ciel* – astronomy's first truly international ventures – were conceived in 1887 to record the whole sky, an international project of such grand proportions that it absorbed and nearly exhausted the astronomical resources of some countries, particularly France, for a substantial time. Although the project was never fully completed (the demands eventually exceeded the committed resources), the original observations still exist, and when fully digitized and aligned with modern observations they will enable the recognition of changes in position or magnitude which can be hard to detect when the temporal base-line is very short.

It was in fact the positive experiences and the communications networks set up through the *Carte du Ciel* that led to the creation of the IAU in 1919. The importance of the roles that surveys play in astronomical research was fully acknowledged when one of the first of the IAU Commissions to be set up was that for the *Carte du Ciel*; it retained its identity as a separate Commission until 2006. Another early Commission (*Instrumentation*) later diversified and sub-divided as technology burgeoned, and in 1991 it updated its *Working Group for Astronomical Photography* to the *Wide-Field Imaging WG*, expressly encompassing Sky Surveys and Patrols and begetting the *Wide-Field Plates Database* [1], astronomy's virtual catalogue of photographic wide-field images.

Other surveys with parallel purposes, though smaller in terms of observatory involvement, also started to flourish in the late 19th and early 20th centuries. Probably the best known is the collection of Harvard College Observatory, featuring over half a million images and objective-prism spectra dating back ~130 years. The on-going programme, *DASCH* [2], to digitize the entire collection is making encouraging headway, and uncovering vast numbers of variable stars. Next in size, the Sonneberg *Sky Patrol*, initiated in 1926 with relatively modest equipment, delivered some 240,000 plates, led to the detection of over 10,000 variable stars, and laid foundations for 20th-century stellar physics. Smaller contemporary surveys, many made in Eastern Europe, provided essential back-up data or filled in gaps in the larger surveys.

Those early observations, while surely not as sophisticated or detailed as the output from modern surveys, or as accessible (unless or until transformed into electronic output), nevertheless offer a base-line extending to a century or beyond, whereas recent surveys cannot by definition encompass more than a handful of years. Observational selection is often a crippling restriction to results from limited surveys, and fascinating new science can be uncovered when its effects are sufficiently reduced. A fine example described by [1] (see Fig. 1) demonstrates how the results of a radial-velocity programme that commenced in 1975 show the actual distribution of binary-star orbital amplitudes and periods to be substantially different from what could previously be deduced from even the most comprehensive catalogue ([2]).

3 The Present

There is nothing magical about "the present". It represents neither a particularly advanced nor an especially sophisticated state, since those are relative properties that are closely tied to, and consequently constrained by, the evolution of relevant technology. The present is merely where we now happen to be along the continuum that connects what was with what will be; the Future is some unknown and undefinable state towards which astronomers will always strive but will never know in absolute terms when it is reached. What they can do is to ensure maximal use and application of all the materials to hand at any point, whether recently deposited or in a long-term archive. All of those materials are complementary, unique, and irreproducible.

[1] `http://www.wfpdb.org`
[2] `http://dasch.rc.fas.harvard.edu/project.php`

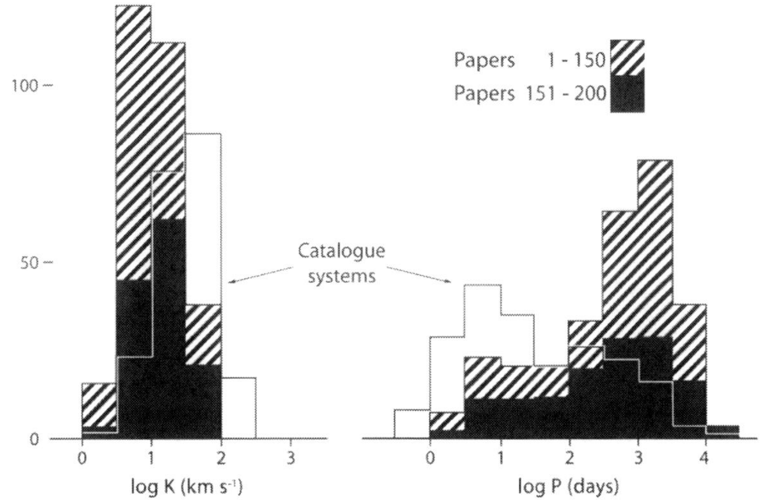

Figure 1. What was believed to be sure knowledge regarding the distribution of amplitudes (K, left) and periods (P, right) for tolerably-well determined binary-star orbits listed in the *Seventh Catalogue of the Orbital Elements of Spectroscopic Binary Systems* [2] (no in-filling) has been substantially altered simply by measuring enough radial velocities over a considerably longer time-span than was customary. Griffin's Papers 1–200 [1] containing the analyses were published at intervals of 2 months. The distributions of K and P are markedly different when the base-line is increased to (a) 25 years (cross-hatching) and (b) 33 years (black). The changes will have had a non-negligible influence on theories of binary-star formation. (*Reproduced by kind permission of Observatory Magazine.*)

However, where "the present" *does* offer a once-only chance it is the opportunity to digitize those archived heritage materials *before it is too late* – before the knowledge about their acquisition and specific properties becomes lost or the data themselves – the photographic plates – destroyed, or discarded in ignorance of their scientific potential. Custom-built digitizing equipment is now available, but will not out-last the plates, and the training and expertise will evaporate even sooner unless passed on to younger members of the community. With those losses will go all the new science whose discovery was the ultimate objective of storing the data for so long.

There is also something important which all astronomers can do to assist, and that is to give support whenever possible to requests for resources to carry out digitizing programmes. Heritage materials are dogged by the stigma that old ≡ old-fashioned, though when they produce evidence of (say) the precursor of a supernova, opinions can change. It should also be pointed out that these heritage materials only need to be digitized once.

Acknowledgments: I am grateful to the Canadian Astronomy Data Centre for a travel grant, and to the conference organizers for additional financial assistance.

References

[1] Griffin, R. F., Observatory, **128**, 448 (2008)

[2] Batten, A. H., Fletcher, J. M., & Mann, P. J., *Seventh Catalogue of the Orbital Elements of Spectroscopic Binary Systems*, PDAO, **15**, 121, (1978)

On the building blocks of the M31 and Milky Way halos

Matteo Monelli[1,2,★]

[1] *Instituto de Astrofísica de Canarias (IAC), E-38205 La Laguna, Tenerife, Spain*
[2] *Universidad de La Laguna (ULL), Dpto. Astrofísica, E-38206 La Laguna, Tenerife, Spain*

Abstract. We discuss the formation of the halo of M31 and the Milky Way as traced by the population of RR Lyrae stars, in comparison with the population of such stars preent in satellite dwarf galaxies. We find that both halos and the massive dwarf host a population of high amplitude short period RRab stars, absent in low-mass dwarfs. These stars are explained as the metal-rich tail of the RR Lyrae distribution ([Fe/H] ~ -1.5), and thus their existence imply fast chemical enrichment in the host system. Their presence in both halos implies that massive building blocks had an important rôle in their formation.

1 Introduction

RR Lyrae stars (RRL) are excellent tracers of the old population, and bring direct insights on the environmental conditions of the host stellar systems at early epochs. For this reason, they are important tracers of the mass accretion history of the halo of spiral galaxies such as M31 and the Milky Way (MW). [6] presented a comparison of the properties of RRL stars in the MW halo and in its system of globular clusters and satellite dwarfs. In particular, they detected the existence of a family or RRab type stars characterized by high amplitudes and short periods ($P < 0.48$ d, $A_V > 0.75$ mag), hence the name, HASP). These stars are associated to a population of stars as metal-rich as [Fe/H] ~ -1.5. [6] discusses the existence of such stars in massive MW satellites such as the LMC and the Sagittarius dSph, in metal-rich globular clusters but not in lower mass galaxies. This indicates that *i)* low-mass dwarfs did not experience a chemical evolution fast enough to produce HASP, and *ii)* to explain the existence of HASP in the MW halo one has to accept that massive building blocks, already enriched at the time of accretion, had to participate in the halo construction.

2 On the M31 halo building blocks

We have performed a similar analysis in the case of M31. Figure 1 compares the period distributions of RRL stars in the M31 and MW halos, and with their satellite systems, as detailed in the caption. The plot shows that the normalized period distribution of both halos is very similar to the corresponding sample of bright satellites. The histograms relative to the RRL stars from faint galaxies are characterized by a rising branch and the main peak occurring at longer period. This reflects the absence of HASP stars in these systems. The plot clearly shows that to repodure the period distribution of both

★`monelli@iac.es`

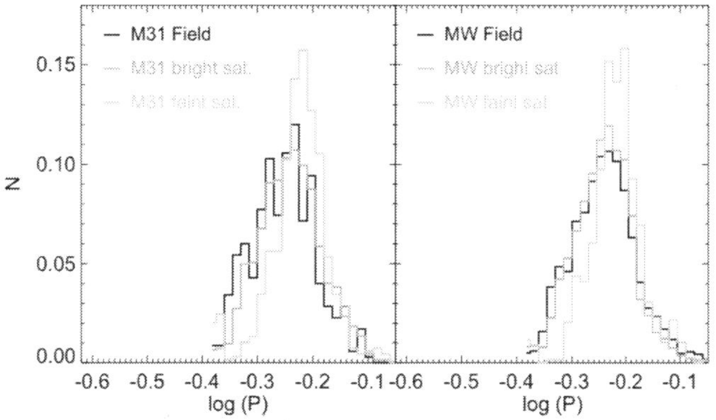

Figure 1. Period distribution of RRab star in different samples of galaxies. The adopted data are *M31 halo:* [7] and Field 2 from [11]. *MW halo:* same collection as in [6]. *M31 bright satellites:* dwarfs with $M_V < -12$, namely: And VII, NGC 147, and NGC 185 from Monelli et al. (2017, submitted); M32: [5]; And II, from the compilation of Martínez-Vázquez et al. (2017, submitted). *MW bright satellites:* OGLE-III data for the LMC, the SMC and Sagittarius dwarf spheroidal ([13–15]); *M31 faint satellites:* And I, And III, And XV, And XVI and And XXVIII from Martínez-Vázquez et al. (2017, op. cit.); And VI ([10]); And XI and And XIII ([18]); And XIX ([2]), And XXI ([3]) and And XXV ([4]) observed with the Large Binocular Telescope; *faint MW satellites:* Carina ([1]); Leo II ([12]); Draco ([8]); Sculptor ([9]); Leo I ([17]).

the MW and the M31 halo the contribution of massive dwarf galaxies is required. A Kolomogorov-Smirnov test performed on the period distributions supports this conclusion, as the probability that the the M31 (MW) halo distribution is drawn from the same one as the bright satellite's is 46% (99%).

References

[1] Coppola, G., Marconi, M., Stetson, P. B., et al., ApJ, **814**, 71 (2015)
[2] Cusano, F., Clementini, G., Garofalo, A., et al., ApJ, **779**, 7 (2013)
[3] Cusano, F., Garofalo, A., Clementini, G., et al., ApJ, **806**, 200 (2015)
[4] Cusano, F., Garofalo, A., Clementini, G., et al., ApJ, **829**, 26 (2016)
[5] Fiorentino, G., Contreras Ramos, R., Tolstoy, E., et al., A&A, **539**, A138 (2012)
[6] Fiorentino, G., Bono, G., Monelli, M., et al., ApJL, **798**, L12 (2015)
[7] Jeffery, E. J., Smith, E., Brown, T. M., et al., AJ, **141**, 171 (2011)
[8] Kinemuchi, K., Harris, H. C., Smith, H. A., et al., AJ, **136**, 1921 (2008)
[9] Martínez-Vázquez, C. E., Monelli, M., Bono, G., et al., MNRAS, **454**, 1509 (2015)
[10] Pritzl, B. J., Armandroff, T. E., Jacoby, G. H., & Da Costa, G. S., AJ, **124**, 1464 (2002)
[11] Sarajedini, A., Mancone, C. L., Lauer, T. R., et al., AJ, **138**, 184 (2009)
[12] Siegel, M. H., & Majewski, S. R., AJ, **120**, 284 (2000)
[13] Soszyński, I., Udalski, A., Szymański, M. K., et al., AcA, **60**, 165 (2010)
[14] Soszyński, I., Udalski, A., Szymański, M. K., et al., AcA, **59**, 1 (2009)
[15] Soszyński, I., Dziembowski, W. A., Udalski, A., et al., AcA, **61**, 1 (2011)
[16] Soszyński, I., Udalski, A., Szymański, M. K., et al., AcA, **65**, 233 (2015)
[17] Stetson, P. B., Fiorentino, G., Bono, G., et al., PASP, **126**, 616 (2014)
[18] Yang, S.-C., Wagner-Kaiser, R., Sarajedini, A., Kim, S. C., & Kyeong, J., ApJ, **784**, 76 (2014)

An LBT view of the Andromeda's satellite galaxies

Felice Cusano[1,*], *Alessia* Garofalo[1,2], *Gisella* Clementini[1], and *Michele* Cignoni[3]

[1] *INAF - Osservatorio Astronomico di Bologna, Via Ranzani 1, I - 40127 Bologna, Italy*
[2] *Dipartimento di Fisica e Astronomia, Universitá di Bologna, viale Berti Pichat, 6/2, I - 40127 Bologna, Italy*
[3] *Dipartimento di Fisica, Universitá di Pisa, Largo Bruno Pontecorvo, 3, 56127 Pisa, Italy*

Abstract. Results are presented on deep ($V \sim 26.5$ mag) time series observations of four dwarf spheroidal galaxies (dSphs) in the Andromeda (M31) complex, namely, And XIX, And XXI, And XXV and And XXVII, that we have observed with the Large Binocular Telescope (LBT). We discovered in these galaxies a total of over 200 RR Lyrae stars and 19 Anomalous Cepheids. We also characterised the stellar populations and the spatial distributions of these dSphs.

1 Introduction

Λ-CDM theory predicts the formation of massive galaxies like the Milky Way and M31 to occur by accretion and merging of smaller structures. However, some issues challenge this scenario like for instance, the missing satellites problem ([8]) or the observed alignment of satellite galaxies ([7]). The characterization of the resolved stellar populations and variable stars of the M31 satellites may allow to trace merging and accretion episodes occurred and still occurring in the M31 environment, thus helping to constrain the origin and the fate of Andromeda. We obtained B and V time-series photometry with the Large Binocular Cameras (LBC) of the LBT of four M31 dwarf satellite galaxies, namely, And XIX ([3]), And XXI ([4]), And XXV ([5]) and And XXVII ([6]).

2 Observations and data reduction

Observations of the four galaxies were obtained between 2010 and 2011. The LBCs were used in binocular mode and B images were acquired with the Blue camera of the LBC, whereas the V images with the Red LBC. PSF photometry was performed using the DAOPHOT-ALLSTAR-ALLFRAME package ([10]). Variable stars were identified using the variability index computed in DAOMASTER. The light curves of the candidate variables were then analyzed with the Graphical Analyzer of Time Series (GRATIS, [2]). Examples of light curves for RR Lyrae stars and Anomalous Cepheids (ACs) that we discovered in these galaxies are shown in Figure 1.

* felice.cusano@oabo.inaf.it

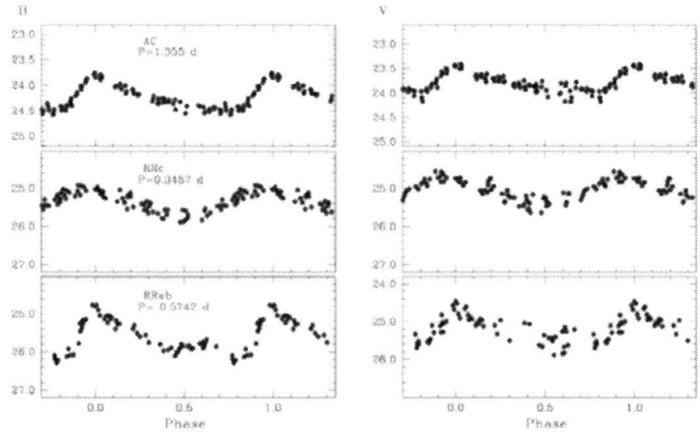

Figure 1. Examples of variable star light curves in And XXV. From top to bottom an AC, a first-overtone RR Lyrae star (RRc) and a fundamental mode RR Lyrae star (RRab).

3 Results

We discovered a total of 218 RR Lyrae stars and 19 ACs in the four M31 satellites observed within our program. The properties of the variable stars are summarized in [6]. The M31 satellite galaxies that we imaged can be classified as Oosterhoff Intermediate (Oo-Int) systems ([1, 9]). There is a slight tendency towards Oo-I type with decreasing the distance to the M31. Besides identifying variable stars we also built color-magnitude diagrams (CMDs) to characterize the stellar populations of these M31 satellite galaxies. We found evidence for three different stellar populations in And XIX and And XXI: (1) an old ($\sim$ 12 Gyr) and metal-poor ([Fe/H] $= -1.7$ dex) component traced by the RR Lyrae stars; (2) a slightly younger (10-6 Gyr) and more metal-rich ([Fe/H] $= -1.5$ dex) component populating the red horizontal branch, and (3) an intermediate age ($\sim$ 1 Gyr) component with the same metallicity that produced the ACs. Instead, for And XXV and And XXVII we found indication of a single old star formation burst ($\sim$ 12 Gyr, [Fe/H] $= -1.9$ dex). The large field of view of the LBC allowed to study the spatial distribution of the different stellar populations in each galaxy. Isodensity contours coupled with the location of the RR Lyrae stars helped us to trace signatures of past and/or ongoing gravitational interactions with M31 or other dwarf galaxies. In particular: (a) we found evidence for And XXI to be the result of a past minor merger between two dwarf galaxies; (b) we discovered a spherically-shaped high density of stars near the And XXV center ([5]). This structure appears to be at a distance consistent with And XXV and we suggest it could either be a star cluster or the nucleus of And XXV; (c) we confirmed that And XXVII is a completely disrupted galaxy.

References

[1] Catelan, M., Ap&SS, **320**, 261 (2009)
[2] Clementini, G., Di Tomaso, S., Di Fabrizio, L., et al., AJ, **120**, 2054 (2000)
[3] Cusano, F., Clementini, G., Garofalo, A., et al., ApJ, **779**, 7 (2013)
[4] Cusano, F., Garofalo, A., Clementini, G., et al., ApJ, **806**, 200 (2015)
[5] Cusano, F., Garofalo, A., Clementini, G., et al., ApJ, **829**, 26 (2016)
[6] Cusano, F., Garofalo, A., Clementini, G., et al., ApJ, submitted (2017)
[7] Ibata, R. A., Lewis, G. F., Conn, A. R., et al., Nature, **493**, 62 (2013)
[8] Moore, B., Ghigna, S., Governato, F., et al., ApJL, **524**, L19 (1999)
[9] Oosterhoff, P. Th., The Observatory, **62**, 104 (1939)
[10] Stetson, P. B., PASP, **106**, 250 (1994)

Little Bear's pulsating stars: Variable star census of UMi dSph Galaxy

K. Kinemuchi[1,2,⋆], *E.* Jeffery[3,⋆⋆], *C.* Kuehn[4,⋆⋆⋆], *K.* Grabowski[1,2,⋆⋆⋆⋆], and *J.* Nemec[5,†]

[1]*Apache Point Observatory, P.O. Box 59, Sunspot, NM 88349, USA*
[2]*New Mexico State University, Department of Astronomy, Las Cruces, NM 88001, USA*
[3]*Brigham Young University, Provo, UT, USA*
[4]*University of Northern Colorado, Greeley, CO, USA*
[5]*Camosun College, Victoria, B.C., Canada*

Abstract. Recent observations and a photometric search for variable stars in the Ursa Minor dwarf spheroidal galaxy (UMi dSph) are presented. Our observations were taken at Apache Point Observatory in 2014 and 2016 using the 0.5m ARCSAT telescope and the West Mountain Observatory (WMO) 0.9m telescope of Brigham Young University in 2016. Previously known RR Lyrae stars in our field of view of the UMi dSph are identified, and we also catalog new variable star candidates. Tentative classifications are given for some of the new variable stars. We have conducted period searches with the data collected with the WMO telescope. Our ultimate goal is to create an updated catalog of variable stars in the UMi dSph and to compare the RR Lyrae stellar characteristics to other RR Lyrae stars found in the Local Group dSph galaxies.

1 Introduction

The Ursa Minor dwarf spheroidal galaxy (UMi dSph) is a Local Group member with a rich variable star population. RR Lyrae (RRL), anomalous Cepheids (AC), and eclipsing binary stars have been found and identified ([1, 4, 7]). Two different epochs of star formation are indicated via the RRL (older population) and the ACs (intermediate age). The UMi dSph is a metal-poor galaxy, and from the analysis of the RRL, it has been classified as an Oosterhoff-II (OoII) type object. UMi is one of a handful of known dwarf galaxies with a clear Oosterhoff classification.

The goals of this project are to revisit the UMi dSph galaxy and reinvestigate the variable star population. We present our variable star census based on the northeast portion of the galaxy. With our photometric data, we wil reevaluate and recalculate many of the stellar characteristics, in particular the pulsational characteristics.

⋆kinemuchi@apo.nmsu.edu
⋆⋆ejeffery@byu.edu
⋆⋆⋆charles.kuehn@unco.edu
⋆⋆⋆⋆kgrabowski@apo.nmsu.edu
†jmn@isr.bc.ca

DOI: 10.1051/epjconf/201715201016

2 Data acquisition and reduction

Data sets were collected at the 0.5m Astrophysical Research Consortium Small Aperture Telescope (ARCSAT) at Apache Point Observatory and at the 0.9m telescope at the West Mountain Observatory (WMO) of Brigham Young University. Due to the field coverage of the telescopes, we focused our survey on the northeast portion of the galaxy, centered at α = 15:09:11.34 and δ = 67:15:51.7. With the ARCSAT telescope, we obtained ~ 100 epochs, and at WMO, over 200 epochs were observed.

Standard data reduction techniques were implemented, and to standardize the magnitudes, Landolt standards ([2]) were observed. Photometry was performed using DAOPHOT/ALLSTAR packages ([6]), and period analysis was done primarily with the dataset collected at WMO. Period solutions were obtained using the Supersmoother program ([5]), which uses a running linear regression algorithm. For amplitude, we used a basic measurement of the difference between maximum and minimum, omitting outliers 2.5 σ from the mean.

Table 1. Identified variable stars in the north portion of UMi dSph.

Star	α_{2000} (deg)	δ_{2000} (deg)	V	Period (d)	type
V5	227.48149	67.474992	19.69	0.766705	ab
V6	227.49143	67.464966	18.17	0.725528	AC
V7	227.44961	67.454614	19.61	0.690328	ab
V8	227.294	67.451651	19.70	0.553879	ab
V9	227.53442	67.419453	19.59	0.356460	c?
V10	227.58564	67.405209	19.69	0.617496	ab
V11	227.63938	67.417687	18.88	0.673431	AC
V12	227.63475	67.395875	19.65	0.771735	ab
V13	227.46745	67.403259	19.66	0.646050	ab
V19	227.53903	67.357643	19.67	0.341565	c
V24	227.54186	67.310734	19.83	0.301081	c?
V28	227.6731	67.278736	19.70	0.308163	c?
V29	227.62463	67.252691	19.73	0.409747	c?
V41	227.46443	67.232054	19.84	0.490488	ab
V42	227.59471	67.209778	19.74	0.646473	ab
V45	227.35121	67.207436	19.02	0.216662	?
V48	227.18208	67.218741	19.69	0.687258	ab
V49	227.18687	67.212797	19.73	0.415464	c?
V78	227.29384	67.38137	19.69	0.372640	c?
V79	227.38367	67.328828	19.70	0.313023	c?
V80	227.41875	67.3245	18.74	0.512738	?
v1540*	227.88001	67.426652	19.79	0.714490	ab
v1678*	227.19679	67.413053	19.75	0.336900	c?
v1895*	227.09781	67.386081	19.45	0.490887	ab
v2078*	227.83088	67.361948	19.73	0.719031	ab
v3176*	227.32619	67.236128	19.81	0.327060	c?

* Newly identified star.

3 RR Lyrae results

From the 543 objects identified and photometered, we find 22 RR Lyrae candidates. Cross-matching with the finding charts and variable star catalog of [4], we find 12 ab-type and 10 c-type RRL stars. We additionally find 2 ACs, V6 and V11 (using Nemec's IDs). In Table 1, we provide our identified variable stars' IDs, equatorial coordinates (as determined from `astrometry.net`, [3]), mean V magnitude, period solution, and variable star type. The star ID is the same as in [4], unless it is denoted with an asterisk, then it is a newly identified star.

4 Curious case of V80

The variable star identified as V80 in [1, 4, 7] has an unclear classification. This star's variability appears real in our datasets from 2016. We find the amplitude of this star to be closer to the published value in [4], but the periodicity is longer than in any of the published literature. The period solution for V80 was 0.512738 days. We cannot conclusively classify this variable star at this time. Applying frequency analysis will help understand the nature.

5 Future Work

Additional data are expected to be collected in March 2017 and will cover the southern regions of UMi dSph that has been missed. With the fairly complete coverage of the pulsation cycle to many of the RRab stars, we will calculate the [Fe/H] metallicities through Fourier decomposition parameters. This work will allow us to obtain an independent measurement of the dSph galaxy's metallicity.

Acknowledgments: K.K. would like to thank Suzanne Hawley, director of ARCSAT for the generous time granted for this project. Professional development funds through the Astrophysical Research Consortium allowed K.K. to present this work at this conference and is gratefully acknowledged.

References

[1] Kholopov, P. N., Peremennye Zvezdy, **18**, 117 (1971)
[2] Landolt, A. U., AJ, **104**, 340 (1992)
[3] Lang, D., Hogg, D. W., Mierle, K., Blanton, M., & Roweis, S., AJ, **139**, 1782 (2010)
[4] Nemec, J. M., Wehlau, A., & Mendes de Oliveira, C., AJ, **96**, 528 (1988)
[5] Reimann, J. D., Ph.D. Thesis (1994)
[6] Stetson, P. B., PASP, **99**, 191 (1987)
[7] van Agt, S. L. T. J., Bulletin of the Astronomical Institutes of the Netherlands, **19**, 275 (1967)

Finding beat Cepheids in M33 and their use in measuring the metallicity gradient

Lawrence Short[1,*] and *David* Bersier[1]

[1] *Astrophysics Research Institute, Liverpool John Moores University, IC2, Liverpool Science Park, 146 Brownlow Hill, Liverpool, L3 5RF, UK*

Abstract. The period ratio between the two different pulsation modes of beat Cepheids is directly affected by the metallicity of the star. Our aim is to find all of the beat Cepheids in M33 and determine their metallicities hence allowing us to derive the metallicity gradient. To achieve this we have M33 data from the INT and the CFHT variability survey conducted by Hartman et al. (2006, [1]).

1 Introduction

Metallicity measurements are traditionally achieved by spectroscopic analysis but this can lead to considerable uncertainties. However, beat Cepheids can be used to trace metallicity because the period ratio, P_1/P_0, is sensitive to metal abundance.

2 Data

2.1 CFHT

We have data from the M33 variability survey conducted by Hartman et al. (2006, [1]) in which they used the MegaCam instrument on the 3.6-m Canada-France-Hawaii Telescope (CFHT) with a field of view of 1 deg^2. Observations were taken in the g', r', and i' Sloan bands covering 27 nights over 17 months giving at least 33 measurements for each filter. They identified more than 36000 variable sources including over 2000 Cepheids. Using these data Beaulieu et al. (2006, [2]) discovered 5 beat Cepheids which they compared to beat Cepheids in the Milky Way, the LMC and the SMC using a Petersen diagram in order to determine their metallicity.

2.2 INT

On top of the CFHT data we have further epochs from the 2.5-m Isaac Newton Telescope (INT), La Palma. The wide field camera allowed us to cover the whole galaxy in 4 pointings. The data set consists of 48 measurements in the r' band taken on 13 nights over 19 months with fewer images in the g' and i' bands. The INT and CFHT data needs cross-calibrating which will give us $\sim$ 80 epochs to find all the variable objects in M33 and produce accurate light curves. A period search can then be applied to the cross-calibrated data in order to determine all of the Cepheids present (we are expecting 2000-3000).

*L.A.Short@2015.ljmu.ac.uk

3 Image subtraction

We have performed image subtraction on a single chip of the 16 from the INT images using the software HOTPANTS[1] based on the ISIS image subtraction package described by Alard & Lupton (1998, [3]) and Alard (2000, [4]). Images with the best seeing are averaged together to build a reference image. Images taken on different nights were then individually subtracted from the reference image so only objects that have changed brightness remain. See Figure 1.

Figure 1. INT images of a small section of M33 in the r' band. Left-hand panel: Image at some epoch. Mid panel: Reference image. Right-hand panel: Resultant subtracted image.

4 Light curves

The flux of the residual pixel counts of a Cepheid found from Figure 1 for each of the subtracted images at every epoch is measured and folded with its period. Figure 2 shows the light curves of a single period Cepheid from both the INT and CFHT data.

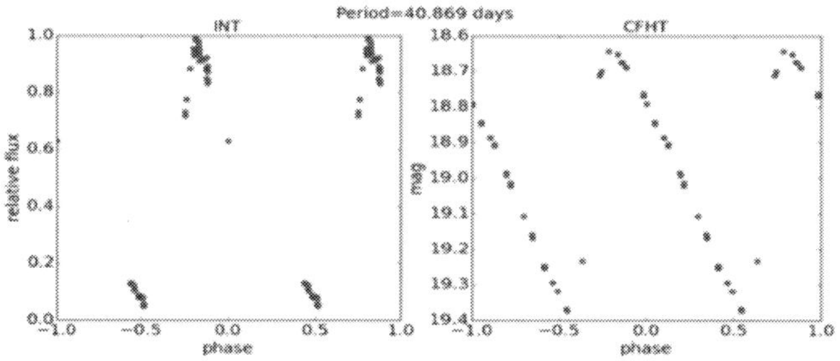

Figure 2. Comparison of folded light curves between the INT subtracted images and the CFHT data of a Cepheid.

References

[1] Hartman, J. D., Bersier, D., Stanek, K. Z., Beaulieu, J.-P., Kaluzny, J., Marquette, J.-B., Stetson, P. B., & Schwarzenberg-Czerny, A., MNRAS, **371**, 1405 (2006)

[2] Beaulieu, J.-P., Buchler, J. R., Marquette, J.-B., Hartman, J. D., & Schwarzenberg-Czerny, A., ApJL, **653**, L101 (2006)

[3] Alard, C., & Lupton, R. H., ApJ, **503**, 325 (1998)

[4] Alard, C., A&AS, **114**, 363 (2000)

[1] `www.astro.washington.edu/users/becker/v2.0`

Infrared excess in semi-regular variables in the Magellanic Clouds

Martin A.T. Groenewegen[1,*] and *Monika I.* Jurkovic[2,3,**]

[1]*Koninklijke Sterrenwacht van België (KSB), Ringlaan 3, B-1180 Brussels, Belgium*
[2]*Astronomical Observatory of Belgrade (AOB), Volgina 7, 11 060 Belgrade, Serbia*
[3]*Konkoly Observatory, Research Centre for Astronomy and Earth Sciences, Hungarian Academy of Sciences, H-1121 Budapest, Konkoly Thege Miklós út 15-17, Hungary*

Abstract. Following up on our recent work we are investigating if Type II Cepheids could be "hiding" in the sample of stars classified by the OGLE team as long-period variables.

1 Introduction

In two recent papers, [1, 2] (and see these proceedings, [3]) studied the 246 Type II Cepheids (T2Cs) and 89 Anomalous Cepheids (ANCEPs) discovered in the OGLE-III survey in the SMC and LMC ([5, 7, 8]). T2C are pulsating stars associated with globular clusters and dwarf spheroidal galaxies and considered to be useful distance indicators for objects where classical Cepheids are too few and RR Lyrae too faint. They are typically separated into 3 subgroups according to their pulsation periods: the BL Herculis (BLH) stars have periods 1 – 4 days, the W Virginis (WVir) 4 – 20 days and the RV Tauris (RVT) 20 – 70 days pulsation periods. Variables with RVT-like light curves at longer periods may well exist but would appear in the OGLE samples of the long-period variables (LPVs) ([6, 9]).

In [1] the spectral energy distributions (SEDs) were constructed for all these stars, using available photometry from the literature, and fitted with a dust radiative transfer code. In the large majority of cases the SEDs could be fitted by a stellar photosphere model only. However, for $\sim$ 60% of the RVT and $\sim$ 10% of the (p)W Vir objects an infrared excess is detected from the SED fitting. The results of [4] are confirmed that there exist (pulsating) stars with luminosities below that predicted from single-star evolution that show a clear infrared excess, likely related to binarity.

In [2] we used the derived luminosities and effective temperatures to derive radii, and present period – (bolometric) luminosity, and period – radius relations for the T2C and ANCEP.

2 T2C among LPVs/SRs

In this (ongoing) study we want to look for T2C (mostly of the RVT type) in the OGLE catalogs of LPVs. The SMC and LMC sample of the regular survey were considered ([6, 9]) as well as the stars classified as LPVs in the OGLE Shallow Survey in the LMC ([10]) that contains brighter objects.

The initial selection was done on colour and magnitude, and required *I*-band amplitudes > 0.07 mag.

*martin.groenewegen@oma.be
**mojur@aob.rs

DOI: 10.1051/epjconf/201715201018

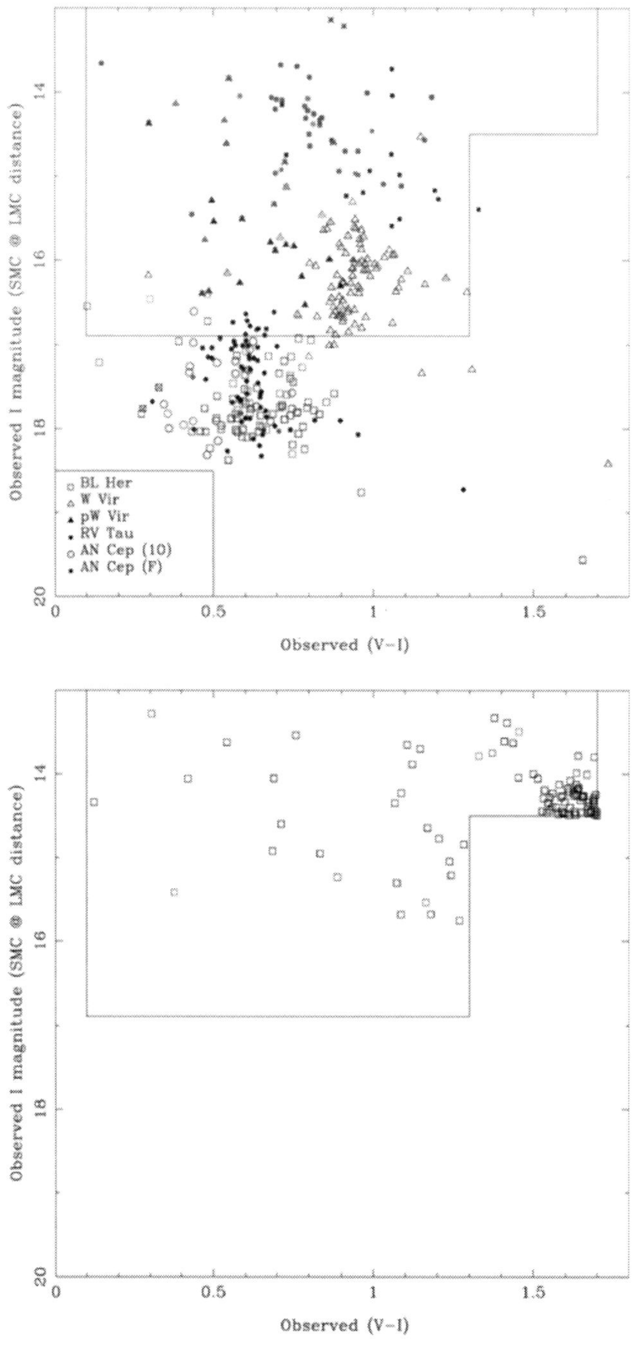

Figure 1. The figures show the observed $I, (V - I)$ CMD. Stars in the SMC are plotted in red, and shifted by −0.43 magnitudes. The top figure shows the sample of T2Cs and ANCEPs studied in [1, 2]. Stars found to have an IR excess are marked by a green plus-sign. Stars which show eclipsing or ellipsoidal variations according to OGLE are marked by a blue cross. The rectangles indicates the colour box used in the present study. The bottom figure shows the stars selected from the OGLE LPV samples and the OGLE Shallow Survey.

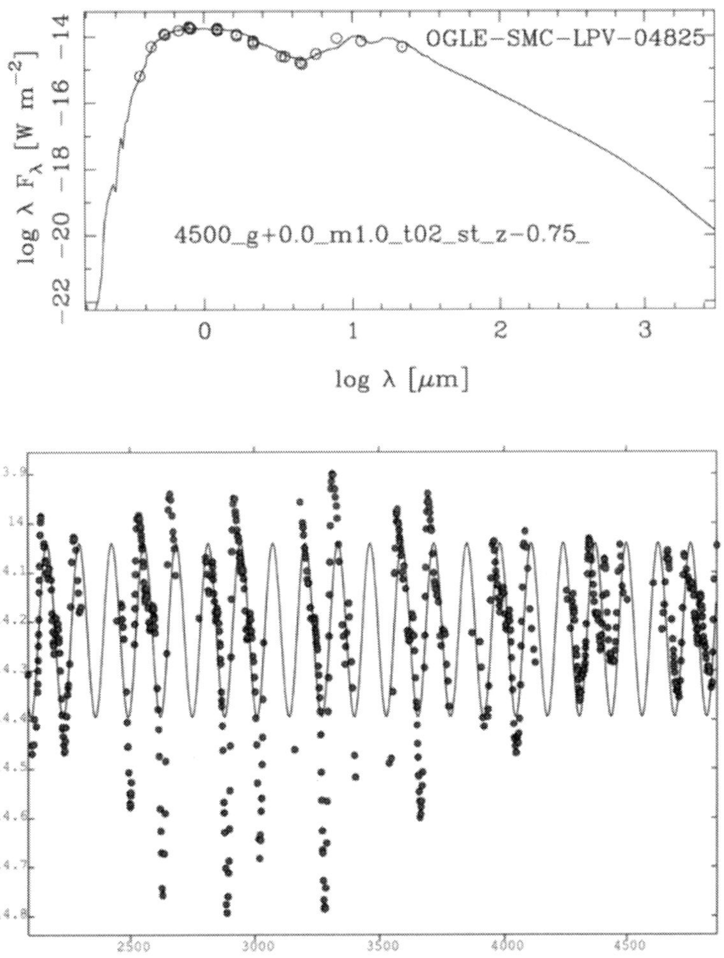

Figure 2. An example: The top panel shows a star with a clear IR excess. The bottom panel shows the OGLE light curve (*I*-band magnitude versus time) of this object, and the best-fitting sine curve which has a period of 130 days.

The colour-magnitude diagram is shown in Figure 1. In the upper panel the selection box is shown together with the T2C and ANCEP sample of [1], and in the bottom panel the initial sample of 5 Miras, 78 SRV, 35 OSARGS, and 16 bright targets is displayed.

Phased light curves were constructed and inspected, also considering MACHO and EROS data when available. For the moment 29 stars have already been removed from the sample: 2 are also classified as T2C and are therefore already considered in [1], 13 are eclipsing binaries or show ellipsoidal variations, 3 of the bright stars have amplitudes smaller than the 0.07 magnitude limit, 3 have R CrB type light curves, and 8 have clear Mira like light curves. *As this is an ongoing study, further reduction of the sample is very likely when considering light curve shape, period (range), luminosity (range), and the period-luminosity relation in more detail!*

The SEDs were constructed and fitted as in [1]. Most stars can be fitted by a stellar photosphere alone, but other SEDs clearly show IR excess; see the example in Figure 2.

References

[1] Groenewegen, M. A. T., & Jurkovic, M. I., A&A, in press (arXiv:1705.00886) (2017a)
[2] Groenewegen, M. A. T., & Jurkovic, M. I., A&A, in press (arXiv:1705.04487) (2017b)
[3] Jurkovic, M. I., & Groenewegen, M. A. T., these proceedings (2017)
[4] Kamath, D., Wood, P. R., Van Winckel, H., & Nie, J. D., A&A, **586**, L5 (2016)
[5] Soszyński, I., Udalski, A., Szymański, M. K., et al., AcA, **58**, 293 (2008)
[6] Soszyński, I., Udalski, A., Szymański, M. K., et al., AcA, **59**, 239 (2009)
[7] Soszyński, I., Poleski, R., Udalski, A., et al., AcA, **60**, 17 (2010)
[8] Soszyński, I., Udalski, A., Szymański, M. K., et al., AcA, **60**, 91 (2010)
[9] Soszyński, I., Udalski, A., Szymański, M. K., et al., AcA, **61**, 217 (2011)
[10] Ulaczyk, K., Szymański, M. K., Udalski, A., et al., AcA, **63**, 159 (2013)

A search for low-metallicity pulsating B stars

Chris Engelbrecht[1,*], *Refilwe* Kgoadi[1,2], and *Fabio* Frescura[3]

[1]*Department of Physics, University of Johannesburg, PO Box 524, Auckland Park, Johannesburg 2006, South Africa*
[2]*College of Science and Engineering, James Cook University, Douglas, Townsville 4810, Queensland, Australia*
[3]*School of Physics, University of the Witwatersrand, Private Bag 3, WITS, Johannesburg 2050, South Africa*

Abstract. We report on some recent results from a long-term *UBVI* survey of various fields in the Large Magellanic Cloud (LMC), which is aimed at identifying and classifying pulsating B stars in the selected LMC fields. Difference Imaging Analysis shows a clear advantage over conventional PSF fitting. Tentative indications have been found of a varying incidence of pulsation amplitudes (and, by inference, of metal content of the pulsators) across the LMC bar.

1 Introduction

Until about a decade ago, models of main-sequence B stars set a lower threshold of $Z = 0.01$ on the metallicity required to enable β Cephei-type pulsation through the kappa mechanism (e.g., [9]). (Slightly lower metallicity thresholds were found for excitation of SPB-type pulsation.) The subsequent firm discovery of three β Cephei (BCep) stars in the Large Magellanic Cloud (LMC, [10]) using OGLE data, and the reporting of 92 BCep candidates in the LMC (and 6 BCep candidates in the Small Magellanic Cloud, SMC) by [6], further bolstered by the results of [4] and [5], prompted more detailed work on models of main-sequence B stars, as well as the exploration of the conditions required for pulsation instability in these stars. As the BCep pulsations are driven by the iron-group (Fe/Ni/Mn/Cr) opacity bump at 160 000 K (through the kappa mechanism), plausible reasons for increasing either the abundances of these elements, or the opacities due to them (or both), were explored. Some workers ([7] and [11]) were able to lower the threshold for pulsation excitation to $Z = 0.007$ by using new opacity calculations. Subsequently, [12] showed model calculations which generated BCep pulsations at both LMC and SMC metallicities (i.e., down to $Z = 0.004$) through an ad hoc amplification of, specifically, the Ni opacity at the bump. Recently, [8] followed up on the report by [2] on direct laboratory measurements of iron opacity in stellar conditions, showing a 75% enhancement of iron opacity, to re-calculate the BCep and SPB instability strips. A modest expansion of these strips was found.

Despite these efforts, the claimed detections of BCep and SPB stars in the LMC and SMC remain inexplicable without invoking ad hoc measures. Since this conundrum might be resolved by ascertaining the actual pulsational properties and physical parameters of both confirmed and suspected BCep

*engelbrecht.chris@gmail.com

stars in the LMC and SMC, a ground-based observing campaign was constructed to do exactly this. The LMC was selected as the target for this campaign as it would be far more likely to produce a positive result within the project's design constraints than the SMC.

2 Project details and results

Five distinct 2.6 arcmin × 2.6 arcmin fields in the LMC were observed in Johnson *UBVI*, producing about 600 useful time series points per filter, per field. The fields were selected based on a Fourier transform run through OGLE data, looking for power in the BCep frequency domain. We will apply for time on the new High-Resolution Spectrograph (HRS) on SALT to obtain spectra of all pulsators found in our sample of the LMC, with the express purpose of determining elemental abundances. We initially used PSF fitting in IRAF to identify potential B stars. Standard calibrations of Johnson photometric indices were then used to estimate key physical parameters of those stars. Lomb-Scargle periodograms and associated statistical significance estimates were calculated for those stars that occupy the expected instability strip, based on calibrations of the photometry. In general, the IRAF-based photometry data produced many marginal/dubious cases of pulsation detection. A greater deal of success is now being achieved with the pyDIA code ([1], [3]). Preliminary analysis of the time series obtained suggests that pulsations appear more frequently – and also with higher amplitudes – as one moves to the eastern side of the bar of the LMC. Final results of pulsation frequencies, amplitudes and modal types will be presented once this analysis is complete.

Acknowledgments: CE and RK acknowledge financial support from the University of Johannesburg and the National Reserarch Foundation of South Africa.

References

[1] Albrow, M. D., Horne, K., Bramich, D. M., et al., MNRAS, **397**, 2099 (2009)
[2] Bailey, J. E., Nagayama, T., Loisel, G. P., et al., Nature, **517**, 56 (2015)
[3] Bramich, D. M., Horne, K., Albrow, M. D., et al., MNRAS, **428**, 2275 (2013)
[4] Diago, P. D., Gutiérrez-Soto, J., Fabregat, J., & Martayan, C., A&A, **480**, 179 (2008)
[5] Diago, P. D., Gutiérrez-Soto, J., Fabregat, J., & Martayan, C., CoAst, **158**, 184 (2009)
[6] Kolaczkowski, Z., Pigulski, A., Soszyński, I., et al., Mem. S. A. It., **77**, 336 (2006)
[7] Miglio, A., Montalbán, J., & Dupret, M.-A., MNRAS, **375**, L21 (2007)
[8] Moravveji, E., MNRAS, **455**, L67 (2016)
[9] Pamyatnykh, A., Acta Astr., **49**, 119 (1999)
[10] Pigulski, A., & Kolaczkowski, Z., A&A, **388**, 88 (2002)
[11] Salmon, S., Montalbán, J., Miglio, A., Dupret, M. -A., Morel, T. & Noels, A., preprint (arXiv:1007.3088) (2010)
[12] Salmon, S., Montalbán, J., Morel, T., Miglio, A., Dupret, M.-A., & Noels, A., MNRAS, **422**, 3460 (2012)

A new search for R Coronae Borealis stars in the SMC

Fatemeh Nikzat[1,2,⋆] and *Márcio* Catelan[1,2,⋆⋆]

[1] *Pontificia Universidad Católica de Chile, Facultad de Física, Instituto de Astrofísica, Av. Vicuña Mackenna 4860, 782-0436 Macul, Santiago, Chile*
[2] *Millennium Institute of Astrophysics, Santiago, Chile*

Abstract. R Coronae Borealis (RCB) stars are rare, and their evolutionary origin is not well understood. Since they are obscured due to formation of carbon dust around the star during their mass loss events, RCB stars can be classified as *self-eclipsing variable stars*. The purpose of this work is to present a new search for RCB stars in the Small Magellanic Cloud (SMC), by analysing *VI* data from the OGLE project.

1 Introduction

R Coronae Borealis (RCB) stars are red supergiant carbon stars presenting dramatic dimming episodes at irregular intervals. The dimming episodes are caused by self-obscuration by dust, occurring as a consequence of mass loss events (e.g., [1–3, 6] and references therein). A subclass of RCB stars called DY Persei also exists. The DY Per stars tend to be cooler and have more symmetrical drops in comparison to RCBs (e.g., [1, 12], and references therein). Our goal was to search for new RCB stars in the Small Magellanic Cloud (SMC), using *VI* data from the OGLE project ([10]). The *VI* light curves of all SMC red variable stars were visually inspected, and compared against templates from the literature. New detected RCB and DY Per candidates were presented in Nikzat & Catelan (2016, [7]), which had previously been classified as semi-regular or Mira variables. If confirmed, these detections would lead to a significant increase in the number of known RCB + DY Per stars in the SMC.

2 Data

Observations were made with the 1.3 m Warsaw telescope at Las Campanas Observatory, northern Chile, in the course of the OGLE-III campaigns ([10]). The catalog data for red variable stars in the SMC are available online from the OGLE project.[1] All constructed *VI* light curves were examined, in a search for deep, non-periodic declines that might be indicative of RCB-like behavior.

3 Results

We found three new RCB and 63 new DY Per candidates ([7]). An additional spectroscopic RCB candidate, MSX-SMC-014 ([5]), three confirmed DY Per stars and three DY Per candidates from the

⋆ fnikzat@astro.puc.cl
⋆⋆ mcatelan@astro.puc.cl
[1] ftp://ftp.astrouw.edu.pl/ogle/ogle3/OIII-CVS/smc/lpv/

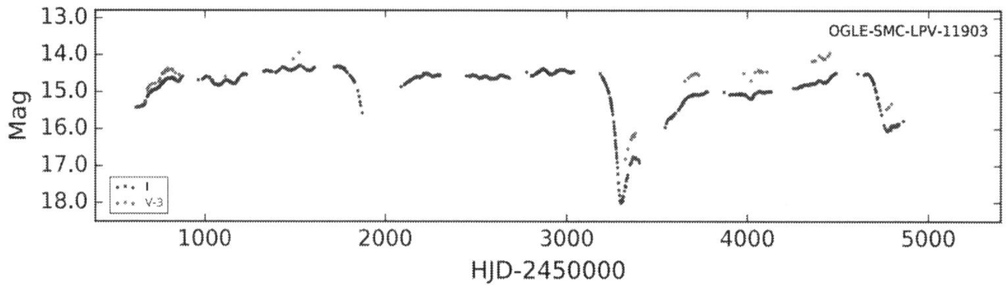

Figure 1. Light curves in *I* (*blue*) and *V* (*green*) of a known DY Per star in the SMC ([11]), identified using OGLE data ([7]), strongly reminiscent of an RCB star, and which we accordingly classify as a "transitional" DY Per/RCB star (see [7] for more details).

EROS2 project ([11]) were also analyzed. Their names, coordinates and light curves can be found in [7]. In addition, we noted the possible existence of a "borderline" DY Per-like star and a "transitional" DY Per/RCB star (Fig. 1). The latter might be consistent with the presence of an evolutionary sequence among H-deficient carbon stars, as suggested by [8] and supported by [4] and [9]. We point out that we have only used light curve morphology as indicative of potential RCB/DY Per status. Follow-up observations are required to conclusively confirm the nature of our candidates.

Acknowledgments: Support for this project is provided by the Ministry for the Economy, Development, and Tourism's Millennium Science Initiative through grant IC 120009, awarded to the Millennium Institute of Astrophysics (MAS); by Proyecto Basal PFB-06/2007; by CONICYT's PCI program through grant DPI20140066; and by FONDECYT grants #1141141 and #1171273. FN is grateful for financial support by Proyecto Gemini CONICYT grants #32130013 and #32140036.

References

[1] Catelan, M., & Smith, H. A., *Pulsating Stars* (Wiley-VCH, Weinheim, 2015)
[2] Clayton, G. C., PASP, **108**, 225 (1996)
[3] Clayton, G. C., JAAVSO, **40**, 539 (2012)
[4] De Marco, O., Clayton, G. C., Herwig, F., Pollacco, D. L., Clark, J. S., & Kilkenny, D., AJ, **123**, 3387 (2002)
[5] Kraemer, K. E., Sloan, G. C., Wood, P. R., Price, S. D., & Egan, M. P., ApJ, **631**, L147 (2005)
[6] Lambert, D. L., & Rao, N. K., JApA, **15**, 47 (1994)
[7] Nikzat, F., & Catelan, M., IBVS, **6190**, 1 (2016)
[8] Saio, H., & Jeffery, C. S., MNRAS, **333**, 121 (2002)
[9] Schaefer, B. E., MNRAS, **460**, 1233 (2016)
[10] Soszyński, I., Udalski, A., Szymański, M. K., et al., AcA, **61**, 217 (2011)
[11] Tisserand, P., Wood, P. R., Marquette, J. B., et al., A&A, **501**, 985 (2009)
[12] Začs, L., Mondal, S., Chen, W. P., Pugach, A. F., Musaev, F. A., & Alksnis, O., A&A, **472**, 247 (2007)

Catalogue of variable stars in Milky Way globular clusters

http://www.astro.utoronto.ca/~cclement/read.html

Christine Clement[1],*

[1] *Department of Astronomy & Astrophysics, University of Toronto, 50 St. George St., Toronto, ON, M5S 3H4, Canada*

Abstract. Globular cluster variable stars have been studied for more than a century. In the early investigations, more than 90% of the known variables were of the RR Lyrae type. However, in the interim, technological advances have facilitated the discovery of other types of variables. As a result, although RR Lyrae stars still dominate, they now constitute less than 70% of the known variables.

1 Introduction

A summary paper based on this catalogue was published in the Astronomical Journal by Clement et al. (2001) [1]. Since 2001, the number of known (optical/near infrared) variables has almost doubled because numerous investigations have dealt with variable stars in globular clusters. Combining the new results with summaries published by Helen Sawyer Hogg in 1939, 1955, and 1973 ([2–4]) illustrates how our knowledge of the variable star content of Galactic globular clusters has changed over the past eight decades.

2 Globular cluster variables: 1939-2016

Observations of globular cluster variables in the early 20th century were usually made with photographic plates. This practice continued until the 1980s. Only the brightest variables could be observed and variations of less than ~0.3 mag were not detectable. Consequently, more than 90% of the known variables were of the RR Lyrae type. The rest were Cepheids and long period variables, which were among the brightest stars in clusters. Variables below the horizontal branch in the colour-magnitude (CM) diagram were too faint to be discovered.

The routine use of CCD detectors since the 1980s has changed that. Because CCDs are more efficient than photographic detectors, it is possible to observe fainter stars and to detect smaller variations. Also, since exposure times can be shorter, it is feasible to detect eclipses and other brightness changes on time scales of half an hour or less.

As a result of these developments, numerous SX Phe stars and binaries have been identified. They now constitute more than 15% of the known variables. The number of slow variables has also increased because many stars near the red giant tip in the CM diagram exhibit low level variations.

*cclement@astro.utoronto.ca

In future, the percentage of known binaries will undoubtedly increase because advances in image processing techniques and the use of space telescopes make it possible to study stars in crowded central regions where most binaries are located.

Table 1. Census of globular cluster variable stars.

Variables	1939	1955	1973	2001	2016
RR Lyrae	613	1114	1202	1842	>2925
Type II Cep, RV Tau	27	32	37	60	>85
Slow Vars: L, M, SR	15	53	71	117	>475
SX Phe	0	0	0	117	>325
Eclipsing, ELL	1	2	2	96	>400

The above table lists the number of pulsating and binary variables catalogued at five different epochs. It includes only the stars that were assumed to be cluster members, but membership status can often be ambiguous. In the early days, it was based on apparent magnitude and distance from the cluster centre. Then in the second half of the twentieth century, CM diagrams became available for many clusters so that the magnitude and colour were used. Also, in a few cases, mainly for the brightest stars, radial velocities and proper motions could be derived so that the cluster members could be distinguished from stars in the surrounding field. Now, with improved astrometry, it is becoming routine to derive proper motions for cluster stars.

Acknowledgments: I am pleased to acknowledge the support that I have received from Peter Martin, Howard Yee, and Ray Carlberg, the three people who have served as chairs of the University of Toronto Astronomy & Astrophysics department since my retirement in 2006. They have all helped me in my research by giving me access to departmental resources. I am also grateful to the worldwide astronomers who have cited the 2001 paper and encouraged me to continue with this project. Finally, I would like to thank Márcio Catelan who inspired me to publish the 2001 paper.

References

[1] Clement, C. M., Muzzin, A., Dufton, Q., Ponnampalam, T., Wang, J., Burford, J., Richardson, A., Rosebery, T., AJ, **122**, 2587 (2001)
[2] Sawyer, H. B., David Dunlap Publ., **1**, No. 4 (1939)
[3] Sawyer, H. B., David Dunlap Publ., **2**, No. 2 (1955)
[4] Sawyer Hogg, H., David Dunlap Publ., **3**, No. 6 (1973)

Variable stars in the VVV globular clusters

Javier Alonso-García[1,2,⋆], *Márcio* Catelan[3,2], *Rodrigo* Contreras Ramos[2,3], *István* Dékány[4], and *Dante* Minniti[5,2,6]

[1]*Universidad de Antofagasta, Unidad de Astronomía, Avda. U. de Antofagasta 02800, Antofagasta, Chile*
[2]*Millennium Institute of Astrophysics, Santiago, Chile*
[3]*Pontificia Universidad Católica de Chile, Instituto de Astrofísica, Av. Vicuña Mackenna 4860, 782-0436 Macul, Santiago, Chile*
[4]*Zentrum für Astronomie der Universität Heidelberg, Astronomisches Rechen-Institut, Mönchhofstr. 12-14, D-69120 Heidelberg, Germany*
[5]*Universidad Andrés Bello, Departamento de Física, Av. Fernández Concha 700, Las Condes, Santiago, Chile*
[6]*Vatican Observatory, V00120 Vatican City State, Italy*

Abstract. The VVV survey observed some of the most crowded and most obscured regions in the inner Milky Way during the last years. A significant sample of the less known globular clusters in our galaxy lie there. Combining the high-resolution, wide-field, near-infrared capabilities of the survey camera, the use of 5 different filters, and multi-epoch observations, we are able to overcome many of the previous challenges that prevented a proper study of these objects. Particularly, the identification of the RR Lyrae stars in these globular clusters is proving to be a fundamental tool to establish accurately their distances and reddenings, and to infer information about the Oosterhoff dichotomy that Galactic globular clusters seem to follow.

1 Introduction

The globular clusters that lie in the inner regions of the Milky Way present important difficulties to extract accurate information about their physical parameters and about their stellar populations. The high extinction perceivable when observing towards the inner Galaxy, the high number of field stars in the region which increases the considerable crowding when observing these compact objects and complicates the proper identification of cluster stars, or the presence of some of the most poorly-populated globular clusters in our Galaxy, are among the most significant of them.

The *Vista Variables in the Vía Láctea* (VVV) survey has been observing the inner regions of the Milky Way during the last 6 years. VVV is one of the ESO Public Surveys conducted with the 4m VISTA telescope at the Paranal Observatory, in Chile. It surveyed a region of 562 square degrees in the region of the Galactic bulge and an adjacent region of the southern Galactic disc ([4]). There are 36 known Galactic globular clusters in the surveyed area, plus a few more new candidates ([3, 5]). The VVV survey allows us not only to observe these globular clusters from their very centers out to their tidal radii and beyond at near-infrared wavelengths were the effects of extinction are highly diminished, but also to search for any variable sources present thanks to its multi-epoch observations.

⋆ jalonso@uantof.cl

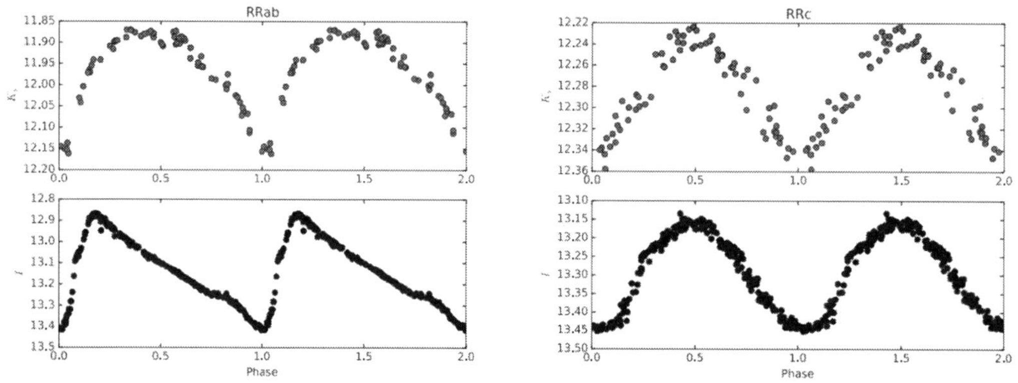

Figure 1. Light-curves of two RR Lyrae stars in NGC 6656 (M 22), one of the globular clusters in our sample. On the left, light-curve of V4, an RRab star; on the right, light-curve of V19, an RRc star. In the upper and lower panels, the K_s- and I-band light-curves are shown, built with VVV and OGLE data ([6]), respectively. Note the change in the amplitudes and shapes between the different wavelengths considered.

2 RR Lyrae in the VVV globular clusters

RR Lyrae constitute a common type of variable stars in Galactic globular clusters. The tight period-luminosity relations they show, especially at near-infrared wavelengths, confer them an additional interest, as they can be used as standard candles to infer information about the physical parameters of the globular clusters they belong to. Their proper identification, however, is complicated by the fact that they show smaller amplitudes and more sinusoidal shapes in their near-infrared light-curves, when compared to the optical ones (see Fig. 1).

We were able to overcome these difficulties and identify a significant number of these variables in two of the most reddened globular clusters in our sample: 2MASS-GC02 and Terzan 10 ([1]). We found both clusters to be further away from us, and much closer to the Galactic center than previously reported. We also found the extinction towards them to be non-canonical, and their Oosterhoff classification uncommon for Galactic globular clusters ([2]). We are currently exploring the cases of several more inner Galactic globular clusters to check the accuracy of their literature physical parameters, and how they fit in the Oosterhoff dichotomy paradigm that seems to be present in our Galaxy.

Acknowledgments: Support is provided by the Ministry for the Economy, Development and Tourism, Programa Iniciativa Científica Milenio grant IC120009, awarded to the Millennium Institute of Astrophysics (MAS). J.A-G. acknowledges support from FONDECYT Iniciación grant 11150916.

References

[1] Alonso-García, J., Dékány, I., Catelan, M., et al., AJ, **149**, 99 (2015)
[2] Catelan, M., Ap&SS, **320**, 261 (2009)
[3] Minniti, D., Hempel, M., Toledo, I., et al., A&A, **527**, A81 (2011)
[4] Minniti, D., Lucas, P. W., Emerson, J. P., et al., New Astronomy, **15**, 433 (2010)
[5] Minniti, D., Palma, T., Dékány, I., et al, ApJL, **838**, L14 (2017)
[6] Soszyński, I., Udalski, A., Szymański, M. K., et al., Acta Astronomica, **64**, 177 (2014)

EPJ Web of Conferences **152**, 01023 (2017) DOI: 10.1051/epjconf/201715201023
Wide-Field Variability Surveys: A 21st Century Perspective

Multiperiodic RR Lyrae stars in the field of NGC 6362

Paweł Moskalik[1,⋆], *Radosław* Smolec[1], *Janusz* Kałużny[1⋆⋆], *Wojciech* Pych[1], *Michał* Różyczka[1], and *Ian B.* Thompson[2]

[1] *Nicolaus Copernicus Astronomical Center, Warsaw, Poland*
[2] *The Observatories of the Carnegie Institution for Science, Pasadena, USA*

Abstract. We present frequency analysis of RR Lyrae stars of globular cluster NGC 6362.

1 Observations and data analysis

NGC 6362 is a nearby globular cluster of Oosterhoff I type. It has a mean metallicity of [Fe/H]= −0.95 and belongs to the old halo population ([1]). 35 RR Lyrae-type stars have been discovered in the cluster ([5]). We revisit these variables, taking advantage of the new CCD photometry accumulated by the Cluster AgeS Experiment (CASE). In our analysis we used 3200 V-band images collected with the 1-m Swope telescope of the Las Campanas Observatory between July 8th, 1999 and September 9th, 2009. The frequency analysis of the data was performed with the standard consecutive prewhitening technique. For full discussion of our results we refer the reader to [6].

2 RR Lyrae variables of NGC 6362

16 RR Lyrae stars of NGC 6362 are fundamental mode pulsators (RRab variables) and 16 are first overtone pulsators (RRc variables). Two objects, previously identified as RRab stars, turned out to be double mode pulsators (RRd stars), with two lowest radial modes simultaneously excited. Finally, in V37 we detect two high-amplitude modes of close frequencies and with lightcurves significantly differing from those of RR Lyrae stars. The nature of V37 is a puzzle. It will not be discussed here.

2.1 Non-radial modes

In 10 out of 16 RRc stars of NGC 6362 (63% of the sample) we detect a secondary short period variability, with the period ratio falling in the range of $P_x/P_1 = 0.60 - 0.65$. The amplitudes of the secondary signals never exceed 6 mmag. Similar pulsators have recently been discovered in many stellar systems ([3], and references therein). Space observations indicate that this form of variability most likely occurs in almost all RRc and RRd stars ([4]). Detecting it in the ground-based data is difficult, however, because of the extremely low amplitude of the secondary periodicities. The incidence rate of 63%, that we report for NGC 6362, is the highest among all stellar systems observed

⋆`mailto:pam@camk.edu.pl`
⋆⋆deceased

from the ground. The properties of the stars identified in NGC 6362 are qualitatively similar to those discovered in the Galactic bulge. In particular, in the period ratio vs. period diagram (so called Petersen diagram) they fall within the three sequences defined by the Galactic bulge stars.

The period ratios of 0.60 – 0.65 cannot be explained by two radial modes. This implies that the secondary periodicities observed in the RRc stars must be due to non-radial modes of oscillation. In the model recently proposed by Dziembowski ([2]), these periodicities are identified with the harmonics of modes of angular degrees of $\ell = 8$ and $\ell = 9$. In this picture, the non-radial modes themselves have amplitudes lower than their harmonics, because of geometric cancelation effects.

2.2 The Blazhko effect

The Blazhko effect is a periodic (or almost periodic) modulation of pulsation amplitudes and phases. In the frequency domain it manifests itself as a presence of equally spaced triplets (or higher multiplets) centered on the pulsation frequency and its harmonics. Sometimes only a doublet is detected, with the other component of the triplet hidden in the noise. The separation of components within the multiplet corresponds to the modulation frequency. We have found the Blazhko effect in 11 RRab and in 3 RRc stars of NGC 6362. This corresponds to incidence rates of 69% and 19%, respectively, which are among the highest reported in ground-based observations. The observed modulation periods range from 17.3 day to 216.4 day for RRab stars and from 8.5 day to 15.5 day for RRc stars. In one RRc and in one RRab star we find a rare double-periodic Blazhko effect. The weakest modulation detected in NGC 6362 has a peak-to-peak amplitude of only 7 mmag (in the RRc star V36).

2.3 Anomalous RRd variables

Two stars previously classified as RRab variables, V3 and V34, turned out to be double mode pulsators. In addition to the dominant radial fundamental mode, we also detect the radial first overtone of very low amplitude. The period ratios for the two modes in V3 and V34 are 0.730 and 0.728, respectively. These values are significantly lower than for the majority of the "classical" RRd variables of the same fundamental mode period. In both stars we detect the Blazhko effect, with either the fundamental mode only (V34) or both radial modes (V3) being modulated. Lower than usual period ratios, strong domination of the radial fundamental mode and presence of a long-period Blazhko modulation all indicate that V3 and V34 are not typical RRd stars. They must be classified as new members of the recently identified group of anomalous RRd variables ([7]).

Acknowledgments: This research is supported by the Polish National Science Center, through grant DEC-2015/17/B/ST9/03421

References

[1] Catelan, M., Ap&SS, **320**, 261 (2009)
[2] Dziembowski, W., Comm. Konkoly Observatory, **105**, 23 (2016)
[3] Netzel, H., Smolec, R., & Moskalik, P., MNRAS, **453**, 2022 (2015)
[4] Moskalik, P., Smolec, R., Kolenberg, K. et al., MNRAS, **447**, 2348 (2015)
[5] Olech, A., Kałużny, J., Thompson, I., Pych, W., Krzemiński, W., & Schwarzenberg-Czerny, A., MNRAS, **321**, 421 (2001)
[6] Smolec, R., Moskalik, P., Kałużny, J., Pych, W., Różyczka, M., & Thompson, I. B., MNRAS, **467**, 2349 (2017)
[7] Soszyński, I., Smolec, R., Dziembowski, W., et al., MNRAS, **463**, 1332 (2016)

Search for variable stars in the young open cluster Stock 18

Tirthendu Sinha[1,*], *Saurabh* Sharma[1,**], *Rakesh* Pandey[1], and *A. K.* Pandey[1]

[1] *Aryabhatta Research Institute of Observational Sciences (ARIES), Manora Peak, Nainital, 263 001, India*

Abstract. We have carried out multi-epoch deep I band photometry of the open cluster Stock 18 to search for variable stars in star forming regions. In the present study, we report identification of 28 periodic and 165 non-periodic variables stars. The periods of most of the periodic variables are in between 2 to 20 hours and their magnitude varies between 0.05 to 0.6 mag. We have derived spectral energy distributions for 48 probable pre-main sequence variables and the average age and mass is found to be 2.7 ± 0.3 Myrs and 2.7 ± 0.2 $M_{\odot}$, respectively.

1 Introduction

Young star clusters are potential sites for studying the characteristics of pre-main sequence (PMS) stars as they are found in plenty of numbers in these regions and they have broad mass range as well as spreads (few Myrs) in their ages. These stars have circumstellar disks around them during their early evolution. The PMS stars usually show a wide range of variability in their luminosity in almost all wavelengths from X-ray to infrared, and the variability time scale ranges from few minutes to years ([1]). Photometric variability results from several facts like irregular distribution of cool spots on stellar photosphere, variable hot spots, obscuration from dust ([3], and reference therein), instability in disk, change in accretion rate, etc. Due to the presence of cool and hot spots on the photosphere, the rotation of the star will give periodic/quasi-periodic variability. The evolution of disks and the accretion rates may also play a prominent role in non-periodic variability ([4–6]). The young open cluster (age ~ 6 Myr, [2]) Stock 18 (RA = 00:01:37, DEC = +64:37:30) is an ideal site to search for PMS variable stars and to study their properties.

2 Observation and data analysis

Optical photometric observations of Stock 18 were taken on the nights of 31/09, 21/10, 11/11, 26/11, 17/12 and 26/12 of year 2016 using the 1.3 m f/4 Devasthal telescope. The telescope has a 2048×2048 pixel square CCD camera covering ~ 18.4 × 18.4 arcmin^2 field of view. In total, about 325 and 13 frames were taken in I and V filters, respectively, with exposure of 180 sec each. The basic image processing such as bias subtraction and flat fielding was done by using tasks available within IRAF. Instrumental magnitude was obtained by PSF fitting using the DAOPHOT package. We have used the DAOMATCH and DAOMASTER routine of DAOPHOT-II to estimate the shifts in the individual

* `tirthendu@aries.res.in`
** `saurabh@aries.res.in`

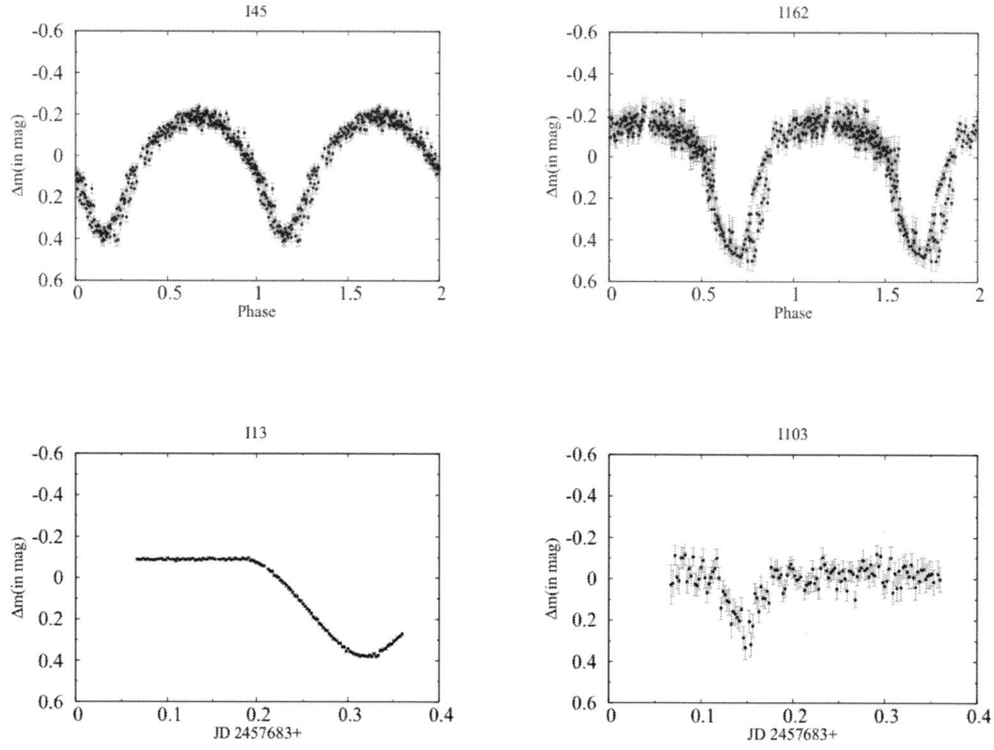

Figure 1. First two panels: Phase folded LC for periodic variables, last two panels: LC for non-periodic variables.

frames with respect to a reference frame and to get magnitudes of stars detected in different frames. Calibration of instrumental magnitude to the standard system was done by using the photometric data published by [2].

3 Result and conclusion

We have calculated the standard deviation (SD) of the difference of magnitude of all the stars in various magnitude bins (e.g., 10-11, 11-12, 12-13, and so on) and the stars in each magnitude bin having smallest SD are taken as comparison stars for that bin. The differential magnitude (Δm) in the sense star – comparison were plotted as a function of Julian date. The probable variables were identified by visual inspection. To determine the probable period of the variable stars we used *Period* software. We have identified 28 periodic and 165 non periodic variables in the region up to a V band magnitude of 20 mag. Light curves (LCs) of a few variables are shown in Figure 1. Most of the periodic variables have periods ranging from 2 – 20 hours and their magnitude varies between 0.05 – 0.6 mag.

The evolutionary status of these variables was explored through their position in the Hertzsprung-Russell diagram (for details, see [7]). Out of the detected variables, 168 have good photometric data (error < 0.1) and 80 of them have ages > 10 Myr. They are most probably main sequence or field stars. The average age of the remaining variables is found to be 4.2 ± 0.4 Myrs (88 stars). We have also derived the spectral energy distribution (SED) for the probable PMS-stars (having ages < 4.6 Myr, i.e., average age + standard error) using the available optical, near-infrared (2MASS) and mid-infrared (WISE) data, with a condition that they have photometric data in at least 5 broad band filters. We derived SEDs for 48 probable PMS variables (5 of them are also showing infrared excess). From the SED fitting analyses (for details, cf. [7]), the average age and mass of probable PMS-variables come out to be 2.7 ± 0.3 Myrs and 2.7 ± 0.2 $M_{\odot}$, respectively which suggests that these stars should be T Tauri stars. Disk/envelope accretion rates and masses of these PMS variable stars are also determined from the SED fitting and detailed analyses is underway.

References

[1] Appenzeller, I., & Mundt, R., A&ARv, **1**, 291 (1989)
[2] Bhatt, H., Sagar, R., & Pandey, J. C., New Astronomy, **17**, 160 (2012)
[3] Herbst, W., Herbst, D. K., Grossman, E. J., & Weinstein, D., AJ, **108**, 1906 (1994)
[4] Hillenbrand, L. A, preprint (arXiv:astro-ph/0210520) (2002)
[5] Percy, J. R., Gryc, W. K., Wong, J. C.-Y., & Herbst, W., PASP, **118**, 1390 (2006)
[6] Percy, J. R., Grynko, S., Seneviratne, R., & Herbst, W., PASP, **122**, 753 (2010)
[7] Sharma, S., Pandey, A. K., Ojha, D. K., et al., MNRAS, **467**, 2943 (2017)

Variable stars around selected open clusters in the VVV area: Young Stellar Objects

Nicolas Medina[1,2,*], *Jura* Borissova[1,2], *Amelia* Bayo[1], *Radostin* Kurtev[1,2], and *Philip* Lucas[3]

[1]*Instituto de Física y Astronomía, Universidad de Valparaíso, Av. Gran Bretaña 1111, Playa Ancha, Casilla 5030, Chile*
[2]*Millennium Institute of Astrophysics, Santiago, Chile*
[3]*Centre for Astrophysics Research, Science and Technology Research Institute, University of Hertfordshire, Hatfield AL10 9AB, UK*

Abstract. Time-varying phenomena are one of the most substantial sources of astrophysical information, and led to many fundamental discoveries in modern astronomy. We have developed an automated tool to search and analyze variable sources in the near infrared Ks band, using the data from the Vista Variables in the Vía Láctea (VVV) ESO Public Survey ([5, 8]). One of our main goals is to investigate the Young Stellar Objects (YSOs) in the Galactic star forming regions, looking for:

- Variability.
- New pre-main sequence star clusters.

Here we present the newly discovered YSOs within some selected stellar clusters in our Galaxy.

1 Methods

- We compute a set of variability indexes: the Stetson J and K indices ([9]), the η index ([10]), the chi square statistic χ^2 ([6]), the small kurtosis κ ([7]), and the ratio between the average Ks magnitude from the light curve over the standard deviation of the data.
- We use an unsupervised algorithm, which identifies patterns among the values of these indices, grouping populations of objects with similar features.
- The periods are determined with Generalized Lomb-Scargle periodograms ([11]), and the Information Potential Metric ([4]).
- The variability class was assigned via comparison with VVV Templates ([2]), the period of the variability and the position on the color-magnitude diagram (CMD).

2 Results

The VVV tiles d 001 and d 002 are centered on RA = 175.8521, DEC = –63.5298 and RA = 179.0511, DEC = –63.8750. Following our algorithm 134 variable sources were discovered, 89 of them projected

*nicomedinap@gmail.com

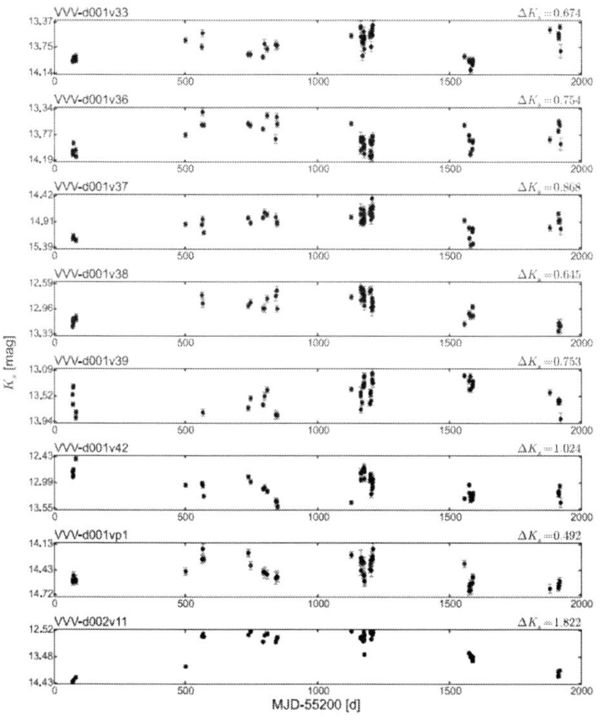

Figure 1. Light curves of the variable YSOs in the vicinities of clusters VVV CL 005, VVV CL 007, LS001, and LS002.

in the d001 area, while 45 belong to d002. Nine Young Star Clusters (YSC) are projected in the field of view of d 001 and d 002, namely VVV CL 005, 007, 008 and 009 ([5]) and La Serena 001, 002, 003, 009 and 015 ([3]). Eight variable stars were discovered in the vicinity of four of the clusters. Based on the shape of the light curves (Fig. 1) and the position on the CMD they are classified as YSO candidates.

Acknowledgments: NM, JB, RK acknowledge support by the Ministry for the Economy, Development, and Tourism's Programa Iniciativa Científica Milenio through grant IC120009, awarded to The Millennium Institute of Astrophysics (MAS). RK acknowledges Fondecyt 1130140, AB acknowledges Fondecyt 11140572.

References

[1] Angeloni, R., Contreras Ramos, R., Catelan, M., et al., A&A **567**, 100 (2014)
[2] Barbá, R. H., Roman-Lopes, A., Nilo Castellón, J. L., et al., A&A, **581**, 120 (2015)
[3] Borissova, J., Bonatto, C., Kurtev, R., et al., A&A, **532**, A131 (2011)
[4] Huijse, P., Estevez, P., Zegers, P., et al., IEEE Signal Processing Letters, **18**, 371 (2011)
[5] Minniti, D., Lucas, P. W., Emerson, J. P., et al., New Astronomy, **15**, 433 (2010)
[6] Rebull, L. M., Stauffer, J. R., Cody, A. M., et al., AJ, **150**, 175 (2015)
[7] Richards, J. W., Starr, D. L., Brink, H., et al., ApJ, **744**, 192 (2011)
[8] Saito, R. K., Hempel, M., Minniti, D., et al., A&A, **537**, 107 (2012)
[9] Stetson, P. B., PASP, **108**, 851 (1996)
[10] von Newmann, J., The Annals of Mathematical Statistics, **12**, 367 (1941)
[11] Zechmeister, M., & Kürster, M., A&A, **496**, 577 (2009)

Variable stars in the Galactic center, as revealed by the VVV Survey

Claudio Navarro Molina[1,2,*], *Jura* Borissova[1,2], *Márcio* Catelan[3,2], *Radostin* Kurtev[1,2], and *Nicolás* Medina[2]

[1] *Universidad de Valparaíso, Instituto de Física y Astronomía, Av. Gran Bretaña 1111, Playa Ancha, Casilla 5030, Valparaíso, Chile*
[2] *Millennium Institute of Astrophysics, Santiago, Chile*
[3] *Pontificia Universidad Católica de Chile, Instituto de Astrofísica, Av. Vicuña Mackenna 4860, 782-0436 Macul, Santiago, Chile*

Abstract. A variability search has been performed in the Galactic center, using the near-infrared images from the Vista Variables in the Vía Láctea (VVV) Survey. Light curves contain 89 epochs in the K_S band. A total of 353 variable stars were found, of which only 47 are already present in the literature.

1 Introduction

The Galactic center (GC) is an exceptional laboratory to test the formation and evolution of stars under extreme conditions. Three massive and young star clusters have been found in this region: the Arches, Quintuplet, and the Nuclear Cluster surrounding Sagittarius A*. The Quintuplet cluster is located at a projected distance of about 30 pc from Sgr A*. It has an age of 3.5 Myr and a mass of $\sim 10^4\ M_\odot$. On the other hand, at only 5 arcmin from Quintuplet, we find Arches. This is the youngest of the three central clusters, with an age of $\sim$ 2.5 Myr.

In order to study the stellar population of this region, we need near-infrared observations, due to the extremely high visual extinction ($A_V \sim 30$ magnitudes) along the line-of-sight. To achieve this goal, we have used the VVV ESO Public Survey ([4]).

Here we present our first results of the variability search around Arches and Quintuplet.

2 Methodology and results

VVV images are obtained with the 4.1-m VISTA telescope at Cerro Paranal, Chile. The total survey area of 562 deg^2 is divided into 348 *tiles*. For this work, we used tile *b333*, located at the GC, in particular in a region covered by a single chip, which is $11.5' \times 11.5'$ in size. Images were automatically reduced by the Cambridge Astronomical Survey Unit (CASU) pipeline ([2]). A total of 84 epochs of K_S images were used, corresponding to observations between 2010 and 2015.

*claudio.navarro@postgrado.uv.cl

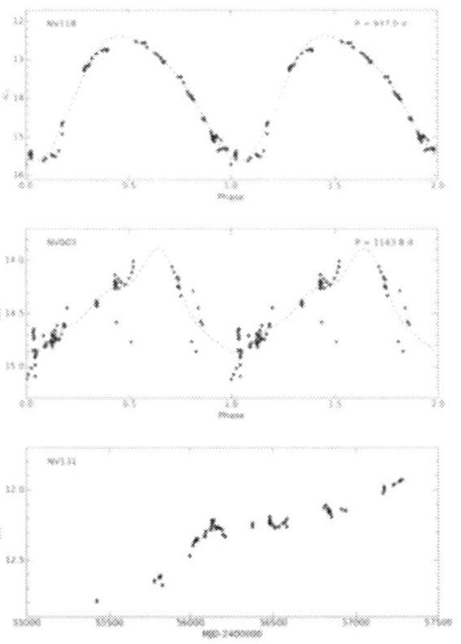

Figure 1. Typical light curves found in this work. Top panel shows a Mira light curve, with a period of 937 days. Middle panel displays a semi-regular variable with a primary period of 1143.8 days. Finally, the bottom panel shows an eruptive variable star candidate, possibly of EXor type.

In order to deal with the extreme crowding present in the images, we performed point-spread function (PSF) fitting photometry of the images. For this, we used DoPHOT ([1]). Further details about the methodology can be found in [5].

After a master catalogue was built from the 89000 sources found on the images, we performed a variability analysis using the Stetson J index ([6]), adapted to single-band time series. 654 candidates were found with this method. A visual inspection of all candidates was performed in order to remove false positives due to blending or image artifacts. Finally, we obtained 353 variable stars. In order to estimate a time interval for their variations, a generalized Lomb-Scargle periodogram was used.

Our 353 variable stars have been cross-matched with the data found in the literature, namely the Mira search carried out by [3]. 47 counterparts were found, meaning that we have discovered 306 new variables in the region. Most of the variable stars correspond to asymptotic giant branch (AGB) stars, like Miras, semi-regulars, and irregulars; in addition, some eruptive variables have been found. Examples of light curves are shown in Figure 1. From the first 50 variables that were analyzed, 4 correspond to Miras, 21 to semi-regulars, while the remaining 25 have irregular light curves. The crossmatch with [3] allowed us to have a long time baseline (over 12 years), enough to check for secular variations in their light curves, which are expected for stars in the AGB phase, both in their amplitude and/or periodicity.

References

[1] Alonso-García, J., Mateo, M., Sen, B., Banerjee, M., Catelan, M., Minniti, D., & von Braun, K., AJ, **143**, 70 (2012)

[2] Irwin, M. J., Lewis, J., Hodgkin, S., et al., in *Optimizing Scientific Return for Astronomy through Information Technologies*, ed. P. J. Quinn & A. Bridger, Procceedings of the SPIE, **5493**, 411 (2004)

[3] Matsunaga, N., Kawadu, T., Nishiyama, S., Nagayama, T., Hatano, H., Tamura, M., Glass, I. S., & Nagata, T., MNRAS, **399**, 1709 (2009)

[4] Minniti, D., Lucas, P. W., Emerson, J. P., et al., New Astron., **15**, 433 (2010)

[5] Navarro Molina, C., Borissova, J., Catelan, M., et al., MNRAS, **462**, 1180 (2016)

[6] Stetson, P. B., PASP, **108**, 851 (1996)

Variable stars in the northern Galactic plane from KISOGP

Noriyuki Matsunaga[1,*] on behalf of KISOGP team

[1] *Department of Astronomy, The University of Tokyo, 7-3-1 Hongo, Bunkyo-ku, Tokyo 113-0033, Japan*

Abstract. We have conducted a large-scale survey of the northern plane using Kiso Wide Field Camera attached to Schmidt telescope at Kiso Observatory. The KISOGP (KWFC Intensive Survey of the Galactic Plane) project has made 40–70 epoch observations in *I* band of about 320 square degrees for 5 years starting in 2012. The limiting magnitude is around 16.5 in *I*. In the data analysis so far, we detected a couple of thousands of variable stars including approximately 100 Cepheids and more than 700 Miras. Roughly 90% of them were not previously reported as variable stars, indicating that there are still many relatively bright variables to be found in the Galactic plane.

A large fraction of stars and gas is concentrated within the Galactic disk. In fact, approximately half of the baryonic mass within our Galaxy, $\sim 4 \times 10^{10}\ M_{\odot}$, is found in the disk ([1]). The stellar populations in a large space of the disk have, however, remained elusive due to the interstellar extinction. Historic surveys of variable stars, mostly done in the optical wavelengths, were far from complete (see, e.g., Fig. 1 in [2]). *Gaia* and other ongoing surveys in the optical are expected to find many new variables, while infrared surveys would be more effective for objects obscured in the disk (see our review [3] in these proceedings). We have conducted an *I*-band survey of variable stars in the northern Galactic disk because it has been less explored compared to the southern part for which OGLE[1] and VVV[2] have discovered a large number of new variables.

The instrument used for our survey is Kiso Wide Field Camera (KWFC) attached to 105-cm Schmidt telescope in Kiso Observatory, Japan ([4]). This camera has been designed for wide-field observations by taking the advantage of a large focal-plane area of the Schmidt telescope. Eight CCD chips with a total of 64 Mpixels cover a field-of-view of $2.2^{\circ} \times 2.2^{\circ}$. The distortion is negligible (at least smaller than $0.3''$) across the entire 4.8 deg^2 field. The survey is named KWFC Intensive Survey of the Galactic Plane (KISOGP) and has been conducted for 5 years starting in 2012. The main goal is to study the Galactic structure using pulsating stars as tracers and we have been also conducting follow-up observations of variable stars we find in both photometric and spectroscopic ways.

Our survey region is composed of 80 KWFC fields-of-view covering $\sim$ 320 deg^2 between 60 and 210 degrees in Galactic longitude (Fig. 1). The photometric observations are only in *I* band and cover the magnitude range of 9.5 to 16.5 magnitude in *I* with signal-to-noise ratio better than 30. For each epoch per each field, an exposure with 5 sec and three exposures with 60 sec are combined. The numbers of epochs for the 5-yr time monitoring ranges from 50 to more than 100, of which 40- or more-epoch datasets are confirmed to be of useful quality.

*matsunaga@astron.s.u-tokyo.ac.jp
[1] http://ogle.astrouw.edue.pl
[2] https://vvvsurvey.org

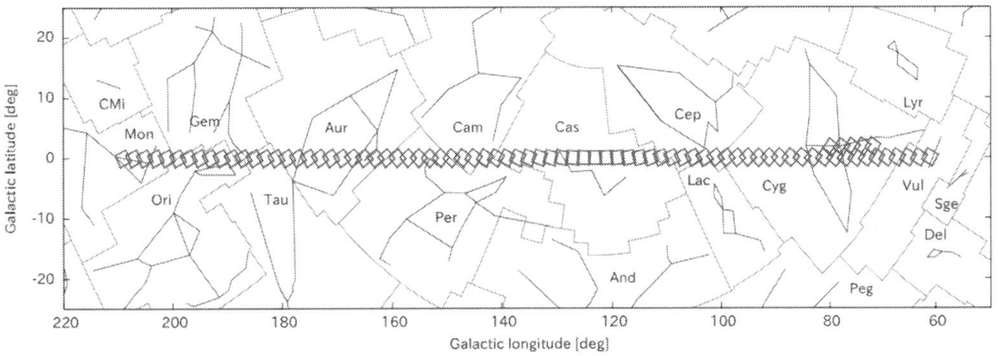

Figure 1. KISOGP survey region.

In our entire survey field, we have detected more than 5 million objects, of which about 50,000 variable star candidates are found, and data analysis is still in progress. So far we have searched for periodic variable stars with relatively large amplitudes and identified $\sim$100 Cepheids and more than 700 Miras. Roughly 90% of them were not reported as variables before. Preliminary estimates of their distances using the period-luminosity relations and 2MASS JHK_s magnitudes indicate that they are scattered across a large space of the northern Galactic disk, from a couple of kpc to more than 10 kpc from the Sun. Their distribution of Galactocentric distances covers ~8 to 15 kpc. They will be important tracers of the outer disk. There are also RR Lyrs, a large number of eclipsing binaries (> 1000), and other types of variables. While we are trying to finalize the data analysis for these variables, there are many other variables with smaller amplitudes and/or irregular variations.

In order to take advantage of the newly found Cepheids and Miras, we have been conducting several follow-up observing programs. For Cepheids, we have used the Subaru telescope to take high-resolution spectra and several telescopes including Las Cumbres Observatory's global network[3] to collect multi-color photometric data. For Miras, we have collected hundreds of low-resolution spectra to classify them into oxygen-rich and carbon-rich using the 188-cm telescope at Okayama Astrophysical Observatory[4], NAOJ, and the 2-m Nayuta telescope at Nishi-Harima Astronomical Observatory[5].

Acknowledgments: We acknowledge a long-term support for the KISOGP project from the staff in Kiso Observatory. NM appreciates financial support from the Japan Society for the Promotion of Science (JSPS) through the Grant-in-Aid, No. 26287028.

References

[1] Bland-Hawthorn, J., & Gerhard, O., ARA&A, **54**, 529 (2016)
[2] Matsunaga, N., Journal of Physics: Conference Series, **372**, 12026 (2012)
[3] Matsunaga, N., these proceedings (2017)
[4] Sako, S., Aoki, T., Doi, M., et al., Proc. SPIE, **8446**, 6L (2012)

[3] https://lco.global
[4] http://www.oao.nao.ac.jp/en/
[5] http://www.nhao.jp/en/

The Large Synoptic Survey Telescope: A summary update on the scientific potential for pulsating star research

Melissa L. Graham[1,*]

[1] *Department of Astronomy, University of Washington, Box 351580, U.W., Seattle WA 98195, USA*

Abstract. The Large Synoptic Survey Telescope (LSST) will provide hundreds of deep images of the southern sky over its 10 year duration, enabling variability studies for an unprecedentedly large and unbiased population of objects. In this proceeding paper I will cover the aspects of the LSST's survey and data products that are most relevant to the study of stellar pulsations (Sect. 1), and provide a directory of pertinent materials for further information. I will also summarize the anticipated variable star sample sizes from the LSST, and highlight recent research from several members of the scientific community which evaluates the scientific potential of the LSST's data products with respect to pulsating stars (Sect. 2).

1 The Large Synoptic Survey Telescope: An overview

The Large Synoptic Survey Telescope (LSST) is currently under construction on Cerro Pachón in Chile. With an 8.4 meter primary mirror, a 3.5 degree radius field of view, and a 3.2 Gigapixel camera of 0.2 arcsecond pixels, the LSST will image the full sky area of 18000 square degrees hundreds of times over its 10 year main "wide-fast-deep" survey (see Table 1 for a link to the project website). These repeated visits allow for the discovery and monitoring of moving objects such as stars with proper motion, and transient and recurrent variables such as pulsating stars. Deep stacks of all images will enable the detection and characterization of faint objects, and the wide unbiased survey area will provide sample populations of unprecedented number. A description of the full range of science drivers and how they have influenced the project's design specifications for the telescope and survey can be found in Ivezić et al. ([1]), and a breakdown of these specifications can be found in the LSST Science Requirements Document ([2]). The LSST Science Book ([3]), prepared by the scientific community, covers many of the anticipated goals in observational astronomy that could be accomplished by the LSST. The scientific community has also collectively produced a comprehensive white paper that analyzes observing strategies for the wide-fast-deep survey and makes recommendations for optimizing a variety of science goals ([4]). Aside from the wide-fast-deep survey, the LSST plans to incorporate "deep drilling fields" (DDF) and "mini-surveys" to attain additional science goals. White papers from the scientific community proposing DDFs were generated in 2011 and four extragalactic fields were chosen (see Table 1 for a link to DDF information). A call for additional fields, cadence proposals for the selected DDFs, and mini-surveys is expected to be released in the next year.

*`mlg3k@uw.edu`

Table 1. Useful links to information about the LSST project and associated science initiatives.

LSST Resource Description	**Weblink** URL
LSST Project Website	http://project.lsst.org
Science Drivers to Reference Design [1]	http://arxiv.org/abs/0805.2366
The LSST Data Management System [5]	http://dm.lsst.org/
	http://arxiv.org/abs/1512.07914
LSST Science Requirements (LPM-17, [2])	http://ls.st/LPM-17
LSST Data Products Definitions (LSE-163, [6])	http://ls.st/LSE-163
LSST DM Applications Design (LDM-151, [7])	http://ls.st/LDM-151
LSST Github Repositories	http://github.com/lsst
LSST Software User Guide (The Stack)	http://pipelines.lsst.io
LSST Community Forums	http://community.lsst.org
LSST Science Collaborations	http://lsstcorporation.org/science-collaborations
The LSST Science Book [3]	http://arxiv.org/abs/0912.0201
	http://lsst.org/scientists/scibook
White Paper: Observing Strategy [4]	http://github.com/LSSTScienceCollaborations/ObservingStrategy
White Papers: Deep Drilling Fields	http://project.lsst.org/content/whitepapers32012
LSST Data Science Fellowship Program	http://ciera.northwestern.edu/Education/LSSTC_DSFPOverview.php
Review: Maximizing Science in the Era of LSST	http://noao.edu/meetings/lsst-oir-study

An overview of the LSST data management system can be found in Jurić et al. ([5]). There are two main data processing pipelines: the Alert Production (AP) and the Data Release Production (DRP). The AP will run in real time, generating and searching difference images and releasing alerts on variable objects within 60 seconds of each observation. Processed images and catalogs from the DRP will be released to the community on an approximately yearly basis. Additional user-defined data processing for science goals not achievable with these products will be developed through the LSST Science Platform. All of the code for processing data from the LSST – collectively referred to as "The Stack" – is open-source and available online through GitHub (see Table 1). Documentation and user guides are included in the hope that the LSST end-users will find the Stack helpful when developing their own added-value processing and analysis through the LSST Science Platform.

In-depth information about the products that will be provided by the LSST data management system, and the processing methods that will be used to generate them, are described in the LSST Data Products Definitions Document ([6]) and the LSST DM Applications Design Document ([7]), respectively. The data products that will be of particular interest to the stellar pulsations community include:

- **Co-added Image Photometry –** Sources will be detected, deblended, measured, and calibrated for all coadd images; these measurements include aperture fluxes. Types of co-adds will include "deep" and "best-seeing", unless they're shown to be one and the same, and "short-period" co-adds, e.g., over one to several years, in order to identify faint long-period variables.
- **Single Visit Photometry –** Sources will be detected, deblended, measured, and calibrated for all single visit images and difference images[1]. Forced photometry on single visit and difference images will be performed at the location of all difference-image sources detected (for AP, this is limited to detections in the previous 12 months; for DRP, there is no limit). All newly-detected difference image sources will have forced photometry generated from the past 30 days of difference images.
- **Variability Characterization –** The exact quantities are to be determined, but the object catalogs will contain commonly-used values that represent variability, such as periodic features extracted from the light-curve based on generalized Lomb-Scargle (similar to, e.g., Table 4 of [8]).

[1]A difference image is a single visit image subtracted from a previously-generated template image.

The author welcomes correspondence with the scientific community and future users regarding additional catalog or image products that would enable science projects with LSST data[2], and/or questions about the current plans for the LSST data management system and its products.

2 The LSST and pulsating star research

Preparations for science with the LSST are occurring mainly within the established Science Collaborations (SC). The two SC most relevant to pulsating star research are Transients and Variable Stars[3] and Stars, Milky Way, and Local Volume[4]. The SC work together through telephone and in-person meetings, "hack-a-thons", and online community forums, and collaborative efforts such as writing the white paper on Observing Strategy, building science roadmaps, and running data challenges are also underway (see Table 1 for links to the forums and the white paper). Eligible members of the scientific community are welcome to submit applications for SC membership (see websites for details; membership requirements vary by SC). Aside from the SC, reviews such as "Maximizing Science in the Era of LSST: A Community-based Study of Needed US OIR Capabilities" are assessing whether the current and planned resources will be able to support LSST-related science, and a fellowship program for students to learn data science techniques that will be applicable to the LSST data products has been established (web links to the review and the fellowship program are in Table 1).

The Science Book ([3]), especially chapters 6 (Stellar Populations), 7 (Milky Way and Local Volume Structure), and 8 (The Transient and Variable Universe), detail how the LSST will advance pulsating star research by providing a larger sample at fainter magnitudes and longer timescales. The predicted yields are $\sim 135 \times 10^6$ variable stars, of which $\sim 59 \times 10^6$ are pulsators ([3]). The Science Book describes how the main science goals to be enabled by these populations include:

- **RR Lyrae:** The LSST will provide $\sim 10^5$ stars in the Magellanic Clouds with well mapped stellar populations for studying relations in brightness, period, age, and metallicity.
- **Cepheid Variables:** The larger samples provided by the LSST can address intrinsic variation, improve calibration, and provide distances for more Type Ia supernova host galaxies for cosmological studies.
- **White Dwarfs:** The LSST's photometric precision can identify new candidate pulsating white dwarfs and provide the ability to find and map (new) instability strips. The LSST's light curve characterizations can constrain, e.g., core composition and physics.
- **Asymptotic Giant Branch Long-Period Variables:** The large sample from LSST can map out the structure of the Milky Way and nearby galaxies in stellar mass, age, and dust production. The LSST's light curve characterization can constrain internal physics and nucleosynthesis.

Also relevant to RR Lyrae science, Sesar et al. ([9]) show that as with the Sloan Digital Sky Survey (SDSS) data, the multi-band aspect of the LSST survey will improve the capability for photometric classification of RR Lyrae stars, which can be difficult to distinguish from, e.g., eclipsing binaries with single-filter observations.

The Observing Strategy white paper on the wide-fast-deep survey has several sections that are relevant to the LSST science capability for pulsating variable stars. The following paragraph is not a complete summary of all pulsating-star science addressed in the Observing Strategy white paper, but intended to highlight and exemplify the type of analysis being done. Section 5.2.1 describes

[2]E.g., see this call for science community feedback on DDF processing and data products: `https://community.lsst.org/t/deep-drilling-fields-and-data-management/1115/1`.
[3]`https://tvs.science.lsst.org/`
[4]`https://milkyway.science.lsst.org/`

metrics for optimizing LSST observations for Cepheid star science, which are also presented in Lund et al. ([10]). Two example metrics are: (1) *periodogram purity function metric*, which represents the fraction of power leaked away from the true frequency due to, e.g., sampling cadence and photometric precision, and (2) the *field star counts metric*, which evaluates the number of LSST filters in which a star would be detected, as a function of stellar mass and distance, for a given observing strategy. This section concludes that the Cepheid stars will be well characterized by the default cadence, but that a more rapid cadence would result in faster discovery and classification, enabling more science in the early years of the survey. Another example relevant to pulsating star research is Section 5.3, written by Keaton Bell, which presents the *VarDepth* metric: the magnitude limit for which 90% of pulsators with a 1% root-mean-square variability will be detected (i.e., a ZZ Ceti star). This work shows that < 100 ZZ Cetis would be discovered and characterized in u-band with the default observing strategy, lower than the sample size of 1000 that would have significant scientific impact. See also the work of Bell et al. ([11]) for a prediction of the number of outbursting cool white dwarf pulsating stars that will be found with LSST. Section 5.5 references the work of Oluseyi et al. ([12]), who predict that at 10 years, the LSST will recover 50% of the RR Lyrae stars at 600 kpc, which is sufficient for the LSST science goals of finding Local Group halo streams and dwarf satellites, estimating their metallicities and placing constraints on formation scenarios. Two of the main conclusions from these and other studies included in the Observing Strategy white paper are that there are two types of cadences that are detrimental to the discovery and period recovery of pulsating stars: exactly uniform visit spacing and extremely non-uniform visit spacing (i.e., concentrating all visits of a field into a few years). As neither of those are likely to happen, the field of periodic variable stars appears poised to meet its LSST science goals for the wide-fast-deep main survey.

As a final note, the recent publication by VanderPlas & Ivezić ([13]), titled "Periodograms for Multi-band Astronomical Time Series", builds on the RR Lyrae predictions of Oluseyi et al. ([12]) with a new technique that models the light curve as a base function common to all bands, plus a residual for each filter. VanderPlas & Ivezić show that their new method outperforms other options, especially for the faintest periodic variable stars: at 5 years, the LSST can reach 100% completeness for RR Lyrae period measurements down to $g = 24.5$ magnitudes, and at 1 year, the LSST can be 50% complete for RR Lyrae to $g \sim 23.8$ magnitudes ([13]). The latter is a remarkable increase in limiting magnitude by > 1.5 magnitudes over previous estimates, increasing the 50% completeness distance from 100 to ~ 250 kpc, so far into new territory in the Milky Way's halo that the expected number counts are unknown.

3 Summary

At the time of writing, both the hardware and software for the LSST are under active construction, and the processes and products of the LSST data management system are becoming solidly defined. This proceedings document has outlined some of the anticipated data products that will be most relevant to the stellar pulsations community. All readers are welcome to contact the author with feedback regarding additional products that would enable science (e.g., light-curve variability parameters, specialized co-adds). This document has also highlighted the significant contributions made by members of the scientific community towards optimizing the LSST's main survey strategy and maximizing science return in the field of pulsating stars. In particular, recently developed analysis tools show a promising ability to enhance the science capability of LSST data products after just one year of the survey, building larger samples of pulsating stars and unlocking LSST science goals earlier than expected.

Acknowledgments: This material is based upon work supported in part by the National Science Foundation through Cooperative Agreement 1258333 managed by the Association of Universities for Research in Astronomy (AURA), and the Department of Energy under Contract No. DE-AC02-76SF00515 with the SLAC National Accelerator Laboratory. Additional LSST funding comes from private donations, grants to universities, and in-kind support from LSSTC Institutional Members.

I thank Željko Ivezić, Keaton Bell, and Paula Szkody for helpful conversations and for their contributions to the content of my oral presentation at the meeting, and I thank the conference organizing committee for extending the invitation to speak and for covering my accommodations and registration. I very much appreciate the many good conversations I had with other attendees throughout the meeting, and I'd like to extend a special thanks to the local organizers for arranging the unique opportunity to visit the ALMA site.

References

[1] Ž. Ivezić, J.A. Tyson, B. Abel, E. Acosta, R. Allsman, Y. AlSayyad, S.F. Anderson, J. Andrew, R. Angel, G. Angeli et al., ArXiv e-prints (2008), 0805.2366
[2] Ž. Ivezić et al., *LSST Science Requirements Document*, LSST Project Management LPM-17 (2011), http://ls.st/srd
[3] LSST Science Collaboration, P.A. Abell, J. Allison, S.F. Anderson, J.R. Andrew, J.R.P. Angel, L. Armus, D. Arnett, S.J. Asztalos, T.S. Axelrod et al., ArXiv e-prints (2009), 0912.0201
[4] P. Marshall et al., *Science-Driven Optimization of the LSST Observing Strategy*, in preparation (2017), https://github.com/LSSTScienceCollaborations/ObservingStrategy
[5] M. Jurić, J. Kantor, K. Lim, R.H. Lupton, G. Dubois-Felsmann, T. Jenness, T.S. Axelrod, J. Aleksić, R.A. Allsman, Y. AlSayyad et al., ArXiv e-prints (2015), 1512.07914
[6] M. Jurić et al., *LSST Data Products Definition Document*, LSST Data Management LSE-163 (2016), http://ls.st/LSE-163
[7] M. Jurić et al., *LSST Data Management Applications Design*, LSST Data Management LDM-151 (2013), http://ls.st/LDM-151
[8] J.W. Richards, D.L. Starr, N.R. Butler, J.S. Bloom, J.M. Brewer, A. Crellin-Quick, J. Higgins, R. Kennedy, M. Rischard, ApJ **733**, 10 (2011)
[9] B. Sesar, Ž. Ivezić, S.H. Grammer, D.P. Morgan, A.C. Becker, M. Jurić, N. De Lee, J. Annis, T.C. Beers, X. Fan et al., ApJ **708**, 717 (2010)
[10] M.B. Lund, R.J. Siverd, J.A. Pepper, K.G. Stassun, PASP **128**, 025002 (2016)
[11] K.J. Bell, J.J. Hermes, M.H. Montgomery, D.E. Winget, N.P. Gentile Fusillo, R. Raddi, B.T. Gänsicke, *The First Six Outbursting Cool DA White Dwarf Pulsators*, in *Astronomical Society of the Pacific Conference Series*, edited by P.E. Tremblay, B. Gaensicke, T. Marsh (2017), Vol. 509 of *Astronomical Society of the Pacific Conference Series*, p. 303
[12] H.M. Oluseyi, A.C. Becker, C. Culliton, M. Furqan, K.L. Hoadley, P. Regencia, A.J. Wells, Ž. Ivezic, R.L. Jones, K.S. Krughoff et al., AJ **144**, 9 (2012)
[13] J.T. VanderPlas, Ž. Ivezić, ApJ **812**, 18 (2015)

Pulsating star research and the Gaia revolution

Laurent Eyer[1,*], *Gisella* Clementini[2], *Leanne P.* Guy[3], *Lorenzo* Rimoldini[3], *Florian* Glass[1], *Marc* Audard[1,3], *Berry* Holl[1,3], *Jonathan* Charnas[3], *Jan* Cuypers**[4], *Joris* De Ridder[5], *Dafydd W.* Evans[6], *Gregory* Jevardat de Fombelle[3,7], *Alessandro* Lanzafame[8,9], *Isabelle* Lecoeur-Taibi[3], *Nami* Mowlavi[1,3], *Krzysztof* Nienartowicz[3,7], *Marco* Riello[6], *Vincenzo* Ripepi[10], *Luis* Sarro[11], and *Maria* Süveges[3,12]

[1] *Department of Astronomy, University of Geneva, Ch. des Maillettes 51, 1290 Versoix, Switzerland*
[2] *INAF - Osservatorio Astronomico di Bologna, Via Gobetti 93/3, 40129 Bologna, Italy*
[3] *Department of Astronomy, University of Geneva, Ch. d'Ecogia 16, 1290 Versoix, Switzerland*
[4] *Royal Observatory of Belgium, Ringlaan 3, 1180 Brussels, Belgium*
[5] *Institute of Astronomy, KU Leuven, Celestijnenlaan 200D, 3001 Leuven, Belgium*
[6] *Institute of Astronomy, University of Cambridge, Madingley Road, Cambridge CB3 0HA, UK*
[7] *SixSq, Rue du Bois-du-Lan 8, 1217 Geneva, Switzerland*
[8] *Universita di Catania, Dipartimento di Fisica e Astronomia, Sezione Astrofisica, Via S. Sofia 78, 95123 Catania, Italy*
[9] *INAF-Osservatorio Astrofisico di Catania, Via S. Sofia 78, 95123 Catania, Italy*
[10] *INAF-Osservatorio Astronomico di Capodimonte, Via Moiariello 16, 80131 Napoli, Italy*
[11] *Dpto. Inteligencia Artificial, UNED, c/ Juan del Rosal 16, 28040 Madrid, Spain*
[12] *Max Planck Institute for Astronomy, Koenigstuhl 17, 69117 Heidelberg, Germany*

Abstract. In this article we present an overview of the ESA Gaia mission and of the unprecedented impact that Gaia will have on the field of variable star research. We summarise the contents and impact of the first Gaia data release on the description of variability phenomena, with particular emphasis on pulsating star research. The Tycho-Gaia astrometric solution, although limited to 2.1 million stars, has been used in many studies related to pulsating stars. Furthermore a set of 3,194 Cepheids and RR Lyrae stars with their times series have been released. Finally we present the plans for the ongoing study of variable phenomena with Gaia and highlight some of the possible impacts of the second data release on variable, and specifically, pulsating stars.

1 Introduction

The variable phenomena of celestial objects present very diverse behaviour. An attempt to visualise this zoo of variability types was presented in [1] and shows that pulsating stars form an abundant group.

Gaia, as a global and comprehensive survey, will have an unprecedented impact on the field of pulsating star research and more generally on variable star research. Gaia will be able to detect

*`Laurent.Eyer@unige.ch`
**Deceased on February 28, 2017.

many/most of the variability types presented in the variability tree of [1]. Furthermore, this mission will help to characterise most if not all stars that are members of a variability type presented in this tree.

More concretely, we can enumerate some examples of the foreseen impact of Gaia:

1. The discovery of an enormous number of new variable stars,
2. The definition of the precise location of instability strips in the Colour Magnitude Diagram (CMD),
3. The calibration of standard candles,
4. The comparison of results with those obtained from asteroseismology,
5. The description of structures of the Milky Way and other galaxies with variable stars used as tracers of certain galactic populations.

Gaia's three instruments observe objects near-simultaneously to collect astrometric, photometric and spectroscopic data, covering the entire sky. This multi-messenger aspect of Gaia is a strong asset for the study of variability as compared with single band surveys. We can remark that currently the number of stars known with multiple pieces of information (good astrometric precision, photometric and radial velocity time series) is very low. There are only 30 pulsating stars known with a relative parallax precision of better than 10% (from Hipparcos, [2]), Hp-band photometric and Coravel ([3]) radial velocity time-series. It is clear that Gaia will increase these figures by several orders of magnitude.

Gaia will be, for example, particularly interesting when we find pulsating a star in a binary system that is a member of a cluster. Gaia may be able to characterise the "different" objects, i.e. the cluster, the binary system, the star and the pulsation. The comparison of these different and independent pieces of information can provide stringent tests for our understanding of pulsation and stellar evolution.

Obviously Gaia won't solve all problems; there are limitations on the precision of measurements and also in the density of the time sampling. The combination of Gaia with other surveys, with other types of measurements (e.g., high cadence photometry, high resolution spectroscopy, interferometry) and other methods of determining distance will be another way to significantly increase the scientific impact of Gaia on pulsating star research.

Within the Gaia consortium, a dedicated group is charged with the variability processing and analysis, see [4]. This analysis is mostly based on the work of the group responsible for the photometric reduction ([5]). The results of this variability processing were and will be made public across official ESA data releases. In the two following sections we explain the content of the first and second data releases.

2 Gaia Data Release 1

2.1 The contents of Gaia Data Release 1

The contents of the first Gaia data release were based on the accumulation of 14 months of Gaia observations, collected between 25 July 2014 and 16 September 2015. During these 14 months, two different scanning laws were employed. For an initial period of 28 days, the Ecliptic Poles Scanning Law (EPSL, which repeatedly covered the ecliptic poles on every spin) was used, producing high-cadence measurements in these regions. During the subsequent 13 months the Nominal Scanning Law (NSL) was used.

This 14-month data set comprises more than half a trillion observations that were processed and used to derive the following contents of this first Gaia public release:

- Positions and mean G-band photometry for 1.1 billion stars,
- Parallaxes and proper motions for 2.1 million stars,
- Light curves of 3,194 Cepheids and RR Lyrae stars,
- Positions and magnitudes of 2,191 quasars.

The Gaia consortium produced comprehensive documentation to accompany the first data release, which can be found on the ESA website.[1] A series of articles were published in the journal Astronomy & Astrophysics. The mission is described in [6] and the results of the first release are summarised in [7]. For the topics related to variable stars and stellar pulsation, the interested reader can find specific information in the following three articles:

1. On the general approach of the variability analysis developed by the Gaia consortium and its application to G-band time series, [4]. Note that this analysis, aimed at extracting a list of Cepheid and RR Lyrae candidates, was restricted to objects within 38 deg of the South Ecliptic Pole.

2. On the specific processing applied to the Cepheid and RR Lyrae candidates, to derive parameters from a Fourier decomposition of the light curve, [8].

3. On the period luminosity relation of Cepheids and RR Lyrae stars ([9]) derived from the Tycho-Gaia Astrometric Solution (TGAS), [10].

For a mission of this breadth and complexity, to have produced a first data release of this quality only 2 years following the beginning of spacecraft operations is a real achievement. Obviously, there are some limitations, for which the reader should refer to the official website for the first Gaia data release.[2] We list here a few of the limitations which could be pertinent to pulsating star research:

- Some filtering has been applied, e.g. cuts on the number of observations, cuts on colour, cuts on objects with "too large" uncertainties, cuts on sources lacking astrometry and/or photometry, cuts on sources with an upper limit on errors in parallax, position and photometry, etc.
- There are no high-proper-motion stars (i.e. with $\mu > 3.5$ arcsec/year),
- All sources are treated as single stars (so the parallax of some astrometric binary stars may obviously be erroneous),
- Various un-modelled effects are present in the data (chromaticity, micro-meteoroid hits, micro-clanks, ...),
- Variations of the basic angle were derived from the on-board metrology,
- There are some spurious sources.

One consequence of these limitations is that your favorite source may be missing from the data release. It is important to bear in mind that the processing of the Gaia data by the DPAC consortium is iterative; with each subsequent data release the quality of the data improves as additional effects are taken into account or effects are better modelled.

[1] `https://gaia.esac.esa.int/documentation/GDR1/`
[2] `http://www.cosmos.esa.int/web/gaia/dr1`

2.2 The processing and analysis of Cepheids and RR Lyrae stars

The classification selection of RR Lyrae stars and Cepheids was done using supervised classification algorithms ([4]). These classifiers were expressly tuned to be permissive. We remark that for the South Ecliptic Pole region, the completeness is very high. A comparison with the OGLE survey on RR Lyrae stars and Cepheids showed that all stars observed by the OGLE survey have also been observed by Gaia. Following the application of various selection and quality criteria, including *(i)* a cut on the number of measurements to be greater than 20, *(ii)* a requirement that the star be processed in two different reduction cycles, *(iii)* the removal of stars for which an unambiguous definition of the period and a robust classification as Cepheid or RR Lyrae was not possible, and *(iv)* the removal of multimode stars, we arrive at a completeness of 58% for RR Lyrae stars and 67% for Cepheids. These numbers are similar to those found by [11].

The inclusion of Cepheids and RR Lyrae stars in the first Gaia data release should be considered a bonus as it was originally not scheduled until much later in the DPAC processing cycle. Our primary goal was to showcase a selection of stars somewhat similar (in terms of the number of measurements) to what we expect in the final data release.

The quality of the photometry allows a more detailed description of the light curve. An approach to describe light curve morphology was done through a Fourier series and was published in the first data release, see [8]. We can remark on the quality of the photometry when we present for example the Hertzsprung progression ([12]). This phenomenon, the period-dependence of the location of a bump in certain Cepheid light curves, is shown in Figure 1.

2.3 The structure of Magellanic Clouds thanks to pulsating stars

It is well known that some pulsating stars can be used as standard candles and tracers of certain galactic populations. Among them are the Cepheids, RR Lyrae stars and Mira stars.

From the first Gaia data release, it could be remarked that the new Large Magellanic Cloud RR Lyrae stars discovered in the South Ecliptic Pole region seem to trace a more extended halo than previously thought ([8]).

The OGLE team produced 2 articles about the structure of the Magellanic Clouds ([13, 14]). The second article was contemporaneous to a study made by [15] which used the mean photometry of the first Gaia data release to extract RR Lyrae star candidates in the neighbourhood of the Magellanic Clouds. [15] made comparisons between the different studies and remark that a bridge of such stars exists between the Large and Small Magellanic Clouds and conclude that it is a stellar tidal tail. Using a similar method with Gaia data but using Mira stars instead of RR Lyrae stars, [16] are unveiling the outskirts of the Large Magellanic Cloud. Although based on a small fraction of the Gaia data, such studies demonstrate the power of using certain groups of pulsating stars to describe the structures of two of our neighbouring galaxies.

2.4 Tests of distances and parallaxes

There have been several comparisons between the Gaia parallaxes and distances obtained from pulsating/variable stars.

For example, Cepheids were used by [10], to select far away objects to test the zero point of the Gaia parallaxes and the north-south asymmetry. Both are found to be unsignificant. A series of comparisons was performed by [9] with Hipparcos and HST parallaxes. Furthermore, Clementini et al. (this volume and [9]) derived new period-luminosity relations. Alternatively [17] used 212 Galactic

DOI: 10.1051/epjconf/201715202002

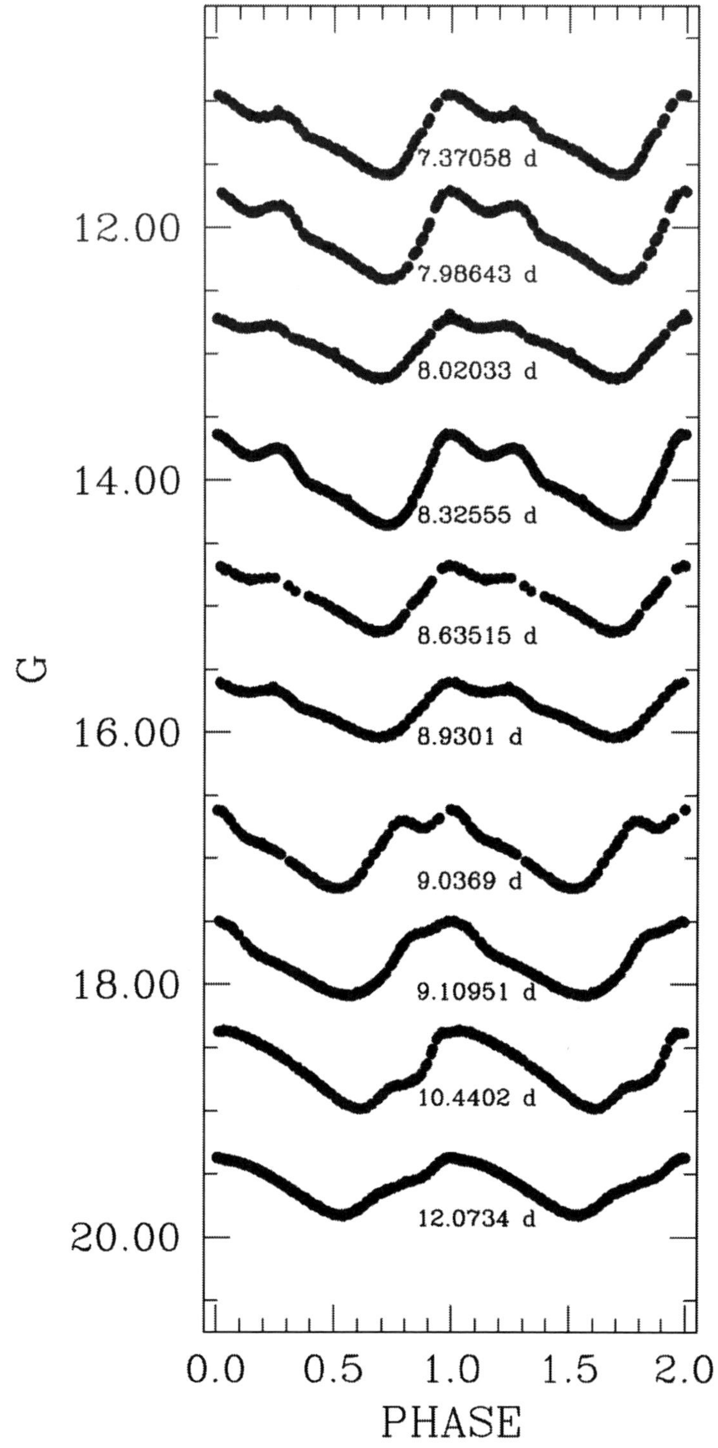

Figure 1. The Hertzsprung progression of classical Cepheids with Gaia data.

Cepheids to test the Gaia parallaxes. They conclude that the Gaia parallax uncertainties seem to be overstimated.

A series of articles used asteroseismic results to derive distances that can be compared to Gaia parallaxes. Here come some statistical difficulties to compare on one side parallaxes and distances, especially when $\mid \frac{\sigma_\pi}{\pi} \mid$ becomes large. It should be noted that such comparison, when concordant, validates both the asteroseismic distance and parallax determinations. Such studies can be found in the articles [18–20]. When statistically significantly different, a detailed and thorough study should be done to understand the discrepancy. Currently the second data release will be much more precise than the TGAS parallaxes, so that we are eagerly awaiting it to understand some of these discrepancies.

Variable stars other than pulsating stars can be used as tests of the Gaia parallaxes. For example, [21] used eclipsing binaries.

2.5 Position of variable stars in the CMD

We used the TGAS parallaxes ([10]) as well as the $B_T - V_T$ Tycho colours ([22]) to construct a CMD. We selected stars that have $\mid \frac{\sigma_\pi}{\pi} \mid < 20\%$ and with a standard deviation on the colour of less than 0.1 mag. We took care not to exclude some large amplitude variables that could have large colour variations and would therefore inflate the value of the standard deviation. We used the AAVSO catalogue of variable stars ([23]) to cross-match with the Tycho identifiers.

The quality of the AAVSO catalogue, being a compilation of many heterogeneous sources, can vary considerably. Many cross-match problems and incorrect class assignments resulting from single-band photometric surveys, especially near the Galactic plane, were identified and removed from the used data. We present the resulting CMD in Figure 2. Such a diagram, when fully cleaned using data from future Gaia data releases, will be most useful in determining regions of variability of certain variability types and can be used to compare theoretical predictions of the instability strips. Working on this subject, with the encountered problems, we can state that we eagerly await future Gaia data releases, which will reduce many of the problems we have had to face here. Indeed Gaia has the advantage of having one set of instruments observing the whole sky and therefore producing an homogeneous and coherent data set.

2.6 Calibration of luminosities of standard candles

Gaia will play a crucial role in the calibration of different kinds of standard candles.

For the first Gaia data release a study was done by Clementini et al. ([9], see also this volume). Given the precision of the TGAS astrometry, the best improvement was seen for RR Lyrae star luminosity calibrations. Clementini et al. also showed the differences between different statistical methods used to calibrate the RR Lyae stars, Classical Cepheids and Type II Cepheids. The three different methods are: a direct least squares applied to the log(period)-magnitude diagram, the ABL method ([24]) and a Bayesian approach. The first method is known to introduce biases in the estimates of luminosity calibrations when $\mid \sigma_\pi/\pi \mid$ is large. Biased estimators can occur because of many different reasons. One way to get a biased estimator is to estimate a quantity from censored samples, here for example due to the removal of sources with negative measured parallaxes, for which no distance or absolute magnitude can be inferred. Another way to get a biased estimator is when uncertainties have distributions which differ (asymmetric and/or non Gaussian distribution) from the desired ones.

Clementini et al. (this volume and [9]) are showing that the direct “naive” least squares method gives a distance shorter than more elaborated methods.

We can predict that the second Gaia data release will have a profound impact on any of the pulsating stars which are standard candles (LPVs, OSARGs, Cepheids, RR Lyrae stars, etc.).

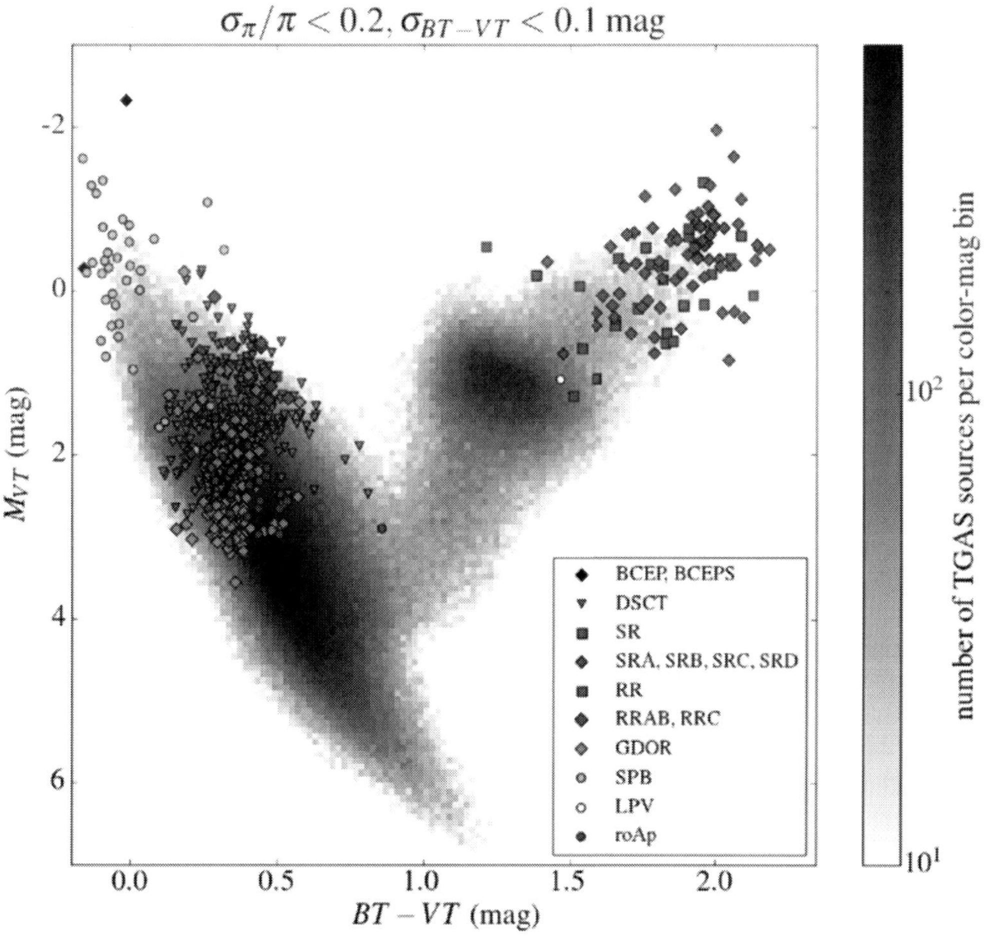

Figure 2. CMD showing the location of several pulsating stars.

3 Future Data Releases

In 2012, a release scenario was presented in a Gaia Consortium internal document by Wil O'Mullane. In that scenario five data releases were planned, and were presented for example in [25]. After the success of the first data release, and its associated knowledge, the long term planning of the future releases for the Gaia archive is currently under revision. If the number of releases is under discussion, the goal for the content of the final release is still a target.

The second Gaia Data Release is planned for April 2018. This new date, compared to the initial planned date, was driven by the desire of the Gaia consortium to have a self-consistent data release in which every delivered component makes use of the same reduction cycle data. The content of this second Data Release is described in `https://www.cosmos.esa.int/web/gaia/release` and reproduced below:

- Five-parameter astrometric solutions for all sources with acceptable formal standard errors (> 10^9 anticipated), and positions (α, δ) for sources for which parallaxes and proper motions cannot be derived.
- G and integrated BP and RP photometric mean fluxes and magnitudes for all sources.
- Median radial velocities for sources brighter than $G_{\rm RVS}$ = 12 mag.
- For stars brighter than G = 17 mag estimates of the effective temperature and, where possible, line-of-sight extinction will be provided, based on the above photometric data.
- Photometric epoch data for a sample of variable stars.
- Epoch astrometry for a pre-selected list of > 10,000 asteroids.

The actual data release content may, however, deviate from the above target content.

3.1 Variability outputs

The Gaia variability team has set ambitious goals for the upcoming data releases. First, we aim to increase significantly the number of released objects compared to the first Data Release, up to two order of magnitudes and to increase the complexity of the data processing. For the second data release, we aim to use the Gaia astrometric data and the BP and RP integrated photometry, and to release the G, BP, and RP photometric time series. Several work packages, corresponding to distinct variability types, are activated for the second data release.

Apart from identifying different types of variability in Gaia data, one major goal is to identify at least RR Lyrae stars over the entire celestial sphere. Using data from the 14-month solution, we can achieve total coverage with less than 10 measurements.

3.2 Variability types as tracers of galactic structure

Pulsating stars can be, and have been used to study galactic structures (see Sect. 2.3). Figure 3 shows a colour-magnitude diagram of 71,464 known fundamental mode ("ab" Bailey type) RR Lyrae stars.

These known RR Lyrae stars are coming from different surveys: OGLE ([26, 27]), the Catalina survey ([28–31]) and the LINEAR survey ([32]). We cross-matched these catalogues with the Gaia catalogue, allowing us to obtain an homogeneous set of magnitudes and colours from G, BP, and RP photometry.

This diagram of apparent magnitude versus colour reveals very interesting structures. We discern in the diagram some over-densities, sub-structures. It would be most interesting to see if these over-densities are due to streams or are spatially clumped. We can see that:

- The black and brown points (Catalina and LINEAR surveys) produce a vertical structure that traces the Galactic halo.
- The diagonal structure in red is caused by interstellar extinction, which attenuates and reddens the light of (RR Lyrae) stars in the direction of the Bulge. These stars are also affected by metallicity, expected to increase the spread of apparent magnitudes by about 0.2 – 0.4 mag ([33]). The small clump in red (with G ~ 18 mag) includes primarily RR Lyrae stars from the Sagittarius dwarf spheroidal galaxy at a heliocentric distance of about 25 kpc.
- The blue and green points correspond to RR Lyrae stars in the Large and Small Magellanic Clouds. Notice that the Large Magellanic Cloud is closer than the Small Magellanic Cloud, with a difference in distance modulus of about 0.5 mag, and both galaxies show a larger spread in colour indices (partially due to the photon noise).

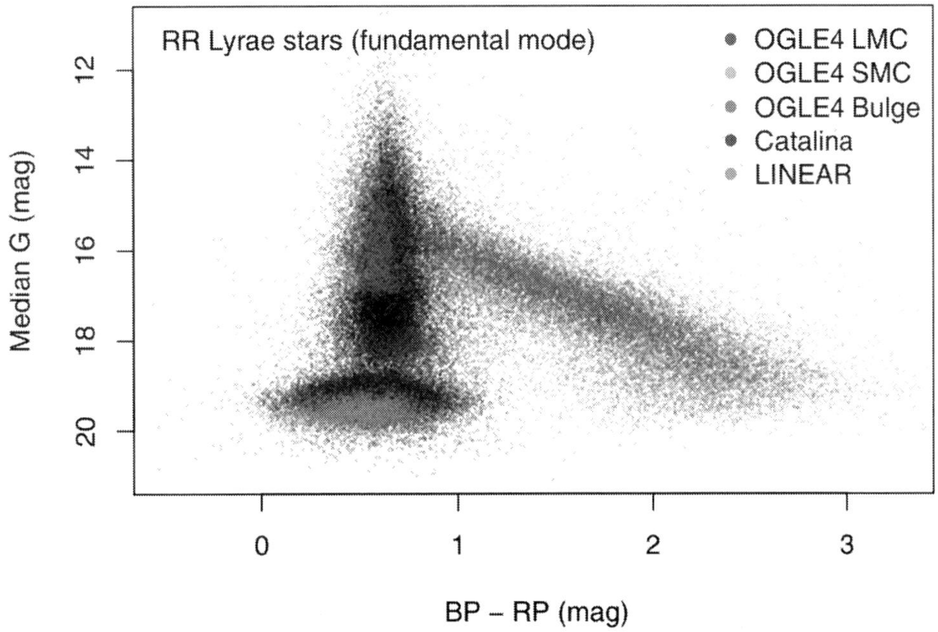

Figure 3. Colour-(apparent) magnitude diagram, in Gaia pass-bands of known fundamental mode (ab-type) RR Lyrae stars. Magnitudes and colour are not corrected for extinction (from the Gaia Image of the Week of 24/03/2017).

In Figure 4, we present the sky distribution of the cross-matched RR Lyrae stars, colour-coded with apparent magnitude. We see clearly the Magellanic clouds, and the Sagittarius stream, and also the clump of distant and RR Lyrae stars below the bulge at Galactic latitude −14 deg and longitude +6 deg.

Certainly a dedicated study should be done, in order to reach a precise description in the three spatial dimensions. The addition of kinematical information would be even more interesting to decipher the origin of these structures. The question of how far in time a halo configuration can be rolled back to understand in detail the formation process of the Milky Way is under debate. [34] showed the potential of astrometric missions to unravel the formation history of our Galaxy. However, [35] argued that the rolling back cannot be done too far in the past, because of the non-conservation of the integral of motions (probably due to dynamical friction) and the pollution from the stars born in the disk which were diffused later by the accreted satellite. Whatever the result of such theoretical studies, the detailed description of the Milky Way halo should be done with great detail. It is very important to gather individual chemical abundances to add to the positional and kinematical descriptions. The Gaia-ESO ([36]), Apogee ([37]), and future spectroscopic surveys are very important in this respect.

3.3 The Gaia mission extension

Initially the mission was planned for half a year of commissioning and 5 years of nominal science operations plus possibly one year of extension. Obviously a longer mission means more data and more precise results. For example, the precision on parallax and proper motion for Gaia is proportional to

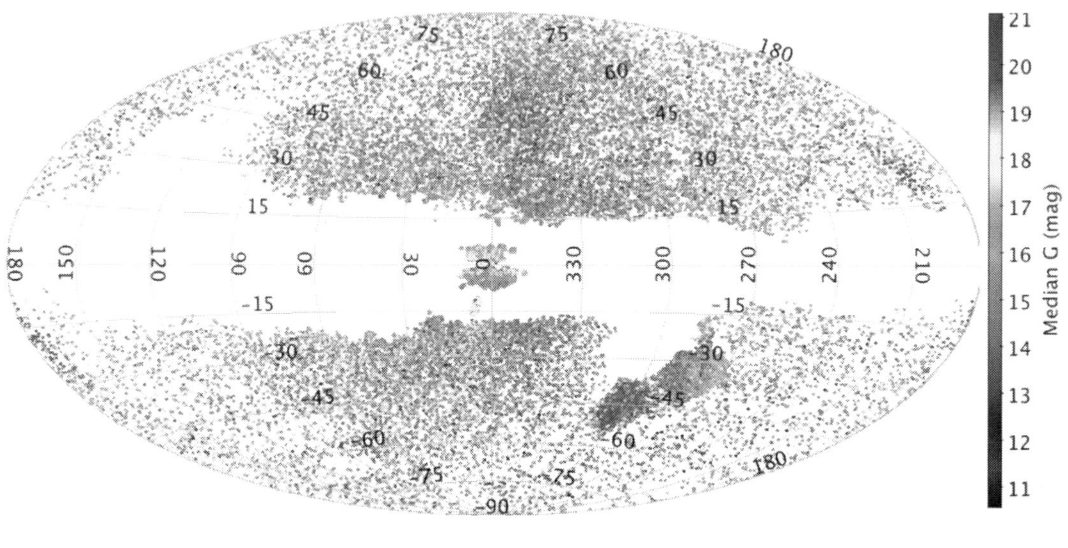

Figure 4. Sky distribution in Galactic coordinates of Gaia cross-matched RR Lyrae stars coloured in apparent magnitude (from the Gaia Image of the Week of 24/03/2017).

$1/\sqrt{T}$, and $1/\sqrt{T^3}$ respectively, where T is the length of the mission. This improvement of the precision is obviously not applicable when a floor of systematic effects is reached, which will most likely be the case for the brightest (~1%) subset of the targets.

A document was coordinated by the Gaia project scientist, Timo Prusti, to make the case of a longer mission, i.e., an extension up to 3-5 years. While the mission lifetime limitation is due most likely to fuel for the micro-propulsion system, a 5-year extension could be achievable with the current fuel consumption and sunshield ageing conditions. These coming years, the mission will go every second year through the ESA usual procedure to decide about the iterative 2-year extension of space missions.

4 Conclusions

In this article, we presented the Gaia mission and the content of the first data release. We also explored some aspects of the future Gaia data releases in the perspective of pulsating star research.

The intent of this first release for variable stars was more to present a showcase of some capabilities of Gaia. The peculiar ecliptic pole sampling of Gaia produces a number of measurements which is similar to the one reached in the nominal scanning law for 5 years. Even if the number of released light curves was small we had to go through all the mechanisms and procedures of an official release of the Gaia consortium and ESA. This work was very instructive and demonstrates that the system (DPAC/ESA) is flexible enough to allow some results to be available publicly even though it was not initially planned. Furthermore, as presented in the previous sections, there were already numerous articles published related to variable stars.

Obviously the first data release is just a very tiny tip of the iceberg, an iceberg that will reveal itself through the future releases. We can confidently conclude that Gaia will be an exceptional survey for pulsating star research.

Acknowledgments: On February 28 2017, Jan Cuypers passed away unexpectedly. As co-authors we are filled with sadness, and we will miss our colleague dearly. At the same time we feel grateful for all the valuable contributions he made to this work, which will undoubtedly leave a lasting mark on all achievements of CU7 we can expect in the coming years. The authors would therefore like to dedicate this paper to Jan. We thank Jos de Bruijne for valuable comments, Yves Revaz and Misha Haywood for discussions and Lennart Lindegren for a list of limitations of first Gaia data release. This work has made use of data from the ESA space mission *Gaia*, processed by the *Gaia* Data Processing and Analysis Consortium (DPAC). We acknowledge with thanks the variable star observations from the AAVSO International Database contributed by observers worldwide and used in this research.

References

[1] L. Eyer, N. Mowlavi, Journal of Physics Conference Series **118**, 012010 (2008), 0712.3797
[2] M.A.C. Perryman, L. Lindegren, J. Kovalevsky, E. Hoeg, U. Bastian, P.L. Bernacca, M. Crézé, F. Donati, M. Grenon, M. Grewing et al., A&A **323** (1997)
[3] A. Baranne, M. Mayor, J.L. Poncet, Vistas in Astronomy **23**, 279 (1979)
[4] L. Eyer, N. Mowlavi, D.W. Evans, K. Nienartowicz, D. Ordonez, B. Holl, I. Lecoeur-Taibi, M. Riello, G. Clementini, J. Cuypers et al., ArXiv e-prints (2017), 1702.03295
[5] F. van Leeuwen, D.W. Evans, F. De Angeli, C. Jordi, G. Busso, C. Cacciari, M. Riello, E. Pancino, G. Altavilla, A.G.A. Brown et al., ArXiv e-prints (2016), 1612.02952
[6] Gaia Collaboration, T. Prusti, J.H.J. de Bruijne, A.G.A. Brown, A. Vallenari, C. Babusiaux, C.A.L. Bailer-Jones, U. Bastian, M. Biermann, D.W. Evans et al., A&A **595**, A1 (2016), 1609.04153
[7] Gaia Collaboration, A.G.A. Brown, A. Vallenari, T. Prusti, J.H.J. de Bruijne, F. Mignard, R. Drimmel, C. Babusiaux, C.A.L. Bailer-Jones, U. Bastian et al., A&A **595**, A2 (2016), 1609.04172
[8] G. Clementini, V. Ripepi, S. Leccia, N. Mowlavi, I. Lecoeur-Taibi, M. Marconi, L. Szabados, L. Eyer, L.P. Guy, L. Rimoldini et al., A&A **595**, A133 (2016), 1609.04269
[9] Gaia Collaboration, G. Clementini, et al., Submitted to A&A (2017)
[10] L. Lindegren, U. Lammers, U. Bastian, J. Hernández, S. Klioner, D. Hobbs, A. Bombrun, D. Michalik, M. Ramos-Lerate, A. Butkevich et al., A&A **595**, A4 (2016), 1609.04303
[11] A. Udalski, I. Soszyński, D.M. Skowron, J. Skowron, P. Pietrukowicz, P. Mróz, R. Poleski, M.K. Szymański, S. Kozłowski, Ł. Wyrzykowski et al., Acta Astornomica **66**, 433 (2016), 1611.08300
[12] E. Hertzsprung, Bulletin of the Astronomical Institutes of the Netherlands **3**, 115 (1926)
[13] A.M. Jacyszyn-Dobrzeniecka, D.M. Skowron, P. Mróz, J. Skowron, I. Soszyński, A. Udalski, P. Pietrukowicz, S. Kozłowski, Ł. Wyrzykowski, R. Poleski et al., Acta Astronomica **66**, 149 (2016), 1602.09141
[14] A.M. Jacyszyn-Dobrzeniecka, D.M. Skowron, P. Mróz, I. Soszyński, A. Udalski, P. Pietrukowicz, J. Skowron, R. Poleski, S. Kozłowski, Ł. Wyrzykowski et al., ArXiv e-prints (2016), 1611.02709
[15] V. Belokurov, D. Erkal, A.J. Deason, S.E. Koposov, F. De Angeli, D.W. Evans, F. Fraternali, D. Mackey, MNRAS (2016), 1611.04614
[16] A.J. Deason, V. Belokurov, D. Erkal, S.E. Koposov, D. Mackey, ArXiv e-prints (2016), 1611.04600
[17] S. Casertano, A.G. Riess, B. Bucciarelli, M.G. Lattanzi, A&A **599**, A67 (2017), 1609.05175

[18] J. De Ridder, G. Molenberghs, L. Eyer, C. Aerts, A&A **595**, L3 (2016), 1609.08945

[19] V. Silva Aguirre, M.N. Lund, H.M. Antia, W.H. Ball, S. Basu, J. Christensen-Dalsgaard, Y. Lebreton, D.R. Reese, K. Verma, L. Casagrande et al., ApJ **835**, 173 (2017), 1611.08776

[20] G.R. Davies, M.N. Lund, A. Miglio, Y. Elsworth, J.S. Kuszlewicz, T.S.H. North, B. Rendle, W.J. Chaplin, T.S. Rodrigues, T.L. Campante et al., A&A **598**, L4 (2017), 1701.02506

[21] K.G. Stassun, G. Torres, ApJ **831**, L6 (2016)

[22] E. Høg, C. Fabricius, V.V. Makarov, S. Urban, T. Corbin, G. Wycoff, U. Bastian, P. Schwekendiek, A. Wicenec, A&A **355**, L27 (2000)

[23] C.L. Watson, Society for Astronomical Sciences Annual Symposium **25**, 47 (2006)

[24] F. Arenou, X. Luri, *Distances and absolute magnitudes from trigonometric parallaxes*, in *Harmonizing Cosmic Distance Scales in a Post-HIPPARCOS Era*, edited by D. Egret, A. Heck (1999), Vol. 167 of *Astronomical Society of the Pacific Conference Series*, pp. 13–32, astro-ph/9812094

[25] L. Eyer, L. Rimoldini, B. Holl, P. North, S. Zucker, D.W. Evans, D. Pourbaix, S.T. Hodgkin, W. Thuillot, N. Mowlavi et al., *The Gaia Mission, Binary Stars and Exoplanets*, in *Living Together: Planets, Host Stars and Binaries*, edited by S.M. Rucinski, G. Torres, M. Zejda (2015), Vol. 496 of *Astronomical Society of the Pacific Conference Series*, p. 121, 1502.03829

[26] I. Soszyński, A. Udalski, M.K. Szymański, Ł. Wyrzykowski, K. Ulaczyk, R. Poleski, P. Pietrukowicz, S. Kozłowski, D.M. Skowron, J. Skowron et al., Acta Astronomica **66**, 131 (2016), 1606.02727

[27] I. Soszyński, A. Udalski, M.K. Szymański, P. Pietrukowicz, P. Mróz, J. Skowron, S. Kozłowski, R. Poleski, D. Skowron, G. Pietrzyński et al., Acta Astronomica **64**, 177 (2014), 1410.1542

[28] A.J. Drake, M. Catelan, S.G. Djorgovski, G. Torrealba, M.J. Graham, V. Belokurov, S.E. Koposov, A. Mahabal, J.L. Prieto, C. Donalek et al., ApJ **763**, 32 (2013), 1211.2866

[29] A.J. Drake, M. Catelan, S.G. Djorgovski, G. Torrealba, M.J. Graham, A. Mahabal, J.L. Prieto, C. Donalek, R. Williams, S. Larson et al., ApJ **765**, 154 (2013), 1301.6168

[30] A.J. Drake, M.J. Graham, S.G. Djorgovski, M. Catelan, A.A. Mahabal, G. Torrealba, D. García-Álvarez, C. Donalek, J.L. Prieto, R. Williams et al., ApJS **213**, 9 (2014), 1405.4290

[31] G. Torrealba, M. Catelan, A.J. Drake, S.G. Djorgovski, R.H. McNaught, V. Belokurov, S. Koposov, M.J. Graham, A. Mahabal, S. Larson et al., MNRAS **446**, 2251 (2015), 1410.7653

[32] L. Palaversa, Ž. Ivezić, L. Eyer, D. Ruždjak, D. Sudar, M. Galin, A. Kroflin, M. Mesarić, P. Munk, D. Vrbanec et al., AJ **146**, 101 (2013), 1308.0357

[33] G. Clementini, R. Gratton, A. Bragaglia, E. Carretta, L. Di Fabrizio, M. Maio, AJ **125**, 1309 (2003)

[34] A. Helmi, P.T. de Zeeuw, MNRAS **319**, 657 (2000), astro-ph/0007166

[35] I. Jean-Baptiste, P. Di Matteo, M. Haywood, A. Gomez, M. Montuori, F. Combes, B. Semelin, ArXiv e-prints (2016), 1611.07193

[36] G. Gilmore, S. Randich, M. Asplund, J. Binney, P. Bonifacio, J. Drew, S. Feltzing, A. Ferguson, R. Jeffries, G. Micela et al., The Messenger **147**, 25 (2012)

[37] C. Allende Prieto, S.R. Majewski, R. Schiavon, K. Cunha, P. Frinchaboy, J. Holtzman, K. Johnston, M. Shetrone, M. Skrutskie, V. Smith et al., Astronomische Nachrichten **329**, 1018 (2008), 0809.2362

Gaia's Cepheids and RR Lyrae stars and luminosity calibrations based on Tycho-Gaia Astrometric Solution

Gisella Clementini[1,*], *Laurent* Eyer[2], *Tatiana* Muraveva[1], *Alessia* Garofalo[1,3], *Vincenzo* Ripepi[4], *Marcella* Marconi[4], *Luis* Sarro[5], *Max* Palmer[6], *Xavier* Luri[6], *Roberto* Molinaro[4], *Lorenzo* Rimoldini[7], *Laszlo* Szabados[8], *Richard I.* Anderson[9], and *Ilaria* Musella[4]

[1]*INAF - Osservatorio Astronomico di Bologna, Via Gobetti 93/3, 40129 Bologna, Italy*
[2]*Department of Astronomy, University of Geneva, Ch. des Maillettes 51, 1290 Versoix, Switzerland*
[3]*Dipartimento di Fisica e Astronomia, Università di Bologna, Via Gobetti 93/2, 40129 Bologna, Italy*
[4]*INAF-Osservatorio Astronomico di Capodimonte, Via Moiariello 16, 80131 Napoli, Italy*
[5]*Dpto. Inteligencia Artificial, UNED, c/ Juan del Rosal 16, 28040 Madrid, Spain*
[6]*Institut de Ciències del Cosmos, Universitat de Barcelona (IEEC-UB), Martí Franquès 1, E-08028 Barcelona, Spain*
[7]*Department of Astronomy, University of Geneva, Ch. d'Ecogia 16, 1290 Versoix, Switzerland*
[8]*Konkoly Observatory, Research Centre for Astronomy and Earth Sciences, Hungarian Academy of Sciences, Konkoly Thege Miklós út 15-17, 1121 Budapest, Hungary*
[9]*Department of Physics and Astronomy, The Johns Hopkins University, 3400 N Charles St, Baltimore, MD 21218, USA*

Abstract. Gaia Data Release 1 contains parallaxes for more than 700 Galactic Cepheids and RR Lyrae stars, computed as part of the Tycho-Gaia Astrometric Solution (TGAS). We have used TGAS parallaxes, along with literature (V, I, J, K_s, W_1) photometry and spectroscopy, to calibrate the zero point of the period-luminosity and period-Wesenheit relations of classical and type II Cepheids, and the near-infrared period-luminosity, period-luminosity-metallicity and optical luminosity-metallicity relations of RR Lyrae stars. In this contribution we briefly summarise results obtained by fitting these basic relations adopting different techniques that operate either in parallax or distance (absolute magnitude) space.

1 Introduction

Cepheids and RR Lyrae stars are primary standard candles of the cosmological distance ladder and excellent tracers of young (the classical Cepheids) and old (the RR Lyrae stars and the Type II Cepheids) stellar populations. Gaia will be an extraordinary discoverer of new Cepheids and RR Lyrae stars in and beyond the Milky Way. Furthermore, the unprecedented accuracy of Gaia's end-of-mission astrometric measurements will allow us to pin down the slope and zero point of the fundamental relations that Cepheids and RR Lyrae stars conform to with the precision required to constrain the Hubble constant to better than 2%.

*gisella.clementini@oabo.inaf.it

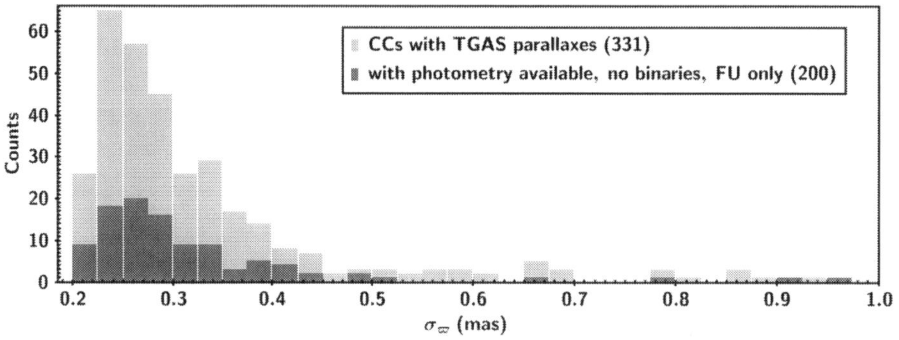

Figure 1. Error distributions of TGAS parallaxes for classical Cepheids. The pink histogram corresponds to the whole sample of 331 Galactic classical Cepheids in the TGAS catalogue, the magenta histogram shows a subsample of 102 fundamental mode classical Cepheids with (V, I, J, K_s) photometry available in the literature after removing the known binaries. The bin size is 0.025 mas.

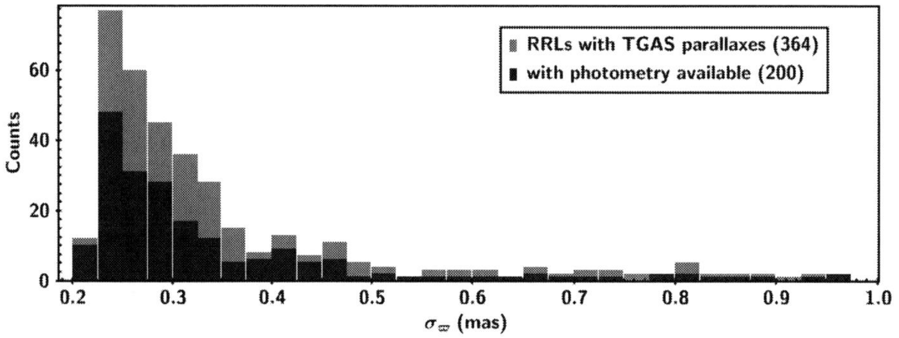

Figure 2. Error distributions of TGAS parallaxes for RR Lyrae stars. The cyan histogram corresponds to the whole sample of 364 RR Lyrae stars in the TGAS catalogue, the blue histogram shows a subsample of 200 stars with (V, K_s, W_1) photometry available in the literature. The bin size is 0.025 mas.

Besides G-band time-series photometry and pulsation characteristics for a few thousand Cepheids and RR Lyrae stars in the Large Magellanic Cloud (LMC; [8], see also Eyer et al., this volume), data for variable stars in Gaia Data Release 1 (DR1) include parallaxes for 331 classical Cepheids, 31 Type II Cepheids and 364 RR Lyrae stars, in common between Gaia and the Hipparcos and Tycho-2 catalogues. They were computed as part of the Tycho-Gaia Astrometric Solution (TGAS, [10]).

The error distributions of the TGAS parallaxes for classical Cepheids and RR Lyrae stars are shown in Figures 1 and 2, respectively. Errors range from 0.2 to about 1 milliarcsecond (mas) and peak around ~ 0.25 mas for both types of pulsating stars[1]. Although this is not comparable to the final Gaia precision, it already represents a significant general improvement with respect to Hipparcos parallaxes ([15]). Additionally, as described in [10], there could still be some systematic effects at a typical level of ± 0.3 mas depending on the sky position and the colour of the source. However, these systematic effects are spatially correlated and become negligible for RR Lyrae stars and Type II Cepheids that are evenly distributed on the sky, while they may be relevant for classical Cepheids,

[1]After publication of the Gaia DR1 catalogue, a number of authors have suggested that the standard errors of TGAS parallaxes may be overestimated (e.g., [5] and Michael Feast's concluding comments, this volume).

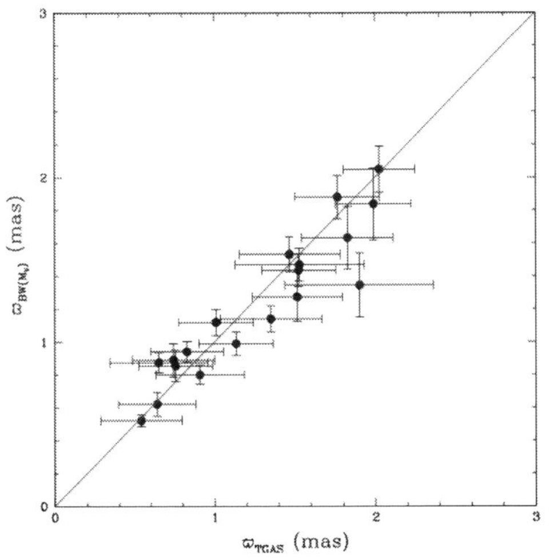

Figure 3. Comparison between the TGAS parallaxes and photometric parallaxes inferred from the absolute visual magnitude (M_V) estimated via the Baade-Wesselink (B-W) technique for 19 Galactic RR Lyrae stars (see Table 2 in [11] and references therein). The blue line represents the bisector.

which are mainly concentrated in the Galactic disk. We also note that the TGAS samples are the result of several processing stages and truncations (e.g., removal of sources brighter than a certain limit and bluer/redder than a certain colour interval, truncation of sources with less than 5 transits, or with large astrometric uncertainties, see [7] for details), each with a different impact on the resulting sample. Nevertheless, the very limited size of the Cepheid and RR Lyrae samples and the large relative errors of TGAS parallaxes hide most of these effects.

2 Comparison with trigonometric and photometric parallaxes in the literature

A direct comparison between TGAS and Hipparcos parallaxes ([15]) is possible for 248 classical Cepheids, 31 Type II Cepheids and 188 RR Lyrae stars for which both measurements are available. The number of negative parallaxes is significantly reduced from 32% of the sample of classical Cepheids in Hipparcos to only 4% in TGAS, decreasing from 42% to 16% for the Type II Cepheids and from 32% to 1% for the RR Lyrae stars. This is a clear indication that uncertainties of the TGAS parallaxes are smaller than in Hipparcos. When comparing TGAS and Hipparcos parallaxes in the period-luminosity plane using the K_s-band (PL_{K_s}), the improvement in quality and statistics of the former is impressive, particularly for RR Lyrae stars.

A direct comparison of TGAS versus the *Hubble Space Telescope* (HST) is possible for three classical Cepheids with HST parallaxes from [2],[13] and [4], one Type II Cepheid (VY Pyx) and five RR Lyrae stars with HST parallaxes from [3]. TGAS and HST parallaxes compare favourably for the classical Cepheids and even better for the RR Lyrae stars. On the other hand, the TGAS parallax of VY Pyx is much smaller than the HST value by [3], and nicely places the star on the extrapolation to longer periods of the RR Lyrae star PL_K relation (see Fig. 10 in [7]), while this is not the case with the HST parallax.

Figure 3 shows the comparison between the TGAS parallaxes and photometric parallaxes inferred from the absolute visual magnitude (M_V) estimated via the Baade-Wesselink (B-W) technique for 19 Galactic RR Lyrae stars (see Table 2 in [11] and references therein). The agreement is excellent both in the case of the M_V and the K-band absolute magnitudes (M_K). Similarly, good agreement is found between TGAS parallaxes and the photometric parallaxes inferred from the application of the Infrared Surface Brightness version of the B-W technique to 54 classical Cepheids from the collection

Table 1. PL_{K_s} relations for classical Cepheids, Type II Cepheids and RR Lyrae stars with zero point based on TGAS parallaxes.

	Relation (mag)	r.m.s. (mag)
PL_{K_s} 95 objects (LSQ)	$-3.365 \log P - (2.06 \pm 0.08)$	0.74
PL_{K_s} 102 objects (ABL)	$-3.365 \log P - (2.63 \pm 0.10)$	0.88
PL_{K_s} 102 stars (BA)	$-3.365 \log P - (2.60^{+0.11}_{-0.15})$	1.33
PL_{K_s} 22 objects (LSQ)	$-2.385 \log P - (1.18 \pm 0.12)$	0.81
PL_{K_s} 26 objects (ABL)	$-2.385 \log P - (1.58 \pm 0.17)$	1.10
PL_{K_s} 26 objects (BA)	$-2.385 \log P - (1.51^{+0.23}_{-0.22})$	1.14
PL_{K_s} 195 stars (LSQ)	$-2.73 \log P - (1.06 \pm 0.04)$	0.84
PL_{K_s} 200 stars (ABL)	$-2.73 \log P - (1.26 \pm 0.04)$	0.90
PL_{K_s} 200 stars (BA)	$-2.73 \log P - (1.24 \pm 0.05)$	1.02

in [9]. Pulsation parallaxes for a few classical Cepheids and RR Lyrae stars in our samples have been estimated from the theoretical modelling of the stars' multi-band light curves through nonlinear convective pulsation models (Marconi, this volume) and compared with the corresponding TGAS parallaxes. Agreement is generally satisfactory except for RS Cas, a classical Cepheid for which the TGAS parallax is likely incorrect as it places the star about 2 magnitudes below the *PL* relation.

3 Luminosity calibrations based on different fitting approaches

A procedure often used to calibrate the period-luminosity (*PL*), period-Wesenheit (*PW*), period-luminosity-metallicity (*PLZ*) or luminosity-metallicity (M_V – [Fe/H]) relations is the direct transformation to distance (hence, absolute magnitude) by parallax inversion and then the least squares fit (LSQ) of the derived parameters. However, symmetrical errors in parallaxes translate into asymmetric errors in the magnitudes, an effect that becomes specially problematic for parallaxes with large relative errors that introduces an overall bias. Furthermore, this method does not allow us to use negative parallaxes, thus biasing the samples against more distant sources. On the contrary, in methods that operate in parallax space such as the Astrometric Based Luminosity (ABL, [1]) and Bayesian approaches (hereinafter, BA), the parallaxes are used directly, thus maintaining the errors symmetrical and allowing negative parallaxes to be used. We applied all three methods (LSQ, ABL and BA) to fit the canonical relations that Cepheids and RR Lyrae stars conform to. In Table 1, we summarise the TGAS-based PL_{K_s} relations of Cepheids and RR Lyrae stars obtained with the three different fitting approaches and adopting the slope from [9] for classical Cepheids, from [14] for the Type II Cepheids and from [11] for the RR Lyrae stars. Table 1 shows that the ABL and Bayesian approaches are generally in good agreement with each other and provide brighter absolute magnitudes (hence longer distances) than the direct transformation of parallaxes and the LSQ fit. Differences are larger (of about 0.5-0.6 mag, on average) for the classical Cepheids, reduce to 0.4-0.5 mag for the Type II Cepheids, and are the smallest ones, 0.2 mag, for the RR Lyrae stars. However, we note that the r.m.s. scatter of all relations is very large, due to the large parallax uncertainties.

Figure 4 shows the results obtained for the LMC distance modulus by applying the TGAS-based relations of the RR Lyrae stars to the LMC variable stars studied by [6]. The results obtained for the RR Lyrae stars show a much better agreement among the three methods and also a reasonably good agreement with the currently adopted LMC distance modulus from [12] (solid, vertical line).

The TGAS parallaxes represent a significant improvement upon the previous Hipparcos estimates. However, the TGAS-based luminosity calibrations presented in this study have to be considered pre-

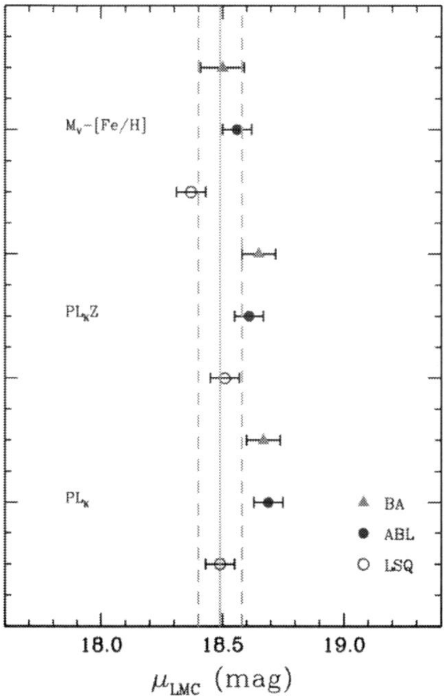

Figure 4. LMC distance moduli obtained by fitting the TGAS-based PL_K, PL_KZ and M_V − [Fe/H] relations of RR Lyrae stars with the LSQ (blue open circle), the ABL (blue filled circles) and the Bayesian approaches (magenta filled triangles), respectively. From bottom to top: PL_{K_s} relations for 195/200/200 stars (LSQ, ABL, BA methods) with slope from [11]; $PL_{K_s}Z$ relations with slope of the dependence on period from [11]; M_V − [Fe/H] relations with slope from [6]. A solid vertical line shows the LMC distance modulus from [12].

liminary and to be superseded by new more accurate relations calibrated on Gaia-only parallaxes that will be published in future releases, to start with Gaia Data Release 2 in 2018.

Acknowledgments This work has made use of data from the ESA space mission Gaia, processed by the Gaia Data Processing and Analysis Consortium (DPAC), and of the SIMBAD database, operated at CDS, Strasbourg, France.

References

[1] F. Arenou, X. Luri, in *Harmonizing Cosmic Distance Scales in a Post-Hipparcos Era*, ed. D. Egret & A. Heck, ASP Conference Series, **167**, 13 (1999)
[2] Benedict, G. F., McArthur, B. E., Feast, M. W., et al., AJ, **133**, 1810 (2007)
[3] Benedict, G. F., McArthur, B. E., Feast, M. W., et al., AJ, **142**, 187 (2011)
[4] Casertano, S., Riess, A. G., Anderson, J., et al., ApJ, **825**, 11 (2016)
[5] Casertano, S., Riess, A. G., Bucciarelli, B., & Lattanzi, M. G., A&A, **599**, A67 (2017)
[6] Clementini, G., Gratton, R. G., Bragaglia, A., et al., AJ, **125**, 1309 (2003)
[7] Gaia Collaboration, Clementini, G., et al., A&A, in press (arXiv:1705.00688) (2017)
[8] Clementini, G., Ripepi, V. Leccia, S., et al., A&A, **595**, A133 (2016)
[9] Fouqué, P., Arriagada, P., Storm, J., et al., A&A, **476**, 73 (2007)
[10] Lindegren, L., Lammers, U., Bastian, U., et al., A&A, **595**, A4 (2016)
[11] Muraveva, T., Palmer, M., Clementini, G., et al., ApJ, **807**, 127 (2015)
[12] Pietrzynski, G., Graczyk, D., Gieren, W., et al., Nature, **495**, 76 (2013)
[13] Riess, A. G., Casertano, S., Anderson, J., MacKenty, J., & Filippenko, A. V., ApJ, **785**, 161 (2014)
[14] Ripepi, V., Moretti, M.-I., Marconi, M., et al., MNRAS, **446**, 3034 (2015)
[15] van Leeuwen, F., A&A, **474**, 653 (2007)

Cepheids with the eyes of photometric space telescopes

László Molnár[1,⋆], *András* Pál[1,2], and *Emese* Plachy[1]

[1] *Konkoly Observatory, MTA CSFK, Konkoly Thege Miklós út 15-17, H-1121 Budapest, Hungary*
[2] *Eötvös Loránd University, H-1117 Pázmány Péter sétány 1/A, Budapest, Hungary*

Abstract. Space photometric missions have been steadily accumulating observations of Cepheids in recent years, leading to a flow of new discoveries. In this short review we summarize the findings provided by the early missions such as *WIRE*, *MOST*, and *CoRoT*, and the recent results of the *Kepler* and *K2* missions. The surprising and fascinating results from the high-precision, quasi-continuous data include the detection of the amplitude increase of Polaris, and exquisite details about V1154 Cyg within the original *Kepler* field of view. We also briefly discuss the current opportunities with the *K2* mission, and the prospects of the *TESS* space telescope regarding Cepheids.

1 Introduction

By instinct, Cepheid variables might not seem to be ideal targets for photometric space missions. Their variations are known to be rather simple: they pulsate in one or a few radial p-modes in a rather clockwork-like manner. They often have long pulsation periods that may extend to several weeks. And we have excellent ground-based observations for many of them, particularly from the OGLE project, that made several interesting discoveries (see, e.g., [1, 2]). The expanding capabilities of time-domain astronomy provide us with ample data about Cepheid stars from the ground.

But is observing Cepheids from space really a moot point? We now have had missions like *CoRoT* and *Kepler* that observed long enough to cover several pulsation cycles of a typical Cepheid. Although long-term ground-based observations can reach very high precision for periodic signals, the per point accuracy of space telescopes is still unmatched. And as the results below illustrate, these benefits allowed us to uncover new, unexpected phenomena in Cepheids that have been not or are barely observable from the ground.

Before we delve into the details, it is worth pointing out that the *Hubble Space Telescope* itself accumulated a large amount of photometry about Cepheids, mostly of extragalactic ones. However, the main goal of these observations was to pin down the period-luminosity relations for the various galaxies, not asteroseismology (for a summary of these observations, see [3]). Therefore we won't discuss these observations in detail here.

⋆molnar.laszlo@csfk.mta.hu

2 Small(ish) space telescopes

2.1 Early efforts

Space-based photometric telescopes were envisaged for a long time. The first instrument dedicated to variable stars was the French *EVRIS* telescope that was flying—briefly—on the failed *Mars-96* mission two decades ago ([4]). Interestingly, the first actual photometric mission was born from another mishap: the small optical star tracker camera on the otherwise unusable *WIRE* infrared space telescope was utilized to monitor bright stars. The well-known Cepheid Polaris (α UMi) was observed with it multiple times in 2004-5, complementing the more extended, but significantly lower-quality data set of the *SMEI* all-sky camera aboard the *Coriolis* satellite ([5]). These observations revealed that Polaris is not switching off and its pulsation amplitude started to increase after reaching a minimum around 2000. Weak excess variation in the low-frequency regime was tentatively connected to granulation patterns on the surface of the star ([5]).

2.2 The first photometric missions

Meanwhile, missions dedicated to stellar variability were also launched into space. The Canadian *MOST* and the French-led *CoRoT* missions utilized larger telescopes, compared to the ones mentioned so far. While the former telescope is more or less limited to one month long observations, the latter was able to perform long runs lasting for 150 days. Unfortunately, almost all stars (a single type II and seven classical Cepheids) that fell into the observing fields of *CoRoT* turned out to be very stable, with almost no signs of variations from one cycle to the next ([8]). There was one star though, CoRoT 0223989566, that revealed an unexpected pulsation mode. This O1/O2 (first- and second-overtone) beat Cepheid also features a longer-period mode at $P_1/P_3 = 0.68$ ratio that has been also identified in a number of first-overtone RR Lyrae stars, possibly suggesting a common origin for the mode in the two classes ([9, 10]).

The venerable *MOST* space telescope observed four Cepheids during its lifetime, all of which belonged to different subgroups. First, it surveyed RT Aur and SZ Tau, a fundamental-mode and a first-overtone star, and looked for signs of irregularity in the pulsations. The results suggested that overtone stars may have less stable light curves ([11]). Later, the telescope serendipitously observed the beat Cepheid U TrA, and carried out a targeted campaign on the modulated, second-overtone star V473 Lyr. The latter observations led to the first discovery of period doubling in a classical Cepheid, hinting that similar dynamical processes may occur in modulated Cepheid and RR Lyrae stars ([12]).

More recently, the BRITE-Constellation mission also covered some bright Cepheids. Although *BRITE* utilizes multiple small satellites to observe the same field simultaneously in blue and red colors, Cepheids are too faint for them in the blue band. Red light curves have been obtained for seven targets so far, with varying success. Nevertheless, the observations revealed a possible modulation in T Vul, and signs of additional modes in DT Cyg and V1334 Cyg ([6, 7]).

3 The *Kepler* and *K2* years

3.1 Four years with a Cepheid

The field-of-view of the original *Kepler* mission included a single Cepheid only, V1154 Cyg. The first months of data indicated that V1154 Cyg is a regular Cepheid pulsating only in the fundamental mode ([13]). More extended observations then revealed that the star is, in fact, not an entirely regular clock. Various measures of the shape, amplitude, and length of the pulsation cycles showed irregular jitter around the mean values that averaged out on longer time scales ([14]).

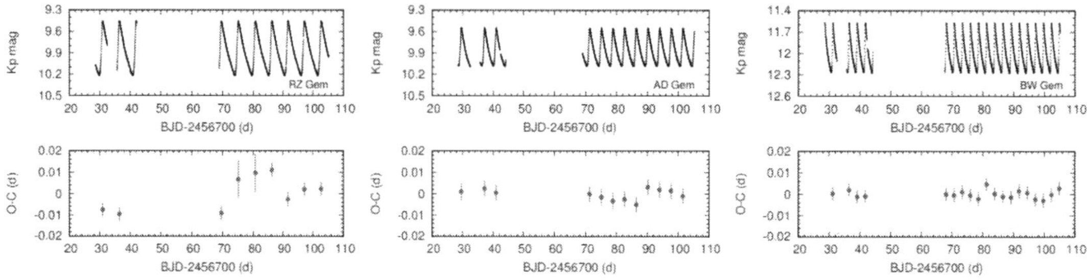

Figure 1. Top: light curves of three fundamental-mode classical Cepheids observed in Campaign 0. Bottom: preliminary O-C values of times of maxima.

And then the four-year-long data set revealed even more. A weak modulation cycle with a period of about 159 d was identified. But the length and precision of the data also made it possible to clearly detect granulation noise in a Cepheid for the first time. Interestingly, the star lacked any signs of solar-like oscillations well below the expected amplitudes scaled from red giants ([15]).

3.2 Surveying Cepheids with *K2*

The mechanical issues of the *Kepler* space telescope led to the start of the *K2* mission that opened up new possibilities to observe rare kinds of stars, including Cepheids. Multiple campaigns are aimed towards the vicinity of the Galactic disk where most classical Cepheids reside. Campaigns 9 and 11, in particular, cover the bulge, and provide a unique opportunity to compare the long-term but sparse observations of OGLE with the short, but continuous data from K2.

Cepheids were observed already in Campaign 0 towards the anticenter direction of the Galaxy. The field included three fundamental-mode classical Cepheids, BW Gem, AD Gem, and RZ Gem. Since one of the discoveries of *Kepler* about Cepheids was the detection of pulsation jitter, we examined the light variations of these three stars. Unfortunately, the light curve of the longest-period star, RZ Gem, suffers from jumps caused by the attitude changes of the telescope that need to be mitigated first. Nevertheless, we investigated the O–C values of the brightness maximum timings for all three stars. The scatter for RZ Gem (P = 5.528 d) was in the same range as that of V1154 Cyg (P = 4.925 d), about ± 0.015 d. But as Figure 1 shows, the scatter of maxima timings turned out to be much smaller (± 0.005 d) for the two short-period stars, AD Gem (P = 3.788 d) and RZ Gem (P = 2.635 d). These first results hint that the amount of pulsation jitter might depend on the pulsation period of the star, but more detailed studies with larger sample sizes are required to draw more certain conclusions. Furthermore, observations of first-overtone and beat Cepheids will help us investigate the stability of different modes, as well as to search for low-amplitude additional modes.

The *K2* mission also covers many type II and anomalous Cepheids that earlier missions largely lacked. The first results suggest that W Vir-type stars (the medium-period type II subclass) may experience period doubling and more irregular cycle-to-cycle variations ([16]). The OGLE surveys already suggested that the pulsation of W Vir-type stars is not perfectly repetitive ([2]), but now the *K2* mission may show us how exactly the cycles vary.

Another unexpected capability of the *K2* mission is that it is able to reach some extragalactic pulsating variable stars. First, it observed three RR Lyrae stars in the dwarf spheroidal galaxy Leo IV during Campaign 1 ([18]). Then in Campaign 8 it observed a much larger irregular dwarf galaxy in the Local Group, IC 1613. While it is farther away than Leo IV, IC 1613 contains hundreds of Cepheids,

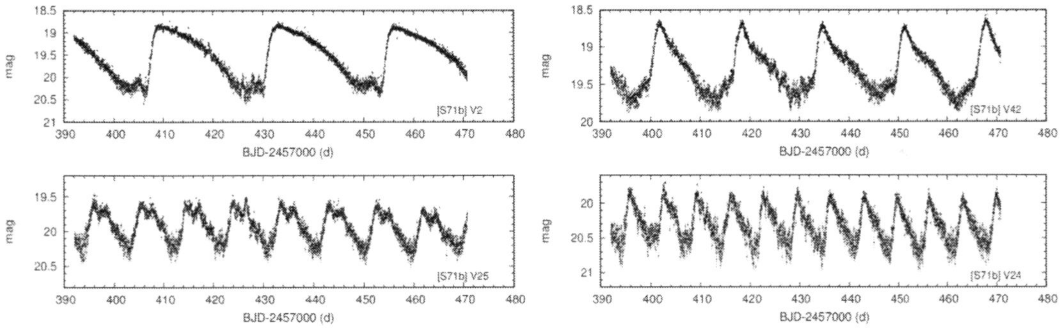

Figure 2. Preliminary *K2* light curves of various Cepheids in the Local Group galaxy IC 1613, obtained during Campaign 8.

among other bright variable stars, that *Kepler* is capable to observe. The low angular resolution makes *Kepler* a less than ideal tool to measure extragalactic variables that suffer from crowding and blending with nearby stars: nevertheless, extracting continuous light curves from even a fraction of the Cepheids in IC 1613 will be a unique result. At the time of writing the photometry is still under way, but we include preliminary light curves of a few bright targets here, in Figure 2, as an appetizer for the complete study. Photometry was carried out using the FITSH package, similar to the method used in the case of Leo IV ([19]). For IC 1613 we also included image oversampling into the pipeline to allow for subpixel-level image shifts that gives us a smoother master image for image subtraction.

4 Upcoming data sources

More missions are under way: the European *Gaia* mission will soon provide accurate geometric parallaxes for a large number of Cepheid stars, as well as sparse photometry for most of them, and potentially discover many, yet unknown members of the class ([20]). The distance measurements of *Gaia* will untangle many, currently uncertain parameters of these stars, including possible companion stars, the amount of infrared excess, and the exact values and possible period dependence of the radial-velocity projection factor ([21]). It will also allow us to properly separate the anomalous Cepheid population in the disk and halo from the long-period RR Lyrae stars, based on their luminosities.

The American *TESS* space telescope will be a new source of continuous photometry from 2018 ([22]). Unlike *Kepler*, it will observe only the brighter stars in the sky, and, for the most part, in 27 d long runs. The short observing time per sector is not favorable for all but the shortest-period Cepheids. Fortunately, the continuous viewing zones (CVZ) of the mission offer one-year-long coverage. The CVZ towards the south ecliptic pole will enable us to observe the brightest, longest-period Cepheids in the Large Magellanic Cloud. The bright Galactic Cepheid β Dor is also included in the CVZ, and the data will potentially be accurate enough to look for the signs of granulation, and compare its properties with that of V1154 Cyg. And in the coming decade, the European *PLATO* mission will expand these kinds of studies even further.

However, single-color photometric observations will not tell us everything about these stars. Targeted efforts to measure accurate line profiles and radial velocities can give us important clues about the physical processes in the atmospheres and the pulsation ([23, 24]). Multicolor photometry from large-étendue surveys also provide important physical information for us, and can be exploited for, e.g., mode identification. And given the usually high apparent brightnesses of Cepheids, surveys that

 DOI: 10.1051/epjconf/201715202004

achieve the large étendue by combining small apertures with extremely large field of view (and high cadence), such as the Fly's Eye system ([25]), are better suited at these tasks than larger telescopes such as PanSTARRS or LSST. The combination of such ground-based observations with space-based photometry may ensure that the golden age for stellar astrophysics continues.

Acknowledgments: L. M. acknowledges with thanks the hospitality and financial support of the organizers of the 22nd Los Alamos Stellar Pulsation Conference Series meeting. L. M. is supported by the János Bolyai Research Scholarship of the Hungarian Academy of Sciences. This project has been supported by the Lendület LP2012-31 and LP2014-17 Programs of the Hungarian Academy of Sciences, and by the NKFIH K-115709, PD-116175, and PD-121203 grants of the Hungarian National Research, Development and Innovation Office. This paper includes data collected by the Kepler and *K2* missions, funded by the NASA Science Mission directorate. This work used *K2* data proposed by WG7 of the Kepler Asteroseismic Science Consortium (GO numbers 0055 and 8041).

References

[1] Soszyński, I., Udalski, A., Szymański, M. K., et al., AcA, **65**, 297 (2015)
[2] Soszyński, I., Udalski, A., Pietrukowicz, P., et al., AcA, **61**, 285 (2011)
[3] Riess, A. G., Macri, L. M., Hoffmann, S. L., et al., ApJ, **826**, 56 (2016)
[4] Buey, J. T., Auvergne, M., Vuillemin, A., & Epstein, G., PASP, **109**, 140 (1997)
[5] Bruntt, H., Evans, N. R., Stello, D., et al., ApJ, **683**, 433 (2008)
[6] Weiss, W. W., Rucinski, S. M., Moffat, A. F. J., et al., PASP, **126**, 573 (2014)
[7] Smolec, R., Moskalik, P., Evans, N. R., Moffat, A. F. J., & Wade, G. A., to appear in Proc. of the 2nd BRITE-Constellation Science Conf., EPJ WoC, preprint (arXiv:1612.02970) (2016)
[8] Poretti, E., Le Borgne, J. F., Rainer, M., Baglin, A., Benkő, J. M., Debosscher, J., & Weiss, W. W., MNRAS, **454**, 849 (2015)
[9] Poretti, E., Baglin, A., & Weiss, W. W., ApJ, **795**, 36 (2014)
[10] Netzel, H., Smolec, R., & Dziembowski, W., MNRAS, **451**, 25 (2015)
[11] Evans, N. R., Szabó, R., Derekas, A., et al., MNRAS, **446**, 4008 (2015)
[12] Molnár, L., Derekas, A., Szabó, R., et al., MNRAS, **466**, 4009 (2017)
[13] Szabó, R., Szabados, L., Ngeow, C.-C., et al., MNRAS, **413** 2709 (2011)
[14] Derekas, A., Szabó, Gy. M., Berdnikov, L., et al., MNRAS, **425**, 1312 (2012)
[15] Derekas, A., Plachy, E., Molnár, L., et al., MNRAS, **464**, 1553 (2017)
[16] Plachy, E., Molnár, L., Jurkovic, M. I., Smolec, R., Moskalik, P. A., Pál, A., Szabados, L., & Szabó, R., MNRAS, **465**, 173 (2017)
[17] Soszyński, I., Udalski, A., Szymański, M. K., et al., AcA, **65**, 1 (2015)
[18] Molnár, L., Pál, A., Plachy, E., Ripepi, V., et al., ApJ, **812**, 2 (2015)
[19] Pál, A., MNRAS, **421**, 1825 (2012)
[20] Clementini, G., Ripepi, V., & Leccia, S., A&A, **595**, 133 (2017)
[21] Kervella, P., Mérand, A., Gallenne, A., et al., in this volume; preprint: (arXiv:1702.07741) (2017)
[22] Ricker, G. R., Winn, J. N., Vanderspeck, R., et al., Proceedings of the SPIE, **9143**, 914320 (2014)
[23] Anderson, R. I., MNRAS, **463**, 1707 (2016)
[24] Nardetto, N., Poretti, E., Rainer, M., et al., A&A, **597**, 73 (2017)
[25] Pál, A., Mészáros, L., Csépány, G., et al., AN, **334**, 932 (2013)

The *Hubble* Catalog of Variables

K. Sokolovsky[1,2,3,*], *A.* Bonanos[1,**], *P.* Gavras[1], *M.* Yang[1], *D.* Hatzidimitriou[4,1], *M. I.* Moretti[5,1], *A.* Karampelas[1], *I.* Bellas-Velidis[1], *Z.* Spetsieri[1,4], *E.* Pouliasis[1,4], *I.* Georgantopoulos[1], *V.* Charmandaris[1], *K.* Tsinganos[1], *N.* Laskaris[6], *G.* Kakaletris[6], *A.* Nota[7,8], *D.* Lennon[9], *C.* Arviset[9], *B.* Whitmore[7], *T.* Budavari[10], *R.* Downes[7], *S.* Lubow[7], *A.* Rest[7], *L.* Strolger[7], and *R.* White[7]

[1]*IAASARS, National Observatory of Athens, Vas. Pavlou & I. Metaxa, 15236 Penteli, Greece*
[2]*Sternberg Astronomical Institute, Moscow State University, Universitetskii pr. 13, 119992 Moscow, Russia*
[3]*Astro Space Center of Lebedev Physical Institute, Profsoyuznaya Str. 84/32, 117997 Moscow, Russia*
[4]*Department of Physics, National and Kapodistrian University of Athens, 15771 Ilissia, Greece*
[5]*INAF-Osservatorio Astronomico di Capodimonte, Salita Moiariello, 16, 80131 Napoli, Italy*
[6]*Athena Research and Innovation Center, Artemidos 6 & Epidavrou, 15125 Maroussi, Greece*
[7]*Space Telescope Science Institute, 3700 San Martin Drive, Baltimore, MD 21218, USA*
[8]*European Space Agency, Research and Scientific Support Department, Baltimore, MD 21218, USA*
[9]*European Space Astronomy Centre, Camino bajo del Castillo, Urbanizacion Villafranca del Castillo, Villanueva de la Cañada, 28692 Madrid, Spain*
[10]*The Johns Hopkins University, Baltimore, MD 21218, USA*

Abstract. We aim to construct an exceptionally deep ($V \lesssim 27$) catalog of variable objects in selected Galactic and extragalactic fields visited multiple times by the *Hubble Space Telescope* (HST). While HST observations of some of these fields were searched for specific types of variables before (most notably, the extragalactic Cepheids), we attempt a systematic study of the population of variable objects of all types at the magnitude range not easily accessible with ground-based telescopes. The variability timescales that can be probed range from hours to years depending on how often a particular field has been visited. For source extraction and cross-matching of sources between visits we rely on the Hubble Source Catalog which includes 10^7 objects detected with WFPC2, ACS, and WFC3 HST instruments. The lightcurves extracted from the HSC are corrected for systematic effects by applying local zero-point corrections and are screened for bad measurements. For each lightcurve we compute variability indices sensitive to a broad range of variability types. The indices characterize the overall lightcurve scatter and smoothness. Candidate variables are selected as having variability index values significantly higher than expected for objects of similar brightness in the given set of observations. The *Hubble* Catalog of Variables will be released in 2018.

1 Introduction

The *Hubble Space Telescope* (HST) is great for deep imaging thanks to the unique combination of low sky background, sharp point spread function (PSF) and wide field of view (compared to ground-

*`kirx@noa.gr`
**`bonanos@astro.noa.gr`

based adaptive optic systems; [13]). It is sensitive to ultraviolet light not accessible from the ground. Some fields were imaged by the HST multiple times opening a window to time-domain studies at faint magnitudes and high spatial resolution. High resolution imaging is important to overcome confusion in dense star fields, like the ones found in nearby galaxies and globular cluster cores. A number of dedicated variability studies were conducted with the HST including [1, 2, 5, 9, 10, 17, 25].

While the HST has superb internal astrometric precision, obtaining accurate absolute astrometry has been a challenge in the previous years. The positional information provided by the observatory's attitude control system is limited by the position accuracy of individual Guide Star Catalog ([14]) stars. The *Hubble* Source Catalog (HSC, [27]) solves this problem by cross-matching sources detected in individual HST visits ([3]) and matching the brightest detected sources to deep reference catalogs: PanSTARRS, SDSS and 2MASS. The Gaia catalog will be used as reference in future HSC versions.

The HSC provides access to a uniform reduction of the majority of publicly available images obtained with the WFPC2, ACS/WFC, WFC3/UVIS, and WFC3/IR instruments. The HSC is very inhomogeneous: it includes 112 instrument-filter combinations; some filters are more popular than others. Most fields were observed only few times and the time between visits varies. The current HSC version 2 has 89 fields containing 1.7 million sources that were visited at least 5 times while only 7 fields containing 290000 sources visited $\geq$ 25 times. The HSC is based on visit-combined images that are deeper than individual exposures and mostly clean of cosmic rays. However, that comes at a price: information about changes on timescales shorter than one visit (that may last a few hours) is averaged-out and the resulting number of independent measurements is by a factor of a few smaller than the number of HST exposures of the field.

We aim to define a set of algorithms that will detect and validate candidate variable sources within the HSC, producing the *Hubble* Catalog of Variables (HCV). This is a work in progress that will lead to the release of the first version of the HCV in 2018. The initial project overview is given by [8].

2 Variability search problem

It is common in optical photometry that measurement errors are not accurately known. The contribution of random background variations and photon noise to the brightness measurements uncertainty can be estimated easily for CCD observations. However, the hard-to-quantify residual systematic effects (pixel-to-pixel sensitivity variations, charge transfer inefficiency, blending with nearby stars) limit the photometric accuracy for brighter objects. Some measurements get corrupted by cosmic ray hits, CCD cosmetic defects, incorrect background estimation near frame edges and other undetected data processing anomalies. Since the majority of field stars are not variable at a few per cent level, we utilize them to estimate typical photometric accuracy in a given dataset as a function of magnitude.

Ground-based photometry was used by [22] to compare a number of variability indices that characterize "how variable" a lightcurve appears by quantifying its scatter and smoothness. We extend this work by comparing the variability indices listed in Table 1 to simulations based on HSC data. We inject variability with random amplitude into HSC lightcurves of non-variable objects using the technique described by [22]. The simulations confirm that the interquartile range (IQR) of the measured magnitudes, m_i together with the inverse von Neumann ratio that characterizes the lightcurve smoothness $1/\eta = \sum_{i=1}^{N}(m_i - \bar{m})^2 / \sum_{i=1}^{N-1}(m_{i+1} - m_i)^2$ (where $\bar{m}$ is the mean magnitude and N is the number of measurements) can recover a broad range of variability patterns and are robust against individual outlier measurements. These indices do not depend on the estimated errors (Table 1) that may be unreliable. Figure 1 presents IQR and $1/\eta$ indices as a function of magnitude in one of the simulations. Figure 2 illustrates how the variability detection efficiency of the indices changes as a function of the number of points in a lightcurve.

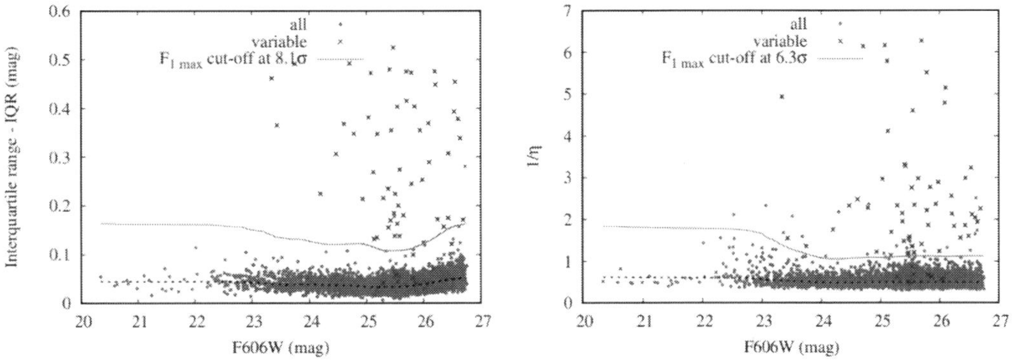

Figure 1. Variability indices characterizing scatter (left panel) and lightcurve smoothness (right panel) plotted as a function of magnitude. The indices are computed for HSC lightcurves of non-variable objects in the M31 halo field ([2]). Crosses mark the objects in which artificial non-periodic variability was injected as described in [22]. Selecting objects above the solid line as candidate variables results in the maximum F_1-score. Dotted line shows the median value of a variability index as a function of magnitude.

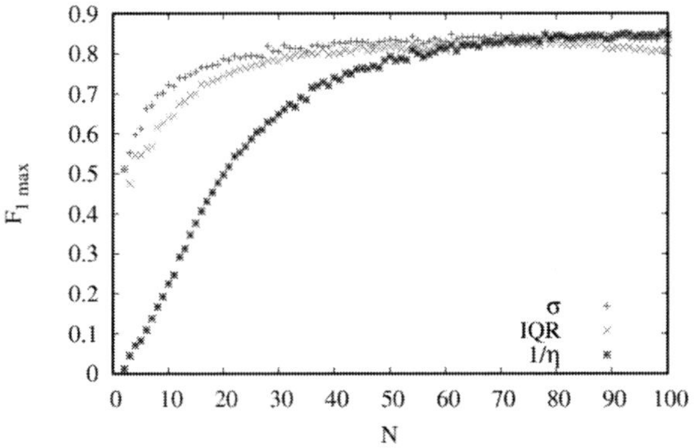

Figure 2. The efficiency of variable objects selection characterized by the maximum (over all possible cut-off thresholds) F_1-score as a function of the number of lightcurve points, N, for three variability indices: σ, IQR and $1/\eta$ (see the references in Table 1). Based on simulated non-periodic variability injected in HSC lightcurves (M4 field, [17]). For the definition of F_1-score, see `https://en.wikipedia.org/wiki/F1_score`.

Table 1. Variability indices computed by the HCV pipeline.

Index	Errors	Ref.	Index	Errors	Ref.
Indices quantifying lightcurve scatter			time-weighted Stetson's $J_{\rm time}$	✓	[7]
reduced χ^2 statistic – $\chi^2_{\rm red}$	✓	[4]	clipped Stetson's $J_{\rm clip}$	✓	[22]
weighted std. deviation – σ	✓	[12]	Stetson's L index	✓	[24]
median abs. deviation – MAD		[29]	time-weighted Stetson's $L_{\rm time}$	✓	[7]
interquartile range – IQR		[22]	clipped Stetson's $L_{\rm clip}$	✓	[22]
robust median stat. – RoMS	✓	[19]	consec. same-sign dev. – *Con.*		[28]
norm. excess variance – $\sigma^2_{\rm NXS}$	✓	[16]	excursions – E_x	✓	[18]
norm. peak-to-peak amp. – υ	✓	[21]	autocorrelation – l_1		[11]
Indices quantifying lightcurve smoothness			inv. von Neumann ratio – $1/\eta$		[20]
Welch-Stetson index – I	✓	[26]	excess Abbe value – $\mathcal{E}_{\mathcal{A}}$		[15]
Stetson's J index	✓	[24]	S_B statistic	✓	[6]

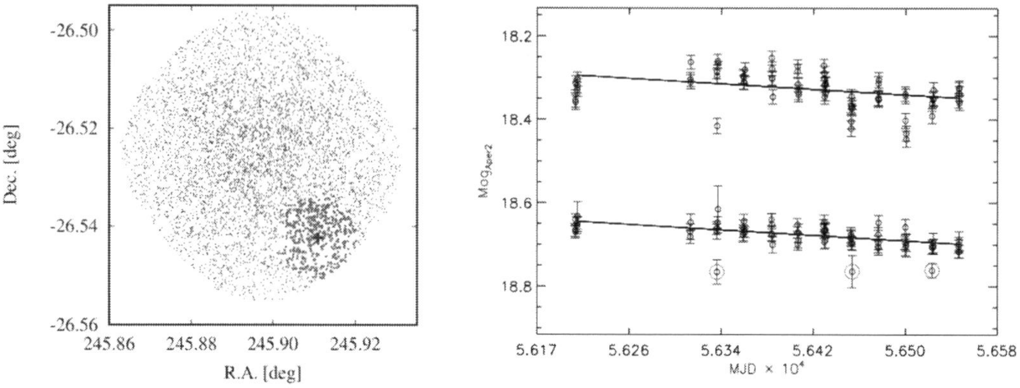

Figure 3. Lightcurve pre-processing. Left: spatial distribution of HSC objects in the M4 field ([17]). Highlighted are the objects used to determine local magnitude zero-point correction for the object marked with the cross. Right: lightcurve of that object before (top) and after (bottom) applying the local correction. The circles mark identified outlier points. The two lightcurves are shifted along the magnitude axis for clarity.

3 HCV data pre-processing and variability detection pipeline

The HSC data are imported and grouped according to the observed field. The measurements of objects near frame edges or obtained with an old version of the image processing pipeline are excluded as unreliable. For each object all its measurements in a given filter that pass the above selection are collected to construct a lightcurve. Each lightcurve is fitted with a straight line using robust regression and outliers from the fit are identified. If a high percentage of measurements obtained during some visit are identified as outliers in the corresponding lightcurves, all measurements associated with this visit are discarded (bad image). For each of the remaining visits, for each object a local zero-point correction is computed as the mean difference between the measured magnitudes and the ones predicted by the robust line fit for all objects within a specified radius around the object being corrected (Fig. 3). This should compensate for the residual large-scale sensitivity variations across the image.

A set of corrected HSC lightcurves of objects observed in a given field with the same instrument-filter combination is the basic unit of variability search. For each lightcurve in a set we compute variability indices and select as candidate variables the objects that have variability index values significantly higher than the typical value for their brightness. Magnitude-dependent cuts in robust indices IQR and $1/\eta$ are currently used to select candidate variables (Fig. 1). We are investigating ways to efficiently combine information captured by all indices listed in Table 1 by means of the principal component analysis and machine learning.

The HCV pipeline is implemented in Java and parallelized using the Apache Spark framework. The variability indices implementation in the pipeline is consistent with their implementation in VaST ([23]) which is also used for lightcurve visualization and testing while a dedicated HCV visualization software is being developed.

4 Summary

- The HCV is a catalog of variable objects derived from the HSC. It will be released in 2018.
- The catalog is very heterogeneous due to the nature of the HSC dataset. It covers selected fields in the Galaxy, the Local Group and beyond.

- The HCV is very deep, it ventures into poorly explored region of variability parameter space.
- Data pre-processing and variability detection techniques used for HCV are applicable to other multi-epoch surveys.

Acknowledgments: This work is supported by ESA under contract No. 4000112940.

References

[1] Bernard, E. J., et al., MNRAS, **432**, 3047 (2013)

[2] Brown, T. M., Ferguson, H. C., Smith, E., Kimble, R. A., Sweigart, A. V., Renzini A., & Rich, R. M., AJ, **127**, 2738 (2004)

[3] Budavári, T., & Lubow, S. H., ApJ, **761**, 188 (2012)

[4] de Diego, J. A., AJ, **139**, 1269 (2010)

[5] Dolphin, A. E., et al., ApJ, **550**, 554 (2001)

[6] Figuera Jaimes, R., Arellano Ferro, A., Bramich, D. M., Giridhar, S., & Kuppuswamy, K., A&A, **556**, A20 (2013)

[7] Fruth, T., et al., AJ, **143**, 140 (2012)

[8] Gavras, P., et al., preprint (arXiv:1703.00258) (2017)

[9] Hoffmann, S. L., et al., ApJ, **830**, 10 (2016)

[10] Jeffery, E. J., et al., AJ, **141**, 171 (2011)

[11] Kim, D.-W., Protopapas, P., Alcock, C., Byun, Y.-I., & Khardon, R., ASPC, **442**, 447 (2011)

[12] Kolesnikova, D. M., Sat, L. A., Sokolovsky, K. V., Antipin, S. V., & Samus, N. N., AcA, **58**, 279 (2008)

[13] Lanzerotti, L. J., *Assessment of options for extending the life of the Hubble Space Telescope: final report* (National Academies Press, Washington D.C, 2005)

[14] Lasker, B. M., et al., AJ, **136**, 735 (2008)

[15] Mowlavi, N., A&A, **568**, A78 (2014)

[16] Nandra, K., George, I. M., Mushotzky, R. F., Turner, T. J., & Yaqoob, T., ApJ, **476**, 70 (1997)

[17] Nascimbeni, V., et al., MNRAS, **442**, 2381 (2014)

[18] Parks, J. R., Plavchan, P., White, R. J., & Gee, A. H., ApJS, **211**, 3 (2014)

[19] Rose, M. B., & Hintz, E. G., AJ, **134**, 2067 (2007)

[20] Shin, M.-S., Sekora, M., & Byun Y.-I., MNRAS, **400**, 1897 (2009)

[21] Sokolovsky, K. V., Kovalev, Y. Y., Kovalev, Y. A., Nizhelskiy, N. A., & Zhekanis G. V., AN, **330**, 199 (2009)

[22] Sokolovsky, K. V., et al., MNRAS, **464**, 274 (2017)

[23] Sokolovsky, K. V., & Lebedev, A. A., preprint (arXiv:1702.07715) (2017)

[24] Stetson, P. B., PASP, **108**, 851 (1996)

[25] Villforth, C., Koekemoer, A. M., & Grogin N. A., ApJ, **723**, 737 (2010)

[26] Welch, D. L., & Stetson, P. B., AJ, **105**, 1813 (1993)

[27] Whitmore, B. C., et al., AJ, **151**, 134 (2016)

[28] Wozniak, P. R., AcA, **50**, 421 (2000)

[29] Zhang, M., Bakos, G. Á., Penev, K., Csubry, Z., Hartman, J. D., Bhatti, W., & de Val-Borro, M., PASP, **128**, 035001 (2016)

Progress report on the K2 RR Lyrae survey

Róbert Szabó[1,*]

[1]*Research Centre for Astronomy and Earth Sciences, Konkoly Observatory, H-1121, Budapest, Konkoly Thege Miklós út 15-17., Hungary*

Abstract. We have initiated a large survey with K2, to observe thousands of RR Lyrae stars along the ecliptic. The high photometric precision and the 70-90-day continuous coverage will allow us to investigate the light variation of these galactic structure tracer variable stars with unprecedented detail. The survey will help us to conduct a thorough statistical study of RR Lyrae pulsation dynamics including old and recently discovered dynamical phenomena, like resonances, non-radial modes, period-doubling and the Blazhko effect. In this contribution I present a status report of the survey.

1 Introduction

With the failure of the second reaction wheel of *Kepler*, a new mission, dubbed K2, was conceived ([4]). The new mission observes ecliptic fields for 70–90 days with a precision which is comparable to the original mission. K2 is a bona fide community endeavour: without a central, pre-determined research plan, exoplanets, stellar (asteroseismic and other variable) sources, extragalactic objects, microlensing events, and even Solar System objects have been proposed by community members.[1] This opened up new possibilities in RR Lyrae research, too, since compared to the original *Kepler* field, where roughly 50 RR Lyrae stars were found ([1, 6, 7]), thousands of RR Lyrae stars can be observed with K2. Members of the *Kepler* Asteroseismic Science Consortium (KASC) RR Lyrae and Cepheids Working Group initiated a large survey to ensure that as many RR Lyrae stars will be observed with K2 as possible.[2] This contribution discusses the present status of this undertaking.

2 The K2 RR Lyrae survey

As Figure 1 demonstrates, our group was successful in proposing RR Lyrae targets throughout the whole K2 Mission. From Campaign 4 the proposals are submitted through the NASA NSPIRES system.[3] Figure 1 also shows that RR Lyrae stars are measured in very different environments: in the Galactic halo and halo sub-structures (Sagittarius-stream), in the Galactic bulge, in globular clusters (M4, M80), and even in external galaxies (e.g., the Leo IV dwarf spheroidal galaxy; [5]). We used all possible variable star catalogs for target selection, and heavily relied on the Catalina Sky Survey

*szabo.robert@csfk.mta.hu

[1]https://keplerscience.arc.nasa.gov/k2-approved-programs.html

[2]The same strategy applies to Cepheids, both classical and Type II ones (see Molnár et al., these proceedings).

[3]https://nspires.nasaprs.com/

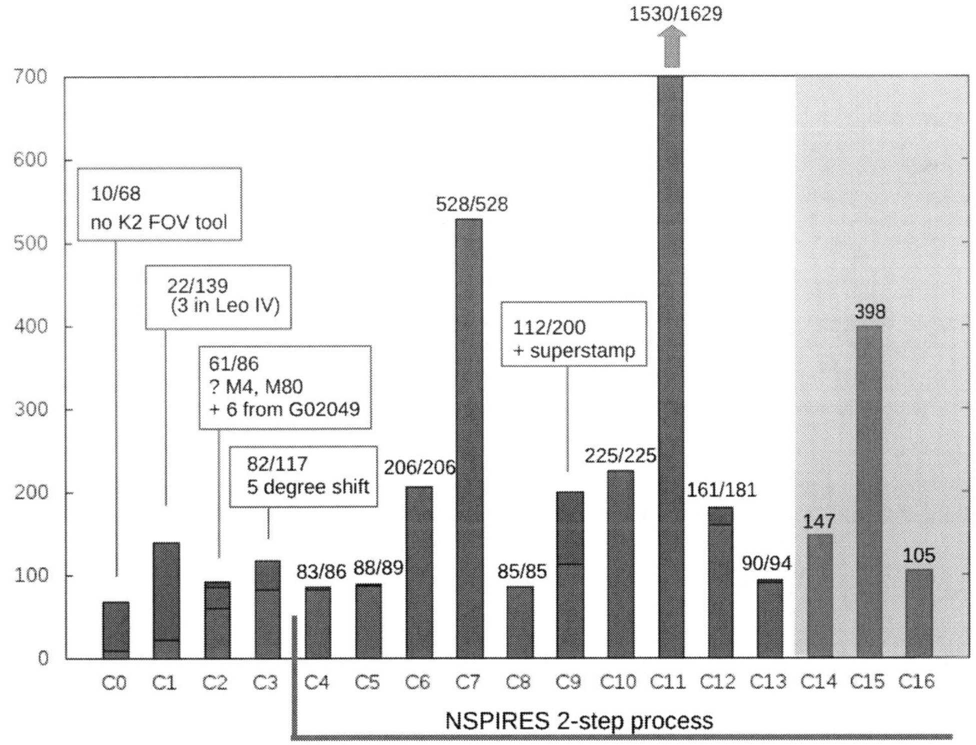

Figure 1. Statistics of the number of proposed (blue) and observed (green) RR Lyrae targets in the K2 campaigns. The K2 RR Lyrae survey has been largely successful in securing most of the known RR Lyrae stars for K2 observations. The grey area shows those campaigns that are scheduled at the time of writing. Two more campaigns (C17, C18) are envisaged, contingent on available fuel on board the *Kepler* spacecraft.

results ([3]). A more detailed description of the target selection process can be found in Plachy et al. (2016, [8]).

By the end of the K2 Mission we expect to have light curves of over 3000 RR Lyrae stars in the *Kepler* photometric band (Kp). The majority of the targets is observed with long cadence (30-min sampling), while a small fraction is observed with short cadence (1-min sampling) mode. The distribution of the K2 RR Lyrae stars on the sky is plotted in Figure 2. As correction maneouvres are necessary to keep the spacecraft on-target, special treatment of the light curves (in particular for high-amplitude variables) is needed. Hence, we developed the Extended Aperture Photometry method ([9]) that delivers optimal results for our targets. This way quasi-continuous light curves covering 70 – 90 days are derived with high photometric precision, which makes the K2 RR Lyrae survey unique. Our project nicely complements other sky surveys, e.g. OGLE, Gaia or LSST.

The main scientific motivation behind our efforts is a statistical study of the occurrence of dynamical phenomena (Blazhko effect, period doubling, nonradial modes, etc.) in RR Lyrae stars as a function of age, metallicity, and Galactic position. By these investigations we hope to achieve a better

DOI: 10.1051/epjconf/201715202006

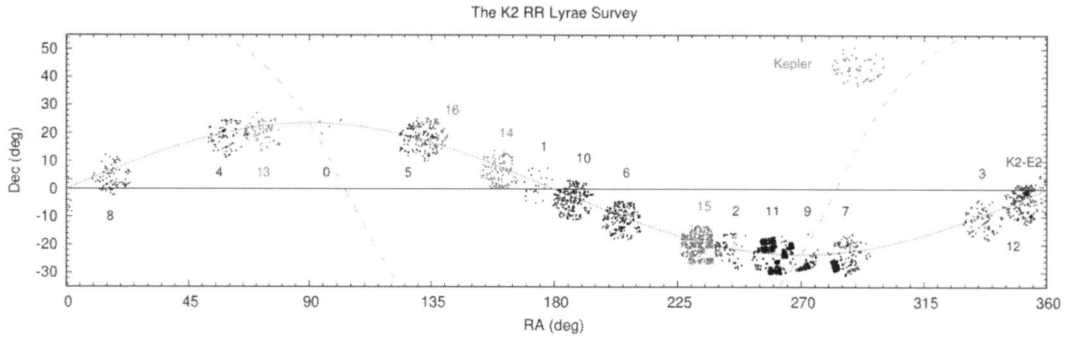

Figure 2. The footprint of the K2 RR Lyrae survey along the Ecliptic. Approved (blue) and observed (black) RR Lyrae targets are marked in the K2 campaign fields. The original *Kepler* field is also plotted for reference (red).

understanding of recently discovered (period doubling, [10]; extra frequencies, [11]) or century-old (like the Blazhko-modulation, [2]) dynamical phenomena occurring in RR Lyrae stars.

3 Summary

We presented the K2 RR Lyrae survey which will be a treasure trove for pulsation and dynamical studies for many years to come. It contains representatives of all currently known subtypes (RRab, RRc, RRd, modulated and non-modulated variants, as well). The corrected, "ready-to-use" RR Lyrae light curves from survey will be made publicly available (Molnár et al. in prep.).

Acknowledgments: This project has been supported by the Hungarian National Research, Development and Innovation Office – NKFIH K-115709, the GINOP-2.3.2-15-2016-00003 grant, and the LP2014-17 Lendület grant of the Hungarian Academy of Sciences. This work includes data collected by the K2 mission. Funding for the K2 mission is provided by the NASA Science Mission Directorate. We acknowledge the *Kepler* team and engineers for their efforts to keep this fantastic instrument alive and design a new mission.

References

[1] Benkő, J., Plachy, E., Szabó, R., Molnár, L., & Kolláth, Z., ApJS, **213**, 31 (2014)

[2] Blazhko, S., AN, **175**, 325 (1907)

[3] Drake, A. J., Graham, M. J., Djorgovski, S. G., et al., ApJS, **213**, 9 (2014)

[4] Howell, S., Sobeck, C., Haas, M. et al., PASP, **126**, 398 (2014)

[5] Molnár, L., Pál, A., Plachy, E. et al., ApJ, **821**, 2 (2014)

[6] Moskalik, P., Smolec, R., Kolenberg, K. et al., MNRAS, **447**, 2348 (2015)

[7] Nemec, J. M., Smolec, R., Benkő, J. M. et al. MNRAS, **417**, 1022 (2011)

[8] Plachy, E., Molnár, L., Szabó, R., Kolenberg, K., & Bányai, E., Communications from the Konkoly Observatory, **105**, 19 (2016)

[9] Plachy, E., Klagyivik, P., Molnár, L., Sódor, Á., Szabó, R., in *Seismology of the Sun and the Distant Stars*, EPJ Web of Conferences, in press (2017)

[10] Szabó, R., Kolláth, Z., Molnár, L. et al. MNRAS, **409**, 1244 (2010)

[11] Szabó, R., Benkő, J., Paparó, M. et al. A&A, **570**, A100 (2014)

RR Lyrae star distance scale and kinematics from inner bulge to 50 kpc

Andrei Dambis[1,⋆], *Leonid* Berdnikov[1,2,⋆⋆], *Eva* Grebel[3,⋆⋆⋆], *Alexey* Kniazev[1,4,5,6,⋆⋆⋆⋆], *Igor* Katkov[1,†], and *Tesfaye* Dagne[2,‡]

[1]*Sternberg Astronomical Institute, M.V.Lomonosov Moscow State University, Universitetskii pr. 13, Moscow, 119992 Russia*
[2]*Astronomy and Astrophysics Research division, Entoto Observatory and Research Center, P.O. Box 8412, Addis Ababa, Ethiopia*
[3]*Astronomisches Rechen-Institut Zentrum für Astronomie der Universität Heidelberg, Germany*
[4]*South African Astronomical Observatory, P.O. Box 9, Observatory, Cape Town, 7935, South Africa*
[5]*Southern African Large Telescope, P.O. Box 9, Observatory, Cape Town, 7935, South Africa*
[6]*Special Astrophysical Observatory of RAS, Nizhnij Arkhyz, Karachai-Circassia 369167, Russia*

Abstract. We use the currently most complete sample of ∼ 3500 type ab RR Lyraes in our Galaxy with available radial-velocity and [Fe/H] measurements to perform a statistical-parallax analysis for a subsample of ∼ 600 type ab RR Lyraes located within 5 kpc from the Sun to refine the parameters of optical and WISE $W1$-band period-metallicity-luminosity relations and adjust our preliminary distances. The new zero point implies the rescaled estimates for the solar Galactocentric distance (R_G = 7.99 ± 0.37 kpc) and the LMC distance modulus ($DM_{\rm LMC}$ = 18.39 ±0.09). We use the kinematic data for the entire sample to explore the dependence of the halo and thick-disk RR Lyrae velocity ellipsoids on Galactocentric distance from the inner bulge out to $R \sim 50$ kpc.

1 Introduction

The high luminosities and old age make RR Lyrae type variables excellent distance indicators and kinematic tracers, in particular, for the halo and the thick disk subsystems of our Galaxy. Hence building a large enough sample of these variables having precise and homogeneous photometry, homogeneous metallicity estimates, and bona fide radial velocities is a task of prime importance. Until recently, only about 400 plus RR Lyraes in the extended solar neighborhood had all these data available ([3, 4]), and other spectroscopically measured Galactic variables of this type have been observed within the framework of SDSS survey and are mostly very distant objects. The situation with spectroscopic observations of RR Lyraes at intermediate distances has improved recently with the public

⋆dambis@yandex.ru
⋆⋆lberdnikov@yandex.ru
⋆⋆⋆grebel@ari.uni-heidelberg.de
⋆⋆⋆⋆akniazev@saao.ac.za
†katkov.ivan@gmail.com
‡tesfayedagne7@gmail.com

DOI: 10.1051/epjconf/201715202007

releases of LAMOST survey. To expand the spectroscopically observed RR Lyrae sample in the extended solar neighborhood we launched a dedicated program with the purpose of acquiring spectra for the greatest possible number of RR Lyraes to complement our long-term program of photometric observations of these variables. Here we briefly describe the data obtained and report some results concerning the RR Lyrae distance scale and halo kinematics based on our measurements combined with published data including those provided by huge SDSS and LAMOST spectroscopic surveys.

2 The sample and data

Our sample consists of all 3444 type ab RR Lyrae variables for which we could find published radial-velocity measurements or measure radial velocities from the spectra we acquired with the Southern African Large Telescope (SALT) in 2015–2016.

2.1 Radial velocities and metallicities

The adopted RR Lyrae radial velocities and metallicities come from a number of sources. First, our own spectroscopic data with 320 measurements for 300 stars including 260 stars with no prior spectroscopic data available. Our observations cover spectral range of 3900-5000 Å, using grating GR1300 and slit with 1.25 arcsec width. It gives us a possibility to detect Balmer lines (Hδ, Hγ, Hβ) and Ca II K (3924 Å) line to use S-method for metallicity determination and produces spectral resolution $R = 1300 - 1500$. All data are very standard long-slit and are reduced automatically with our pipeline based on MIDAS and IRAF programs. Second, the published results of dedicated spectroscopic studies for about 400 stars compiled in our paper ([4]), third, measurements available in public releases of SDSS ([1]), LAMOST ([15]), and BRAVA ([9]) surveys (about 2800 stars). Typical errors of systemic V_R values inferred from these data range from few to 60 km/s (we show their distribution in Fig. 1) with a mean value of ~ 22 km/s. Figure 2 shows the distribution of [Fe/H] values for stars of our sample (transformed to the Zinn & West [16] scale).

2.2 Photometry and photometric distances

We estimated the provisional distances of most of the stars of our sample based on the $W1$-band mid-infrared light curves acquired as a result of spaceborne WISE all-sky photometric survey ([13]) and the corresponding period-luminosity relation of Dambis et al. ([5]):

$$\langle M_{W1} \rangle = -0.814 - 2.381 \cdot \log P_F + 0.106 \cdot [\mathrm{Fe/H}]_{\mathrm{ZW}}. \tag{1}$$

We determined the distances to stars in heavily obscured bulge regions based on OGLE VI photometry ([10]) and the V-band metallicity-luminosity relation of Dambis et al. ([4]) consistent with the above $W1$-band PLZ relation:

$$\langle M_V \rangle = +1.094 + 0.232 \cdot [\mathrm{Fe/H}]_{\mathrm{ZW}}. \tag{2}$$

Finally, we determined the distances to faint stars outside the bulge region with no or unreliable WISE data (discovered mostly as a result of an analysis of Catalina Sky Survey data [6, 11]) from their Gaia G-band photometry using the following PLZ relation derived in the same way as the corresponding $W1$-band relation that we reported in [5]:

$$\langle M_G \rangle = -0.550 - 1.531 \cdot \log P_F + 0.140 \cdot [\mathrm{Fe/H}]_{\mathrm{ZW}}. \tag{3}$$

Interstellar extinction was taken into account either using the extinction map ([7]) or appropriate period-metallicity-color relations.

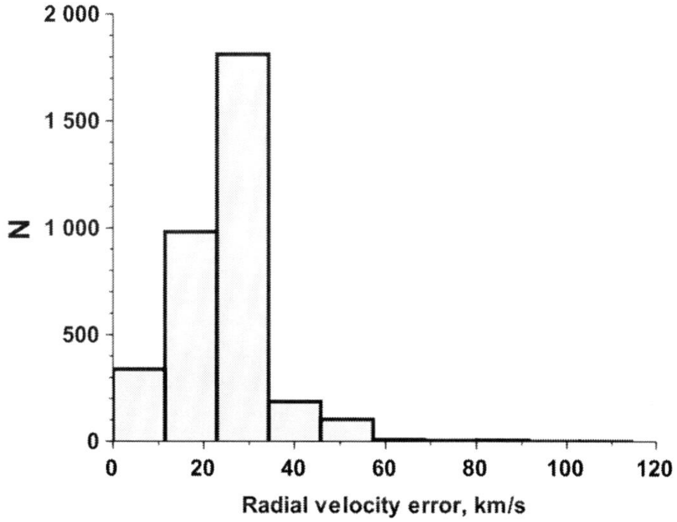

Figure 1. Distribution of errors of systemic V_R values for type ab RR Lyrae variables of our sample.

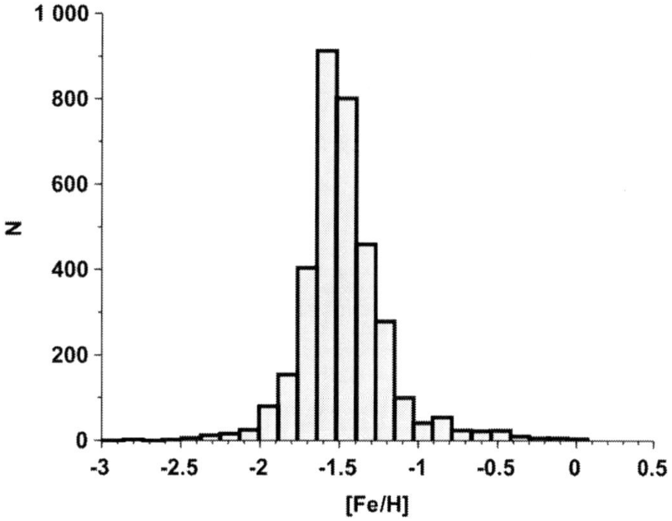

Figure 2. Distribution of [Fe/H] values for type ab RR Lyrae variables of our sample transformed to the Zinn & West scale.

Table 1. Kinematical parameters of RR Lyrae populations in the extended solar neighborhood and the refined WISE $W1$-band PLZ relation.

Population	Sample fraction	U_0	V_0	W_0	σV_R	σV_φ	σV_θ
	(α and 1-α)			km/s			
Halo	0.766	−1.8	−214.6	−6.0	163.6	94.7	88.4
	± 0.028	± 8.0	± 7.5	± 4.4	± 6.9	± 4.7	± 3.8
Disc	0.234	−14.4	−36.5	−18.6	46.5	37.8	27.9
	± 0.028	± 5.5	± 5.4	± 3.6	± 5.5	± 4.9	± 3.4
$M_{W1} = -0.885 + 2.381 \cdot \log P + 0.106 \cdot [\mathrm{Fe/H}] \pm 0.060$							

2.3 Proper motions

Our sources of proper motions are the APOP ([12], used for most of the stars at Galactic latitudes $|b| > 27°$), HSOY ([2]), and UCAC4 ([14]) catalogs.

3 Results

3.1 Distance scale and local kinematics

We first used the subsample of 610 stars located within 5 kpc from the Sun drawn from our initial sample to refine the zero point of our WISE $W1$-band PLZ relation ([5]) and estimate the kinematical parameters (mean velocities and parameters of velocity dispersion ellipsoid) of the halo and thick-disk RR Lyrae populations in the solar neighborhood using the bimodal method of statistical parallax (see [3, 4] for details). The results are summarized in Table 1.

Our refined WISE $W1$-band PLZ zero point is brighter than the value derived in [4, 5] implying the rescaled estimates for the solar Galactocentric distance ($R_G = 7.99 \pm 0.37$ kpc) and the LMC distance modulus ($DM_{\mathrm{LMC}} = 18.39 \pm 0.09$), which agree well with most of the current estimates.

3.2 Velocity ellipsoid of the halo

We then fix the mean velocity components of halo RR Lyrae stars at the values listed in the first row of Table 1 and assume the halo velocity ellipsoid to be two-axial (i.e., we set $\sigma V_\varphi = \sigma V_\theta$) to explore the variation of the radial component of halo velocity ellipsoid (σV_R) with Galactocentric distance R_G. We show the resulting dependence in Figure 3. The main feature of the radial velocity dispersion behavior is that it decreases steadily from $\sim$ 180 km/s to $\sim$ 100 km/s in the Galactocentric distance interval $R_G = 3 - 16$ kpc and remains practically constant at greater distances at least out to $R_G \sim 50$ kpc, which mostly agrees with the results of Kafle et al. ([8]) based on blue horizontal-branch star data.

Acknowledgments: This work makes use of observations from the South African Astronomical Observatory(SAAO) and Southern African Large Telescope (SALT), supported by the National Research Foundation of South Africa, and data from the SDSS and LAMOST projects. The data reduction of all data was supported by the Russian Science Foundation (project no. 14-50-00043), and the light-curve analysis was supported by the Russian Science Foundation (project no. 14-22-00041).

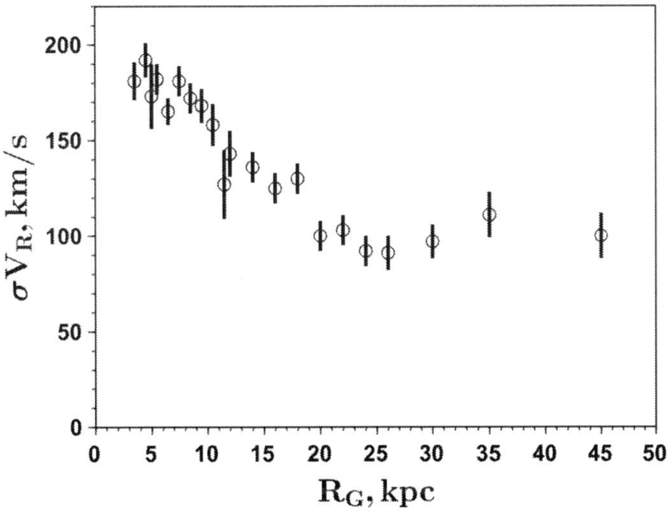

Figure 3. Radial velocity dispersion for halo RR Lyraes as a function of Galactocentric distance.

References

[1] Albareti, F. D., Allende Prieto, C., Almeida, A., et al., preprint (arXiv:1608.02013) (2016)
[2] Altmann, M., Roeser, S., Demleitner, M., Bastian, U., & Schilbach, E., A&A, **600**, L4 (2017)
[3] Dambis, A. K., MNRAS, **396**, 553 (2009)
[4] Dambis, A. K., Berdnikov, L. N., Kniazev, A. Yu., Kravtsov, V. V., Rastorguev, A. S., Sefako, R., & Vozyakova, O. V., MNRAS, **435**, 3206 (2013)
[5] Dambis, A. K., Rastorguev, A. S., & Zabolotskikh, M. V., MNRAS, **439**, 3765 (2014)
[6] Drake, A. J., Catelan, M., Djorgovski, S. G., et al., ApJ, **763**, 32 (2013)
[7] Drimmel, R., Cabrera-Lavers, A., López-Corredoira, M., A&A, **409**, 205 (2003)
[8] Kafle, P. R., Sharma, S., Lewis, G. F., & Bland-Hawthorn, J., ApJ, **761**, 98 (2012)
[9] Kunder, A., Rich, R. M., Koch, A., et al., ApJL, **821**, L25 (2016)
[10] Soszyński, I., Udalski, A., Szymański, M. K., et al., AcA, **64**, 177 (2014)
[11] Torrealba, G., Catelan, M., Drake, A. J., et al., MNRAS, **446**, 2251 (2015)
[12] Qi, Z., Yu, Y., & Bucciarelli, B., AJ, **150**, 137 (2015)
[13] Wright, E. L., Eisenhardt, P. R. M., Mainzer, A. K., et al., AJ, **140**, 1868 (2010)
[14] Zacharias, N., Finch, C. T., Girard, T. M., et al., AJ, **145**, 44 (2013)
[15] Zhao, G., Zhao, Y.-H., Chu, Y.-Q., Jing, Y.-P., & Deng, L.-C., RA&A, **12**, 723 (2012)
[16] Zinn, R., & West, M. J., ApJS, **55**, 45 (1984)

Faint (and bright) variable stars in the satellites of the Milky Way

A. Katherina Vivas[1,*]

[1] *Cerro Tololo Inter-American Observatory, Casilla 603, La Serena, Chile*

Abstract. I describe two ongoing projects related with variable stars in the satellites of the Milky Way. In the first project, we are searching for dwarf Cepheid stars (a.k.a δ Scuti and/or SX Phe) in some of the classical dwarf spheroidal galaxies. Our goal is to characterize the population of these variable stars under different environments (age, metallicity) in order to study their use as standard candles in systems for which the metallicity is not necessarily known. In the second project we search for RR Lyrae stars in the new ultra-faint satellite galaxies that have been discovered around the Milky Way in recent years.

1 Introduction

The last two years have revealed a completely new picture in our understanding of the Galactic neighborhood with the discovery of $\sim$ 20 dwarf systems surrounding the Milky Way (MW), doubling the number of satellites known in 2014 ([1–9]), and narrowing the discrepancy between the predicted number of satellites in the standard Lambda Cold Dark Matter (LCDM) cosmological framework and observations, a problem known for many years as the missing satellites problem. Many of these new satellites are extremely low-luminosity systems and are opening research in a new low mass regime for galaxies. The role of these small ultra-faint satellites in the formation of galaxies like our own remains to be understood. Key to such characterization is accurate distance measurements toward the new galaxies.

Variable stars have a long tradition of being an excellent tool for studying the content and structure of stellar systems. RR Lyrae stars are not only excellent standard candles, but they also trace the old population of these systems. Anomalous Cepheids (which are brighter than RR Lyrae stars) are usually interpreted as belonging to an intermediate-age population (e.g., [10]). Below the horizontal branch, dwarf Cepheids may have been produced by different stellar populations and are potentially another good standard candle.

Here I discuss two ongoing projects to study and characterize variable stars in satellites of the Milky Way.

2 Faint variable stars: dwarf Cepheid stars in the Sextans dSph galaxy

δ Scuti and SX Phe stars (which I will collectively call dwarf Cepheid stars here) co-exist in the same region of the color-magnitude diagram (CMD) of dwarf spheroidal (dSph) galaxies. The region of

*kvivas@ctio.noao.edu

the instability strip below the horizontal branch may contain not only main sequence pulsating stars (δ Scuti) from young and intermediate-age stellar populations, but also pulsating blue stragglers (SX Phe) belonging to the old stellar population. If a galaxy contains both types of populations, it is usually not possible to separate them and that is the reason why a collective name for both types of stars is useful.

The search for dwarf Cepheid stars in extragalactic systems has been difficult because of observational limitations. First, dwarf Cepheid stars are faint ($\sim$ 1.5 – 2.5 magnitudes below the horizontal branch, which for distant systems are beyond the limit of most of the existing variability surveys, except for OGLE/MACHO in the Magellanic Clouds (see [11]). Second, dwarf Cepheids have very short periods ($<$ 0.1 d) which in practice means that short exposure times ($\sim$ a few minutes) need to be used in order to correctly sample the light curves. This imposition on the exposure times requires the use of medium/large aperture telescopes in order to reach the magnitude of dwarf Cepheid stars in distant systems. Finally, several of the dSph galaxies around the MW extend over large areas of the sky and a complete census is observationally expensive unless a large field of view (FOV) imager is availiable at the telescope. It is important to note that all these difficulties will be naturally overcome by the LSST project ([12]).

The importance of finding and characterizing dwarf Cepheid stars resides in their potential use as standard candles. They are known to follow a clear period-luminosity (PL) relationship ([13]). It has been suggested that such PL relationship does not depend on color and/or metallicity ([14]). However, more systems with dwarf Cepheid stars coming from different environments (age, metallicity, star formation histories) are needed in order of establishing better constraints on their use as standard candles in systems for which the metallicity is not necessarily known. The dSph galaxies contain stars that inhabit a different age/metallicity range than those found anywhere else in the Galaxy and so offer the chance to study these variables from unique and otherwise hard to study populations.

Dwarf Cepheid stars have been searched so far only in three extra-Galactic systems: the LMC ([11]), Fornax ([15]) and Carina ([16–18]). Our study in Carina ([17]) has raised several interesting points: (i) Carina is very rich in dwarf Cepheid stars; (ii) There seems to be a fundamental difference between the dwarf Cepheid population in dSph galaxies and the Galactic field. While high amplitude dwarf Cepheid stars are extremely rare in the field ([19]), they are at least 100$\times$ more frequent in Carina. (iii) There are important differences observed among the properties of the dwarf Cepheid population of Carina, Fornax and the LMC which may be a reflection of a metallicity spread, depth along the line of sight, and/or different evolutionary paths of the dwarf Cepheid.

In collaboration with Javier Alonso-García (Universidad de Antofagasta, Chile), Mario Mateo (University of Michigan), Alistair Walker and David Nidever (NOAO), we have started a project to look for dwarf Cepheid stars in several satellite galaxies using the large FOV of the Dark Energy Camera (DECam, [20]) on the 4m Blanco telescope at Cerro Tololo Inter-American Observatory, Chile. One of our targets is the Sextans dSph system. Sextans is a low luminosity system located at 86 kpc. It does not have a strong intermediate age population but there is evidence of continuous star formation in the last $\sim$ 8 Gyrs ([21]) and a significant spread in metallicity ([22]). Figure 1 shows the spatial distribution of the variable stars (30 RR Lyrae stars, 2 Sextans dwarf Cepheid stars and 1 foreground dwarf Cepheid) we were able to identify in one of our DECam fields (out of 3 we observed), and the resulting CMD from that field, showing as well the variable stars. All the RR Lyrae stars found were previously known.

The field shown in Figure 1 is not centered at the galaxy but along the semi-major axis and contains stars between 13 and 134 arcmin from the center. We will present a complete coverage of the galaxy in a forthcoming paper. With these preliminary results it is clear that the population of dwarf Cepheid stars in Sextans is not as numerous as in Carina, which is probably a consequence of the small presence of intermediate-age stars in this galaxy. The periods of the two dwarf Cepheid stars in Sextans are

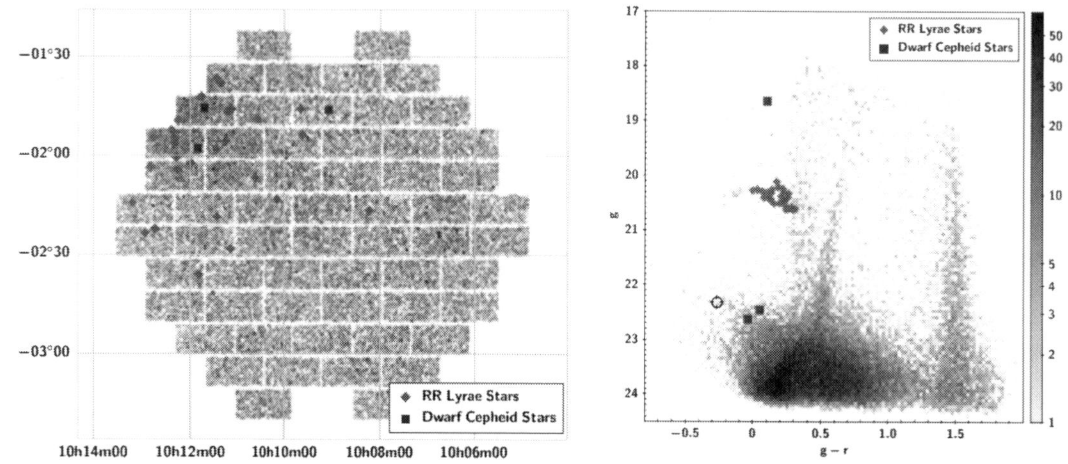

Figure 1. (Left) Spatial distribution of the variable stars in the DECam field. Colored symbols indicate the variable stars found in this field. For reference the center of Sextans is at RA = 10:13:03, DEC = −01:36:53. (Right) Hess diagram of stars closer than 60 arcmin of the center of Sextans. The bright dwarf Cepheid in the diagram is likely a foreground field star in the line of sight of the galaxy. That star is also the more distant from the center of Sextans.

0.0521 and 0.0576 days, which are shorter than the mean period of the dwarf Cepheid stars in Carina (which are shorter than the ones in Fornax and the LMC), following a trend with metallicity already suggested in [17]. This trend needs to be confirmed once the full galaxy has been explored.

3 Bright variable stars: RR Lyrae stars in the ultra-faint satellites of the Milky Way

RR Lyrae stars are excellent standard candles and thus provide an alternative method for determining the distance to galaxies. This is particularly important in distant and low luminosity systems lacking young populations (i.e., systems that cannot host Cepheids that could be used as distance indicators). In addition, estimation of accurate distances by isochrone fitting may be hard because usually there are very few stars in the upper part of the CMD, the contamination by field stars may be important due to the low density of stars in the system, and the main sequence turn-off may not be available in some cases due to their large distances. Improving the distance to the satellites allows a better determination of other parameters such as the physical size (half-light radius) and absolute magnitude of the systems.

As part of the Survey of the Magellanic Stellar History (SMASH, [23]), we obtained time series in *gri* with DECam of the recently discovered Hydra II ultra-faint dwarf ([4]). One RR Lyrae star was found in the system and allowed us to get a distance to Hydra II of 151 ± 8 kpc. With this value, the physical size and absolute magnitude of the galaxy were re-derived to $r_h = 76^{+12}_{-10}$ pc and $M_V = -5.1 \pm 0.3$. Details on this work can be found in [24].

Until now, all dwarf galaxies that have been adequately searched for variable stars have yielded RR Lyrae stars, which confirms that these systems predominantly contain old, metal-poor stellar populations. Figure 2, updated from [24], is a period-amplitude diagram of the RR Lyrae stars known in the Milky Way's ultra-faint satellites. It is clear that while the halo stars (black + symbols) concentrate toward the Oo I locus, most of the RR Lyrae stars in the ultra-faint dwarfs (including the only

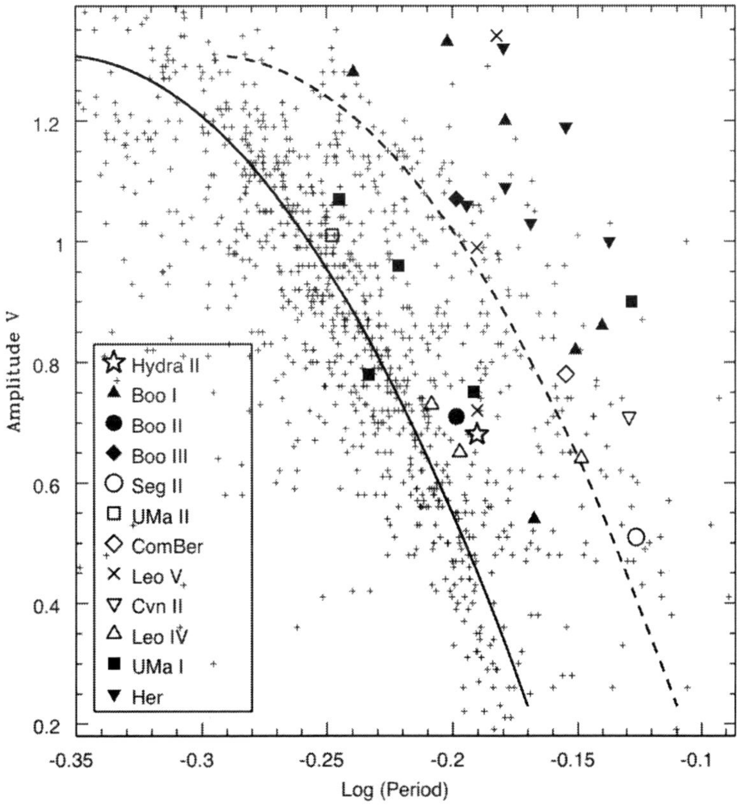

Figure 2. Period-amplitude diagram for fundamental mode RR Lyrae stars in the halo (small + symbols, [25]) and in the Milky Way's ultra-faint satellites. Reference for the later can be found in [24]. Leo V data are from [26]. The amplitudes for the stars in this galaxy are not V but g, but they should not differ by more than 0.2 mags. The locus of the Oosterhoff I and II groups are shown as a solid and dashed line respectively.

star in Hydra II) are mainly located in the region of the Oo-intermediate and Oo II loci. If the Milky Way was formed by the accretion of small sub-structures as expected by the LCDM cosmology, the ultra-faint galaxies may contribute to the Oo II tail of the population of halo RR Lyrae, but they do not seem to be the source of the main population. It seems that mergers with large galaxies such as Sagittarius and Fornax (which have a similar distribution as the halo in a period-amplitude diagram) are needed in order to reproduce the observed RR Lyrae population in the halo of the Milky Way.

4 Conclusions

Variable stars provide important information about the satellites of the Milky Way that goes from the possibility to obtain accurate distances (which allows to determine other physical parameters), to the study of the stellar populations that are present in those systems. They can also provide clues about the formation scenarios of the Milky Way.

Dwarf Cepheid stars are important to study since they are potentially good standard candles that could be used in the Local Group in the LSST era.

References

[1] K. Bechtol, A. Drlica-Wagner, E. Balbinot, A. Pieres, J.D. Simon, B. Yanny, B. Santiago, R.H. Wechsler, J. Frieman, A.R. Walker et al., ApJ **807**, 50 (2015)

[2] S.E. Koposov, V. Belokurov, G. Torrealba, N.W. Evans, ApJ **805**, 130 (2015), 1503.02079

[3] A. Drlica-Wagner, K. Bechtol, E.S. Rykoff, E. Luque, A. Queiroz, Y.Y. Mao, R.H. Wechsler, J.D. Simon, B. Santiago, B. Yanny et al., ApJ **813**, 109 (2015), 1508.03622

[4] N.F. Martin, D.L. Nidever, G. Besla, K. Olsen, A.R. Walker, A.K. Vivas, R.A. Gruendl, C.C. Kaleida, R.R. Muñoz, R.D. Blum et al., ApJL **804**, L5 (2015), 1503.06216

[5] B.P.M. Laevens, N.F. Martin, E.J. Bernard, E.F. Schlafly, B. Sesar, H.W. Rix, E.F. Bell, A.M.N. Ferguson, C.T. Slater, W.E. Sweeney et al., ApJ **813**, 44 (2015), 1507.07564

[6] D. Kim, H. Jerjen, D. Mackey, G.S. Da Costa, A.P. Milone, ApJL **804**, L44 (2015), 1503.08268

[7] D. Kim, H. Jerjen, ApJL **808**, L39 (2015), 1505.04948

[8] G. Torrealba, S.E. Koposov, V. Belokurov, M. Irwin, MNRAS **459**, 2370 (2016), 1601.07178

[9] G. Torrealba, S.E. Koposov, V. Belokurov, M. Irwin, M. Collins, M. Spencer, R. Ibata, M. Mateo, A. Bonaca, P. Jethwa, MNRAS **463**, 712 (2016), 1605.05338

[10] G. Fiorentino, P.B. Stetson, M. Monelli, G. Bono, E.J. Bernard, A. Pietrinferni, ApJL **759**, L12 (2012), 1209.6496

[11] A. Garg, K.H. Cook, S. Nikolaev, M.E. Huber, A. Rest, A.C. Becker, P. Challis, A. Clocchiatti, G. Miknaitis, D. Minniti et al., AJ **140**, 328 (2010), 1004.0955

[12] Z. Ivezic, J.A. Tyson, B. Abel, E. Acosta, R. Allsman, Y. AlSayyad, S.F. Anderson, J. Andrew, R. Angel, G. Angeli et al., ArXiv e-prints (2008), 0805.2366

[13] J.M. Nemec, A.F.L. Nemec, T.E. Lutz, AJ **108**, 222 (1994)

[14] R.E. Cohen, A. Sarajedini, MNRAS **419**, 342 (2012)

[15] E. Poretti, G. Clementini, E.V. Held, C. Greco, M. Mateo, L. Dell'Arciprete, L. Rizzi, M. Gullieuszik, M. Maio, ApJ **685**, 947-957 (2008), 0806.4453

[16] M. Mateo, D. Hurley-Keller, J. Nemec, AJ **115**, 1856 (1998), arXiv:astro-ph/9807233

[17] A.K. Vivas, M. Mateo, AJ **146**, 141 (2013), 1309.2799

[18] G. Coppola, M. Marconi, P.B. Stetson, G. Bono, V.F. Braga, V. Ripepi, M. Dall'Ora, I. Musella, R. Buonanno, M. Fabrizio et al., ApJ **814**, 71 (2015), 1509.02687

[19] L.A. Balona, J.M. Nemec, MNRAS **426**, 2413 (2012)

[20] B. Flaugher, H.T. Diehl, K. Honscheid, T.M.C. Abbott, O. Alvarez, R. Angstadt, J.T. Annis, M. Antonik, O. Ballester, L. Beaufore et al., AJ **150**, 150 (2015), 1504.02900

[21] M.G. Lee, I.S. Yuk, H.S. Park, J. Harris, D. Zaritsky, ApJ **703**, 692 (2009), 0907.5102

[22] G. Battaglia, E. Tolstoy, A. Helmi, M. Irwin, P. Parisi, V. Hill, P. Jablonka, MNRAS **411**, 1013 (2011), 1009.4857

[23] D.L. Nidever, K. Olsen, A.R. Walker, A.K. Vivas, R.D. Blum, C. Kaleida, Y. Choi, B.C. Conn, R.A. Gruendl, E.F. Bell et al., ArXiv e-prints (2017), 1701.00502

[24] A.K. Vivas, K. Olsen, R. Blum, D.L. Nidever, A.R. Walker, N.F. Martin, G. Besla, C. Gallart, R.P. van der Marel, S.R. Majewski et al., AJ **151**, 118 (2016), 1510.05539

[25] R. Zinn, B. Horowitz, A.K. Vivas, C. Baltay, N. Ellman, E. Hadjiyska, D. Rabinowitz, L. Miller, ApJ **781**, 22 (2014), 1312.1602

[26] G. Medina, R. Muñoz, A.K. Vivas, J. Carlin, ApJ (2017), submitted

Pulsating stars and the Virtual Observatory

Juan Carlos Suárez[1,2,*]

[1] *Universidad de Granada (UGR), Dept. Theoretical Physics and Cosmology, 18071, Granada, Spain*
[2] *Instituto de Astrofísica de Andalucía (CSIC), Glorieta de la Astronomía s/n, 18008, Granada, Spain*

Abstract. Virtual Observatory is one of the most used internet-based protocols in astronomy. It has become somewhat natural to find, manage, compare, visualize and download observations from very different archives of astronomical observations with no effort. The VO technology beyond that is now being a reality for asteroseismology, not only for observations but also for theoretical models. Here I give a brief description of the most important VO tools related with asteroseismology, as well as a rough outline of the current development in this field.

1 Why Virtual Observatory?

In the last century astronomy has witnessed an exponential growth in many aspects: number of researchers, ground and space observatories, observation networks, surveys, etc. This has led to a significant advance in many fields of astrophysics. As a consequence an unprecedented improvement of the astronomical instrumentation has occurred. Such an improvement often refers to the quality and also the quantity of the raw data generated. Such an exponential growth (Fig. 1) will soon reach the limits of the data rates exchange and storage capacity of today's technology. The epitome of such saturation scenario is the Square Kilometer Array (SKA) project[1].

Additionally, the increasing number of laboratories and instruments has also led to deal with different types of data structures and data files, each of them requiring a specific interface to be acquired with different computer languages and under different operating systems. Consequently, scientists spend a non-negligible amount of time preparing interfaces and adapting their own codes to be compliant with all different data sources. The Virtual Observatory (VO) was born to solve those problems, seeking to allow the exploration of astronomical data in a single transparent way, providing online powerful analysis and visualisation tools. One of the keys for success of this project is the philosophy of standardisation. This applies to the data, metadata, and the communication protocols among archives and VO tools. As well, the use of a registry. As stated in the International Virtual Observatory Alliance[2] (sic) "The long term vision is not one of a fixed specific software package, but rather one of a framework which enables data centers to provide competing and co-operating data services, and which enables software providers to offer a variety of compatible analysis and visualization tools and user interfaces. The first priority for the VO projects worldwide is to develop the standardized

*`jcsuarez@ugr.es`
[1]`http://skatelescope.org`
[2]The international organism that debates and agrees the technical VO standards.

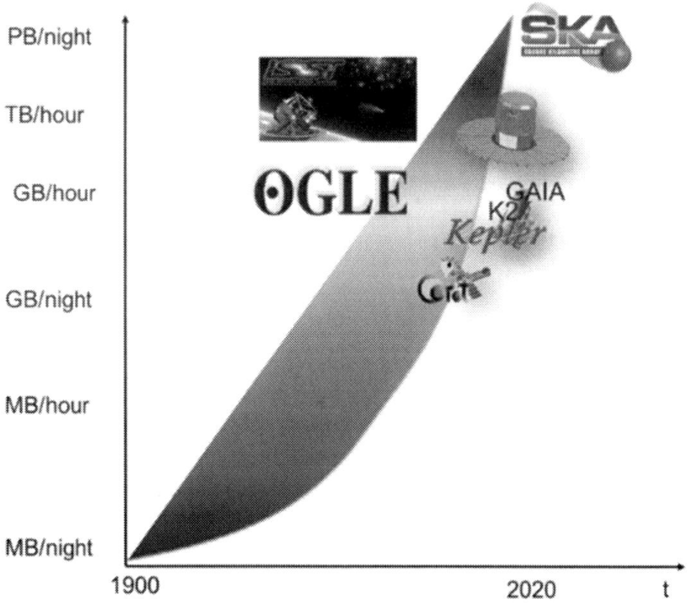

Figure 1. Evolution of the data rates produced by astronomical instrumentation. Notice that SKA in full operation will overwhelm the scale. The use of aperture array radio telescopes in the low and mid frequency ranges will imply data rates to many Petabits (10^{15}) per second, which represents more than 10 times the current global internet traffic.

framework, which will allow such creative diversity". Today, VO projects are being developed worldwide, seeking full interoperability and standardisation among applications, archives and researchers, boosting the time spent by scientists for research ([1]).

Stellar pulsations have become a revolution in the field of stellar physics, and a major contributor to the field of exoplanetary science. This can be understood by the synergies (both instrumental and scientific) between asteroseismology and planet detection and characterization. Space missions like *MOST* ([2]), *CoRoT* ([3]), and *Kepler* ([4]) have provided thousands of light curves (time series of stellar luminosities). Those missions, which were designed mainly to detect planetary systems, use asteroseismology as the main technique to disentangle between signal coming from stellar pulsations and from the planetary transits (or even stellar activity). Space missions require complementary ground-based observations to fully accomplish their scientific objectives. This implies large international networks able to get the extremely competitive observation time in the observatories around the world. All this activity has increased exponentially the amount of data generated in the field of stellar pulsations, scenario that will get worse in the near future with the next-generation missions, namely *CHEOPS* ([5]), *TESS* ([6]), and *PLATO* ([7]).

Likewise, interoperability and data access has also changed dramatically for this field thanks to the VO. This has led to an entry into the era of big data, in which data discovery from automated procedures, statistical inferences and artificial intelligence (AI) have become common practice in the field ([8]). This allows to *connect* the information of many astrophysical surveys whose main scientific case are not necessarily related to stellar pulsations. We find many examples of this at the 22nd Stellar Pulsation Conference (Fig. 2) like the *Gaia* survey in full operation ([9, 10]), which is provid-

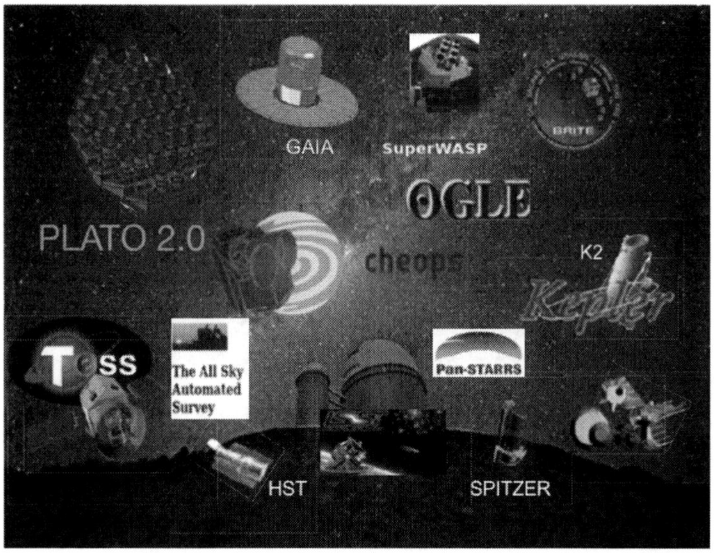

Figure 2. The instruments here schematically shown represent a sample of the current panoply of missions (flying and/or in preparation and/or in legacy mode) related to stellar physics and asteroseismology. Likewise ground-based instrumentation has also improved and grown as a necessary complement to space-based observations. Credits (background image): picture of the 3.6 m telescope at La Silla (ESO), Chile, taken by the author.

ing precise photometric measurements and parallaxes of millions of stars, the Optical Gravitational Lensing Experiment (OGLE) survey ([11, 12]), the Whole Earth Telescope (WET) survey ([13]), the *Spitzer* space survey ([14]), the *Hubble* space telescope ([15]), or the future Large Synoptic Survey Telescope (LSST, [16]), to name a few.

Today, those surveys and their legacy data are essential for the advance in the understanding of Cepheid stars (e.g., the VEGA/CHARA survey, [17]), the asteroseismology of white dwarfs ([18]), the study of Galactic archaeology ([19]), or the study of the stellar interiors of red giants with the combined data of the Apache Point Observatory Galactic Evolution Experiment (*APOGEE*) survey and the *Kepler* mission ([20]).

Today most of the data generated by those missions, instruments and networks are available by interconnected archives throughout the VO technology and standardisation protocols (e.g. *CDS*, *Vizier*, *Simbad*, *Aladin*, *TOPCAT*[3], to name a few). These well-known general purpose VO tools have been used for decades, with no specific run modes or modules for asteroseismology. Today this is changing. Here I will give a brief description of the next generation VO tools focused on asteroseismology: the *Seismic Plus* and *TOUCAN* portals.

2 Next generation tools for stellar pulsations

Next generation is a terminology often used to indicate a leap forward in the technology involved in astronomical instrumentation or even in software application. Here we are dealing with a hybrid

[3] `http://cdsweb.u-strasbg.fr; http://vizier.u-strasbg.fr/viz-bin/VizieR; http://simbad.u-strasbg.fr/simbad/; http://aladin.u-strasbg.fr/aladin.gml; http://www.star.bris.ac.uk/~mbt/topcat/`

of those worlds: the technological development of the instrumentation being employed in asteroseismic studies (see previous section), and the software development under the VO to handle the data generated by that instrumentation.

In 2006 several scientific teams (24 partners) organized a network under the project HELAS (European HELio- and ASteroseismology network), as a network funded by the European Commission as part of the Sixth Framework Programme (FP6). The activities of HELAS structured this European Research Area by bringing together the European groups active in helio- and asteroseismology. Years later, once the HELAS project was finished, the partners got funded again for the Exploitation of Space Data for Innovative Helio- and Asteroseismology (SpaceInn) project[4] with the objective to build on the existing European strength in the field of time-domain stellar physics. Indeed, one of the main tasks of that new project was to build a next generation VO tool able to centralize all the observational data related with stellar variability, helio and asteroseismology. This tool is the *Seismic Plus* portal[5].

Likewise, during the SpaceInn project, ways were explored in which the observational data could efficiently communicate with theoretical models in the era of the Virtual Observatory. The solution was found in the state-of-the art development of theoretical models in the VO for asteroseismology, the *TOUCAN*[6] portal ([21]).

3 Seismic Plus

The asteroseismic data has become the state-of-the-art in the research of stellar interiors, and thereby providing the most updated and unprecedentedly accurate information not only for stellar structure and evolution, but also for planetary research, which requires that the stars that host planets are accurately characterized. The *Seismic Plus* portal is the *place* that gathers an extended list of existing data sources, namely solar and stellar seismic data obtained with space missions (e.g., light curves from *CoRoT Kepler*, etc.) and ground-based facilities (e.g., follow-up observations of those space missions with astrometry, spectroscopy, etc.). Likewise, *Seismic Plus* provides online tools to exploit the data (e.g., combining the data to produce new outputs). The current implemented data sources[7] can be grouped in four main types: (1) Time series. This is the main data type in asteroseismology. It contains not only information of the stellar pulsations but also may show up insights on the stellar activity or even transiting planets; (2) Spectroscopy. Stellar modeling on which asteroseismology relies requires constraints that only spectroscopic analysis (e.g., for [Fe/H] and $T_{\rm eff}$ estimates) can provide with sufficient accuracy to be useful in the field; (3) Stellar parameters. Observational seismic indices, based on large amount of data provided by the space missions like *CoRoT* and *Kepler*, help to characterize large numbers of stars to first order. Together with the classical photometric color indices or parallaxes, those new indices have emerged which characterize the stellar pulsation and thus the stellar structure.

The *Seismic Plus* portal was conceived to remain as simple as possible in all its features, from the query to the presentation of the data, without compromising its main utility, i.e. providing the most relevant information from all the linked sources. To do so, the portal has three main functions: (1) to list and describe the data sources; (2) to cross search for existing data (and data types), and

[4]This project with reference FP7-312844 was ended on December 2016. The HELAS consortium is now applying for European EINFRA project under H2020 program to improve and update the infrastructures related to the asteroseismic observations and data access.

[5]`http://voparis-spaceinn.obspm.fr/seismic-plus/`

[6]`http://www.laeff.cab.inta-csic.es/projects/toucan3/toucan/`

[7]Data sources and data products are in continuous evolution so this list remains open to changes in the observational landscape and in current scientific issues.

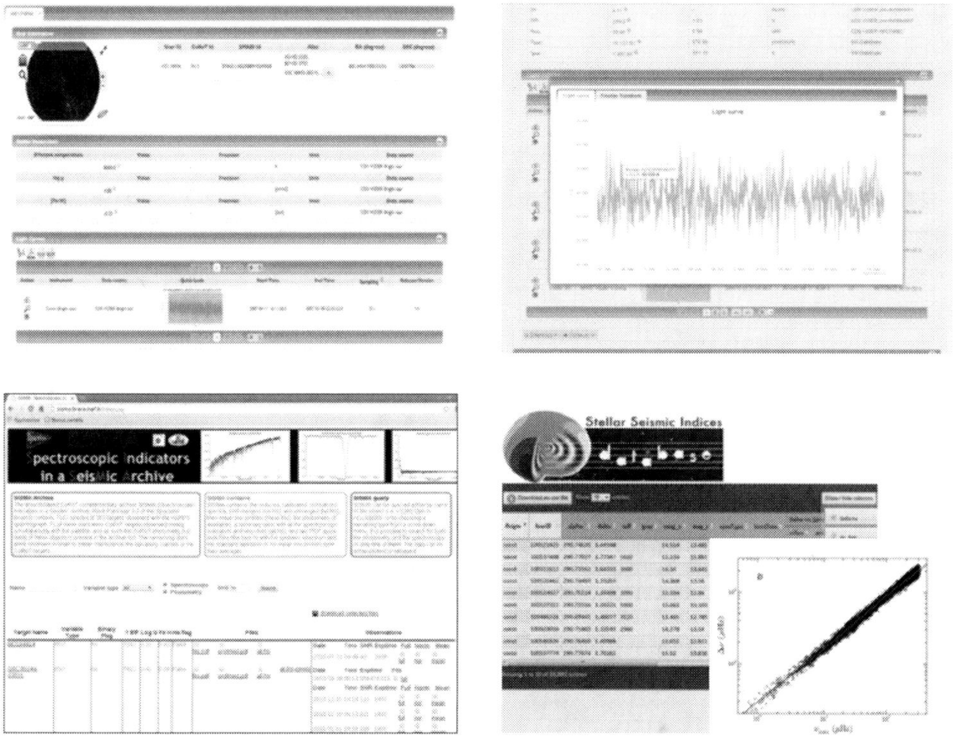

Figure 3. Screenshots of the Seismic Plus' main features: (top, left) first output for a single-star query; (top, right) high-level product consisting in a fast Fourier transform of the selected light curve; (bottom, left) main page of the *SISMA* (Spectroscopic Indicators in a SeisMic Archive) module. This feature calculates indicators from spectra and time series of spectra useful for asteroseismic studies; (bottom, right) main page of the *SSI* (Stellar Seismic Indices) module. The SSI provides asteroseismic indices commonly used in asteroseismology, like large and small separations ($\delta\nu$ and $\Delta\nu$, repectively), the frequency of the mode with observed maximum amplitude ($\nu_{\max}$), etc.

(3) to help in the more complex data queries (cross queries throughout the archives) and in the second level output generated by the tool. Two main tools are now providing second-level outputs within the portal: the *SISMA* ([22]) and the *SSI* tools ([23]) (Fig. 3).

4 Toucan, the VO gateway to stellar models

Nowadays, model databases are extensively used, leading to interpret theoretically large sets of observations, like those provided by the space missions and surveys (e.g., [19, 20, 24, 25]).

Such analyses require different codes, numerical approaches, generally imply different definitions of physical quantities and/or modelling parameters, different output formats, etc. The lack of homogeneity makes it difficult to design automatic tools to simultaneously work with different models and/or applications able to use the models on the fly. The VO (see Sect. 1) is the ideal frame to overcome such difficulties. TOUCAN is the first tool in the VO focused on theoretical asteroseismic models. The portal, which has been developed within the Spanish Virtual Observatory project[8], pro-

[8]`http://svo.cab.inta-csic.es`, supported from the Spanish MINECO through grant AYA2014-55216.

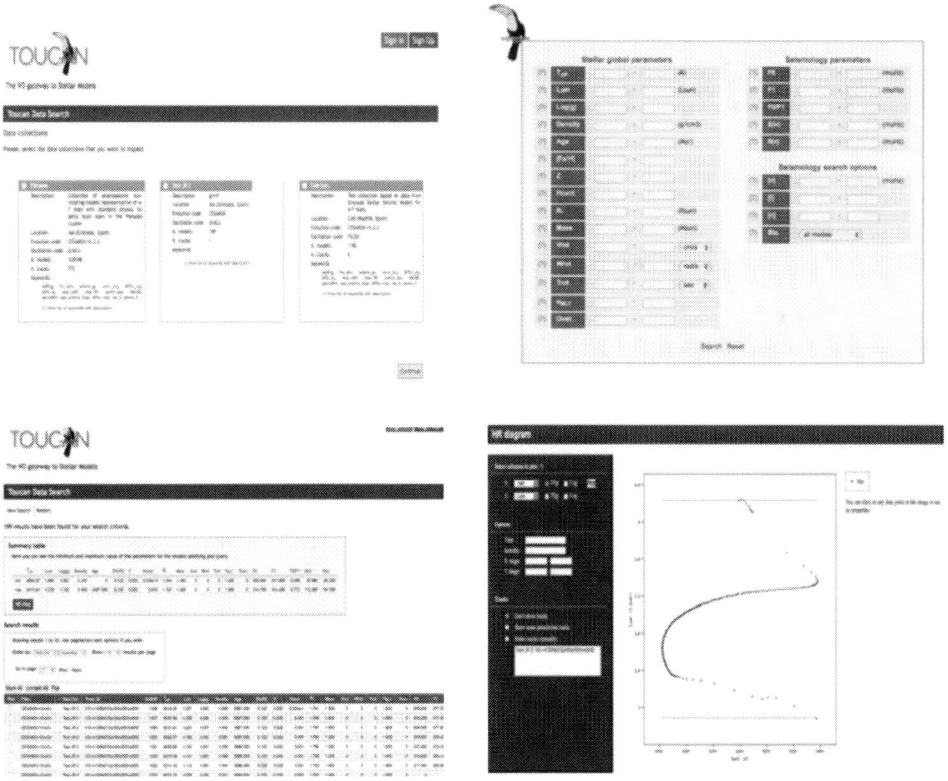

Figure 4. Screenshots of the *TOUCAN*'s main features: (top, left) description of the Data Collections; (top, right) query of models with criteria on both structure/modelling parameters and seismic parameters; (bottom, left) example of the results provided by the *TOUCAN* portal: a summary table indicates the minimum and maximum values of a set of structure and seismic parameters. Below, a complete list of individual models is given; (bottom, right) plots of models selected for comparison. Online plots of both global variables (as illustrated in the panel, with an HR diagram) and shell variables (e.g., the variation of the pressure with the radius) are possible.

vides access to a database of approximately 1.5×10^6 models. Users can search for those models fulfilling physical criteria (structure and seismic properties), visualize the data, get some high-level products, and download models in both original format and VO tables (Fig. 4).

Currently the portal is in beta version[9], running on a local prototype (i.e. asteroseismic models are stored at the portal's site). This prototype has already been used for several scientific works (see, e.g., [21, 26–30]). However, the portal is in continuous development (Fig. 5) in order to adapt the tool to a wider international context and provide access to millions of models distributed in very different and heterogeneous data collections[10].

Likewise, the scientific and technical teams of *Seismic Plus* and *TOUCAN* are performing a feasibility study of the interconnection between the two portals, so that researchers and general users have access to both observations and models in an efficient and smart way, with the objective of maximizing the scientific time in effective research.

[9] `http://svo.cab.inta-csic.es/theory/sisms3/`

[10] There is a prototype of a new version of the tool at `http://svo.cab.inta-csic.es/main/index.php`.

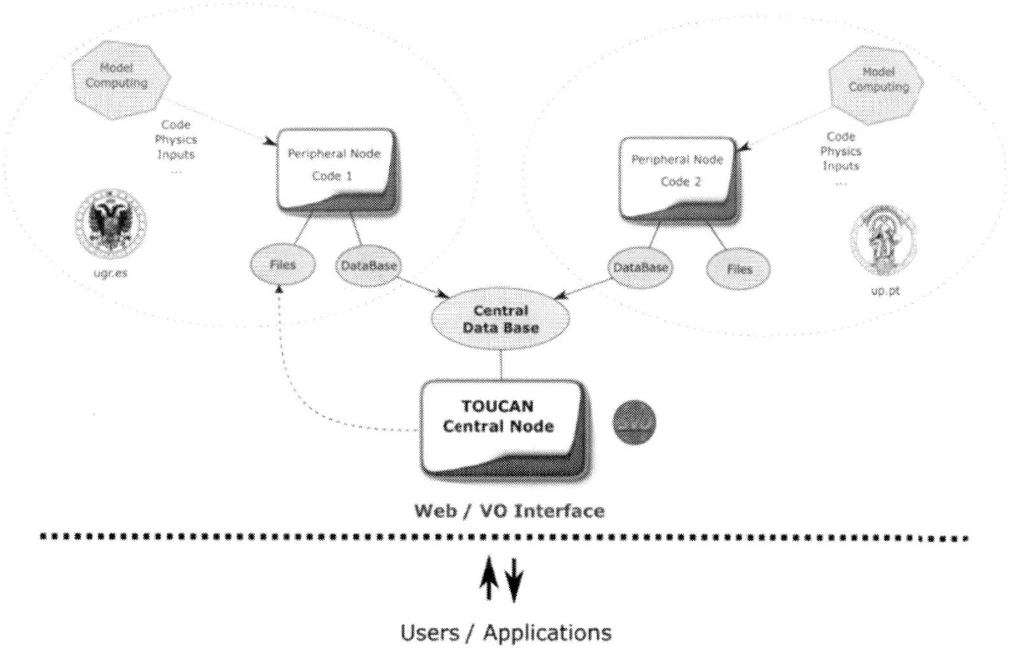

Figure 5. Scheme of the new TOUCAN's client-server web architecture. Current development intends to upgrade the tool to a distributed framework. A TOUCAN central service (server) will communicate with TOUCAN peripheral services (clients). The peripheral services will contain the physically stored Data Collections. The TOUCAN central service will connect to the different peripheral services (or TOUCAN nodes) to get the information queried by the user, and will present it to the users.

Acknowledgments: I acknowledge funding support from Spanish public funds for research under project ESP2015-65712-C5-5-R (MINECO/FEDER), and under the *Ramón y Cajal* research fellowship with reference RYC2012-09913 (MINECO/FEDER).

References

[1] F. Genova, M.G. Allen, C. Arvise, A. Lawrence, F. Pasian, E. Solano, J. Wambsganss, Astronomy and Computing **11**, 181 (2015), 1506.06567

[2] J.M. Matthews, Communications in Astroseismology **150**, 333 (2007)

[3] A. Baglin, E. Michel, M. Auvergne, *The seismology programme of the CoRoT space mission*, in *European Space Agency, (Special Publication) ESA SP* (2006), Vol. 624 SP, ISBN 9290929359, ISSN 03796566, 0811.1080

[4] R.L. Gilliland, T.M. Brown, J. Christensen-Dalsgaard, H. Kjeldsen, C. Aerts, T. Appourchaux, S. Basu, T.R. Bedding, W.J. Chaplin, M.S. Cunha et al., Publications of the Astronomical Society of the Pacific **122**, 131 (2010)

[5] C. Broeg, A. Fortier, D. Ehrenreich, Y. Alibert, W. Baumjohann, W. Benz, M. Deleuil, M. Gillon, A. Ivanov, R. Liseau et al., EPJ Web of Conferences **47**, 03005 (2013), 1305.2270

[6] H. Kjeldsen, *TESS and the Future of Asteroseismology from Space*, in *These Proceedings* (2017)
[7] H. Rauer, C. Catala, C. Aerts, T. Appourchaux, W. Benz, A. Brandeker, J. Christensen-Dalsgaard, M. Deleuil, L. Gizon, M.J. Goupil et al., Experimental Astronomy **38**, 249 (2014)
[8] M. Graham, *Challenges in the Automated Classification of Variable Stars in Large Databases*, in *These Proceedings* (2017)
[9] L. Eyer, *Pulsating Star Research and the Gaia Revolution*, in *These Proceedings* (2017)
[10] T. Metcalfe, *The Impact of Gaia Parallaxes on Asteroseismic Inferences from* Kepler, in *These Proceedings* (2017)
[11] A. Udalski, *OGLE Cepheids in the Milky Way*, in *These Proceedings* (2017)
[12] H. Netzel, *Blazhki Modulation in First-Overtone RR Lyrae Stars from the OGLE Collection*, in *These Proceedings* (2017)
[13] J. Provencal, *A Survey of Pulsating DA and DB White Dwarfs with the Whole Earth Telescope*, in *These Proceedings* (2017)
[14] P. Whitelock., *Spitzer Observations of Large Amplitude Variables in IC1613*, in *These Proceedings* (2017)
[15] K. Sokolovsky, *The Hubble Catalogue of Variables*, in *These Proceedings* (2017)
[16] M. Graham., *LSST and Its Potential for Pulsating Star Research*, in *These Proceedings* (2017)
[17] N. Nardetto, *VEGA/CHARA Interferometric Observations of Cepheids*, in *These Proceedings* (2017)
[18] S. Kepler, *Pulsating White Dwarfs in Large Surveys*, in *These Proceedings* (2017)
[19] V. Silva-Aguirre, *Asteroseismology and Galactic Archaeology in the Era of Large Surveys*, in *These Proceedings* (2017)
[20] M. Pinsonneault, *APOKASC 2.0: Asteroseismology and Spectroscopy of Red Giants*, in *These Proceedings* (2017)
[21] J.C. Suárez, A. García Hernández, A. Moya, C. Rodrigo, E. Solano, R. Garrido, J.R. Rodón, Astronomy & Astrophysics **563**, A7 (2014), 1402.0723
[22] M. Rainer, E. Poretti, A. Mistò, M.R. Panzera, M. Molinaro, F. Cepparo, M. Roth, E. Michel, M.J.P.F.G. Monteiro, AJ **152**, 207 (2016), 1611.02715
[23] R. Samadi, R. De Assis Peralta, E. Michel, K. Belkacem, C. Renie, M. Abed, EPJW, in press (2016)
[24] P. De Cat, *Pulsating Variable Stars and Large Spectroscopic Surveys*, in *These Proceedings* (2017)
[25] I. Soszyński, *The Impact of Large-Scle, Long-Term OPtical Surveys on Pulsating Star Research*, in *These Proceedings* (2017)
[26] A. García Hernández, A. Moya, E. Michel, R. Garrido, J.C. Suárez, E. Rodríguez, P.J. Amado, S. Martín-Ruiz, A. Rolland, E. Poretti et al., Astronomy and Astrophysics **506**, 79 (2009)
[27] E. Poretti, E. Michel, R. Garrido, L. Lefèvre, L. Mantegazza, M. Rainer, E. Rodríguez, K. Uytterhoeven, P.J. Amado, S. Martín-Ruiz et al., Astronomy and Astrophysics **506**, 85 (2009)
[28] L. Mantegazza, E. Poretti, E. Michel, M. Rainer, F. Baudin, A. García Hernández, T. Semaan, M. Alvarez, P.J. Amado, R. Garrido et al., Astronomy & Astrophysics **542**, A24 (2012)
[29] A. García Hernández, A. Moya, E. Michel, J.C. Suárez, E. Poretti, S. Martín-Ruíz, P.J. Amado, R. Garrido, E. Rodríguez, M. Rainer et al., Astronomy & Astrophysics **559**, A63 (2013)
[30] A. García Hernández, S. Martín-Ruiz, M.J.P.F.G. Monteiro, J.C. Suárez, D.R. Reese, J. Pascual-Granado, R. Garrido, ApJL **811**, L29 (2015), 1509.01111

Stellar variability from Dome A, Antarctica

Lingzhi Wang[1,2,3,⋆], *L. M.* Macri[4], *B.* Ma[1,3], *L. F.* Wang[3,4,5], *M. C. B.* Ashley[6], *X.* Cui[3,7], *F. J.* Du[3,7], *J. N.* Fu[8], *L. L.* Feng[3,5], *X.* Gong[3,7], *Y.* Hu[1,3], *G.* Li[8], *X. Y.* Li[3,7], *Z. Y.* Li[3,7], *J. S.* Lawrence[6,9], *D.* Luong-Van[6], *C. R.* Pennypacker[10], *Z.* Shang[3,11], *J. W. V.* Storey[6], *H.* Yang[12], *X.* Yuan[3,7], *D. G.* York[13], *X.* Zhou[1,3], *Z. H.* Zhu[8], *Z. X.* Zhu[3,5], and *J. L.* Zhou[14]

[1] *Key Laboratory of Optical Astronomy, National Astronomical Observatories, Chinese Academy of Sciences, Beijing 100012, China*
[2] *Chinese Academy of Sciences South America Center for Astronomy, China-Chile Joint Center for Astronomy, Camino El Observatorio #1515, Las Condes, Santiago, Chile*
[3] *Chinese Center for Antarctic Astronomy, Nanjing 210008, China*
[4] *Mitchell Institute for Fundamental Physics & Astronomy, Department of Physics & Astronomy, Texas A&M University, College Station, TX 77843, USA*
[5] *Purple Mountain Observatory, Chinese Academy of Sciences, Nanjing 210008, China*
[6] *School of Physics, University of New South Wales, NSW 2052, Australia*
[7] *Nanjing Institute of Astronomical Optics and Technology, Nanjing 210042, China*
[8] *Department of Astronomy, Beijing Normal University, Beijing, 100875, China*
[9] *Australian Astronomical Observatory, NSW 1710, Australia*
[10] *Center for Astrophysics, Lawrence Berkeley National Laboratory, Berkeley, CA, USA*
[11] *Tianjin Normal University, Tianjin 300074, China*
[12] *Polar Research Institute of China, Pudong, Shanghai 200136, China*
[13] *Department of Astronomy and Astrophysics and Enrico Fermi Institute, University of Chicago, Chicago, IL 60637, USA*
[14] *School of Astronomy and Space Science and Key Laboratory of Modern Astronomy and Astrophysics in Ministry of Education, Nanjing University, Nanjing 210093, China*

Abstract. The Antarctic plateau is one of the best observing sites on the surface of the Earth thanks to its extremely cold, dry, stable and transparent atmosphere conditions. Various astronomical activities are underway there and the Chinese Center for Antarctic Astronomy (CCAA) is dedicated to developing Antarctic astronomy at the highest point, Dome A or the Chinese Kunlun station. So far a large number of images have been collected from a 14.5-cm quad-telescope called the Chinese Small Telescope ARray (CSTAR) and the first two of a trio of 50-cm Antarctic Survey Telescopes (AST3-1 and AST3-2).

1 Photometric Data and Stellar Variability

A large number of time-series photometric data in Table 1 had been acquired since 2008 from the first-generation optical telescope CSTAR ([1]) and the second-generation optical telescopes AST3-1, AST3-2 ([2–4]). The high-cadence time-series observations from CSTAR were obtained over 4, 5,

⋆wanglingzhi@bao.ac.cn

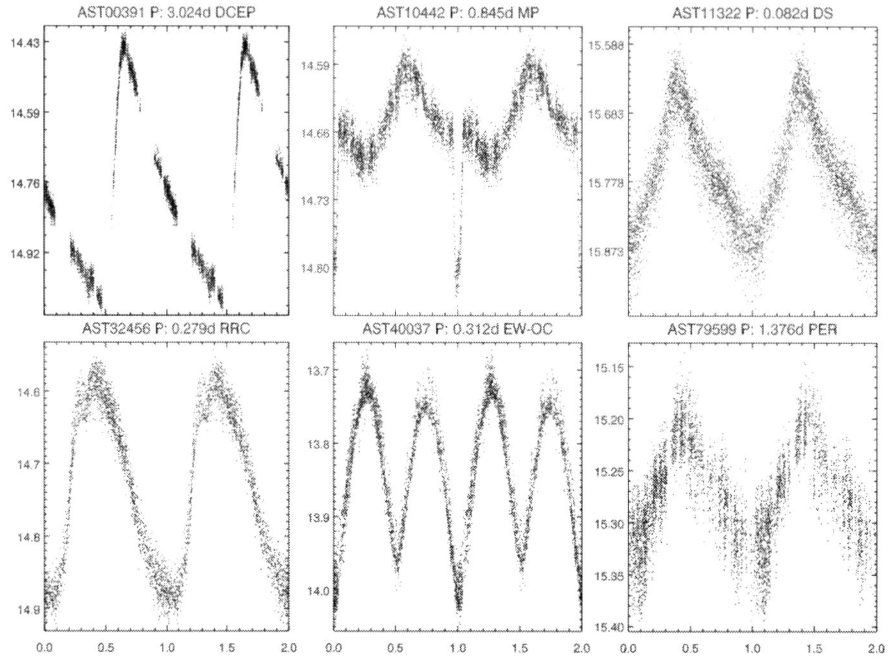

Figure 1. Phased light curves for six typical periodic variable stars from AST3-1 during 8-day observations. The AST IDs, periods, and tentative types are listed above each panel. Top row (from left to right): DCEP, δ Cephei variable; MP, multiperiodic variable; DS, δ Scuti variable. Bottom row (from left to right): RRC, RRc Lyrae variable; EW-OC, W Ursae Majoris-type eclipsing binary (EW) with O'Connell effect; PER, unclassified periodic variable.

6 months of the 2008, 2009, 2010 Antarctic winters, respectively ([5–9]). Those long-term high-cadence data showed that the high-precision long-term photometry is possible at Dome A, Antarctica thanks to its long, continuous, stable and clear atmosphere conditions. Based on time-series photometry of 92583 stars measured by AST3-1, 560 variable stars were detected with i magnitude $\leq$ 16.5 mag during 8 days of observations ([10]). We show the phased light curves of 6 periodic variables stars from AST3-1 in Figure 1.

Table 1. Log of observations for CSTAR and AST3.

Telescope	Filter	Year	# images	Reference
CSTAR#1	*i*	2008	287840	[5, 6]
CSTAR#2	*g*	2009	251310	[7, 8]
CSTAR#3	*clear*	2009	74010	[7]
CSTAR#4	*r*	2009	536383	[7, 8]
CSTAR#1	*i*	2010	342353	[9]
AST3-1	*i*	2012	15783	[10]

Acknowledgments: We thank NSFC grant 11303041, 973 Program (through grant Nos. 2013CB834901, 2013CB834900, 2013CB834903), CHINARE2016-02-03-05, the great efforts made by the 24–32th Dome A expedition teams, who provided invaluable assistance to the astronomers that set up and maintained the AST3 and the PLATO-A system, and the Chinese Academy of Sciences (CAS), through a grant to the CAS South America Center for Astronomy (CASSACA) in Santiago, Chile.

References

[1] Yuan, X., Cui, X., Liu, G., et al., Proc. SPIE, **7012**, 70124G (2008)
[2] Cui, X., Yuan, X., & Gong, X., Proc. SPIE, **7012**, 70122D (2008)
[3] Shang, Z., Hu, K., Hu, Y., et al., Proc. SPIE, **8448**, 26 (2012)
[4] Yuan, X., & Su, D.-q., MNRAS, **424**, 23 (2012)
[5] Wang, L., Macri, L. M., Krisciunas, K., et al., AJ, **142**, 155 (2011)
[6] Wang, S., Zhang, H., Zhou, X., et al., ApJS, **218**, 20 (2015)
[7] Oelkers, R. J., Macri, L. M., Wang, L., et al., AJ, **149**, 50 (2015)
[8] Zong, W., Fu, J.-N., Niu, J.-S., et al., AJ, **149**, 84 (2015)
[9] Wang, L., Macri, L. M., Wang, L., et al., AJ, **146**, 139 (2013)
[10] Wang, L., Ma, B., Li, G., et al., AJ, **153**, 104 (2017)

DOI: 10.1051/epjconf/201715202011

Using APASS and 2GSS for studying variable stars

Arne Henden[1,*], *Stephen* Levine[2], *Doug* Welch[3], *Dirk* Terrell[4], *Ulisse* Munari[5], and *Thomas* Smith[5]

[1]*American Association of Variable Star Observers, Cambridge, MA, USA*
[2]*Lowell Observatory, Flagstaff, AZ, USA*
[3]*McMaster University, Hamilton, ON, Canada*
[4]*Southwest Research Institute, Boulder, CO, USA*
[5]*Dark Ridge Observatory (DRO), Weed, NM, USA*

Abstract. The AAVSO Photometric All-Sky Survey (APASS) provides calibrated magnitudes in the range $7 < V < 17$ for the entire sky, in the $BVugriZ_sY$ bandpasses. While primarily designed for 0.02 mag calibration, it provides photometry over a many-year baseline, and also has near daily cadence in the standard field regions on the equator. Likewise, the Second Generation Synoptic Survey (2GSS) is a follow-on to APASS, and will provide daily cadence in two passbands for the entire sky.

1 Introduction

APASS was begun in 2010 in response to the need for accurate, bright, photometric all-sky calibration. The intent was to extend the Sloan Digital Sky Survey calibration to brighter magnitudes, as well as provide transitional photometry in both the Johnson (B, V) and Sloan ($ugriz$) passbands. This survey has had ten data releases, and covers the entire sky from $7 < V < 17$ mag. The photometry is accurate to about 0.02 mag, and the astrometry is accurate to about 0.15 arcsec.

Data collection is accomplished with telescope systems at CTIO in the south and DRO in the north. Each telescope system uses a single Paramount ME mount, with two bore-sighted optical telescope assemblies (OTAs). One OTA covers the blue photometric bands (B, u, g, Z_s, Y) while the other OTA simultaneously covers the mid-range photometric bands (V, r, i). The OTAs are 20 cm f/3.6 ASA astrographs, along with Apogee U16m CCD cameras. Each OTA covers a 2.9×2.9 degree field of view, with 2.5 arcsec pixels. The survey is designed to cover the entire sky twice with a small amount of overlap between each telescope field. Because there is vignetting in the system and poor image correction in the field corners, a set of secondary field centers, offset from the primary by center-to-corner, is also taken twice. Therefore, the final survey has a minimum of four observations of each object. In reality, many more observations are made, due to non-photometric nights and other processing issues. In addition, nightly standard-star fields are acquired, with several hundred observations for each over the 6-year timespan of the survey. Further information about the survey can be found at `https://www.aavso.org/apass/`.

*`arne@aavso.org`

DOI: 10.1051/epjconf/201715202011

The SHARCnet system of high performance computers (`https://www.sharcnet.ca/my/front/`) is used to find and measure all objects using a combination of Sextractor, DAOPHOT and astrometry.net software. All measures from photometric nights are combined into a master "means" catalog, which is available from the AAVSO or VizieR. We also take all individual measurements with their Julian Date and import them into an Epoch Photometry Database (EPD) that is available to AAVSO members.

2 Variable stars

APASS has been used for variable star studies in five ways. The obvious first use is in providing calibrated secondary comparison stars in the fields of known variables. This was one of the primary reasons for the development of the survey. The means catalog is heavily used by the AAVSO in creating their charts. Second, on non-photometric nights, specific target fields have been observed. These are targets such as recent novae, *Kepler* fields, comets, or fields relating to specific research projects. Third, we also used a couple of dozen nights for a study of the Aquarius Halo Stream ([1]), re-imaging each field often enough to search for RR Lyr variables. Fourth, several researchers have queried the photometric database for stars that lie in a specific locus of color-color space, to be used as a target list for studying a specific class of variable star (such as cataclysmic variables). Finally, the EPD itself contains many fields for which some known variable stars exist, and for which other variables can be discovered (or analysis made if other researchers publish new discoveries).

3 2GSS – the APASS follow-on

Once the APASS survey is complete, we hope to be able to use the hardware to continue observing. The intent is to keep nearly everything the same, except to replace the 20 cm astrographs with 30-40 cm astrographs, and to only image in one filter per OTA (Sloan g and i). This survey technique will optimize both the simultaneous wavelength coverage and the number of fields that can be observed per night. With five such systems, we envision nightly coverage of the available sky. Since the APASS means catalog will be available, no standard fields need to be observed, and each target field will have sufficient secondary standards to zero-point the photometric measures. With sufficient computing power, it should be possible to instrumentally calibrate the images and extract stellar photometry in near-real-time, and provide alerts for those objects exceeding a specified photometric variation. This is similar in many ways to the envisioned use of LSST, with two exceptions: we will have simultaneous 2-color photometry, and the faint limit of 2GSS overlaps slightly with the bright limit of LSST, providing continuous temporal coverage of the sky from 7th magnitude to the LSST limit. A pilot 2GSS system is currently operational at Lowell Observatory.

Acknowledgments: This project was begun with a grant from the Robert Martin Ayers Sciences Fund. APASS is funded in part through NSF grant AST-1412587. We appreciate the help provided by: J. Irwin (Harvard/MEarth) and D. Reichart (UNC); vendors including Software Bisque, DC3 Dreams, Apogee, Astrodon and ASA; and dozens of volunteers.

References

[1] Funari, U., Henden, A., & Frigo, A., NewA, **27**, 1 (2014)

 DOI: 10.1051/epjconf/201715202012

Preliminary results of variable star search using KMTNet data

*Young-Beom Jeon**

[1] *Korea Astronomy and Space Science Institute 776, Daedeokdae-ro, Yuseong-gu, Daejeon, 34055, Republic of Korea*

Abstract. The KMTNet (Korea Microlensing Telescope Network) was completed on October 2nd in 2015. There are three 1.6 m telescopes in Chile (CTIO), Australia (SSO) and South Africa (SAAO). The field of view of the CCD camera is 2 degree × 2 degree. We are monitoring and observing the bulge region for 24-hour continuous observations using the KMTNet telescopes. I present the characteristics of CCD camera and telescope, and the status of observation programs. We are still developing CCD cameras to accurate photometry. From the KMTNet data, we can obtain accurate 24-hour continuous time-series data for many variable stars.

1 Introduction

The Korea Microlensing Telescope Network (KMTNet) is a wide-field photometric system composed by three identical 1.6m telescopes. The most important capability of the KMTNet is 24 hours continuous observations using the three telescopes installed at three different continents; CTIO in Chile, South America, SSO in Australia and SAAO in South Africa. The primary science of the KMTNet is the Galactic bulge monitoring to find extrasolar planets. The monitoring data are very useful to find microlensing events, variable stars, and supernovae, etc. The science programs and the system specifications of the KMTNet are well explained in [2]. Table 1 presents brief specifications for telescopes and CCD camera. The first results for the microlensing effects and a variable star monitoring are published in [3] and [1], respectively.

2 Preliminary results of variable star search

Figure 1 shows sample images and preliminary results of variable star search. There are 1,600 small regions in a N chip. I checked 20 regions at first. From all 20 regions, I found above 100 variable stars including short period δ Sct, eclipsing binary, RR Lyrae, Cepheid and long periodic regular or semi-regular variable stars. We expect to detect planet transits, too. Most images are obtained in the I-band, and also V-band to determine the color, occasionally.

Till now, we are developing image processing procedures and photometry methods. We will use the pieces of 1024 pixels × 1024 pixels for image subtraction processes after 2017.

Acknowledgments: This work was supported by KASI (Korea Astronomy and Space Science Institute) grant 2016-1-832-01.

*ybjeon@kasi.re.kr

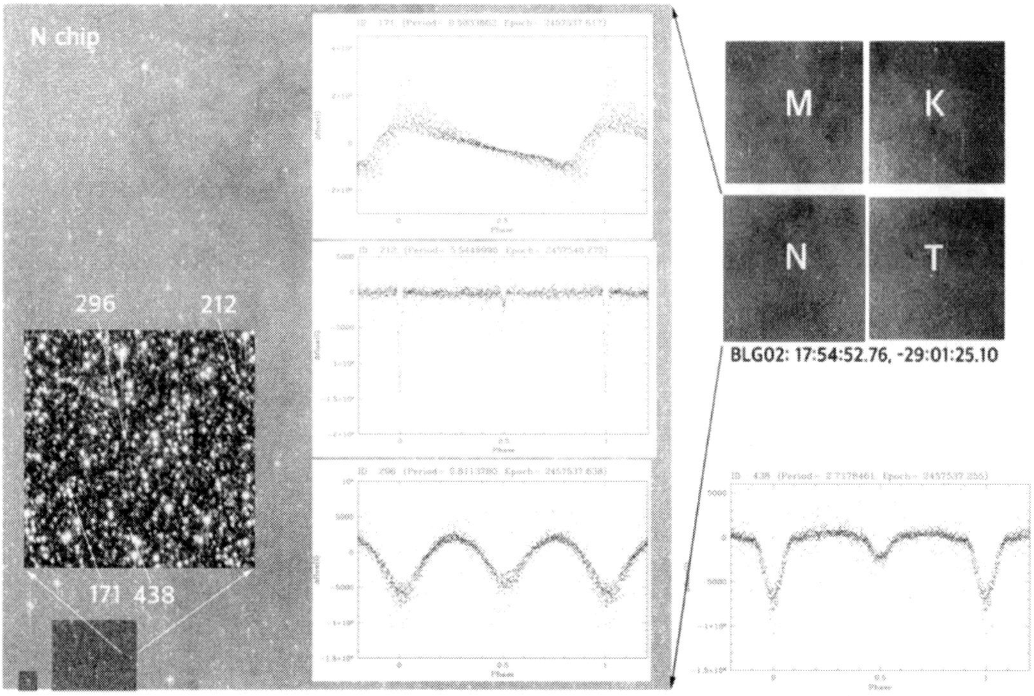

Figure 1. Preliminary results of variable star search using the KMTNet data. The whole K, M, T, N mosaic field is shown on the upper right side. To find variable stars, I selected a few small regions of the N chip field. Four variable stars were found from a small sample region which has 256 pixels × 256 pixels.

Table 1. Specifications of the KMTNet system.

Telescope		
	Location	CTIO Chile, SSO Australia, SAAO South Africa
	Diameter	1.6m, each
	FOV	2 degree × 2 degree
	Focal length	5160mm
	Focus	Prime
CCD camera		
	Manufacture	e2v
	Mosaic	4 ea× 9k × 9k
	Plate scale	10 μm/pixel × 10 μm/pixel, 0.4″/pixel
	Filters	Johnson-Cousins *BVRI*, SDSS *griz*, H_α

References

[1] Kim, S.-L., Cha, S.-M., Lim, B., et al., JKAS, **49**, 199 (2016)
[2] Kim, S.-L, Lee, C.-U., Park, B.-G., et al., JKAS, **49**, 37 (2016)
[3] Shin, I.-G. Ryu, Y.-H., Udalski, A., et al., JKAS, **49**, 73 (2016)

RR Lyrae stars as tracers of halo substructures

Sonia Duffau[1,2,⋆], *A. Katherina* Vivas[3], *Camila* Navarrete[2,1], *Julio* Carballo-Bello[2,1], *Gergely* Hajdu[2,1], and *Márcio* Catelan[2,1]

[1]*Millennium Institute of Astrophysics, Santiago, Chile*
[2]*Pontificia Universidad Católica de Chile, Instituto de Astrofísica, Av. Vicuña Mackenna 4860, 782-0436 Macul, Santiago, Chile*
[3]*Cerro Tololo Interamerican Observatory, Casilla 603, La Serena, Chile*

Abstract. Three RR Lyrae overdensity candidates in the southern sky have been studied using low resolution spectra obtained with the Goodman spectrograph at SOAR. We search for unidentified velocity peaks indicative of the possible presence of streams within the overdensities. Our results suggest that all three overdensities present an excess of radial velocity signal at high mean radial velocities which cannot be explained as a contribution from any known Galactic component.

1 Introduction

We study halo substructure to determine how important a role accretion has played in the formation and evolution of our Galaxy. We do so by using RR Lyrae as tracers, because they are old and bright, and excellent standard candles.

A large effort has been conducted already unveiling substructure in the northern sky, while the southern sky census is still fairly incomplete. From the catalog of RR Lyrae stars presented in [4] we targeted three overdensity candidates. The overdensities in this study are: Hya 2, Crv 1 and Cen 3, having 28, 15 and 20 RR Lyrae members each and with mean heliocentric distances of 9.4, 11.4 and 10.2 kpc, respectively. From low resolution spectra of the member stars, radial velocities have been obtained by cross-correlation with appropriate templates, and the pulsation velocity of these variable stars has been accounted for. As a by-product, mean metallicities have been obtained as well, from the equivalent widths of the Ca II K line and three Balmer lines using Layden's 1994 method ([3]).

2 Observations and results

After obtaining all velocities, we compare the resulting histograms to the expected velocity distribution for the potential galactic contaminants, and check possible associations between our overdensities and previous discoveries in the area. For all three overdensities an excess of high velocity radial velocity signal was found. Figure 1 displays, as an example, the results for Crv 1. In the left panel, the radial velocity histogram (smoothed with a window comparable to the radial velocity error) of the data is presented (in black), along with that for model particles of the Galaxy obtained from the Besançon

⋆ sduffau@astro.puc.cl

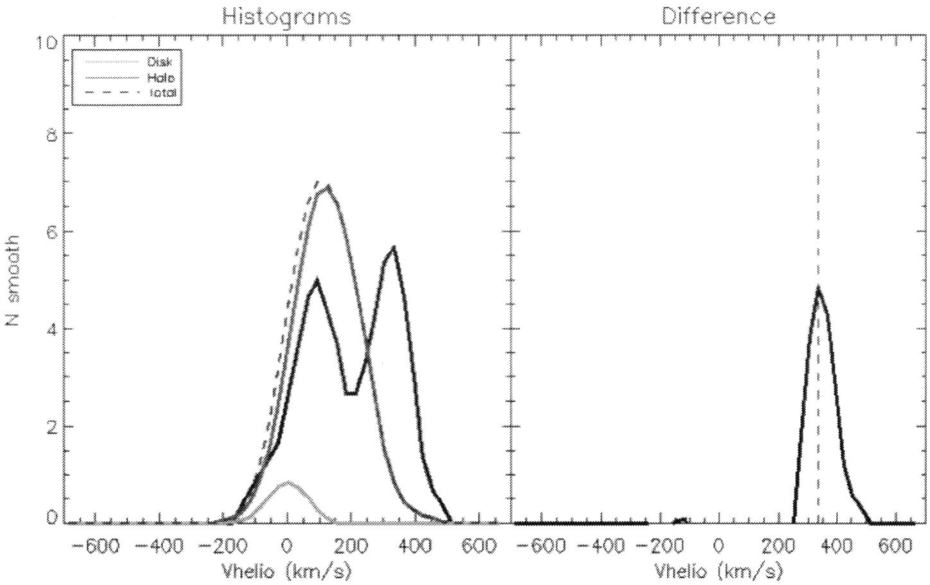

Figure 1. Radial velocity distribution obtained for Crv 1 (black solid line). The dashed blue line indicates the velocity distribution of the Galaxy as predicted by the Besançon model for the same line-of-sight (composed of a halo contribution, in red, and a disk contribution, in green). The vertical line indicates the position of the detected excess signal. Equivalent results were found for Cen 3 and Hya 2 at very similar mean peak velocities.

model for the location of Crv 1 (in red and green). Figure 1 also displays the total contribution from the Galaxy (as a dashed blue line). The potential Galaxy contaminants considered belong to the halo and thick disk population, depicted as red and green smoothed histograms. In the right panel, the excess signal in radial velocity of Crv 1 is displayed. The excess is defined as the difference between the data and the total contribution from the Galaxy, and indicates the presence of a high velocity peak not explained as part of the Milky Way. We interpret this as the original signal from the overdensity, and as evidence of a potential stream candidate.

All three overdensities present an excess of radial velocity signal at high mean radial velocities, as exemplified in Figure 1, which cannot be explained as a contribution from any known Galactic component. Hya 2 is a new detection. Crv 1 seems to be the southern extension to the Virgo Stellar Stream and G4 velocity complex discovered in [5]. Although at a similar location as the expected orbit of the Orphan Stream, Cen 3's distance and velocity do not correspond with the expected values in the area, representing then another new detection as well. Confirmation of the results will be obtained by applying a group finding algorithm to the data, considering both the radial velocity of the stars and the separation between them, and simulating the Galactic contribution to assess the significance of these detections, as done in [1] and [5].

Acknowledgments: We acknowledge financial support from the Ministry for the Economy, Development, and Tourism's Programa Iniciativa Científica Milenio through grant IC 120009, awarded to the Millennium Institute of Astrophysics (MAS), from CONICYT's PCI program through grant DPI20140066, from Proyecto Basal CATA PFB-06/2007, and from FONDECYT grants #1171273 and #3160502.

 DOI: 10.1051/epjconf/201715202013

References

[1] Duffau, S., Vivas, A. K., Zinn, R., Méndez, R. A., & Ruiz, M. T., A&A, **566**, A118 (2014)
[2] Grillmair, C. J., Hetherington, L., Carlberg, R. G., & Willman, B., ApJL, **812**, L26 (2015)
[3] Layden, A. C., AJ, **108**, 1016, (1994)
[4] Torrealba, G., Catelan, M., Drake, A. J., et al., MNRAS, **446**, 2251 (2015)
[5] Vivas, A. K., Zinn, R., Farmer, J., Duffau, S., & Ping, Y., ApJ, **831**, 165 (2016)

Eclipsing binary stars in the era of massive surveys

First results and future prospects

Athanasios Papageorgiou[1,2,*], *Márcio* Catelan[1,2], *Rodrigo* Contreras Ramos[1,2], and *Andrew J.* Drake[3]

[1] *Pontificia Universidad Católica de Chile, Facultad de Física, Instituto de Astrofísica, Av. Vicuña Mackenna 4860, 782-0436 Macul, Santiago, Chile*
[2] *Millennium Institute of Astrophysics, Santiago, Chile*
[3] *California Institute of Technology, 1200 East California Boulevard, CA 91225, USA*

Abstract. Our thinking about eclipsing binary stars has undergone a tremendous change in the last decade. Eclipsing binary stars are one of nature's best laboratories for determining the fundamental physical properties of stars and thus for testing the predictions of theoretical models. Some of the largest ongoing variable star surveys include the Catalina Real-time Transient Survey (CRTS) and the VISTA Variables in the Vía Láctea survey (VVV). They both contain a large amount of photometric data and plenty of information about eclipsing binaries that wait to be extracted and exploited. Here we briefly describe our efforts in this direction.

1 Detached eclipsing binaries in CRTS

Focusing on the eclipsing binaries from CRTS[1] classified as detached eclipsing binaries from [2], a detailed catalog was constructed with the phenomenological properties of the observed light curves (LCs). A total of 4683 LCs were filtered in order to discard erroneous data points, and then a period search was performed to each light curve applying a variety of methods. The phase-folded LCs were visual inspected and the best period values adopted. The final catalog contains revised periods for ~ 10% of the stars. The time of minimum light (MJD_0) was calculated from the phased LCs and a linear ephemeris was constructed for each system. In order to estimate the morphological features of the LCs (e.g., the magnitude at primary and secondary eclipses, and at the quarters out of the eclipses; the difference between eclipse depths; the mean magnitudes and amplitudes; etc.), the total sample was fitted by a chain of second-order polynomials ([4, 6]). In addition, using the Artificial Neural Network (ANN) EBAI ([6]), five fundamental physical parameters can be derived: temperature ratio $\alpha = \frac{T_2}{T_1}$, sum of relative radii $\beta = \rho_2 + \rho_1$, $\gamma = e\cos\omega$, $\delta = e\sin\omega$, and $\epsilon = \sin i$,where e is the eccentricity, ω the argument of periastron, and i the inclination of the orbit. Set of ~ 49,000 LCs and a 20,000 LCs of detached eclipsing binaries (DEB) in the V band were constructed using PHOEBE ([7]) for the training and validation procedure, respectively. In order to optimize the ANN, different topologies (201 input nodes $-n$ hidden nodes -5 output nodes) were tested varying the numbers of hidden nodes n. The best network topologies were found in the range of 100-130 hidden nodes.

* apapageo@astro.puc.cl
[1] http://nesssi.cacr.caltech.edu/DataRelease/

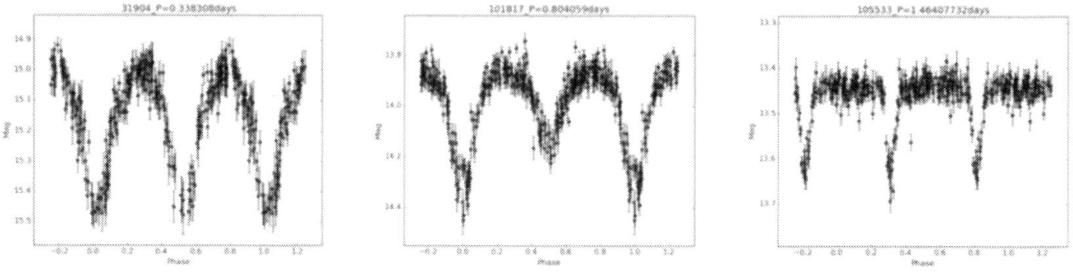

Figure 1. Representative examples of overcontact (left), semi-detached (center), and detached (right) eclipsing binaries in the K_s band from the present work. Each stars' internal id and derived period are given as headers.

2 Eclipsing binaries in VVV

We present a variability search in a $12' \times 12'$ field around the globular cluster NGC 6544 using VVV data ([1]). The observations were carried out during 2010-2015 using the Visible and Infrared Survey Telescope for Astronomy (VISTA) 4.1m telescope located at ESO's Cerro Paranal Observatory in Chile. A total of $\sim$ 160,000 sources were examined for variability, rejecting light curves with less than 10 data points and LCs with standard deviation smaller than 1.5 times the mean photometric error, resulting in 63,133 LCs. Furthermore, σ-clipping was applied to the dataset and a variability search in the K_s band carried out, applying a variety of variability indices to the time series with the python library FATS ([3], and references therein). A total of 790 suspected variables were found, and their periods determined. Among them, 18 EBs were classified as DEBs, 3 as semi-detached, 9 as overcontact, 2 as ellipsoidal, and 8 with multiple classification (Fig. 1). Using the principal component analysis method as implemented in python ([5]), we found in a projection of the variability indices that the suspected variables appeared separately from the non-variable sources of the total sample.

Acknowledgments: Support for this project is provided by the Ministry for the Economy, Development, and Tourism's Millennium Science Initiative through grant IC 120009, awarded to the Millennium Institute of Astrophysics (MAS); by the Basal Center for Astrophysics and Associated Technologies (CATA) through grant PFB-06/2007; by CONICYT's PCI program through grant DPI20140066; and by FONDECYT grants #1171273 and #3160782.

References

[1] Catelan, M., Dékány, I., Hempel, M., & Minniti, D., Boletin de la Asociación Argentina de Astronomía, **56**, 153 (2013)

[2] Drake, A. J., Graham, M. J., Djorgovski, S. G., et al., ApJS, **213**, 9 (2014)

[3] Nun, I., Protopapas, P., Sim, B., et al., preprint (arXiv:1506.00010) (2015)

[4] Papageorgiou, A., Kleftogiannis, G., & Christopoulou, P.-E., Contributions of the Astronomical Observatory Skalnate Pleso, **43**, 470 (2014)

[5] Pedregosa, F., Varoquaux, G., Gramfort, A., et al., Journal of Machine Learning Research, **12**, 2825 (2011)

[6] Prša, A., Guinan, E. F., Devinney, E. J., et al., ApJ, **687**, 542 (2008)

[7] Prša, A., & Zwitter, T., ApJ, **628**, 426 (2005)

Challenges in the automated classification of variable stars in large databases

Matthew Graham[1,2,*], *Andrew* Drake[1], *S. G.* Djorgovski[1], *Ashish* Mahabal[1], and *Ciro* Donalek[1]

[1] *California Institute of Technology, Pasadena, CA 91125, USA*
[2] *National Optical Astronomy Observatory, 950 N. Cherry Ave, Tucson, AZ 85000, USA*

Abstract. With ever-increasing numbers of astrophysical transient surveys, new facilities and archives of astronomical time series, time domain astronomy is emerging as a mainstream discipline. However, the sheer volume of data alone - hundreds of observations for hundreds of millions of sources – necessitates advanced statistical and machine learning methodologies for scientific discovery: characterization, categorization, and classification. Whilst these techniques are slowly entering the astronomer's toolkit, their application to astronomical problems is not without its issues. In this paper, we will review some of the challenges posed by trying to identify variable stars in large data collections, including appropriate feature representations, dealing with uncertainties, establishing ground truths, and simple discrete classes.

1 Introduction

Time domain astronomy is now entering a golden age with an ever-increasing number of new instruments and facilities dedicated to repeated observations of large swathes of sky every few nights or so. This is not just limited to optical astronomy but extends right across the electromagnetic spectrum from radio to gamma-ray wavelengths and now to even more exotic physical domains with the emergence of neutrino (IceCUBE) and gravitational wave (LIGO) observatories. Even though many of these are dedicated to looking for real-time transients (things changing significantly from past observations, if any), they can quickly generate substantial archives of data. Systematic searches of these for particular types of astrophysical source or phenomena require new approaches and in recent years, there have been a number of studies attempting automated classification and outlier detection using machine learning-based techniques (see Table 1). Obviously, these are all with an eye to the next generation of synoptic sky surveys, e.g., Gaia ([1]), ZTF ([2]), and LSST ([3]), which will increase the amount of available data by several orders of magnitude and mandate such approaches.

These types of analyses are not necessarily straightforward, though, and can present a number of issues related to how time series are represented (characterization), what groupings are present (categorization), and how individual time series are assigned to these (classification). The challenges are not just technical but also sociological: the Astronomer's Telegram[1] (ATels) is a popular mechanism for distributing transient alerts but 50% of optical surveys mentioned in them have no public

* `mjg@caltech.edu`
[1] `http://www.astronomerstelegram.org/`

Table 1. A sample of automated classifications of various time domain surveys.

Survey	Passband	Survey area	No. of sources	Refs.
ASAS	V, I	$\delta < 28°$	50000	[4, 5]
CoRoT	Visible	~ 100 deg^2	40000	[6]
Kepler	Visible	105 deg^2	2300	[7]
Hipparcos	Visible	All sky	2 - 4000	[8, 9]
OGLE-II	I	~ 100 deg^2	280000	[10]
Stripe 82	$ugriz$	300 deg^2	300	[11]
LINEAR	V	10000 deg^2	7000	[12]
CRTS	V	33000 deg^2	5×10^8	[13–15]
VVV	$ZYJHK_S$	520 deg^2	10^6	[16, 17]
EROS-2	BR	~ 100 deg^2	3×10^7	[18]
WISE	IR	All sky	6600	[19]

interface to their data holdings. These are also the larger surveys and one wonders how many smaller collections of astronomical observations and time series remain in the private domain. The success of automated classification relies on having access to the greatest amount of information and in this era of data-intensive astronomy, there is no lack of science exploration and discovery to do.

In the paper, we will review the mechanics of automated classification and associated issues that one attempting such an analysis should be cognizant of.

2 How to automatedly classify a data set

Most automated classification work follows the same basic workflow:

1. Astronomical time series are typically irregularly sampled, noisy, and gappy, and even within the same survey, they can vary in terms of number of observations, sampling, signal-to-noise ratios, etc., due to such factors as differing night-to-night observing conditions. These differences obviously make any analysis that is looking for similarities between these time series non-trivial, i.e., it is not just a case of computing a Euclidean distance between them. One way to handle irregular sampling is to interpolate the time series onto a regular temporal grid but this can also introduce additional errors and there may still be issues with missing values or large gaps. A more general approach is to convert the inhomogeneous raw data representation into a homogeneous one through the use of characterizing statistical features. For example, a set of 10,000 light curves, each consisting of between 10 and 50 observations with varying measurement errors can be replaced by a set of 10,000 measurements of median magnitude, interquartile range, and linear trend, say. This can also reduce the dimensionality of the problem: in the example, the raw parameter space is 50-dimensional whereas the feature parameter space has just three dimensions.

2. A *training* data set is defined and associated with a particular ground truth. This normally takes the form of attaching a known label (or set of labels in the multi-label case) to each object in the training set, e.g., class memberships or some categorical variable. If the classes are not known a priori, then clusters may be identified within the training set using unsupervised learning and their labels used. In either case, the training set should be as representative of the parameter space as possible: incompleteness in the training set will lead to inaccuracy in the classifier.

3. A classifier or set of classifiers are trained using the training set. Popular classifiers include random forest, support vector machines, and convolutional neural networks – the latter are a type of deep learning algorithm, which is regarded as the current state-of-the-art for image-based work. *Ensemble* classification methods, such as bagging, boosting, and stacking, use multiple classifiers to improve performance over individual classifiers and generate more certain, precise, and accurate results.

4. A trained classifier is validated on a *test* data set, which is independent from the training set, to assess its strength and utility. Validation using the training set will just result in an overfitted classifier. Performance characteristics such as false negative and positive rates and their ratios to the true positive and negative rates, and summary measures, such as the F-score and Matthews correlation coefficient, are determined from the test set. For binary classifiers, ROC (receiver operating characteristic) curves show the classifier performance as its discrimination threshold is varied and allow optimal models to be selected. If a test data set is not available then *cross validation* can be employed: the training set is partitioned into subsets, which are iteratively used as training and validation sets. Performance characteristics are then calculated by averaging over the subsets.

5. The classifier is applied to the full data set. Population statistics can be determined for each class and outliers identified by poor class membership.

3 Characterizing astronomical time series

The challenge with time series characterization is to represent the time series in such a way that those which are interesting are easily distinguishable from the more run-of-the-mill ones. The most commonly-used discriminative features can be grouped in terms of the type of aspect they try to capture:

- *Location*: Mean, median
- *Scale*: Variance, median absolute deviation
- *Variability*: Stetson JHK indices, von Neumann index
- *Morphology*: Skew, kurtosis, interquartile region, cumulative sum index, ratio of magnitudes brighter/fainter than the mean
- *Timescales*: Lomb-Scargle derived period
- *Trends*: Phase folded slope percentiles, linear regression
- *Model-based*: Fourier amplitude ratios, phase differences, and amplitudes; Shapiro-Wilk normality test

The hope is that such features extracted from the data are informative and non-redundant, although there is a tendency in analyses to employ many features that all aim to broadly capture the same type of information. Unless the time series under analysis are densely sampled, the information content across such features will essentially be uniform.

3.1 Unstated assumptions

There is a danger that feature extraction is applied in a very ad hoc manner without regard for certain statistical aspects of time series. These include:

- *Heteroskedasticity*: Observations are taken on different nights with different seeing, weather conditions, instrumental settings, etc. It is therefore inappropriate to draw measurement errors on flux from the same underlying statistical distribution over the full temporal range of a time series (homoskedasticity), i.e., $\sigma(f_i) \sim \mathcal{N}(0, \sigma^2)$. Features involving errors are instead to be based on $\sigma(f_i) \sim \mathcal{N}(0, \sigma_i^2)$.
- *Non-iid*: Successive data points in a time series cannot be assumed to independently and identically distributed (iid) in the same way that spatial data might be. Temporal data is sequential and so there is an inherent dependence between observations. If the time difference between successive points is small enough then the error measurements can be correlated. Residuals will also carry some correlations, particularly if the time difference is small.
- *Stationarity*: The generating distribution of the time series and its statistical moments, i.e., mean, variance, etc., are assumed to be time independent. There are known astronomical sources for which this is not true: GRS 1915+215 has at least twenty variability states ([20]). Certain classes of time series models can deal with this, for example, generalized autoregressive conditional heteroskedastic (GARCH) models, which are used to describe financial time series, have a variance which is a stochastic function of time. There is also no requirement for nonstationary time series to be stationary in any particular limit. Time series can be made weakly stationary locally by considering first-order differences instead of raw measurements.
- *Ergodicity*: It is assumed that the time average for one sequence is the same as the ensemble average:

$$\hat{f}(x) = \lim_{n\to\infty} \frac{1}{n} \sum_{k=0}^{n-1} f\left(T^k x\right).$$

Observations that are sufficiently far apart in time are assumed to be uncorrelated and new observations therefore give extra information about the underlying process.

3.2 Not all features are equal

Feature relevance is an aspect of machine learning which determines which features contribute the most information to a class label. Several automated classifications ([5, 8, 17, 21, 22]) have analyzed their results in this regard using a number of different techniques: for example, random forest has a feature importance measure or specific iterative approaches successively add/remove the most informative features based on some particular measure, such as class variance or mutual information. These analyses make it clear that most commonly-used features have little importance and are redundant, in part because many of them are highly correlated (all capturing the same information) and so do not offer much discriminative power between them. A practical upshot of this is that only a handful of relevant features need to be derived for a set of time series rather than a bucketful. Which features will, however, depend on the relative classes trying to be identified: there is (as yet) no silver bullet feature that distinguishes all possible classes.

3.3 The most important feature: period

Many features used to characterize light curves rely on a derived period: the time series need to be phase-folded as a precomputation step or the features derive from a periodic basis function decomposition, e.g. Fourier series. The period (or its frequency inverse) is also often identified as the most significant feature in feature relevance analyses. This means that an incorrect period can have

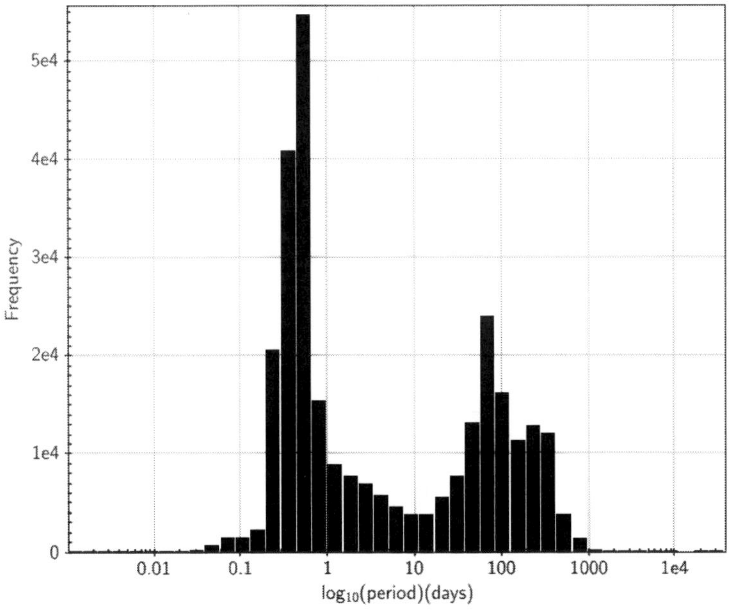

Figure 1. The distribution of 290,000 periods from the Variable Star Index (version 2017-03-20; [24]).

a sizable effect on classification: for example, [8] found a 22% misclassification error rate for non-eclipsing variable stars with a wrong period. Periodic feature routines also account for around 75% of computing time used in feature extraction ([21]) and so a wrong starting value can mean a waste of resources.

More generally, there is the issue of whether one quantity can capture the cyclical aspects of stellar variability. Kepler observations of RR Lyrae stars have shown that about 30% exhibit some form of Blazhko behavior – period and amplitude variation associated with mode switching – and there are also small amplitude cycle-to-cycle modulations in many RRabs. Close binary systems and long period variables are known to have cyclic period changes over multidecade baselines. Semi-regular variable stars commonly exhibit double periods and multiperiodicity. Finally, any variable source that can be described by a (C)ARMA (autoregressive moving average) process can expect to show quasi-periodicity (associated with peaks in the power spectrum).

Of the few hundred thousand periodic variable stars known, we typically have (sparse) temporal data for about a decade or two of coverage. For the majority of sources with periods on the order of a few days or shorter (see Fig. 1), this amounts to at least a couple of thousand cycles (though a much smaller fraction of these will actually have associated data). However, we have very limited information on how stable the measured periods are over these timescales. For decadal length surveys now under way/being planned, this could require a regular retraining of the classifier as class periods drift.

There is also no single algorithm used to find periods. Rather there are a number of different methodologies employed (and described extensively in the literature – see [23] for a review) with the two most commonly-applied ones being: minimizing the fit of a set of basis functions to the time series and minimizing some dispersion measure of the time series in the phase-folded space. The most popular period-finding algorithm, Lomb-Scargle and its variants, belongs to the former category (it uses a least-squares fit of a set of trigonometric functions to the data) as do wavelet-based methods. The latter category includes phase dispersion minimization (PDM; uses means), analysis of variance (AoV; use variance), Lafler-Kinman (uses string length), and entropy-based methods. Other methods employ rank ordering of the time series, Bayesian modeling, neural networks, Gaussian process regression with a periodic kernel, and convolutions.

To understand which period-finding algorithm might be optimal, we undertook an investigation ([23]) into the dependencies of the nine most popular algorithms on signal-to-noise ratio, number of observations, sampling strategy, and variable star class type. We were also interested in the performance characteristics of the algorithms since one that might take 30 s on average to find a period would require 100 CPU years to deal with a 100 million source data set. Our test data set consisted of 67,000 light curves with representation across the eruptive, pulsating, rotating, cataclysmic, and eclipsing classes of variable stars.

We found that no single algorithm was generally better than $\sim$ 60% accurate across the full data set (we assumed that the ground truth was at most a few percent inaccurate – the quoted periods had all been verified by visual inspection by at least three of the authors). All methods are dependent on the quality of the time series and show a decline in period recovery with lower quality time series as a consequence of fewer observations, fainter magnitudes, and noisier data, and an increase in period recovery with higher source variability. The algorithms were stable with a minimum bin occupancy in the phase-folded space of 10 for a bin width of $\Delta\phi = 0.1$. A bimodal observing strategy consisting of pairs (or more) of close-separated observations (short Δt) per night and normal repeat visits (every few nights or so) was better than just regular single visits. The algorithms worked best with pulsating and eclipsing variable classes, perhaps unsurprisingly as these tend to have the most sinusoidal waveforms. Lomb-Scargle (and its variants) was strongly effected by the "half-period issue": for eclipsing binaries, the algorithm tends to return half the true period value. Finally, specific algorithms worked better irregular sampling, bright magnitudes (containing saturated values), or with performance

For a choice of single algorithm, the recommendation was AoV.

3.4 Are we using the best features

As mentioned above, a lot of the standard features used in the literature are correlated, e.g., variance, median absolute deviation, and interquartile range are all estimates of scale. They also focus on capturing morphological aspects of the time series for discriminative purposes - the aim is to create groups of time series that look similar. However, the shape of a time series does not necessarily correlate in any simple way with the underlying physical process(es) that generated it. An alternate approach is to draw on domain knowledge to define phenomenologically motivated features that might be able to group time series based on common physics. For example, chaotic processes can be quantified by the Lyapunov exponent, autocorrelation identified by the Durbin-Watson statistic, and nonlinearity by the Teraesvirta measure.

Using discriminative descriptors also focuses on learning the boundaries between classes whereas generative ones aim to model the individual distributions of classes. Though the former may be easier to compute and be more suitable for classification purposes, the latter offer more potential insight into the physical processes underpinning the time series. In particular, there is interest in modeling time series via generative models that can capture deterministic and stochastic components, such as families of autoregressive moving average models or recurrent neural networks. For example, [25] model the multistate source GRS 1915+105 using an autoregressive model and principal component analysis and can distinguish between purely stochastic states, chaotic behavior contaminated by colored noise, and signatures of deterministic non-linear behavior (chaotic).

4 Which classifier?

Most individual classifiers will give broadly the same results for the same feature set in terms of performance characteristics such as precision and recall. The errors on the characteristics are typically at the few percent level, particularly for classifiers with a strong stochastic element, e.g., random forest, and so there really is no difference between a classifier that is 88% accurate and one that is 90% accurate. As with period finding algorithms, certain classifiers may have slight preferences for certain classes where the discriminating hyperplane has some particular form. The state of the art classifier is random forest (although see also support vector machines, Bayesian networks, and self-organizing maps). Better results can be obtained with an ensemble classifier which incorporates the relative strengths of each individual classifier and thus obtains better predictive performance than any of them (typically accuracy will go from 85-90% to 95-99%) . In general, there are two types of such ensemble methods: averaging methods, such as bagging where the combined (averaged) prediction of the independent constituent methods has reduced variance (random forest actually works along these lines with decision trees as the base classifier); and boosting methods, in which weak component methods are combined incrementally to create a strong method with reduced bias. There are also hierarchical ensemble methods where the class outputs of a base layer of classifiers are used as inputs to a secondary (or subsequent) layer(s).

4.1 Dealing with uncertainties

There are many sources of uncertainty: time series have observation errors in flux (and time, although we rarely consider these); regularization and imputation to deal with irregular sampling and missing values add interpolation uncertainties; and model parameters and hyperparameters have their own uncertainties. Feature representations do not traditionally deal with these (and alternate data representations will also introduce their own uncertainties). Probabilistic classifiers tend only to simulate the effect of observation errors through the choice of priors or parameter space coverage in training sets.

Ideally, though, a full probability density function should be given for any classification: for example, a source is not longer classed as an RR Lyrae or eclipsing binary but has a 62% change of being an RR Lyrae, a 28% change of being an eclipsing binary and a 10% chance of something else. Uncertainty quantification (UQ) formally considers this through both forward UQ with simulations and expansions methods and inverse UQ via Bayesian techniques.

4.2 A word about automated classification

[18] used a random forest classifier to automatically classify periodic variables in the EROS-II LMC data set including 6,607 potential RR Lyrae not previously discovered. [26] reported that 4,408 sources were missed from this data set in their new collection of RR Lyrae in the LMC. 3,234 have a counterpart in the OGLE-IV database and their light curves were carefully analyzed. 149 are probable RR Lyrae, mainly noisy RRc, but the remaining 3,085 are eclipsing binaries, δ Scuti, Cepheids, or constant sources. In other words, the EROS-II classification for 3,085 sources is not confirmed by a subsequent analysis of the same objects from a different survey. Although [26] do not attempt to explain this misclassification, it is most likely due to OGLE-IV data being of higher quality than the EROS-II set. It does illustrate, however, that classifications are not in any sense absolute and reproducibility of classifications across surveys is the only way to validate them.

4.3 Establishing ground truths

It can almost be guaranteed that the assumed ground truth will not be 100% correct and this has a number of consequences. The (automated) classifier will be mistrained and this will raise the false positive/false negative rates in data to which it is applied. It should be noted that a 1% misclassification rate is highly significant when dealing with large data sets, for example, containing 100 million sources.

Currently, there is a lack of a (comprehensive) community-agreed training set for stellar classification, particularly at fainter magnitudes ($V > 16$). Compiled catalogs, such as the General Catalog of Variable Stars (GCVS), the Variable Star Index (VSX), or even SIMBAD, are known to be heterogeneous and have varying inaccuracies in the class flags. The respective population sizes of different classes of objects within these catalogs is also biased towards specific classes, e.g., RR Lyrae and eclipsing binaries, rather than being representative of the true astrophysical populations. For example, MACC ([5]) has $\sim$ 20% of sources as semiregular variables and $\sim$ 20% as small amplitude red giants.

4.4 Class distinctions

The question of exactly how many classes there may in a data set is essentially unanswerable. It will always be dependent on the both the objective state of domain knowledge at the time of classification and subjective choices made by the domain scientist carrying out the classification. There can also be different levels of classification employed, particularly when using a hierarchical system: for example, according to [4] stellar variability may be regarded at a high level as intrinsic or extrinsic to the source in origin or at a lower level originating from eruptive, pulsating, rotating, cataclysmic, or eclipsing processes.

Most classification schemes assume mutually exclusive classes, e.g., a source is either a RR Lyrae or an eclipsing binary but not both. *Fuzzy* clustering employs overlapping class boundaries, described by membership functions, and so multiple class memberships are possible, Sources can thus be designated as 80% RR Lyrae and 20% eclipsing binary, say. This treats class as much of a continuum quantity than a complete discrete one.

Unsupervised learning can determine clusters through different cluster paradigms: partitional, hierarchical, density-based, grid-based, correlation, spectral, gravitational, herd, etc. These all have different assumptions, for example, partitional clustering places all clusters on the same level rather than in a taxonomy, say, and density-based clustering will associate different classes with peaks in the density field of the characterizing feature space. It is also arguable whether classes associated with clusters from an algorithm are physically realistic: does division of a feature space naturally map to a structured view of domain knowledge?

An often overlooked possibility in classification is an "other" class which holds sources that do not fit well into any of the prescribed classes. 60% of objects in MACC [5] are in the "MISC" class and the most accurately classified sources in the CRTS periodic variable catalog ([13]) belong to the aperiodic noise class.

4.5 Extremes

Often the most interesting objects are those classed as outliers and not matching the characteristics of the general population. These can be data artifacts, which help to identify instrumental or reduction problems, or genuine physically unusual sources. [27] defines "an outlier as an observation that differs so much from other observations as to arouse suspicion that it was generated by a different mechanism".

However, there is no reason why the characterized variability of every type of astronomical source in the observable universe over a decadal baseline should follow a Gaussian distribution. For a generic heavy-tailed distribution, defined by

$$\lim_{x\to\infty} P[|X| > x]\, x^{-\alpha} = \lambda,$$

λ and α cannot be estimated from data and so the general statistical significance of a source known. A specific choice has to be made about the distribution, e.g., it is Cauchy or Weibull, for a numerical significance to be evaluated.

There is no formal statistical definition for an outlier but it can be shown that the presence of outliers has no connection with either the existence of heavy tails of an underlying distribution or with experimental errors ([28]). From a topological perspective, a significant outlier is also not necessarily related to an underdensity in any distribution but rather has high persistence over a range of scales and marginal connectivity to the general population.

5 Summary and future work

Automated classification is a necessity when working with large data sets; there are, however, a number of caveats to be aware of. Astronomical time series need to be represented in terms of features but the choice of features is not simple and each feature comes with inherent (statistical) assumptions and dependencies. It needs to be understood how errors in these representations propagate through and combine with others in the classification process to the final class assignment, which should also have an error model associated with it rather than being categorical. The actual choice of classification algorithm is not necessarily as important as is sometimes made out but the choice of training set used with it is. The ground truth associated with it needs to be researched as it it reflects a particular set of assumptions. The available set of class assignments also needs to be looked at and the reality of them considered in terms of current domain knowledge. Finally, if one is interested in outliers, consideration must be given to what an outlier actually is.

There are further challenges for automated classification that we have not considered here. There is obvious overlap between many data sets (the sky is a finite surface), both in similar passbands and also in different wavelength regimes, but how can this data be optimally handled/combined to aid the classification process. Domain adaptation is a branch of machine learning that deals with the transformation of what has worked for one survey to another – this may be particularly appropriate for a new survey which is deeper than any previous one or in a new wavelength regime and so has very or no labels (see the paper by P. Protopapas in this volume for further details). Another issue is the performance and/or hardware requirements for automated classification. An important quantity here is the mean characterization time for a given time series since this will most likely be the top computational cost in the process. With the computational resources available, what is the acceptable range for this given the size of the data set; for example, for 100 million light curves and a 32-core machine, a 1 s characterization requires just over a month to complete. In many analyses, classification will also be an ongoing process rather than one that is performed once a given survey is completed: it is unlikely that the community will wait for a decade before classifying the variable stars in the LSST data set. How do the class assignments deal with new data then? Does the whole process have to be rerun at regular intervals, each data release, say, or are there streaming classification algorithms that can adjust class probability distributions as fresh data arrives?

These issues are not unsurmountable and the promise of automated classification is the generation of statistically significant samples of rare phenomena: a 100 million object data set will contain 10,000 examples of a 1 in 10,000 source and a hundred of a one-in-a-million. The challenge then becomes to understand these.

Acknowledgments: This work was supported in part by the NSF grants AST-1413600 and AST-1518308.

References

[1] Mignard, F., MmSAI, **83**, 918 (2012)
[2] Bellm, E., in *The Third Hot-wiring the Transient Universe Workshop*, ed. P. R. Wozniak et al., p. 27 (2014)
[3] Ivezic, Z., et al., `arXiv:0805.2366` (2011)
[4] Eyer, L., & Blake, C., MNRAS, **358**, 30 (2005)
[5] Richards, J., Starr, D., Miller, A., Bloom, J., Butler, N., Brink, H., & Crellin-Quick, A., ApJS, **203**, 32, (2012)
[6] Debosscher, J., Sarro, L.M., Lopez, M., et al., A&A, **506**, 1 (2009)
[7] Blomme J., Debosscher, J., De Ridder, J., et al., ApJL, **713**, 204 (2010)
[8] Dubath, P., Rimoldini, L., Süveges, M., et al., MNRAS, **414**, 2602 (2011)
[9] Rimoldini, L., Dubath, P., Süveges, M., et al., MNRAS, **427**, 2917 (2013)
[10] Sarro, L. M., Debosscher, J., Aerts, C., & Lopez, M., A&A, **506**, 535 (2009)
[11] Süveges, M., Sesar, B., Varadi, M., et al., MNRAS, **424**, 2528 (2012)
[12] Palaversa, L., Ivezic, Z., Eyer, L., et al., AJ, **146**, 101 (2013)
[13] Drake, A.J., Graham, M.J., Djorgovski, S.G., et al., ApJ, **790**, 3 (2014)
[14] Torrealba, G., Catelan, M., Drake, A.J., et al., MNRAS, **446**, 2251 (2015)
[15] Drake, A.J., Djorgovski, S.G., Catelan, M., et al., MNRAS, submitted (2017)
[16] Catelan, M., Minniti, D., Lucas, P.W., et al., `arXiv:1310.1996` (2013)
[17] Elorrieta, F., Eyheramendy, S., Jordan, A., et al., A&A, **595**, 11 (2016)
[18] Kim, D.-W., Protopapas, P., Bailer-Jones, C. A. L., Byun, Y.-I., Chang, S.-W., Marquette, J.-.B., & Shin, M.-S., A&A, **566**, 43 (2014)
[19] Masci, F., Hoffman, D., Grillmair, C., & Cutri, R., AJ, **148**, 21 (2014)
[20] Hannikainen, D. C., et al., A&A, **435**, 995 (2005)
[21] Richards, J. W., Starr, D. L., Butler, N. R., Bloom, J. S., Brewer, J. M., Crellin-Quick, A., Higgins, J., Kennedy, R., & Riscahrd, M., ApJ, **733**, 10 (2011)
[22] D'Isanto, A., Cavuoti, S., Brescia, M., Donalek, C., Longo, G., Riccio, G., & Djorgovski, S. G., MNRAS, **457**, 3119 (2016)
[23] Graham, M. J., Drake, A. J., Djorgovski, S. G., Mahabal, A. A., Donalek, C., Duan, V., & Maker, A., MNRAS, **434**, 3423 (2013)
[24] Watson, C., Henden, A. A., & Price, A., SASS, **25**, 47 (2006)
[25] Huppenkothen, D., Heil, L. M., Hogg, D. W., & Mueller, A., MNRAS, **466**, 2634 (2017)
[26] Soszynski, I., Udalski, A., Szymanski, M. K., et al., AcA, **66**, 131 (2016)
[27] Hawkins, D. M., *Identification of Outliers* (Springer Netherlands, 1980)
[28] Klebanov, L.B., preprint (`arXiv:1611.05410`) (2016)

A fast template periodogram

John Hoffman[1,*], *Jake* VanderPlas[2,**], *Joel* Hartman[1,***], and *Gáspár* Bakos[1,****]

[1]*Department of Astrophysical Sciences, Princeton University, Princeton NJ 08540*
[2]*eScience Institute, University of Washington, Seattle, WA 98195*

Abstract. This proceedings contribution presents a novel, non-linear extension to the Lomb-Scargle periodogram that allows periodograms to be generated for arbitrary signal shapes. Such periodograms are already known as "template periodograms" or "periodic matched filters," but current implementations are computationally inefficient. The "fast template periodogram" presented here improves existing techniques by a factor of ~a few for small test cases ($O(10)$ observations), and over three orders of magnitude for lightcurves containing $O(10^4)$ observations. The fast template periodogram scales asymptotically as $O(HN_f \log HN_f + H^4 N_f)$, where H denotes the number of harmonics required to adequately approximate the template and N_f is the number of trial frequencies. Existing implementations scale as $O(N_{\rm obs} N_f)$, where $N_{\rm obs}$ is the number of observations in the lightcurve. An open source Python implementation is available on GitHub.

1 Introduction

Template fitting, or periodic matched filtering, is a powerful technique for identifying, classifying, and characterizing periodic variability in noisy time series data. However, template fitting requires several orders of magnitude more computational resources than running a Lomb Scargle periodogram ([1–4]) on the same lightcurve, and the computational requirements scale as $N_{\rm obs} N_f$, where $N_{\rm obs}$ is the number of observations in the lightcurve and N_f is the number of trial frequencies.

For example, [5] used template fitting to model RR Lyrae in the Pan-STARRS DR1 ([6]) and found that, for a given completeness, template fitting produced purer samples of RR Lyrae than Lomb-Scargle or multi-term extensions ([7]). However, [5] found that the computational resources needed to produce template fits (approximately 30 minutes per source per CPU) limited the number of lightcurves for which a full template fit could be performed to $O(10^3)$ out of $O(10^6)$ Pan-STARRS DR1 lightcurves.

Pan-STARRS lightcurves typically have less than 35 observations in 5 photometric filters (~ 7 per filter), but many photometric surveys have orders of magnitude more observations – CoRoT ([8]), SuperWASP ([9]), HATNet ([10]), and *Kepler* ([11]), for example, have $O(10^4 - 10^5)$ observations per lightcurve. Since template fitting scales as $O(N_{\rm obs}^2)$,[1] running template periodograms on these surveys currently requires a prohibitively large amount of computational resources.

*jah5@princeton.edu
**jakevdp@cs.washington.edu
***jhartman@astro.princeton.edu
****gbakos@astro.princeton.edu

[1]The actual scaling is technically $O(N_{\rm obs} N_f)$, however the number of trial frequencies N_f scales linearly with $N_{\rm obs}$ ([12]).

However, if template periodograms were computationally efficient enough to run on entire surveys with $O(10^3 - 10^5)$ observations per lightcurve, they would have the potential to be a powerful tool for automated classification. For example, by selecting a group of templates that capture the variety of periodic signals present in astrophysical data, a template periodogram could provide *simultaneous* detection and (pre-)classification of all lightcurves with higher signal to noise ratio than multi-harmonic periodograms, which are also designed to detect non-sinusoidal signals.

The advantage of templates is that there are only three free parameters (amplitude, phase, and a constant offset), compared to the multi-harmonic periodogram, which has $2H + 1$ free parameters, where H is the number of harmonics. Using a collection of templates encodes "domain knowledge" (see, e.g., [13]) into the periodogram, which helps to reduce the parameter space to a set of physically plausible lightcurve shapes.

2 Extending Lomb-Scargle to arbitrary shapes

The Lomb-Scargle periodogram ([1–4]) is also known as least squares spectral analysis. The formulation of the periodogram given in [1] and [2] fits a linear model

$$\hat{y}(t|\theta,\omega) = \theta_1 \cos\omega t + \theta_2 \cos\omega t \tag{1}$$

to time series data $D = \{t_n, y_n, \sigma_n | n < N\}$ at a number of trial frequencies ω. The Lomb-Scargle periodogram,

$$P_{\rm LS}(\omega) = \frac{1}{2\sigma^2}\left(\frac{\left[\sum_{n=1}^{N}(y_n - \bar{y})\cos\omega t_n\right]^2}{\sum_{n=1}^{N}\cos^2\omega t_i} + \frac{\left[\sum_{n=1}^{N}(y_n - \bar{y})\sin\omega t_n\right]^2}{\sum_{n=1}^{N}\sin^2\omega t_i}\right), \tag{2}$$

is similar to a Fourier intensity for unevenly spaced data. As shown in [2], the periodogram can be derived by minimizing the squared residuals and setting

$$P(\omega) = \frac{\chi_0^2 - \chi^2(\omega)}{\chi_0^2}, \tag{3}$$

where $\chi^2(\omega)$ is the minimum (weighted) sum of squared residuals for a model fit at frequency ω, and χ_0^2 is the weighted sum of squared residuals for a constant fit.

There are many extensions to the Lomb-Scargle periodogram that are useful for time domain astronomy. [14] used *weighted* squared residuals and added a constant offset to $\hat{y}$, [7, 15] further extended the Lomb-Scargle periodogram for models with more than one harmonic,

$$\hat{y}(t|\theta,\omega) = \theta_0 + \sum_{h=1}^{H} \theta_{2h}\cos h\omega t + \theta_{2h-1}\sin h\omega t, \tag{4}$$

and recently [12] provided a formalism for applying periodograms to multi-band time series.

The multiharmonic periodogram is designed to find non-sinusoidal periodic signals in irregularly spaced time series data. However, as shown in [12], increased flexibility comes at a cost – the noise level of the periodogram relative to the maximum peak height tends to increase as the number of free parameters in the model $\hat{y}$ increases, and in some cases the best frequency can tend to a harmonic or a subharmonic of the true frequency.

If the signal shape of interest is known *a priori* and can be expressed as a truncated Fourier series with H harmonics,

$$M(\omega t) = \sum_{h=1}^{H} \psi_{2h} \cos h\omega t + \psi_{2h-1} \sin h\omega t, \tag{5}$$

then a model of the form

$$\hat{y}_M(t|\theta, \omega) = \theta_0 + \theta_1 M(\omega(t - \theta_2)) \tag{6}$$

should provide improved signal to noise over the multiharmonic periodogram, since there are only three free parameters at each trial frequency, regardless of the number of harmonics used to express the template M. The difference between the template periodogram (i.e., the periodogram derived from $\hat{y}_M$) and the multiharmonic periodogram of [7, 15] is that the harmonic amplitudes ψ_i are *not free parameters* in the template periodogram, while they are free parameters for the multiharmonic periodogram.

Unlike the Lomb-Scargle extensions previously mentioned, the template periodogram is *non-linear*. The non-linearity arises from the phase shift parameter θ_2. For Lomb-Scargle and multi-harmonic periodograms, the phase shift is equivalent to changing the relative amplitudes of the cosine and sine terms. This is not true for the template periodogram, since ψ_i terms are deliberately fixed. Instead, by working with $x \equiv \cos \omega\theta_2$, the template can be expressed as:

$$\begin{aligned} M(\omega(t - \theta_2)) = \sum_{h=1}^{H} \Big[& \Big(\psi_{2h} T_h(x) \mp \psi_{2h-1} \sqrt{1 - x^2} U_{h-1}(x)\Big) \cos h\omega t \\ & + \Big(\psi_{2h-1} T_h(x) \pm \psi_{2h} \sqrt{1 - x^2} U_{h-1}(x)\Big) \sin h\omega t \Big], \end{aligned} \tag{7}$$

where T_n and U_n are the Chebyshev polynomials of the first and second kind, respectively, and the $\pm$ sign ambiguity reflects the two possible signs of $\sin \omega\theta_2$.

Detailed derivations of the template periodogram will be presented in a later paper, but the non-linearity of the template periodogram can be reduced to finding the real roots of a polynomial of order $\sim 6H$ at each trial frequency. The root finding step scales as $O(H^3 - H^4)$ at each trial frequency, and the polynomial coefficients can be computed with a single non-equispaced fast Fourier transform (NFFT, [16]) of size $N_f H$. Thus, the template periodogram scales as

$$O(N_f H^4 + N_f H \log N_f H). \tag{8}$$

3 Implementation

A Python implementation of the fast template periodogram is available at

`https://github.com/PrincetonUniversity/FastTemplatePeriodogram`.

Timing comparisons with an existing Python template fitting implementation (`gatspy`, [17]) shown in Figure 1 demonstrate the improved computational efficiency of the fast template periodogram. In terms of accuracy, Figure 1 also shows that the fast template periodogram, which solves for the optimal parameters directly, is usually able to find better fits than the `gatspy` implementation, which uses non-linear optimization tools to find the optimal parameters.

DOI: 10.1051/epjconf/201715203002

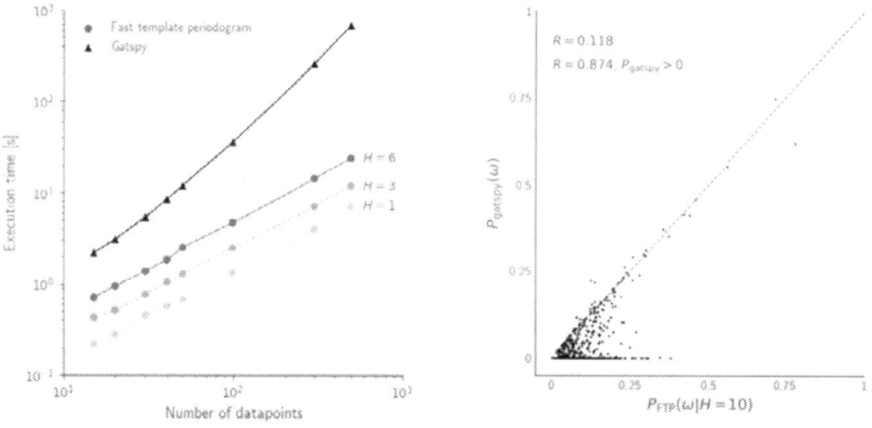

Figure 1. Timing (*left*) and accuracy (*right*) comparisons between the gatspy ([17]) template fitting algorithm and the fast template periodogram. The fast template periodogram improves computational efficiency by a factor of ~a few for small test cases and this improvement factor grows linearly in $O(N_f \sim N_{\text{obs}})$. Though the fast template periodogram has $O(N_f \log N_f)$ asymptotic scaling for a given value of H, the root-finding utilities (which scale as $O(N_f)$) dominate the computational time for all test cases shown here. In most cases, the fast template periodogram finds better template fits than gatspy (points that fall below the dashed line on the *right*-most figure).

4 Conclusion

Finding non-sinusoidal signals in irregularly spaced astrophysical data is a challenging problem that requires balancing model flexibility against the need to use as few model parameters as possible to minimize the relative noise of the periodogram. Template periodograms, or periodic matched filters, allow for the detection of periodic signals with fixed, arbitrary non-sinusoidal shapes without the increase of periodogram noise inherent in the multiharmonic periodogram presented in [7, 15]. However, template periodogram implementations to date ([5, 17]) have relied on non-linear optimization at each trial frequency to obtain the optimal amplitude, phase, and offset of the template. Thus, applying template periodograms to large photometric surveys like Pan-STARRS ([6]) requires prohibitive amounts of computational resources and scale as $O(N_f N_{\text{obs}} \sim N_{\text{obs}}^2)$.

However, template periodograms can be made orders of magnitude more computationally efficient. By using the algorithmic shortcuts described here, template periodograms could potentially prove computationally feasible on large photometric surveys with many ($\gtrsim 10^2$) observations per lightcurve, like CoRoT ([8]), HATNet ([10]), and others. An open source implementation of the fast template periodogram is available on GitHub, and improvements to the existing implementation should continue to improve computational efficiency.

References

[1] Lomb, N. R., Ap&SS, **39**, 447 (1976)
[2] Scargle, J. D., ApJ, **263**, 835 (1982)
[3] Barning, F. J. M., BAIN, **17**, 22 (1963)
[4] Vaníček, P., Ap&SS, **12**, 10 (1971)

DOI: 10.1051/epjconf/201715203002

[5] Sesar, B., Hernitschek, N., Mitrović, S., et al., AJ, **153**, 204 (2017)
[6] Chambers, K. C., Magnier, E. A., Metcalfe, N., et al., preprint (arXiv:1612.05560) (2016)
[7] Palmer, D. M., ApJ, **695**, 496 (2009)
[8] CoRoT Team, *The CoRoT Legacy Book: The Adventure of the Ultra High Precision Photometry from Space* (2016)
[9] Smith, A. M. S., & WASP Consortium, Contr. Astron. Obs. Skalnate Pleso, **43**, 500 (2014)
[10] Bakos, G., Noyes, R. W., Kovács, G., Stanek, K. Z., Sasselov, D. D., & Domsa, I., PASP, **116**, 266
[11] Koch, D. G., Borucki, W. J., Basri, G., et al., ApJL, **713**, L79 (2010)
[12] VanderPlas, J. T., & Ivezić, Ž., ApJ, **812**, 18 (2015)
[13] Yu, T., Simoff, S., & Jan, T., Neurocomputing, **73**, 2614 (2010)
[14] Zechmeister, M., & Kürster, M., A&A, **496**, 577
[15] Schwarzenberg-Czerny, A., ApJL, **460**, L107 (1996)
[16] Keiner, J., Kunis, S., & Potts, D., ACM Trans. Math. Softw., **36**, 19 (2009)
[17] VanderPlas, J., *gatspy: General tools for Astronomical Time Series in Python*, Astrophysics Source Code Library (http://ascl.net/1610.007) (2016)

Finding binaries from phase modulation of pulsating stars with *Kepler*

Hiromoto Shibahashi[1,⋆], *Simon* Murphy[2,3,⋆⋆], and *Tim* Bedding[2,3,⋆⋆⋆]

[1] *University of Tokyo, Department of Astronomy, Tokyo 113-0033, Tokyo, Japan*
[2] *University of Sydney, Sydney Institute for Astronomy, NSW 2006, Australia*
[3] *Aarhus University, Stellar Astrophysics Centre, DK-8000 Aarhus C, Denmark*

Abstract. Binary orbital motion causes a periodic variation in the path length travelled by light emitted from a star towards us. Hence, if the star is pulsating, the observed phase of the pulsation varies over the orbit. Conversely, once we have observed such phase variation, we can extract information about the binary orbit from photometry alone. Continuous and precise space-based photometry has made it possible to measure these light travel time effects on the pulsating stars in binary systems. This opens up a new way of finding unseen brown dwarfs, planets, or massive compact stellar remnants: neutron stars and black holes.

1 Introduction: History of finding unseen stars in binary systems

The history of finding unseen stars in binary systems dates back to 1783 when regularity and periodicity in variation of the magnitude of β Per, called Algol, was noticed visually and interpreted as the interposition of a large body orbiting around the star by Goodricke ([1]). It should be noted that this finding of an eclipsing binary was even before the apparent double stars and the binary stars orbiting under mutual gravitational attraction were first discriminated by William Herschel ([2]). The first visual binary orbits were calculated in the early 19th century, but very few gravitationally bound binary systems were known to science at that time. The best method to detect such binaries was then astrometry. Using this method, Bessel ([3]) discovered that the proper motion of α CMa, Sirius, in right ascension was not constant and he postulated the presence of an unseen companion disturbing Sirius' apparent position.

Development of spectroscopy and astrophotography brought about a revolution in the field of binary stars at the end of the 19th century. In 1890, Vogel ([4]) noticed that the radial velocity of the star Algol, measured from the Doppler shift of the star's spectral lines, varied in accordance with the eclipses. He thus established the first spectroscopic binary (SB1), wherein one set of spectral lines was observed to shift according to the orbital phase. In the same year, Pickering ([5]) detected two sets of spectral lines (i.e. an SB2 binary) in the absorption spectrum of ζ UMa.

Binary orbital motion causes a periodic variation in the path length travelled by light emitted from a star and arriving at Earth. Hence, if the star is regularly pulsating, the observed times of maxima in

⋆shibahashi@astron.s.u-tokyo.ac.jp
⋆⋆simon.murphy@sydney.edu.au
⋆⋆⋆tim.bedding@sydney.edu.au

luminosity vary over the orbit. It was only about 40 years ago that the light-time effect on the pulse timing was first utilized to find unseen binary companions. This won Hulse and Taylor the Nobel prize for their work on a binary pulsar ([6]), and progress was being made on the application to main-sequence stars around the same time. Based on deviation of the photoelectrically observed ephemeris for the times of luminosity maxima from the prediction, a binary nature was first suggested by Barnes and Moffett ([7]) in 1975 for a high-amplitude δ Sct star, SZ Lyn; the suggestion was confirmed correct by radial velocity measurements in 1988 ([8]).

2 New techniques: The FM method and the PM method

The pulse timing method may work very well in the case of stars pulsating with a single mode, because the intensity maxima from the single mode, particularly in the case of sharp maxima, are easy to track and then any deviations from precise periodicity are easy to detect. However, when the pulsating star is multiperiodic the situation becomes complex. It is more desirable to extract phase information for multiple modes systematically. This problem was first discussed by Woltjer ([9]) and later by Irwin ([10]). However, the required photometric and timing precision was not reachable at that time, hence application of their methods was not practical.

In contrast, recent space-based photometry has made it possible to measure the orbital phase variation of pulsating stars in binary systems with extremely high precision over long time spans. Shibahashi and Kurtz ([11]) resorted to the frequency domain to analyse the photometry data, and showed that the phase variation of each mode caused by the light-time effect manifests itself, in the Fourier transform, as a multiplet with spacing equal to the orbital frequency as far as pulsations are coherent over the orbital period as in the case of δ Sct variables. Their method to obtain an orbital solution, called the Frequency Modulation (FM) method, was later improved ([12]) and the validity of the FM technique was verified by the consistent results obtained from it when compared to a traditional eclipsing binary light-curve analysis ([13]).

Soon thereafter, Murphy et al. ([14]) developed a complementary method to be applied in the time domain, called the Phase Modulation (PM) method, which works by converting the phase shifts into the light arrival-time delay for many oscillation frequencies simultaneously ([15]). If multiple oscillation frequencies show the same periodic time delays, the origin can be attributed to binarity. The PM method was further strengthened by adapting automation with a Markov-chain Monte Carlo (MCMC) algorithm to obtain robust uncertainties in the determined orbital elements ([16]). In some binary systems, two series of time delays are identifiable with the same period but opposite phase. This indicates that both binary components pulsate. In such a case, the mass ratio of these systems can be obtained by the PM method without dependence on the inclination $\sin i$, making them very attractive to study. We refer to these as "PB2" systems by analogy with the SB2 spectroscopic binaries.

3 Application of the PM method to exoplanet hunting

It took 18 years to detect Sirius' white-dwarf companion directly ([17]), and Algol's companion is now known to be a cool subgiant. However, some companions are not directly detectable because they do not emit their own visible light. Planets fall into this category. It was the pulse timing method applied to a millisecond pulsar, PSR1257+12, that led in 1992 to the first success in detecting an exoplanetary system ([18]). The discovery of a giant planet orbiting a hot subdwarf B star, V391 Peg, was also made by the pulse timing method ([19]).

Detection of an exoplanet orbiting around a solar-type star was first made in 1995 from spectroscopic observations of the star's radial velocity ([20]). High-precision photometric measurements also led to the success in detecting partial eclipses due to planetary transits across a sun-like star in 2000 ([21]). The former is called the RV method in the field of exoplanet hunting, while the latter is called the transit method. Ultra-high precision, high-cadence photometric measurements from space missions such as the *Kepler* Space Telescope has rapidly accelerated progress in exoplanet searches.

Most of the thousands of planets (and planet candidates) detected by *Kepler* have been found around cool stars ([22]), while exoplanets orbiting A stars remain difficult to find both by the RV method and the transit method. This is because the RV method is hindered by significantly broadened spectra due to rapid rotation of A stars and a lack of suitably shallow absorption lines, while the transit method is hindered by the luminosity variations caused by stellar pulsations, which amount to several mmag. Fortunately, pulsations that limit RV and transit surveys of A stars can themselves be used as precise clocks for the FM or the PM methods. Both the FM and the PM methods are capable of detecting companions down to brown dwarfs or even planetary masses when applied to the *Kepler* data of δ Sct stars. They also complement the spectroscopic radial velocity measurement with sensitivity at long periods of several hundred days, where radial velocity amplitudes tail off. By applying the PM method, Murphy et al. ([23]) succeeded in detecting a 12 $M_{\rm Jup}$ planet in an 840-day orbit around a main-sequence A star. This is the first case (and only one yet) of an A star known to host a planet within the habitable zone.

4 Future application of the PM and FM methods to search for stellar-mass black holes

The contrast between bright intermediate-mass stars and planets is so great that the planets are essentially invisible. The same is true when the optical counterpart is a neutron star or black hole. We then plan to search for stellar remnants, namely white dwarfs, neutron stars and black holes as binary components of pulsating stars by using the PM or the FM methods. If we consider the case of δ Sct stars as the pulsating stars, they are intermediate-mass stars (1.4 to 2.4 $M_{\odot}$) on or near the main sequence that lie within the classical instability strip. The lower limit for the mass of the binary companion is obtained by either the PM or the FM method. Thus, any luminous companions that have dynamic masses above around 1.4 $M_{\odot}$ should pulsate, or should be more massive than the instability strip boundary (i.e., $M > 2.4\,M_{\odot}$) and dominate the spectrum at optical wavelengths. Otherwise, the companions are non-luminous and are therefore compact objects. Specifically, if the dynamical masses of the companions exceed the Chandrasekhar limit (1.4 $M_{\odot}$), the companions should be neutron stars or black holes. If the masses of the companions exceed the mass limit for neutron stars ($\sim 3\,M_{\odot}$), the companions should be black holes.

Only around 20 stellar-mass black holes have been found to date ([24, 25]), and they have all been found through their X-ray emission and high energy physics. However, stellar evolution theory tells us they should be ubiquitous, and hundreds of their progenitors within 1 kpc of the Sun are known. The discovery of black holes in the optical through their gravitational interactions would be a major scientific breakthrough, and is well within the capabilities of the PM and FM methods. Stellar evolution theory also predicts that many main-sequence stars should have neutron stars as companions, to which the PM and FM methods are also expected to be sensitive, and which are even more common. We will conduct a thorough search for such objects.

References

[1] Goodricke, J., Phil. Trans. Roy. Soc. London, **73**, 474 (1783)
[2] Herschel, W., Phil. Trans. Roy. Soc. London, **92**, 477 (1802)

[3] Bessel, W. F., MNRAS, **6**, 136 (1844)
[4] Vogel, H. C., AN, **123**, 289 (1890)
[5] Pickering, E. C., MNRAS, **50**, 296 (1890)
[6] Hulse, R. A., & Taylor, J. H., ApJL, **195**, L51 (1975)
[7] Barnes, T. G., & Moffett, T. J., AJ, **80**, 48 (1975)
[8] Moffett, T. J., Barnes, T. G., Fekel, F. C., Jr., Jefferys, W. H., & Achtermann, J. M., AJ, **95**, 1534 (1988)
[9] Woltjer, J., Jr., BAN, **1**, 93 (1922)
[10] Irwin, J. B., ApJ, **116**, 211 (1952)
[11] Shibahashi, H., & Kurtz, D. W., MNRAS, **422**, 738 (2012)
[12] Shibahashi, H., Kurtz, D. W., & Murphy, S. J., MNRAS, **450**, 3999 (2015)
[13] Kurtz, D. W., Hambleton, K. M., Shibahashi, H., Murphy, S. J., & Prša, A., MNRAS, **446**, 1223 (2015)
[14] Murphy, S. J., Bedding, T. R., Shibahashi, H., Kurtz, D. W., & Kjeldsen, H., MNRAS, **441**, 2515 (2014)
[15] Murphy, S. J., & Shibahashi, H., MNRAS, **450**, 4475 (2015)
[16] Murphy, S. J., Shibahashi, H., & Bedding, T. R., MNRAS, **461**, 4215 (2016)
[17] Bond, G. P., AN, **57**, 131 (1862)
[18] Wolszczan, A., & Frail, D. A., Nature, **355**, 145 (1992)
[19] Silvotti, R., Schuh, S., Janulis, R., et al., Nature, **449**, 189 (2007)
[20] Mayor, M., & Queloz, D., Nature, **378**, 355 (1995)
[21] Charbonneau, D., Brown, T. M., Latham, D. W., & Mayor, M., ApJL, **529**, L45 (2000)
[22] Coughlin, J. L., Mullarly, F., Thompson, S. E., et al., ApJS, **224**, 12 (2016)
[23] Murphy, S. J., Bedding, T. R., & Shibahashi, H., ApJL, **827**, L17 (2016)
[24] Cowley, A. P., ARA&A, **30**, 287 (1992)
[25] Casares, J., in *Black Holes from Stars to Galaxies – Across the Range of Masses*, Proc. of IAU Symposium 238, ed. V. Karas & G. Matt, p. 3 (2007)

KIC 2831097

A short orbital-period candidate RR Lyrae binary

Ádám Sódor[1,*], *Marek* Skarka[1,**], *Jiří* Liška[2,3,***], and *Zsófia* Bognár[1,****]

[1]*Konkoly Observatory, MTA Research Centre for Astronomy and Earth Sciences, Konkoly Thege út 15–17, H–1121 Budapest, Hungary*
[2]*Department of Theoretical Physics and Astrophysics, Masaryk University, Kotlářská 2, CZ–611 37 Brno, Czech Republic*
[3]*CEITEC BUT, Purkyňova 656/123, CZ–612 00 Brno, Czech Republic*

Abstract. KIC 2831097 was discovered to be a first-overtone RR Lyrae pulsator based on 4-year *Kepler* photometry (Sódor et al. 2017, MNRAS, 465, L1). The data show strong, 0.1 d amplitude systematic phase variations that can be explained by light travel-time effect caused by an about 2-year period orbital motion in a binary system, superimposed on a linear pulsation-period decrease. To verify the binary hypothesis, several well-timed radial-velocity observations will be sufficient.

1 Introduction

KIC 2831097 is the fifth first-overtone *Kepler* ([2]) RR Lyrae (RRc) star with detailed analysis published (first four: [3]; KIC 2831097: [1]). The pulsation properties of the star were found to be typical compared with the four previously known *Kepler* RRc stars (see details in [1]).

The $O - C$ variations are shown in Figure 1. We modelled the $O - C$ data by the superposition of a constant period chane and the orbital light-travel-time effect ([1]).

2 Verification of the binarity

Even though almost two hypothetical orbital revolutions appear in the $O - C$ data, the intrinsic origin of these phase variations cannot be excluded (see details and discussion in [1]). Therefore, radial-velocity (RV) measurements are needed in order to confirm or reject the binarity. Since the hypothetical orbit is short and the radial-velocity amplitude is large, the verification can be attained in two consecutive observing seasons with high confidence, according to Figure 2.

The organization of the observations are underway. Meanwhile, we are also collecting new $O - C$ data to extend the *Kepler* baseline.

*sodor@konkoly.hu
**marek.skarka@csfk.mta.hu
***jiriliska@post.cz
****bognar@konkoly.hu

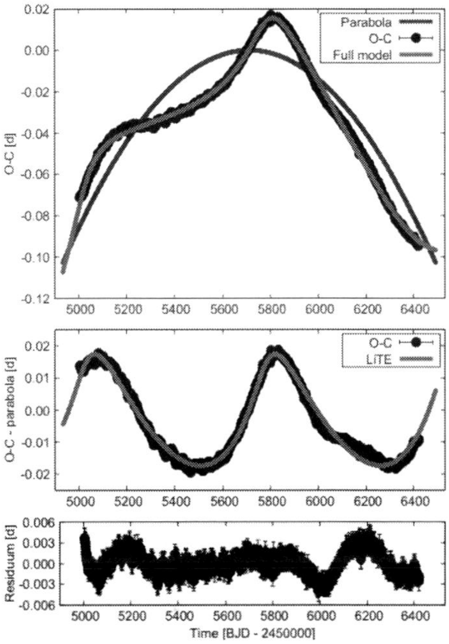

Figure 1. $O - C$ data of KIC 2831097, and the constant period change (blue) and orbital light-travel-time (red) models.

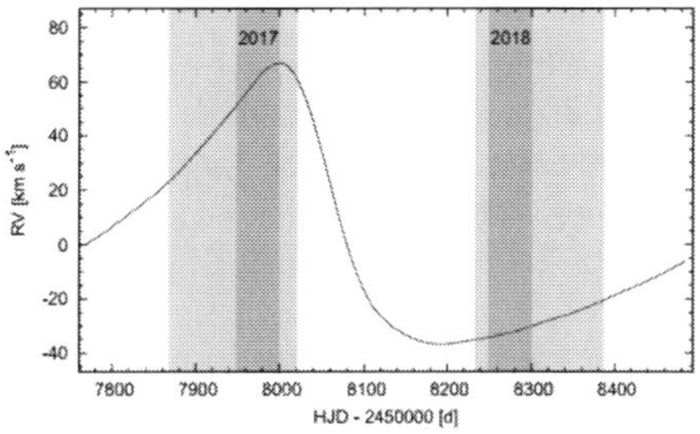

Figure 2. Predicted RV curve of KIC 2831097 for the 2017–2018 observing seasons, based on the orbital solution.

Acknowledgments: The financial support of the Hungarian NKFIH Grants K-115709, K-113117 and K-119517 are acknowledged. ÁS was supported by the János Bolyai Research Scholarship of the Hungarian Academy of Sciences. MS acknowledges the support of the postdoctoral fellowship programme of the Hungarian Academy of Sciences at the Konkoly Observatory as a host institution.

References

[1] Sódor, Á., Skarka, M., Liška, J., & Bognár, Zs., MNRAS, **465**, L1 (2017)
[2] Borucki, W. J., Koch, D., & Basri, G., Science, **327**, 977 (2010)
[3] Moskali P., Smolec, R., & Kolenberg, K., MNRAS, **447**, 2348 (2015)

RR Lyrae binary systems in the Galactic bulge

Gergely Hajdu[1,2,★] and *Márcio* Catelan[1,2,★★]

[1] *Pontificia Universidad Católica de Chile, Facultad de Física, Instituto de Astrofísica, Av. Vicuña Mackenna 4860, 782-0436 Macul, Santiago, Chile*
[2] *Millennium Institute of Astrophysics, Santiago, Chile*

Abstract.
The possibility of a direct mass measurement of RR Lyrae variable stars in binary systems has remained elusive for many years, due to the apparent paucity of such systems. Motivated by our recent success of finding a population of high confidence candidates, we have been continuing the search for RR Lyrae variables in binary systems towards the Galactic bulge. We describe the preliminary results of our project in these proceedings.

1 Introduction

Direct mass measurements of variable stars provide an indispensable tool to test the models of stellar pulsation and evolution. Previously only one RR Lyrae star was known to be in a binary system with high confidence ([5, 10]). This has changed when we announced 20 candidate binary systems, using OGLE observations of RR Lyrae stars towards the bulge ([1]). Furthermore, an independent analysis of field RR Lyrae has also found a few plausible binary systems ([6]). Finding new binary RR Lyrae is paramount to understand their paucity, as well as to select suitable systems for follow-up observations.

2 Analysis

We are continuing our search for variations caused by the light-time effect ([2]) in the O−C diagrams ([9]) of the combined I band light curves of RR Lyrae stars (RRab subtype) of the OGLE-II, III and IV projects ([7, 8]) towards the Galactic bulge, utilizing the same method as described in [1]. Compared to the limited sample of the previous study (1952 stars), we have analyzed more objects (5948), by relaxing the quality requirements of the analyzed light curves.

During our analysis, we have inspected the generated O−C diagrams and folded light curves of the stars one by one. In many cases, the O−C diagrams show periodic signals similar to those caused by the light-time effect, but inspection of the folded light curve revealed the presence of the Blazhko effect ([4]), usually with the same, long-term period as the O−C oscillations. In some cases the amplitude modulation caused by the Blazhko effect is on the level of ~ 0.01 mag, despite the thousand-day long periods. Based on this, we warn against the *overinterpretation* of O−C diagrams derived from timing of the maxima of RR Lyrae stars alone. Unfortunately, this could be a plausible explanation for some of the bright RR Lyrae binary candidates published by [6].

★`ghajdu@astro.puc.cl`
★★`mcatelan@astro.puc.cl`

3 Binary candidates

We have found ~35 new good binary candidates in the analyzed sample, after discarding all variables showing the Blazhko effect. Preliminary analysis indicates the following results:

- The distribution of orbital periods is similar to that in our original study ([1]); the bulk of the candidates is in the 8 – 12 yr range. Binaries with longer periods remain undetectable with the light-time effect method due to the limited time coverage of the observations.
- There are no short-period (< few yr) RR Lyrae binaries; this is probably a side effect of binary co-evolution prohibiting the formation of normal RR Lyrae variables in binaries with short orbital periods.
- The frequency of RR Lyrae binaries is compatible with a lower limit of 4% as in [1], when taking the lower average quality of the data into consideration.
- The majority of the candidates have low minimum companion masses, making radial velocity follow-up observations challenging; the expected radial velocity signal is only a few km s^{-1}, which is superposed on the signal of the pulsation with tens of km s^{-1}.

We are organizing follow-up observations of our best binary candidates. An interesting possibility of binary stellar evolution is that for a subset of these binaries the secondary body in the system could be the degenerate remnant of an originally higher-mass star. If the star went through the asymptotic giant branch (AGB) phase, it could have polluted the photospehere of the RR Lyrae with *s*-process elements, producing a so-called CEMP (Carbon-Enhanced Metal-Poor) star. Up to this date, several RR Lyrae variables have been found with such elemental enhancements ([3]); however, none of them has been shown to still reside in a physical binary with the white dwarf remnant of the AGB star. The detection of carbon enhancement by observing the CN band using low-resolution spectroscopy can easily prove the existence of degenerate companions for our binary candidates, at a moderate expense of telescope time on 8 m-class telescopes. In turn, these stars would be the very first RR Lyrae variables spectroscopically confirmed to reside in a binary system.

Acknowledgments: Support for this project is provided by the Ministry for the Economy, Development, and Tourism's Programa Iniciativa Milenio through grant IC 120009; by Proyecto Basal PFB-06/2007; by Proyectos Fondecyt #1141141 and #1171273; and by CONICYT-PCHA/Doctorado Nacional grant #2014-63140099.

References

[1] Hajdu, G., Catelan, M., Jurcsik, J., Dékány, I., Drake, A. J., & Marquette, J.-B., MNRAS, **449**, L113 (2015)
[2] Irwin, J. B., ApJ, **116**, 211 (1952)
[3] Kennedy, C. R., Stancliffe, R. J., Kuehn, C., et al., ApJ, **787**, 6 (2014)
[4] Kovács, G., in *Stellar Pulsation: Challanges for Theory and Observation*, ed. J. A. Guzik & P. A. Bradley, AIP Conference Proceedings, **1170**, 261 (2010)
[5] Liška, J., Mikulášek, Z., Zejda, M., & Chrastina, M., A&A, **589**, 94 (2016)
[6] Liška, J., Skarka, M., Zejda, M., Mikulášek, Z., & de Villiers, S. N., MNRAS, **459**, 4360 (2016)
[7] Soszyński, I., Dziembowski, W. A., Udalski, A., et al., AcA, **61**, 1 (2011)
[8] Soszyński, I., Udalski, A., Szymański, M. K., et al., AcA, **64**, 177 (2014)
[9] Sterken C., in *The Light-Time Effect in Astrophysics*, ed. C. Sterken, ASP Conference Series, **335**, 3 (2005)
[10] Wade, R. A., Donley, J., Fried, R., White, R. E., & Saha, A., AJ, **118**, 2442 (1999)

The Blazhko RR Lyrae variables and phase modulation in binary systems

Hiromoto Shibahashi[1,*]

[1] *University of Tokyo, Department of Astronomy, Tokyo 113-0033, Tokyo, Japan*

Abstract. With the help of the Fourier transform, I verify that some stars classified as Blazhko variables are actually regularly pulsating RR Lyrae stars in binary systems.

1 Introduction and summary

The Blazhko phenomenon is the periodic variation in pulsation amplitude or phase (or equivalently, frequency) seen in a large fraction of RR Lyrae stars ([1, 2]), and it still remains a long-lived enigma to be explained. However, phase modulation is also seen in pulsating stars in binary systems, due to the light-travel time effect ([3, 4]). Some stars classified as Blazhko variables might be then explained as binary systems. We should carefully discriminate between intrinsic phase modulation in RR Lyrae stars and phase modulation due to binarity. Hajdu et al. ([5]) adopted the O-C method to detect binarity. Another way to discriminate these two causes more clearly is the Fourier transform ([6, 7]). By carefully investigating the multiplets seen in the Fourier spectra of RR Lyrae stars, I verify that some of the stars classified as Blazhko variables are actually non-Blazhko RR Lyrae stars in a binary system, and conversely show that some stars classified as binary systems are actually not binaries.

2 The light-travel time effect and evidence for binarity

Let us consider a star pulsating with a frequency ν_0 and its harmonics in a binary. The observed luminosity variation at time t is then given by

$$\Delta L(t) = \sum_k A_k \cos 2\pi \left\{ k\nu_0 \left[t - \frac{1}{c} \int_0^t v_{\rm rad}(t')\,dt' \right] + \phi_k \right\}, \tag{1}$$

where c is the speed of light and $v_{\rm rad}(t)$ denotes the radial velocity, due to the orbital motion, of the pulsating star at the time t', where the epoch is the time at which the star passes the nodal point directed towards us.

The Fourier transform of eq. (1) leads to ([7]) (i) multiplets with an equal spacing in frequency, equal to the orbital frequency, (ii) the amplitude ratio between the first sidelobe and the central component of each multiplet being proportional to the oscillation frequency, $k\nu_0$, and (iii) the phase difference between the central component and the first sidelobes of each multiplet being $\pi/2$. These features discriminate the regularly pulsating RR Lyrae stars in a binary system from the Blazhko RR Lyrae variables.

*shibahashi@astron.s.u-tokyo.ac.jp

3 Stars showing features of binarity

A good example is OGLE-BLG-RRLYR-06498. Figure 1(a) shows the amplitude spectrum around the highest peak, ν_0. The conspicuous side peaks are caused by a year-gap of observations appearing in the window spectrum. After pre-whitening ν_0, we see clearly a pair of sidelobes of a triplet around ν_0 (Fig. 1(b)). As seen in Figure 1(c), the frequency spacings between members of this and other multiplets are equal. The amplitude ratio between the sidelobe and the central component of each triplet, shown in Figure 1(d), is proportional to frequency, being consistent with binarity. Figure 1(e) shows that the phase difference between the mean of the sidelobes and the central component of triplets is $\pi/2$, again consistent with binarity. Hence, we conclude that the phase modulation seen in this star is caused by orbital motion rather than an intrinsic Blazhko effect.

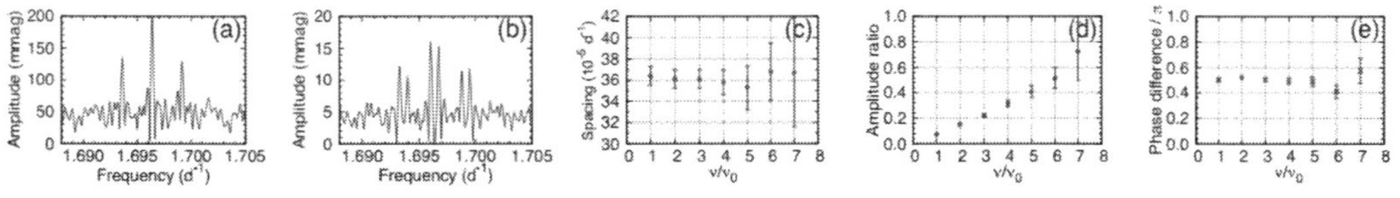

Figure 1. The amplitude spectrum of OGLE-BLG-RRLYR-06498 and its features, showing binarity.

4 Stars showing features of the Blazhko cycle

An example is KIC 10789273. Figure 2(a) shows the amplitude spectrum around the highest peak, ν_0. Pre-whitening of ν_0 makes a triplet visible as shown in Figure 2(b). Figure 2(c) shows that the frequency spacings among multiplets are equal. The amplitude ratio between the sidelobe and the central component of triplets, shown in Figure 2(d), is, however, *not* proportional to frequency, being *inconsistent* with binarity. Figure 2(e) shows that the phase difference between the mean of the sidelobes and the central component of triplets is *not* $\pi/2$ at all, being *inconsistent* with binarity.

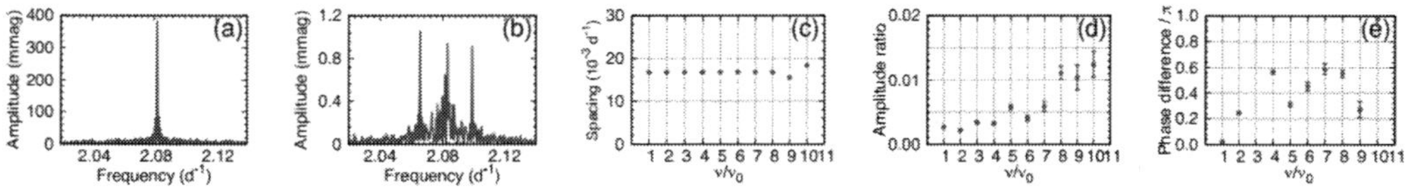

Figure 2. The amplitude spectrum of KIC 10789273 and its features are *inconsistent* with binarity.

References

[1] Blažko, S., AN, **175**, 325 (1907)
[2] Shapley, H., ApJ, **43**, 217 (1916)
[3] Murphy, S. J., Bedding, T. R., Shibahashi, H., Kurtz, D. W., & Kjeldsen, H., MNRAS, **441**, 2515 (2014)
[4] Murphy, S. J., & Shibahashi, H., MNRAS, **450**, 4475 (2015)
[5] Hajdu, G., Catelan, M., Jurcsik, J., Dékány, I., Drake, A. J., & Marquette, J.-B., MNRAS, **449**, L113 (2015)
[6] Shibahashi, H., & Kurtz, D. W., MNRAS, **422**, 738 (2012)
[7] Shibahashi, H., Kurtz, D. W., & Murphy, S. J., MNRAS, **450**, 3999 (2015)

Four years' interferometric observations of Galactic binary Cepheids

A. Gallenne[1], A. Mérand[1], P. Kervella[2], N. R. Evans[3], C. Proffitt[4], G. Pietrzyński[5], and W. Gieren[6]

[1]European Southern Observatory, Alonso de Córdova 3107, Casilla 19001, Santiago 19, Chile
[2]LESIA, Obs. de Paris, CNRS UMR 8109, UPMC,Univ. Paris 7, 5 Pl. Jules Janssen, 92195 Meudon, France
[3]Smithsonian Astrophysical Observatory, MS 4, 60 Garden Street, Cambridge, MA 02138, USA
[4]Space Telescope Science Institute, 3700 San Martin Drive, Baltimore, MD 21218, USA
[5]Nicolaus Copernicus Astro. Centre, Polish Academy of Sci., Bartycka 18, PL-00-716 Warszawa, Poland
[6]Universidad de Concepción, Departamento de Astronomía, Casilla 160-C, Concepción, Chile

Abstract. We give an update on our long-term program of Galactic Cepheids started in 2012, whose goal is to measure the visual orbits of Cepheid companions. Using the VLTI/PIONIER and CHARA/MIRC instruments, we have now detected several companions, and we already have a good orbital coverage for several of them. By combining interferometry and radial velocities, we can now derive all the orbital elements of the systems, and we will be soon able to estimate the Cepheid masses.

1 Introduction

Cepheids are powerful astrophysical laboratories providing fundamental clues for studying the pulsation and evolution of intermediate-mass stars. However, one of the most critical parameters, the mass, is a long-standing problem because of the 10-20% difference between masses predicted from stellar evolution and pulsation models. Cepheids in binary systems are the only tool to constrain models and make progress on this mass discrepancy. Studying Cepheids' companions can also provide insight on the impact of binarity on the calibration of the period-luminosity relation from the Baade-Wesselink technique and the IR surface brightness method. As many of the companions are blue main-sequence stars, the flux contribution in the near-IR is often negligible (< 1-2%), but can be as large as 10% in V.

Most of the companions are too close to the Cepheid (< 40 mas) to be resolved with single-dish 8-meter class telescopes, and the high contrast between the companions and the Cepheids makes the detection even more difficult. But long-baseline interferometry (LBI) is able to reach high-spatial resolution and high-dynamic range.

2 Update and results

Since 2012 we have observed some binary Cepheids in both the northern and southern hemispheres with the interferometric instruments CHARA/MIRC and VLTI/PIONIER. They are bright Cepheids ($H < 7$ mag), and single-line spectroscopic binaries with long orbital periods ($\sim 1 - 40$ yr). Because

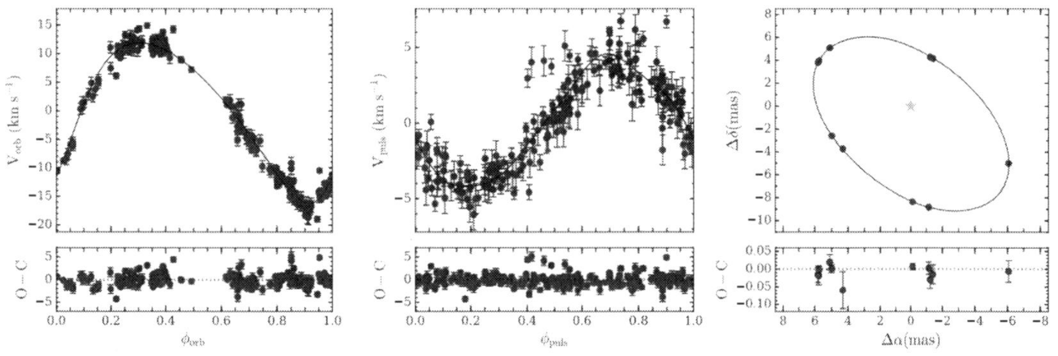

Figure 1. Combined fit of single-line velocities with astrometry for the Cepheid V1334 Cyg.

of the brightness of the Cepheid, it is difficult to spatially/spectrally detect companions using ground-based telescopes. However, the companion causes the Cepheid to wobble (as the two components orbit their center of mass), which is more easily detectable from radial velocities.

Using LBI, we detected companions separated by less than 50 mas from the Cepheids with contrast as high as 6.5 mag in H. For this purpose, we have created a dedicated tool, named `CANDID`, which searches for high-contrast companions from interferometric data ([1]). It allows a systematic search on a $N \times N$ grid, whose minimum needed resolution is estimated a posteriori. The tool delivers, among other things, the flux ratio f, the astrometric separation (α, δ), and the (non-)detection level of the companion based on χ^2 statistics. So far, we have detections for six binary Cepheids, with projected separations ranging from 1.5 to 40 mas and flux ratios (in H) from 0.8 to 4%. The astrometric positions can then be combined with the radial velocities of the primary, as shown in Figure 1 for V1334 Cyg, to provide the full set of orbital elements, including a, i and Ω previously unknown ([2, 3]). However, the distance and masses are still degenerate, and unfortunately no accurate parallaxes exist for these binary Cepheids.

From space, ultraviolet observations from *HST*/STIS will provide the spectra of the companions. However, broad features (some of them are fast rotators) and blended lines complicate the analysis, and can prevent the determination of accurate radial velocities (C. Proffitt 2017, these proceedings).

With the current upgrades on the instruments and facilities, we will soon be able to observe fainter Cepheids with dimmer orbiting companions. The coming Gaia parallaxes will allow us to break the degeneracy between mass and distance, and we will then be able to combine interferometry with single-line velocities to provide dynamical mass measurements of Cepheids.

Acknowledgments: W.G. and G.P. acknowledge support from the BASAL Centro de Astrofísica y Tecnologias Afines PFB-06/2007. This research has received funding from the European Research Council under the European Union's Horizon 2020 research and innovation program (grant No 695099). PK acknowledges support from the French Agence Nationale de la Recherche (ANR), under grant ANR-15-CE31-0012-01.

References

[1] Gallenne, A., Mérand, A., Kervella, P., et al., A&A, **579**, A68 (2015)
[2] Gallenne, A., Mérand, A., Kervella, P., et al., A&A, **561**, L3 (2014)
[3] Gallenne, A., Monnier, J. D., Mérand, A., et al., A&A, **552**, A21 (2013)

Spectro-Photo-Interferometry of Stellar Pulsation (SPIPS)

A more precise and more accurate implementation of the parallax of pulsation method

Antoine Mérand[1,⋆], *Pierre* Kervella[2,3,4], *Alexandre* Gallenne[1], *Joanne* Breitfelder[1], *Boris* Trahin[2,3,4], and *Marion* Villenave[5]

[1] *European Southern Observatory, Alonso de Córdova 3107, Casilla 19001, Santiago 19, Chile*
[2] *LESIA (UMR 8109), Observatoire de Paris, PSL, CNRS, UPMC, Univ. Paris-Diderot, 5 place Jules Janssen, 92195 Meudon, France*
[3] *Unidad Mixta Internacional Franco-Chilena de Astronomía, CNRS/INSU, France (UMI 3386), Departamento de Astronomía, Universidad de Chile, Camino El Observatorio 1515, Las Condes, Santiago, Chile*
[4] *Universidad de Concepción, Departamento de Astronomía, Casilla 160-C, Concepción, Chile*
[5] *Institut Supérieur de l'Aéronautique et de l'Espace, Toulouse, France*

Abstract. We present our implementation of the parallax of pulsation method which integrates all observables and physical modelling of the photosphere to get the best statistical precision and controlled biases. This method has been validated on well known stars and used to estimate observationally the projection factor of the *HST*-FGS sample. Our future developments include application to the Gaia Cepheids and modelling of the spectrum.

1 Why revisit the parallax of pulsation method?

The parallax of pulsation (PoP) method compares the integrated pulsation velocity and angular diameter variations of pulsating stars, in order to derive the distance. The PoP method is the "most geometrical" way, apart from measuring the direct parallax, to estimate distance to Cepheids. As such, the method is central in calibrating the Leavitt's (period-luminosity) law.

Traditional implementation (so called Baade-Wesselink) have lots of limitations which affect the accuracy and precision of the derived distances.

First, pulsation velocity measured using spectroscopy does not represent exactly the pulsation velocity: the use of an ad-hoc projection factor to estimate the pulsation velocity from the radial velocity introduces a bias to the distance. This p-factor depends on the extraction method of radial velocity, which makes the bias complicated to account for in practice.

Secondly, angular diameters are usually derived using 2 bands photometry (such as V and K): two parameters ($T_{\rm eff}$, diam) are actually adjusted from two measurements, which is an under-use of multiple bands usually available (up to a dozen different). Surface brightness laws need to be calibrated, and use a simplistic linear dependency with the color. Moreover, reddening and reddening law need to be known. Note that so-called Wesenheit relations (a linear combination of 2 magnitudes) is not immune to that since the linear relation is derived from reddening laws.

⋆ amerand@eso.org

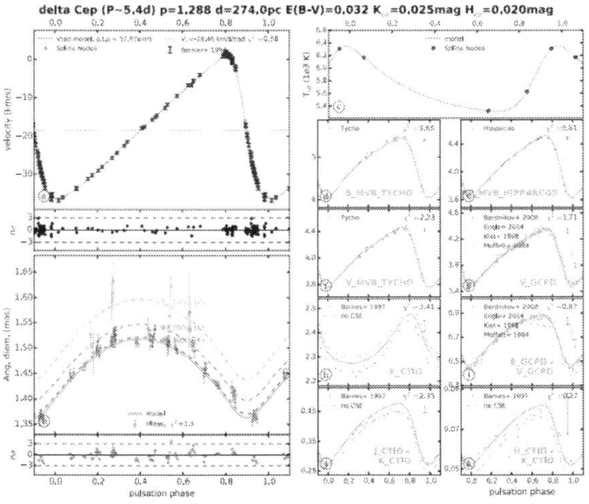

Figure 1. SPIPS fit for δ Cep: velocity, angular diameter and photometric fit. Note the infrared excess (in K band photometry) as well as the bias in the apparent angular diameter.

Finally, the traditional Baade-Wesselink implementation uses non optimal data combination schemes. Usually, data are fitted in various stages, i.e. not at once, which render the final uncertainties too optimistic. Too often, low order interpolation is used which leads to poor statistical significance and biases to the distance. Additionally, only all-at-once data combination allows to efficiently combine data with poor phase coverage and/or account for period change.

2 A better approach

The PoP method is not geometric. We embrace its complexity and dependence on the pulsation's physics. Our approach to improve precision and accuracy is as follows. **Combination of all data possible:** spectroscopic radial velocities; high resolution spectra; any number of photometric measurements (not limited to 2 bands); optical interferometric measurements. **Simultaneous fit of ALL the parameters:** 2 profiles as function of time for `Vpuls(t)` and `Teff(t)`; period and period change parameters; reddening $E(B-V)$; distance or p-factor. **Use of atmospheric models** to derive synthetic photometry from the effective temperature. Inclusion of morphological and photometric modelling of **infrared excesses**.

This approach allowed us to successfully model all observables from 2 stars in great details (δ Cep and η Aql, the first of which can be seen on Fig. 1) which proves the validity of the approach [1]. Our future work includes high resolution spectroscopic modelling, as well as extension of SED modelling beyond the K-band. Other uses of SPIPS include the modelling and derivation of p-factor for the Type II Cepheid κ Pav ([2]), the *HST*-FGS Cepheids sample ([3]) and for the long period Cepheid RS Pup ([4]). All these examples of SPIPS modelling of Cepheids with available distances point toward a constant p-factor of the order of 1.25 to 1.30, regardless of the pulsation period (see Kervella et al., these proceedings).

References

[1] Mérand, A., Kervella, P., Breitfelder, J., et al., A&A, **584**, 10 (2015)
[2] Breitfelder, J., Kervella, P., Mérand, A., et al., A&A, **576**, 64 (2015)
[3] Breitfelder, J., Mérand, A., Kervella, P., et al., A&A, **587**, 117 (2016)
[4] Kervella, P., Trahin, B., Bond, H. E., et al., A&A , **600**, A127 (2017)

Between Cepheids and the Cosmos

Nancy Remage Evans[1,*], *Scott* Engle[2,**], *Ed* Guinan[2,***], *Hilding* Neilson[3,****], *Massimo* Marengo[4,†], *Lynn* Matthews[5,‡], and *Moritz* Guenther[6,§]

[1] *Smithsonian Astrophysical Observatory, 60 Garden St., Cambridge MA 02138, USA*
[2] *Villanova University, Department of Astronomy, Villanova, PA 19085, USA*
[3] *University of Toronto, Dept of Astronomy & Astrophysics, 60 St. George St. Toronto, ON M5S 3H4, Canada*
[4] *Iowa State University, A313E Zaffarano, Dept. of Physics and Astronomy, Ames, IA 50010, USA*
[5] *MIT Haystack Observatory, Westford MA, 01886, USA*
[6] *MIT Kavli Institute for Astrophysics and Space Research, 70 Vassar St., Cambridge MA 02139, USA*

Abstract. The X-ray observations of three classical Cepheids produce a surprising result. At approximately the phase of maximum radius there is a sharp increase in X-ray flux above the normal "quiescent" level. The relation of this new upper atmosphere diagnostic to other phenonena above the photosphere is discussed.

1 Introduction

We have a number of tantalizing clues about material above the photosphere in classical Cepheids: spatially extended infrared emission seen with *Spitzer*, including an apparent bow shock associated with δ Cephei itself, extended nebulosity detected in the H I 21-cm line, and evidence of circumstellar shells based on near-infrared interferometry ([3]). The understanding of how these are related is patchy. There are a number of photospheric/chromospheric diagnostics from *Hubble Space Telescope* ultraviolet (UV) spectra, confirming pulsation related activity immediately following minimum radius as discussed in [1]. The recent addition to these diagnostics is a brief X-ray increase at the surprising phase of maximum radius (when the photospheric UV lines are quiescent). This has been seen in two cycles of δ Cep, and three of β Dor. In additon, a number of random phase observations of Cepheids have been made with the XMM-Newton satellite ([2]), summarized in Figure 1. The only *Cepheid* detected is V473 Lyr, again shortly after maximum radius. The emerging picture is that at the phase when the pulsation wave passes through the photosphere and chromosphere (minimum radius) many features are seen at those levels – UV emission lines and an increase in microturbulence. There is, however, no increase in X-rays flux above the level typically produced in a star with a convective envelope ("quiescent" flux). However, at maximum radius as the star starts to contract the X-ray

*nevans@cfa.harvard.edu
**scott.engle@villanova.edu
***edward.guinan@villanova.edu
****neilson@astro.utoronto.ca
†mmarengo@iastate.edu
‡lmatthew@haystack.mit.edu
§hgunther@mit.edu

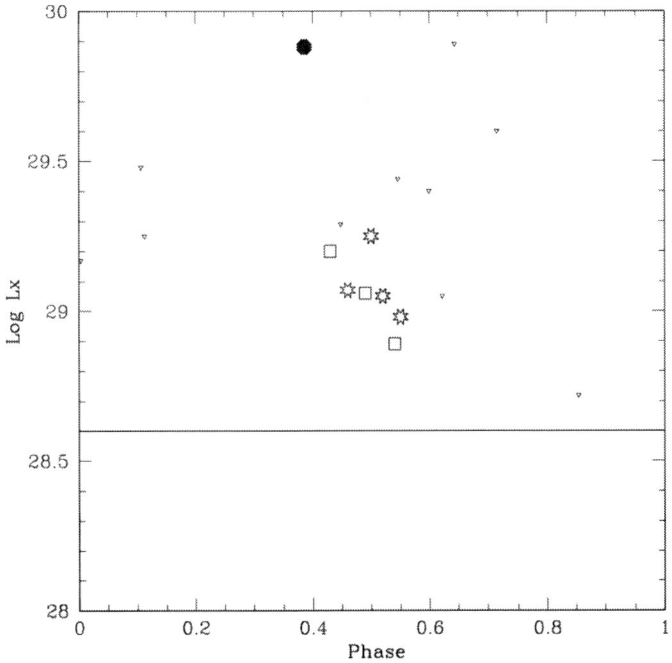

Figure 1. Random phase XMM observations of Cepheids, compared with observations of δ Cep (asterisks) and β Dor (squares). Upper limits (triangles) were found for all Cepheids in the XMM observations, except V473 Lyr (filled circle), which was observed at maximum radius, the same phase as the increased flux in δ Cep. The horizontal line is the quiescent level of X-rays for δ Cep from 9 observations at other phases. L_X is in ergs sec^{-1}.

flux increases dramatically by nearly a factor of five. In contrast, photospheric and chromospheric indicators are at a low level in this "coasting phase". The morphology of the X-ray increase is a rapid rise, a short duration and a rapid fall, reminiscent of a flare or a pulsation driven shock. Thus, the "working model" is that the X-ray activity occurs above the photosphere and chromosphere, and something triggers the X-ray event as the atmosphere starts to collapse. Importantly, this ties the X-ray increase to the pulsation process.

Are the X-rays (at about 0.1 AU) related to the circumstellar envelopes (CSEs, at about 1 AU) and the possible bow shock seen in δ Cep (at 10^4 to 10^5 AU)? One possible scenario would be for the atmospheric collapse to trigger a coronal loop reconnection event. This might in turn result in a coronal mass ejection event. Altogether this might produce mild outflow similar to the solar wind. If this scenario proves correct, it would have important implications for mass loss in Cepheids (related to evolutionary calculations) as well as the influence of the CSE on the Cepheid Leavitt law (period-luminosity relation).

Acknowledgments: Support was provided by Chandra X-ray Center NASA Contract NAS8-03060.

References

[1] Engle, S., Guinan, E. F., Harper, G. M., Neilson, H. R., & Evans, N. R., ApJ, **794**, 80 (2014)
[2] Evans, N. R., Pillitteri, I., Wolk, S., et al., AJ, **151**, 108 (2016)
[3] Marengo, M., Evans, N. R., Matthews, L. D. et al., ASSP, **31**, 99 (2013)

Eclipsing binary stars with extreme light curve asymmetries mined from large astronomical surveys

Athanasios Papageorgiou[1,2,*] , *Georgios* Kleftogiannis[3,**] and *Panagiota-Eleftheria* Christopoulou[3,***]

[1]*Pontificia Universidad Católica de Chile, Instituto de Astrofísica, Av. Vicuña Mackenna 4860, 782-0436 Macul, Santiago, Chile*
[2]*Millennium Institute of Astrophysics, Santiago, Chile*
[3]*University of Patras, Department of Physics, 26500, Patra, Greece*

Abstract. The O'Connell effect is one of the most perplexing challenges in binary studies as it has not been convincingly explained. Furthermore, a simple method to obtain essential parameters for eclipsing binaries exhibiting this effect and to extract information describing the asymmetry in the light curve maxima is needed. We have developed an automated program that characterizes the morphology of light curves by depth of both minima, height of both maxima and curvature outside the eclipses.

1 Introduction

In the last few decades, due to the automatic ground-based and space telescopes, the amount of data has increased dramatically. In order to search for peculiar light curves, automatic pipelines are needed to deal with this amount of data. Further to our search ([4]) we apply an improved version of our automated method in several databases in order to detect Light Curve Asymmetries (LCA). The program is being developed in Python (`https://www.python.org/`).

2 Method – results

The program reads from a given catalog (name, period P, $V_{\max}$, etc.) of a specific database and applies two different sets of constraints ([4]) which were improved as follows. Each data point is examined to be valid according to the local variance of its neighbors in a certain phase bin. Combining this method with σ-clipping, the outliers were better defined and the performance of fitting improved accordingly. Then the program constructs the phase-magnitude diagram in the range -0.25 – 1.25 where a second order polynomial is fitted (Fig. 1), using the NumPy package (`http://www.numpy.org/`). Finally, fitting alert flags were included in order to control the goodness of fitting.

The method was applied in 13987 eclipsing binaries (EB) light curves from ASAS, OGLE II, OGLE III, NSVS and OMC databases. Of those 345 were found with strong out of eclipse variations

*apapageo@astro.puc.cl
**gkleftis@upatras.gr
***pechris@physics.upatras.gr

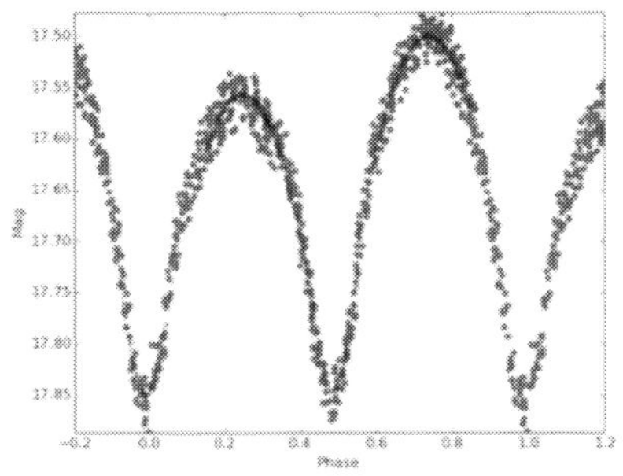

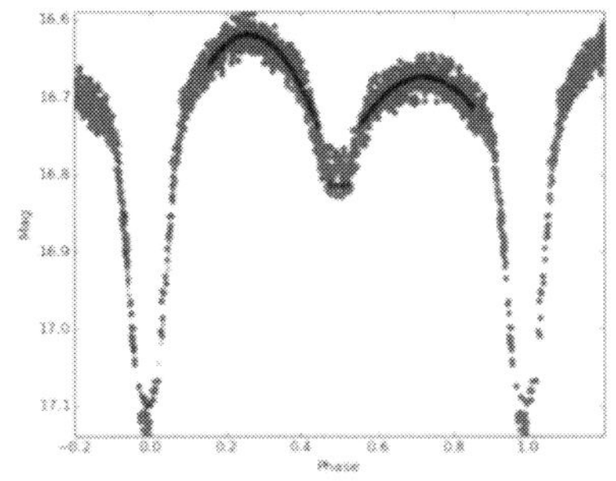

Figure 1. Contact EB OGLE-GD-ECL-07791 (left) and semi-detached EB OGLE-GD-ECL-03789 (right) with light curve asymmetries. Black points denote the polynomial fitting.

Table 1. A sample of 5 eclipsing binary stars from the new catalogue with strong LCA

ID	$\|MaxII - MaxI\|$ (mag)	P (d)	RA (h:m:s)	Dec (d:m:s)
OGLE-GD-ECL-09205	0.0743	0.278594	13:15:52.74	-64:34:52.1
OGLE-GD-ECL-10742	0.0751	0.291847	13:31:28.61	-64:08:21.0
OGLE-GD-ECL-08412	0.0757	0.396209	11:55:07.87	-61:45:31.8
OGLE-GD-ECL-07791	0.0567	0.363248	11:33:42.20	-60:36:44.9
OGLE-GD-ECL-03789	0.0547	0.490414	10:51:12.11	-60:48:38.7

($|MaxI - MaxII| > 0.025$ mag) as shown in Table 1. From ASAS, 120 EBs were classified as Eclipsing Contact type (EC) and 23 had multiple classification ([7, 8]); from OGLE II, 3 EBs were classified as Eclipsing W UMa type (EW) and 2 had multiple classification ([9, 10]); from OGLE III, 189 EBs were classified as EC ([2, 5, 6]); from NSVS, 4 EBs were classified as W UMa type ([3]); and from OMC, 4 EBs had multiple classification ([1]).

Acknowledgments: A.P. gratefully acknowledges the support provided by Fondecyt through grant #3160782.

References

[1] Alfonso-Garzon, J., Domingo, A., Mas-Hesse, J.M., & Gimenez, A., A&A, **548**, A79 (2012)
[2] Graczyk, D., Soszyński, I., Poleski, R., et al., Acta Astron., **61**, 103 (2011)
[3] Hoffman D. I., Harrison T. E., & McNamara B. J., AJ, **138**, 2 (2009)
[4] Papageorgiou, A., Kleftogiannis, G., & Christopoulou, PE., Contrib. Astron. Obs. Skalnate Pleso, **43**, 3 (2014)
[5] Pawlak, M., Graczyk, D., Soszynski, I., et al., Acta Astron., **63**, 323 (2013)
[6] Pietrukowicz, P., Mroz, P., Soszynski, I., et al., Acta Astron., **63**, 115 (2013)
[7] Pojmanski, G., & Maciejewski, G., Acta Astron., **54**, 153 (2004)
[8] Pojmanski, G., Pilecki, B., & Szczygiel, D., Acta Astron., **55**, 275 (2005)
[9] Wyrzykowski, L., Udalski, A., Kubiak, M., et al., Acta Astron., **53**, 1 (2003)
[10] Wyrzykowski, L., Udalski, A., Kubiak, M., et al., Acta Astron., **54**, 1 (2004)

Machine learning techniques to select variable stars

Alejandro García-Varela[1,*], *Muriel* Pérez[1,2,**], *Beatriz* Sabogal[1,***], and *Adolfo* Quiroz[2,****]

[1] *Universidad de los Andes, Departamento de Física, Cra.1 No.18A-10, Edificio Ip, Bogotá, Colombia*
[2] *Universidad de los Andes, Departamento de Matemáticas, Cra.1 No.18A-10, Edificio H, Bogotá, Colombia*

Abstract. In order to perform a supervised classification of variable stars, we propose and evaluate a set of six features extracted from the magnitude density of the light curves. They are used to train automatic classification systems using state-of-the-art classifiers implemented in the R statistical computing environment. We find that random forests is the most successful method to select variables.

1 Introduction

Machine learning techniques have proved to be quite useful in classification of variable stars. In the existing literature, quantities related to the magnitude density of the light curves and their Fourier coefficients are chosen as features. However, the calculation of Fourier coefficients is computationally expensive for large data sets. In order to perform a supervised classification, we propose and evaluate a set of six robust descriptive statistics that can be calculated efficiently and do not need to be checked externally. We calculate this set of features for OGLE-III variables belonging to the Milky Way and the LMC and SMC galaxies, classified as Cepheids (Ceph), δ Scuti (δ-Sct), Eclipsing Binaries (EBS), Long Period Variables (LPV), RR Lyræ (RRLyr), Type 2 Cepheids (T2Ceph) and a set of Be Star Candidates (BeSC) reported in the literature. We evaluate the performance of the features over the following classifiers: K-nearest neighbors, classification trees, random forests, support vector machines, and gradient boosted trees. We use 10-fold cross-validation to estimate the recall and precision of these classifiers ([1]).

2 Selection of variable stars

In order to avoid quantities sensitive to outliers, we choose six robust estimators. To visualise how data look in the six-dimensional feature space, we use the t-distributed Stochastic Neighbour Embedding (t-SNE) visualisation technique ([2]). Figure 1 shows a t-SNE plot, where the different variability classes are separated in the six-dimensional feature space. Some overlapping of the classes is also evident. After the training process, we validate the classifiers on a data set from the OGLE-IV database. Our classifiers yield correct classifications with high probability, which shows that our proposed set of features can be used to characterise different variability types. We find that the random forest classifier produces the best results on the validation samples (see Table 1).

* josegarc@uniandes.edu.co
** mf.perez648@uniandes.edu.co
*** bsabogal@uniandes.edu.co
**** aj.quiroz1079@uniandes.edu.co

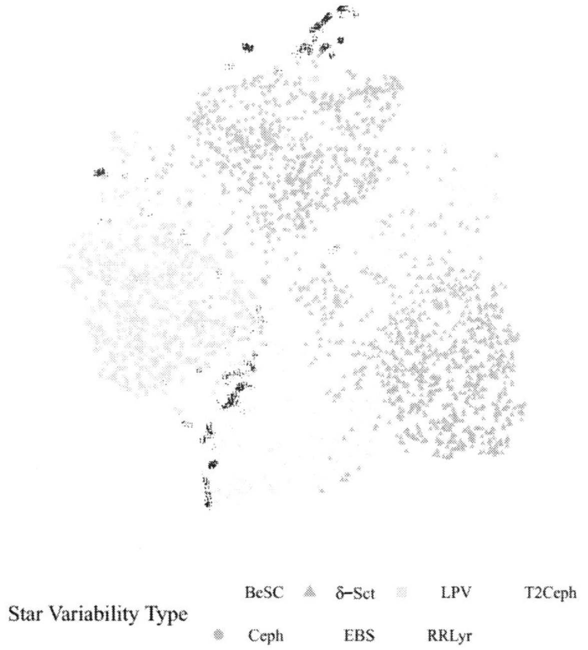

Figure 1. Visualisation of the six-dimensional feature space by t-SNE. The axes are omitted because the scale of this embedding carries no meaning.

Table 1. Confusion matrix. Classification results of the random forest method on the OGLE-IV data set.

Prediction	Ceph	δ-Sct	EBS	LPV	RRLyr	T2Ceph	Other
BeSc	1		21	3			108
Ceph	**126**	1	42		10	1	52
δ-Sct		**146**	209		9		316
EBS		2	**1110**	2	13		676
LPV	1		105	**2790**		1	226
RRLyr	3	10	19		**652**		86
T2Ceph	4		26	4	2	**3**	9
Total	135	159	1532	2799	686	5	1473

Acknowledgments: Authors acknowledge support from Departamento de Física and Facultad de Ciencias, Universidad de los Andes, through CENIF program.

References

[1] Pérez-Ortiz, M. F., García-Varela, A., Sabogal, B., & Quiroz, A., A&A, submitted (2017)
[2] Van der Maaten, L., & Hinton, G., The Journal of Machine Learning Research, **9**, 85 (2008)

Ultra-precise analysis of the light curves of CoRoT and *Kepler* δ Scuti stars

Javier Pascual-Granado[1,⋆], *Juan Carlos* Suárez[2], and *Rafael* Garrido[1]

[1] *Instituto de Astrofísica de Andalucía, Glorieta de la Astronomía 18008, Granada, Spain*
[2] *Universidad de Granada, Facultad de Ciencias, Departamento de Física Teórica y del Cosmos, 18071, Granada, Spain*

Abstract. It is well known that the residuals of the multifrequency analysis of δ Scuti stars are correlated, giving rise to challenging features such as a plateau (HD 50844, HD 50870, HD 49434, ...) of non-resolved frequencies with amplitude higher than the expected noise level. Several hypotheses have been advanced to explain these features: effective convection, rotation, non-linear interactions, etc. We have recently demonstrated that in some cases the underlying function describing the light variations of δ Scuti stars has not the property of being analytic. The strong consequence of this result is that the Fourier expansion on which the harmonic analysis is based could be not justified.
In order to know the extension of this phenomenon among δ Scuti stars, we have used photometric data from CoRoT seismofield and a set of *Kepler* stars. The results show that this inconsistency in the application of harmonic analysis is almost ubiquitous to the δ Scuti pulsating stars.

1 Context

To perform seismological studies of the stars it is essential to detect pulsational frequencies in their photometric or radial velocity time series. A periodogram ([1]) is a well defined and easy to calculate estimator of the spectral density of a time series. Therefore, it is widespread in the asteroseismology literature. However, it is not always guaranteed that a periodogram is an unbiased estimator of the spectral density, i.e. the discrete Fourier transform converges to the real frequency content of the time series. Indeed, it is only guaranteed that a function has a convergent Fourier expansion when the function is analytic.

In [2] we proposed a method to study the analyticity of the underlying function of light curves through the "connectivities" in order to test whether the periodogram is an unbiased estimator of the spectral density. Connectivities are calculated by fitting an analytic model to the time series. In [2] we used an L2 metric to estimate the best fitting model, here instead we make use of an L1 optimization since this proved to be superior when the data has a non-Gaussian distribution ([3]).

We have applied this test to a set of 10 light curves from CoRoT and *Kepler* observations (5 each). *Kepler* light curves were selected from the list of δ Scuti stars characterized in [4] and CoRoT light curves were selected from the seismo targets in the literature taking into account the classification during the preparatory phase of the mission ([5]).

⋆ javier@iaa.es

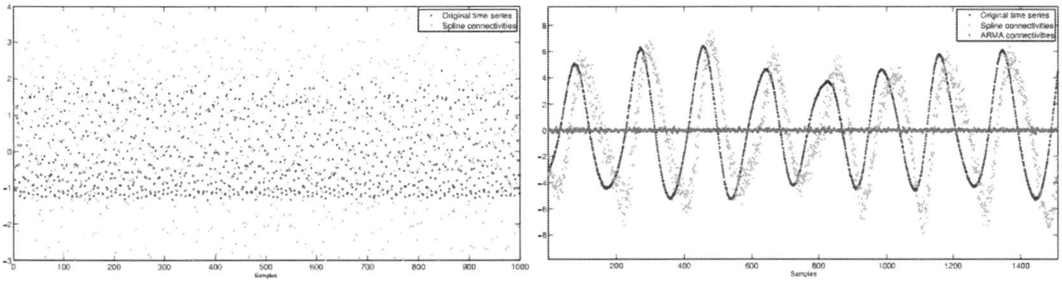

Figure 1. Analyticity test for KIC 09408694 (left panel) and KIC 010273384 (right panel). Light curves are shown in blue, connectivities calculated by a spline fitting (analytic) in green and by an ARMA model (non-analytic) in red. Correlation indicates a non-analytic underlying function.

2 Analyticity test and results

We have found that 8 of the 10 δ Scuti stars analyzed show light curves with a non-analytic underlying function. Figure 1 shows the result of the analyticity test for the light curves of two *Kepler* stars: one which has an analytic underlying function (KIC 9408694) and one which does not (KIC 10273384). Notice that splines connectivities are correlated with the light curve of KIC 10273384[1] while for KIC 9408694 there is no apparent correlation and the Pearson correlation coefficient is -3.9×10^{-5}. Correlation with the original time series indicates that the underlying function is non-analytic according to the test (for more details, see [2]).

Our results suggest that the non-analyticity of the underlying function is an almost ubiquitous phenomenon in the photometric time series of δ Scuti stars as observed by the space satellites CoRoT and *Kepler*. In consequence it is not guaranteed that the periodogram is an unbiased estimator of the frequency content of δ Scuti stars.

A massive study of *Kepler* and CoRoT stars is in progress to shed more light on the origin of the non-analyticity of the underlying function.

References

[1] Scargle, J., ApJ, **263**, 835 (1982)

[2] Pascual-Granado, J., Garrido, R., Suárez, J. C., A&A, **581**, A89 (2015)

[3] Scargle, J., IEEE Trans. Inf. Theor., **23**, 140 (1977)

[4] Uytterhoeven, K., Moya, A., Grigahcène, A., et al., A&A, **534**, A125 (2011)

[5] Michel, E., Baglin, A., Auvergne, M., et al., in *The CoRoT Mission Pre-Launch Status – Stellar Seismology and Planet Finding*, ed. M. Fridlund and A. Baglin, ESA Special Publication, **1306**, 39

[1] ARMA connectivities are shown here to demonstrate that a non-analytic approximation can fit the data.

Core overshoot and convection in δ Scuti and γ Doradus stars

Catherine Lovekin[1,*] and *Joyce A.* Guzik[2]

[1] *Physics Department, Mount Allison University, Sackville, NB, E4L 1C6, Canada*
[2] *XTD-NTA, MS T-082 Los Alamos National Laboratory, Los Alamos, NM, 87545, USA*

Abstract. The effects of rotation on pulsation in δ Scuti and γ Doradus stars are poorly understood. Stars in this mass range span the transition from convective envelopes to convective cores, and realistic models of convection are thus a key part of understanding these stars. In this work, we use 2D asteroseismic modelling of 5 stars observed with the *Kepler* spacecraft to provide constraints on the age, mass, rotation rate, and convective core overshoot. We use Period04 to calculate the frequencies based on short cadence *Kepler* observations of five γ Doradus and δ Scuti stars. We fit these stars with rotating models calculated using MESA and adiabatic pulsation frequencies calculated with GYRE. Comparison of these models with the pulsation frequencies of three stars observed with *Kepler* allowed us to place constraints on the age, mass, and rotation rate of these stars. All frequencies not identified as possible combinations were compared to theoretical frequencies calculated using models including the effects of rotation and overshoot. The best fitting models for all five stars are slowly rotating at the best fitting age and have moderate convective core overshoot. In this work, we will discuss the results of the frequency extraction and fitting process.

1 Introduction

One of the theoretical challenges presented by δ Scuti stars is determining which stars should be pulsating. Although all the stars in the δ Scuti instability strip are predicted to show pulsations, observations show that less than half of the stars in this region actually pulsate ([1, 2]). This difference between theory and observation is not presently understood. It has been suggested that nonlinear mode coupling may stabilize the pulsations ([7]), or the opacity driving mechanism could be saturated ([13]). Unfortunately, testing these predictions requires nonlinear models of nonadiabatic nonradial pulsations, which do not currently exist.

Additionally, space-based observations, such as with CoRoT ([11]) and *Kepler* (e.g., [5, 9]), have found that a number of δ Scuti stars are observed to pulsate with γ Dor-type frequencies, and these are usually described as hybrid stars. Hybrid stars are particularly interesting asteroseismic targets, as the g modes sample deeper regions of the star than the p modes. Hybrid stars, γ Dor and δ Scuti stars collectively span the mass range $1.4 - 2.5\,M_{\odot}$, where the interior structure transitions from a convective envelope and radiative core in the lower mass stars to higher mass stars with convective cores and radiative envelopes. Studies of stars in this mass range with *Kepler* have found more hybrid star candidates than expected by current stellar pulsation theory ([18]). The leading explanation for

*clovekin@mta.ca

Table 1. Literature parameters. All parameters are taken from the KIC.

KIC #	T_{eff} (K)	log g (± 0.4)	Magnitude (K_P)
5466537	6979 ± 221	4.141	10.34
6131093	6629 ± 180	4.358	9.25
9962653	7254 ± 220	4.277	10.11
10451090	7577 ± 152	4.134	9.21
11649699	7182 ± 202	4.211	10.14

this discrepancy is that the interaction between convection and pulsation in both the deeper layers (γ Dor, g modes) and upper layers (δ Scuti, p modes) is not fully understood.

In this work, we study the pulsation characteristics of 5 γ Doradus and δ Scuti stars, including candidate hybrid stars. The frequency analysis described here will provide the foundation for modelling these stars, providing us with constraints on their interior convection and rotational properties. In Section 2 we describe the observations, and in Section 3 we describe the results of our frequency extraction. Our preliminary conclusions are summarized in Section 4.

2 Observations

Observations were made with the *Kepler* spacecraft ([3]) as part of a Guest Observer project (GO 20022, PI J. A. Guzik) to observe δ Scuti, γ Doradus, and solar-type variables. Based on previous observations, 14 bright stars were chosen with masses between 1.0 and 1.8 $M_{\odot}$. Bright targets were selected to give the best quality data possible, and the resulting signal-to-noise ratio is very high. These targets were then observed in short-cadence mode (1 point integrated every minute) for 30 days. All of these stars have been previously observed through other programs, and all available short-cadence data were used to extract frequencies. For most of the stars, this consisted of two sets of observations separated by several hundred days. Two of the stars (KIC 9962653 and KIC 10451090) have been observed for 4 and 5 sets of 30 days, respectively. The available data covers a total of about 600 days for each star, although the coverage is incomplete. Processed light curves were downloaded from the Mikulski Archive for Space Telescopes (MAST). Stellar parameters for these stars are primarily taken from the *Kepler* Input Catalogue (KIC, [6]), with updates for some of the parameters for three of the stars ([4, 16, 18]).

3 Results

All five stars proved to have very rich Fourier spectra, with close to 100 frequencies with SNR > 4 in all cases. The details of the frequency analysis for each star are described below. The literature parameters for each star are summarized in Table 1.

3.1 KIC 5466537

KIC 5466537 has 132 identified frequencies, of which 23 frequencies were not considered to be combinations. One of these 23 frequencies is 1.39 c/d, consistent with γ Doradus pulsations, while the remaining 22 frequencies are between 9 and 28 c/d, as expected for δ Scuti variables. This is consistent with the previous identification of this star as a hybrid ([5]).

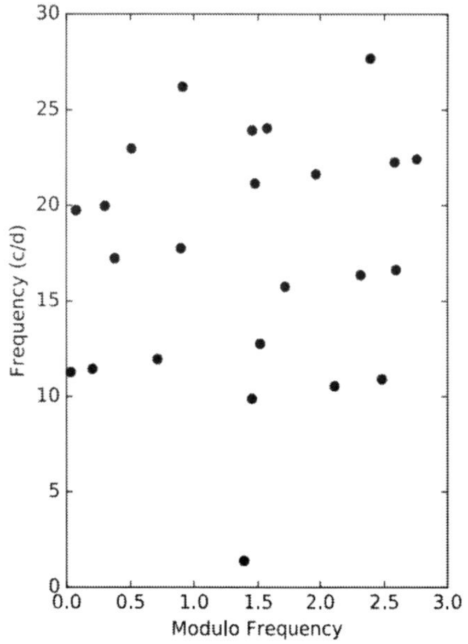

Figure 1. An echelle diagram for observed frequencies in KIC 5466537. The large spacing was found to be 2.81 c/d using the Dirac's comb method ([8]) based on 23 frequencies.

Although δ Scuti stars are not expected to obey an asymptotic frequency spacing, there are patterns in the frequency spectrum that can be detected ([8, 14]). We used the Dirac's comb method ([8]) to look for periodic spacing in the detected frequencies. Although we only used 23 frequencies, we were able to detect a spacing pattern at 2.81 c/d. This results in the echelle diagram shown in Figure 1. Since only one of the non-combination frequencies was in the g-mode region, no attempt to analzye patterns in the g modes has been made.

3.2 KIC 6131093

KIC 6131093 has been previously identified as a flare star with a rotation period of 8.028 d ([2]). We identified 65 signficant frequencies in this star, mainly g modes expected in a γ Doradus star. We used the techniques for determining significance described in [19] rather than a strict signal-to-noise ratio criterion. We looked for patterns in the period spacings and found upward trends, as shown in Figure 2. This is consistent with an identification of $m = -1$ for these modes ([19]). We also observe dips in the period spacing pattern which indicates there is a chemical discontinuity within the core of the star ([12]). To determine the location of the chemical discontinuity, we need to identify the radial orders of the modes, which has not been done at this time. The effects of rotation will further complicate the frequency spectrum, making it more difficult to interpret. Even at relatively low rotation rates, the resulting splitting can be comparable in size to the period spacing. Spectroscopic mode identification may be required in order to interpret the observed frequency spectrum.

3.3 KIC 9962653

KIC 9962653 was previously identified as a γ Doradus star ([1]). We identified nearly 100 significant frequencies in this star, primarily in the range 1-4 c/d, consistent with the identification as a γ Doradus star. We find multiple lines in the period spacing pattern, all with the upward slope expected for

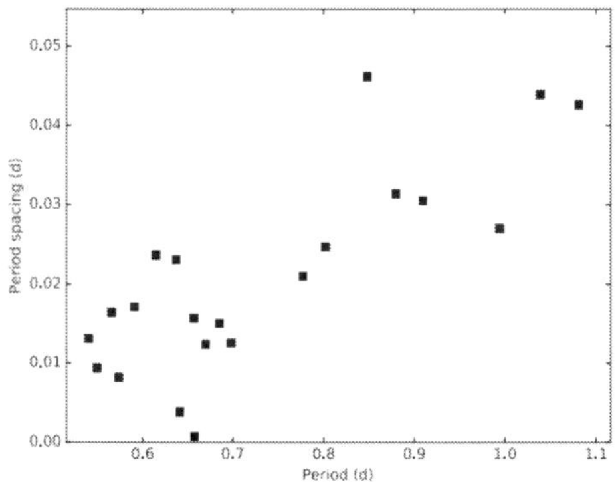

Figure 2. A section of the period spacing diagram for KIC 6131093. The upward trend suggests the majority of the observed frequencies are retrograde ($m = -1$) modes. The dips in the period spacing at 0.65 d and 1 d indicate that there are chemical discontinuities in the star.

retrograde ($m = -1$) modes. As for KIC 6131093, there are visible dips in this pattern, indicating a chemical discontinuity within the star.

3.4 KIC 10451090

KIC 10451090 was previously identified as a δ Scuti star based on long cadence observations, which identified many frequencies in the range 10-20 c/d ([5]). In our short cadence data, we found a large number of frequencies (108), primarily in the range 29-38 c/d, with no detected frequencies between 10 and 20 c/d. The frequencies detected in the long cadence data were probably reflections of frequencies above the Nyquist frequency (24.5). This effect has been shown to occur in other stars ([10]). We also detected a few very low frequencies (0.03-0.05 c/d), but these are relatively low amplitude and are likely the result of combinations of real frequencies. Using the same techniques as for KIC 5466537, we were able to find a large frequency separation of 2.27 c/d, resulting in two ridges in the echelle diagram.

3.5 KIC 11649699

No previous studies of this star have been performed. We identified nearly 100 significant frequencies in this star. As for KIC 6131093 and KIC 9962653, the modes are tentatively identified as $m = -1$ based on the patterns visible in the period spacings. The majority of the detected frequencies in this star are low, consistent with γ Doradus pulsation. However, a few low-amplitude frequencies have been detected between 5-10 c/d. These are likley combination frequencies, but may suggest that KIC 11649699 is also a hybrid star.

4 Conclusions

We present preliminary frequency analysis for 5 δ Scuti and γ Doradus stars observed with *Kepler*. A large number of significant frequencies has been detected in each of the five stars. Two of the stars (KIC 5466537, KIC 10451090) have been identified as δ Scuti stars based on the range of observed frequencies. For these two stars, we have been able to identify the large frequency separation by taking the Fourier transform of the Dirac comb produced by the observed frequencies. In both cases, the large separation can be used to make an echelle diagram showing two ridges. We have not yet identified the ℓ of these ridges.

The other three stars in our sample are all consistent with γ Doradus pulsations. Plotting the period separation against the observed period produces regular patterns, and these have been used to tentatively identify the modes as primarily retrograde ($m = -1$) modes. All three stars also show dips in the period spacing pattern, which indicates the presence of a chemical discontinuity inside the star.

Now that the observed frequencies and patterns have been extracted, we will compare the observed frequencies to theoretical models calculated with MESA ([15]) and GYRE ([17]). These theoretical models will be used to constrain the interior rotation and convection properties of the stars.

Acknowledgments: This work was supported in part by NASA ATP grant 12-ATP12-0130. This paper includes data collected by the *Kepler* mission as part of the Guest Observer program (GO20022). Funding for the *Kepler* mission is provided by the NASA Science Mission Directorate. This research has made use of NASA's Astrophysics Data System and Astropy, a community developed core Python package for astronomy.

References

[1] Balona, L. A., & Dziembowski, W. A., MNRAS, **417**, 591 (2011)
[2] Balona, L. A., Daszyńska-Daszkiewicz, J., & Pamyatnykh, A. A., MNRAS, **452**, 3073 (2015)
[3] Basri, G., Borucki, W.J., & Koch, D., New Astron. Rev., **49**, 478 (2005)
[4] Bastien, F. A., Stassun, K. G., Basri, G., & Pepper, J., ApJ, **818**, 43 (2016)
[5] Bradley, P. A., Guzik, J. A., Miles, L. F., Uytterhoeven, K., Jackiewicz, J., & Kinemuchi, K., ApJ, **149**, 68 (2015)
[6] Brown, T. M., Latham, D. W., Everett, M. E., & Esquerdo, G. A., AJ, **142**, 112 (2011)
[7] Dziembowski, W., Krolikowska, M., & Kosovichev, A., AcA, **38**, 61 (1988)
[8] García Hernández, A, Moya, A., Michel, E, et al., A&A, **506**, 79 (2009)
[9] Grigahcène, A., Antoci, V., Balona, L., et al., ApJL, **713**, L192 (2010)
[10] Guzik, J. A., Bradley, P. A., Jackiewicz, J., et al., Astronomical Review, **11**, 1 (2015)
[11] Hareter, M., Reegen, P., Miglio, A., et al., preprint (arXiv:1007.3176) (2010)
[12] Miglio, A., Montalbán, J., Noels, A., & Eggenberger, P., MNRAS, **386**, 1487 (2008)
[13] Nowakowski, R. M., AcA, **55**, 1, (2005)
[14] Paparó, M., Benkő, J. M., Hareter, M., & Guzik, J. A., ApJ, **822**, 100 (2016)
[15] Paxton, B., Bildsten, L., Dotter, A., et al., ApJS, **192**, 3 (2011)
[16] Pinsonneault, M. H., An, D., Molenda-Żakowicz, J., et al., ApJS, **199**, 30 (2012)
[17] Townsend, R. H. D., & Teitler, S. A., MNRAS, **435**, 3406 (2013)
[18] Uytterhoeven, K., Moya, A., Grigahcène, A., et al., A&A, **534**, A125 (2011)
[19] Van Reeth, T., Tkachenko, A., Aerts, C., et al. A&A, **574**, A17 (2015)

DOI: 10.1051/epjconf/201715203014

"Dipper" stars in the Upper Sco and ρ Oph star forming regions identified from the K2 mission

Yapeng Zhang[1,⋆] and *Valentin* Ivanov[2,3,⋆⋆]

[1] *School of Astronomy and Space Science, Nanjing University, Nanjing, 210046, China*
[2] *European Southern Observatory, Ave. Alonso de Córdova 3107, Vitacura, Santiago, Chile*
[3] *European Southern Observatory, Karl-Schwarzschild-Str. 2, 85748 Garching bei München, Germany*

Abstract. We report the discovery of 24 new "dippers" – young stars with disks that show 1-3 day long abrupt drops in their light curves. The search was based on the K2 Campaign 2 observations of the Upper Sco and ρ Oph star forming regions.

1 Introduction

Young stars are typically irregular variables, and they vary on different time scales, due to a number of mechanisms, such as rotational modulation by cold spots on the surface of the star or hot spots related to accretion and variations in the accretion rate from circumstellar disks.

The *Spitzer* YSOVAR (Young Stellar Objects VARiability) project ([4]) was the first to report "dippers" in the infrared: 41 objects in their sample showed "brief, sharp drops in their flux density with durations of one to a few days", superimposed on otherwise smooth or nearly flat light curves (LCs). Similar behaviour was seen earlier in the optical LCs of individual protostars and T Tau stars by [3].

Campaign 2 of the K2 mission utilized the *Kepler* spacecraft ([2]) to observe the Upper Scorpius and the ρ Ophiuchus star-forming regions. [1] announced 25 dippers. Here we report a thorough search aimed at finding more dippers in the same K2 campaign data.

2 Search and results

First, we filtered the normalized LCs of [5] through a high-pass filter with a cut-on frequency of 1 day^{-1}, and estimated the standard deviation σ of each filtered LC using a biweight function. We considered as candidate dippers the objects with $\geq 3\sigma$ dipping event(s) on the filtered LCs. Dips with only one data point were discarded. This filtering removes periodic variations, because if the rotation-related variability is not removed, it can mask some of the dips.

Following [1], we parametrized dippers' behavior: (1) number of dips in the filtered LC N_{dip} – a lower limit to N_{dip} excludes spurious events; (2) average depth of the three deepest dips in normalized LC D_{dip} – it characterizes the dips' strength, as a sample; (3) the ratio R_{dip} of the average depth of the three deepest dips D_{dip} to σ – it measures the significance of the dips, again as a sample.

⋆chrisamo@163.com
⋆⋆vivanov@eso.org

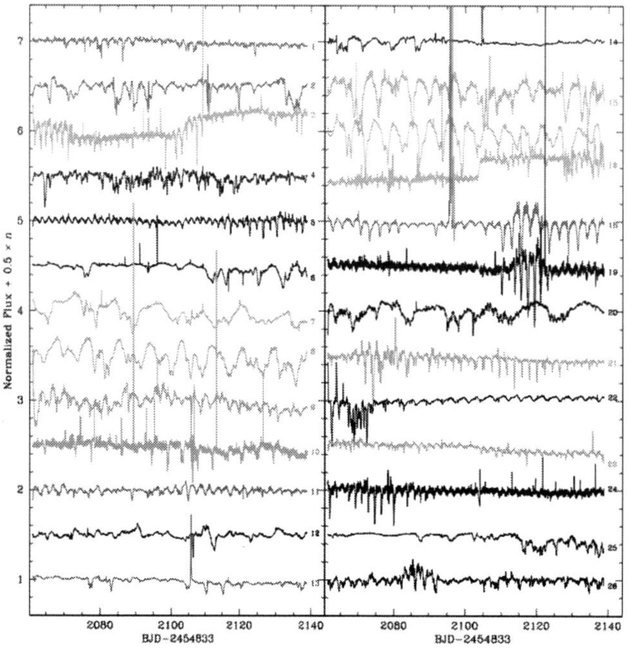

Figure 1. Light curves of a subset of the selected dippers: newly identified objects from this work are plotted in black; objects previously reported in [1] are plotted in red and green, depending on whether they are listed in their Tables 1 or 5, respectively.

In our survey we set the limits as follows: $N_{dip} > 3$, $D_{dip} > 0.07$ and $R_{dip} > 5$, somewhat more relaxed than the values adopted by [1]. These numbers were selected after inspecting the LCs, and the distributions of the parameters for the entire Campaign 2. We found no significant correlations between any of these three parameters, because they appear to constrain different aspects of the objects. The automatic identification yielded 475 dipper candidates. Next, we conservatively ignored some LCs that show extremely unreasonable artifacts, most likely due to failure of the de-trending. We removed sources with periodic dips, probably related to stellar spots, and objects that suffer contamination from nearby sources.

Our approach works well – we recovered all dippers listed by [1] and 24 new ones (Fig. 1). Further details are available in a forthcoming paper (Zhang & Ivanov, in preparation).

Acknowledgments: YZ is grateful to the Summer program supported by School of Astronomy and Space Science, Nanjing University for financing her visit to ESO Garching.

References

[1] Ansdell, M., Gaidos, E., Rappaport, S. A., et al., ApJ, **816**, 69 (2016)
[2] Borucki, W. J., Koch, D., Basri, G., et al., Science, **327**, 977 (2010)
[3] Herbst, W., & Shevchenko, V. S., AJ, **118**, 1043 (1999)
[4] Morales-Calderón, M., Stauffer, J. R., Hillenbrand, L. A., et al., ApJ, **733**, 50 (2011)
[5] Vanderburg, A., & Johnson, J. A., PASP, **126**, 948 (2014)

Searching for Be star candidates within large databases

Beatriz Sabogal[1,*] and *Alejandro* García-Varela[1,**]

[1]*Universidad de los Andes, Departamento de Física, Cra.1 No.18A-10, Ed. Ip, A.A.4976, Bogotá, Colombia*

Abstract. Long time surveys yield many light curves among their byproducts, useful to identify candidate Be stars that can be confirmed by spectroscopic follow-ups. We present in this work different techniques we have used to find Be star candidates in the Galaxy, the Magellanic Clouds and the OGLE-IV Gaia South Ecliptic Pole Field.

1 Introduction

Be stars are non-supergiant very rapid rotators, with spectral types between late O and early A, whose spectra show or have shown at some time one or more Balmer lines in emission ([1]). The photometric variations shown by these stars can exist in different time scales depending on their causes ([5]). Photometric searches for Be star candidates (BeSC hereafter) are useful in order to have larger samples of Be stars to analyze their dependency on metallicity and evolutionary state. A photometric search for BeSC within the Small Magellanic Cloud (SMC) OGLE-II catalog of variable stars was performed by [3]. Those authors found that, contrary to the Galactic Be stars, many light curves of BeSC had jumps of brightness and quasi-periodic variations.

2 Techniques used in Searches for BeSC

After the work of [3] we searched for BeSC within the Large Magellanic Cloud (LMC) OGLE-II catalog of variable stars, using photometric criteria. We looked for stars in the range of magnitudes, colors and amplitudes of Galactic Be stars. From a list of about 5000 stars, we selected 2446 BeSC that were visually inspected. Comparing these stars with those BeSC of the SMC, we found an important dependency of the amplitude and duration of outbursts with the metallicity: they are brighter, longer and have a slower decline at the lower metallicity of the SMC ([6]).

We also searched for BeSC using photometric criteria on 48 fields of the Galactic bulge within the OGLE-II catalog of stars. To determine if the selected stars were BeSC, we needed to inspect visually more than one hundred thousand light curves. This motivated us to implement a statistical criterion to select only the stars with light curves similar to those of the LMC, SMC and Galactic BeSC. A correlation between skewness and kurtosis was very useful to select the stars, guaranteeing the recovery of BeSC at the 70% level. This process conducted us to a catalog of about 29000 BeSC. We found that 198 of these stars exhibit a periodic behavior ([7]).

*bsabogal@uniandes.edu.co
**josegarc@uniandes.edu.co

We also developed a search for BeSC within the ASAS catalog of variable stars ([8]). These Galactic stars are located mainly in the Galactic disk at different distances. For this reason, we could not use photometric criteria. Instead, we used the statistical criterion introduced by [7], using the ranges of skewness and kurtosis found in that work for BeSC in the I-band. Once a sample of stars was selected by this criterion, we applied morphological and photometric constraints and found 213 BeSC, showing outbursts and stochastic variability. We did not find stars with jumps in brightness in their light curves, suggesting that, for the detection range of ASAS telescopes, these stars do not exist in the Galaxy.

The advent of machine learning techniques to classify light curves of variable stars ([2, 9]) motivated us to implement a useful technique to select BeSC. We first propose and evaluate a set of six features extracted from the light curves. We used as training sample the following OGLE-III variable stars: Cepheids, RR Lyrae, δ Scuti, long-period variables, type II Cepheids, eclipsing binaries, a set of OGLE-III BeSC reported in the literature, and samples of BeSC reported in previous works. The best classifier to select BeSC was random forests ([4]). We used this classifier, and later color criteria, to search for BeSC within the OGLE-IV Gaia South Ecliptic Pole Field data. The result was a sample of 50 new BeSC, besides 16 BeSC previously reported by [10]. We also found stars with jumps of brightness that could belong to the LMC, due to the proximity of these fields to the center of that galaxy.

3 Conclusions

In this work we have shown the usefulness of combining photometric criteria with statistical criteria to select BeSC. The random forest method implemented with the features found in this work will be a very useful tool to select automatically BeSC from large databases.

Acknowledgments: Authors acknowledge support from Departamento de Física and Facultad de Ciencias, Universidad de los Andes, through CENIF program.

References

[1] Collins, G. W., in *Physics of Be stars*, Proc. IAU Coll. 92, ed. A. Slettebak & T. P. Snow, Cambridge University Press (1987)

[2] Debosscher, J. Sarro, L. M., Aerts, C., Cuypers, J., Vandenbussche, B., Garrido, R., & Solano, E., A&A, **475**, 1159 (2007)

[3] Mennickent, R. E., Pietrzyński, G., Gieren, W., & Szewczyk, O., A&A, **393**, 887 (2002)

[4] Pérez-Ortiz, M.F., García-Varela, A., Sabogal, B., & Quiroz, A., A&A, submitted (2017)

[5] Rivinius, Th., Carciofi, A. C., & Martayan, Ch., A&ARv, **21**, 69 (2013)

[6] Sabogal, B., Mennickent, R., Pietrzyński, G., & Gieren, W., MNRAS, **361**, 1055 (2005)

[7] Sabogal, B., Mennickent, R., Pietrzyński, G., García, J.A, Gieren, W., & Kolaczkowski, Z., A&A, **478**, 659 (2008)

[8] Sabogal, B., García-Varela, A., & Mennickent, R., PASP, **126**, 219 (2014)

[9] Sarro, L. M., Debosscher, J., López, M., & Aerts, C., A&A, **494**, 739 (2009)

[10] Soszyński, I., Udalski, A., Poleski, R., et al., Acta Astron., **62**, 219 (2012)

Pulsating variable stars and large spectroscopic surveys

Peter De Cat[1,*]

[1]*Royal Observatory of Belgium, Ringlaan 3, 1180 Brussels, Belgium*

Abstract. In the past decade, the research of pulsating variable stars has taken a giant leap forward thanks to the photometric measurements provided by space missions like MOST, CoRoT, *Kepler*/K2, and BRITE. These missions have provided quasi uninterrupted photometric time-series with an ultra-high quality and a total length that is not achievable from Earth. However, many of the success stories could not have been told without ground-based spectroscopic follow-up observations. Indeed, spectroscopy has some important assets as it can provide (more) accurate information about stellar parameters (like the effective temperature, surface gravity, metallicity, and abundances that are mandatory parameters for an in-depth asteroseismic study), the radial velocity (that is important for the detection of binaries and for the confirmation of cluster membership, if applicable), and the projected rotational velocity (that allows the study of the effects of rotation on pulsations). Fortunately, several large spectroscopic surveys are (becoming) available that can be used for these purposes. For some of these surveys, sub-projects have been initiated with the specific goal to complement space-based photometry. In this review, several spectroscopic surveys are introduced and compared with each other. We show that a large amount of spectroscopic data is (becoming) available for a large variety of objects.

1 Introduction

How can large spectroscopic surveys help the research of pulsating variable stars? Or, in other words, *are they useful for the gathering of the required input for in-depth asteroseismic studies?* A first requirement for asteroseismology is the observation of a large number of pulsation frequencies. Each pulsation frequency reveals information of a specific internal layer. So the more pulsation frequencies can be observed, the more complete the fingerprint of the stellar interior will be. Therefore, there is a need for time-series of high-quality data (to detect low amplitude variations) with a sufficiently long time-base (to cover the beat periods) in photometry (brightness variations) and/or in spectroscopy (radial velocity and/or line-profile variations). Space-based observations from past and ongoing missions like MOST ([1]), CoRoT ([2]), *Kepler*/K2 ([3, 4]), and BRITE ([5]) have already proven their usefulness and we still have some high hopes for (future) space-missions like GAIA ([6]), TESS ([7]), and PLATO ([8]). GAIA is the only mission from this list that also provides spectroscopic observations.

The second requirement is that the observed pulsation modes need to be identified. It means that their degree ℓ and azimuthal number m should be known. In the most fortunate cases, quasi-regular spacings in either frequency (for pressure modes; [9]) or in period (for gravity modes; [10]) can be detected, making it possible to read the value of ℓ from the ridges in the corresponding Échelle

*Peter.DeCat@oma.be

diagram (e.g., [10]). However, for most types of pulsating stars, the mode identification needs to be done by using techniques in multi-colour photometry (determination of ℓ only; [11]) and/or high-resolution, high-quality spectroscopy (full identification; [12, 13]). Unfortunately, the data provided by large spectroscopic surveys are in general not appropriate for this purpose.

The third requirement is that accurate values for stellar parameters like the effective temperature ("$T_{\rm eff}$"), surface gravity ("$\log g$"), metallicity ("[M/H]"), projected rotational velocity ("$v \sin i$"), and chemical abundances are needed to pinpoint the best model amongst those that are fitting the observed frequency spectrum. All these parameters can already be derived from one spectrum of sufficient resolution and quality. Large spectroscopic surveys can be extremely important in this respect. The resulting accurate position in the Hertzsprung-Russell (H-R) diagram can also be used for the confirmation of pulsation class membership or for the detection of new types of pulsating stars (like the mysterious star OGLE-GD-DSCT-0058; [14]). For some of the pulsation mechanisms, the driving of the modes heavily depends on the amount of metals that are present in the stellar interior. This is nicely illustrated by the B8.3 v star KIC10526294 for which only half of the observed gravity modes are driven in models with standard metal mixtures ([15]). Therefore, the importance of accurate values for especially the metallicity and chemical abundances must not be underestimated for making progress in the understanding of the physics of the stellar interior.

Finally, values of the radial velocity ("v_{rad}") can give extra useful input for asterseismic modeling. A few v_{rad} values could already be sufficient to detect multiple systems. Knowledge of their orbit can lead to constraints on the mass, the inclination and the distance of the components. A single v_{rad} value could already be sufficient to confirm cluster membership. As cluster stars are formed from the same molecular cloud, they all have a common age, composition and distance that can be used in the asteroseismic investigation. In this respect, large spectroscopic surveys can be a goldmine too.

In what follows, a brief overview of large spectroscopic surveys is given. In this context, a spectroscopic survey is considered as large when at least 10 000 objects are observed. Surveys only focusing on galaxies were omitted. We first present past and currently ongoing surveys (Sect. 2) before introducing a few surveys that are planned for the coming years (Sect. 3). We end with the summary and conclusions in Section 4.

2 Past and ongoing large spectroscopic surveys

2.1 Xinglong Observatory (China)

LAMOST

The Large Sky Area Multi-Object Fiber Spectroscopic Telescope (LAMOST; Table 1; `http://www.lamost.org`; [16, 17]) is a new 4-m telescope located in the Xinglong Observatory (China) in the northern hemisphere. It is a unique instrument that combines a large aperture (4 meter) with a wide field-of-view ("FoV"; circular region with a diameter of 5 degrees on the sky) that is covered with 4 000 optical fibers. It is the ideal tool to obtain spectra with a low spectral resolution ($R \sim 1\,800$) in the optical (370-900 nm) for a large number of objects in a homogeneous and efficient way. The LAMOST ExtraGAlactic Survey (LEGAS) and the LAMOST Experiment for Galactic Understanding and Exploration (LEGUE; [18]) were the two initial scientific driving forces for the LAMOST project. The survey aims to observe more than 5 000 000 objects (stars, galaxies and quasars) for which $r < 17.8$ mag is a realistic magnitude limit. However, it was soon realized that the observation of the FoV of the nominal *Kepler* mission with the LAMOST would be a pure scientific goldmine, being a win-win opportunity for both communities: it would provide the *Kepler* community with the data needed for a homogeneous spectroscopic determination of stellar parameters for objects observed by *Kepler* while

Table 1. Characteristics of the large spectroscopic survey conducted at the Xinglong Observatory (Xing; China) in the northern hemisphere with the 4.0-m Large Sky Area Multi-Object Fiber Spectroscopic Telescope (Lamost).

Name	Lamost
Telescope	4.0-m Xing
Fibers	#4 000
Wavelengths	370-900 nm
Resolution	~1 800 (fixed)
Targets	>5 000 000 stars, galaxies & quasars (r < 17.8 mag)
Date range	2011-2017

the Lamost community could benefit from high-precision values for stellar parameters for *Kepler* objects derived from data obtained elsewhere to calibrate the Lamost results. Therefore, the proposal of the Lamost-*Kepler* project ([19]) was well received in 2010 and the first observations for the project were already done during the test phase of the Lamost in early 2011. The scientific observations of the pilot survey began on October 24, 2011 while the regular survey started about one year later, on September 28, 2012. So far, observations on 45 nights have led to 198 727 Lamost spectra of stars within the *Kepler* FoV. The 101 086 spectra obtained before October 2014 have been analyzed by three groups leading to atmospheric parameters and spectral classification for 51 406 and 63 149 stars, respectively ([20–22]). The accuracy of the derived $T_{\rm eff}$, $\log g$, and [Fe/H] values amounts to ~3.5 %, ~0.3 dex and ~0.2 dex, respectively. Since the end of 2015, five K2 fields have been observed with the Lamost. Up to now, 96 233 Lamost spectra have been collected and they were used to derive the stellar atmospheric parameters for 59 213 stars ([23]). The first regular survey of the Lamost will end in June 2017. For the next phase, one of the options is to equip the Lamost with medium resolution spectrographs.

2.2 Apache Point Observatory (New Mexico, USA)

There are several large spectroscopic surveys for which the observations were/are being done from the Apache Point Observatory (New Mexico, USA; Table 2) in the northern hemisphere. They all use the 2.5-m Sloan Foundation Telescope having a circular FoV with a diameter of 3 deg (~7 deg^2).

Segue-1 & Segue-2

The Sloan Extension for Galactic Understanding and Exploration (Segue; Table 2, top left; `http://www.sdss.org/surveys/segue/`) was the first large spectroscopic survey that used this telescope. It was combined with a spectrograph with 640 fibers to obtain spectra with a resolution of ~2000 with a typical signal-to-noise ratio ("SNR") of 25 in the optical wavelength range (385-920 nm). The focus of the first phase (Segue-1; [24]) was on the exploration of the structure of the Milky Way by observing ~240 000 stars with a g-magnitude below 19 and covering a surface area of 1 438 deg^2. These observations were done in dark and gray time from the winter of 2004 until the winter of 2008. During the second phase (Segue-2), the focus shifted to stars in the halo with distances from 10 to 60 kpc. A total of 118 151 stars with g < 19 mag were observed from the winter of 2008 to the fall of 2009.

Table 2. Characteristics of the large spectroscopic surveys conducted at the Apache Point Observatory (Apo; New Mexico, USA) in the northern hemisphere with the 2.5-m Sloan Foundation Telescope and at the Las Campanas Observatory (Lco; Atacama, Chile) in the southern hemisphere with the 2.5-m Irénée du Pont Telescope.

Name	Segue-1	Segue-2	Marvels
Telescope	2.5-m Apo		2.5-m Apo
Fibers	#640		#60
Wavelengths	385-920 nm (*H*-band)		range of 90 nm (center at ~540 nm)
Resolution	~2 000		~5 100
Survey Area	1438 deg^2	1317 deg^2	
Targets	240 000 stars ($g < 19$ mag)	118 151 stars ($g < 19$ mag)	~11 000 main-sequence stars ~1 000 giant stars ($8 < V < 12$ mag)
Date range	2004-2008	2008-2009	2008-2014

Name	Apogee	Apogee-2	
Telescope	2.5-m Apo	2.5-m Apo	2.5-m Lco
Fibers	#300	#300	
Wavelengths	1 510-1 700 nm (*H*-band)	1 510-1 700 nm (*H*-band)	
Resolution	~22 500	~22 500	
Targets	>100 000 giant stars ($H < 12.2$ mag)	>300 000 stars ($H < 13.5$ mag)	
Date range	2011-2014	2014-2017	2017-2020

Marvels

The Multi-object Apo Radial Velocity Exoplanet Large-area Survey (Marvels; Table 2, top right; `http://www.sdss.org/surveys/marvels/`; [25]) combined the 2.5-m Sloan Foundation Telescope with two interferometric spectrographs with 60 fibers each. It obtained spectra with a spectral resolution $R \sim 5\,100$ with a wavelength range of 90 nm centered at ~540 nm for about 11 000 main sequence stars and about 1 000 giant stars with a V-magnitude between 8 and 12 mag. For each star, a time-span of 1.5 years was covered with 25 to 35 spectra from which the v_{rad} could be determined with a typical error of 10 m s^{-1} at $V = 10$ mag. Such observations are ideal to detect gas giant planets with orbital periods ranging from several hours up to two years. The observations were done in bright time from the fall of 2008 until the spring of 2014.

Apogee & Apogee-2

For the Apache Point Observatory Galactic Evolution Experiment (Apogee; Table 2, bottom left; `http://www.sdss.org/surveys/apogee/`; [26]), the 2.5-m Sloan Telescope was combined with a spectrograph with 300 fibers that is capable of obtaining spectra with a resolution of $R \sim 22\,500$. It observed the region around the H-band in the near infrared (1.51-1.70 μm). These observations in the northern hemisphere were done in bright time from the spring of 2011 until the spring of 2014. In total, more than 160 000 red giant stars across the full range of the Galactic bulge, bar, disk, and halo with a H magnitude below 12.2 were observed. The SNR of the Apogee spectra is typically above 100. This is sufficient to derive the stellar parameters ($T_{\rm eff}$, $\log g$, [Fe/H]), the abundances with a precision of 0.1 dex for 15 species, and v_{rad} with an error below 100 m s^{-1}. Apogee also has a sub-project, ApoKasc ([27]), that focuses on stars within the *Kepler* FoV. Apart from the Apogee spectroscopic parameters, the ApoKasc catalog also lists asteroseismic values for log g, masses, radii, and mean densities. It is often used as reference work for the quality check of the results of other surveys.

The second phase of the Apache Point Observatory Galaxy Evolution Experiment (Apogee-2; Table 2, bottom right; `http://www.sdss.org/surveys/apogee-2`; [28]) aims to study the assembly and formation history of the Milky Way. Spectra with the same characteristics and of similar quality will be obtained for in total more than 300 000 stars in the entire Milky Way. Compared to Apogee, the chemical abundance analysis will be extended with 10 species. For planet-hosting stars, the goal is to find out if their chemical composition differs from the one of stars without planets. Apogee-2 will last for 6 years and only observes during bright time. The observations started in the northern hemisphere in the fall of 2014. After three years, the observations will turn to the south by using the 2.5-m Irénée du Pont Telescope of the Las Campanas Observatory (Atacama, Chile) with a circular FoV with a diameter of 2.1 deg (3.5 deg^2).

2.3 Siding Spring Observatory (Australia)

At the Siding Spring Observatory (Australia) in the southern hemisphere, two telescopes have been used for large spectroscopic surveys (Table 3): the 1.2-m UK Schmidt Telescope (Uks) with a circular FoV with a diameter of 5.7 deg (~25 deg^2) and the 3.9-m Anglo-Australian Telescope (Aat) with a circular FoV with a diameter of 2 deg (~3 deg^2).

Rave

The Radial Velocity Experiment (Rave; Table 3, left; `https://www.rave-survey.org/project`; [29]) combined the 1.2-m UK Schmidt Telescope with the multi-object spectrograph 6dF (6 degree Field) that utilizes interchangeable field plates with 150 fibers each. It enables to observe the wavelength region 841.0– 879.5 nm (containing the Ca II triplet) with $R \sim 7\,500$. The observations were done from April 2003 to July 2005 during 7 nights per lunation, and from August 2005 to April 2013 during 20-25 nights per lunation. In total 574 630 spectra were gathered for 483 330 stars with an I-magnitude between 9 and 13 mag. Apart from the basic stellar parameters ($T_{\rm eff}$, log g, and [M/H]), also the v_{rad}, distances and elemental abundances ([Fe/H] and six α-elements) were determined with an accuracy of ~1.5 km s^{-1}, ~20 % and ~0.2 dex, respectively. Given the large number of observed stars and the big overlap with the Tycho-Gaia asterometric solution catalog ([30]), it was the first spectroscopic survey that allowed to study the Milky Way morphology and history by complementing the Rave v_{rad} values with proper motions listed in astrometric databases.

Argos

The Abundances and Radial Velocity Galactic Origins Survey (Argos; Table 3, middle; [31–33]) used the 3.9-m Anglo-Australian Telescope in combination with the AAOmega fibre spectrometer having 400 fibers of which 350 are used for stars. About 28 000 red giants with an I-magnitude between 13 and 16 mag and a Galactic latitude near +10 deg or between −10 and −5 deg were observed in two wavelength ranges: a first region including the Ca II triplet (840-885 nm) with $R \sim 11\,000$ and a typical SNR of 75, and a second region including the Mgb/MgH feature (500-560 nm) with a lower spectral resolution of $R \sim 3\,000$. The latter was included to be able to reject foreground dwarf stars from the sample of more distant giants. In total, 28 fields were covered with 1 000 spectra each (except for one with only 600 spectra) during 42 nights in the months May-August of the years 2008-2011. The

Table 3. Characteristics of the large spectroscopic surveys conducted at the Siding Spring Observatory (Sso; Australia) in the southern hemisphere with the 1.2-m UK Schmidt Telescope (Uks) and the 3.9-m Anglo-Australian Telescope (Aat).

Name	Rave	Argos		Galah
Telescope	1.2-m Uks@Sso	3.9-m Aat@Sso		3.9-m Aat@Sso
Fibers	#150	#400		#400
Wavelengths	841.0-879.5 nm	840-885 nm	500-560 nm	471.8-490.3 nm
	(Ca II triplet)	(Ca II triplet)	(Mgb/MgH)	564.9-587.3 nm
				648.1-673.9 nm
				759.0-789.0 nm
Resolution	~7 500	~11 000	~3 000	~28 000
Targets	483 330 (stars)	~28 000 red giants		1 000 000 stars
	$(9 < I < 13\,\mathrm{mag})$	$(13 < I < 16\,\mathrm{mag})$		$(12 < V < 14\,\mathrm{mag})$
Date range	2003-2013	2008-2011		2012-2017

main reason to study these stars in the Galactic bulge and inner disk was to find out more about the formation processes of the bulge.

Galah

The Galactic Archaeology with Hermes Survey (Galah; Table 3, right; `https://galah-survey.org/`; [34–37]) combines the 3.9-m Anglo-Australian Telescope with the High Efficiency and Resolution Multi-Element Spectrograph (Hermes). It has 400 fibers and observes in four wavelength regions: 471.8-490.3 nm (blue), 564.9-587.3 nm (green), 648.1-673.9 nm (red), and 759.0-789.0 nm (IR). The spectra with a resolution of $R \sim 28\,000$ and a required SNR of 100 are being used to derive the chemical abundances of 30 elements with an accuracy of 0.05 dex. In total, about 1 000 000 stars with a V-magnitude between 12 and 14 mag and a declination below 10 deg are being observed. They originate from different parts of our galaxy: 77 % thin-disk stars, 22 % thick-disk stars, 0.8 % bulge stars, and 0.2 % halo stars. The observations started in 2012 and will end in 2017.

2.4 Gaia related

Gaia

The Global Astrometric Interferometer for Astrophysics (Gaia; Table 4, left; (`http://www.esa.int/Our_Activities/Space_Science/Gaia`; [6]) is a space mission with a $1.45 \times 0.5\,\mathrm{m}^2$ size telescope on-board. It performs photometric, astrometric and spectroscopic observations of objects passing through its FoV of $0.22 \times 0.39\,\mathrm{deg}^2$. Gaia was launched on December 19, 2013 and started its scientific observations in sky scanning mode on August 22, 2014. During the 5.5 years of the nominal mission, it is expected to gather data for approximately 150 000 000 objects (but rumour has it that the mission will be extended). For the spectroscopic part, the Radial Velocity Spectrometer (Rvs) is used to obtain spectra with $R \sim 11\,500$ in the wavelength range surrounding the Ca II triplet (847-874 nm). The aimed precision of the v_{rad} determinations ranges from $1\,\mathrm{km\,s^{-1}}$ for stars with $V = 11.5$ mag to $30\,\mathrm{km\,s^{-1}}$ for stars with $V = 17.5$ mag. For stars with a V-magnitude up to 15 mag, also the atmospheric parameters will be provided in the output catalogue.

DOI: 10.1051/epjconf/201715204001

Table 4. Characteristics of the large spectroscopic surveys related to the GAIA space-mission conducted in space with the Radial Velocity Spectrometer (RVS) and at the European Southern Observatory (ESO; Paranal, Chile) with the 8.2-m Very Large Telescope (VLT/UT2) in the southern hemisphere.

Name	GAIA	GAIA-ESO	
Telescope	1.45 m × 0.5 m	8.2-m VLT@ESO	
Instrument	RVS	GIRAFFE	UVES
Fibers		#132	#8
Wavelengths	847-874 nm	370-950 nm	300-1100 nm
	(Ca II triplet)		
Resolution	~11 500	7 500-30 000	~40 000
Targets	~150 000 000 stars	>10 000 stars	
	(V < 17.5 mag)	(V < 19 mag)	(V < 16.5 mag)
Date range	2014-2020	2012-2016	

GAIA-ESO

The GAIA-ESO survey (Table 4, right; `https://www.gaia-eso.eu/`; [38]) is a ground-based spectroscopic support effort for the GAIA mission. It is a large observing program based in the southern hemisphere at the European Southern Observatory (ESO) that runs on the second unit (UT2) of the 8.2-m Very Large Telescope (VLT) in Paranal (Chile). This telescope only has a small, circular FoV with a diameter of 25 arcmin (~0.135 deg^2). The observations are done with the Fibre Large Array Multi Element Spectrograph (FLAMES) feeding two spectrographs covering the whole visual spectral range. GIRAFFE is equipped with 130 fibers and is used to obtain spectra with a spectral resolution in the range of 7 500 to 30 000 for stars with $V < 19$ mag. Spectra with $R \sim 40\,000$ are collected for up to 8 stars with $V < 16.5$ mag simultaneously with the Ultraviolet and Visual Echelle Spectrograph (UVES). The observations of the GAIA-ESO survey started on December 31, 2011 and ended recently after a period of 5 years. In total, more than 10 000 Milky Way stars, both in the field and in open clusters, were observed. Their astrophysical parameters, including detailed abundances for at least 12 elements (Na, Mg, Si, Ca, Ti, V, Cr, Mn, Fe, Co, Sr, Zr, Ba) and v_{rad} values with a precision between 0.1 and 5 km s^{-1}, are being derived as we speak.

3 Future large spectroscopic surveys

In the next few years, there are some large spectroscopic surveys planned. They all serve as ground-based spectroscopic support for current and/or future space missions. In this section, three of them are briefly introduced.

WEAVE

The WHT Enhanced Area Velocity Explorer (WEAVE; `http://www.ing.iac.es/weave/about.html`; [39]) is designed to use an upgraded version of the 4.2-m William Herschell Telescope (WHT), located in the Observatorio Roque de los Muchachos (ORM; La Palma, Canary Islands, Spain) in the northern hemisphere. The new optics that will be installed will increase its circular FoV to a diameter of 2 deg (~3 deg^2). A total 1000 fibers will be installed to obtain spectra with $R \sim 5\,000$ in the full optical range (370-1000 nm) or with $R \sim 20\,000$ with a limited wavelength coverage. WEAVE aims to deliver more accurate v_{rad} values than those that will be delivered by GAIA for some 1 000 000 stars to investigate how the Milky Was was assembled. It will also observe distant galaxies to deduce their

Table 5. Characteristics of the large spectroscopic surveys that will be conducted at the Observatorio Roque de los Muchachos (Orm; La Palma, Canary Islands, Spain) in the northern hemisphere with the 4.2-m William Herschel Telescope (Wht) and at the European Southern Observatory (Eso; Paranal, Chile) in the southern hemisphere with the 8.2-m Very Large Telescope (Vlt) and the 4.1-m Vista telescope.

Name	Weave	Moons	4Most	
Telescope	4.2-m Wht@Orm	8.2-m Vlt@Eso	4.1-m Vista@Eso	
Hemisphere	north	south	south	
FoV	~3 deg^2 (2 deg; circular)	~0.135 deg^2 (25 arcmin; circular)	~4.1 deg^2 (hexagonal)	
Fibers	#1000	#1024	#1624	#812
Wavelengths	370-1000 nm	640-1800 nm	370-950 nm	392.6-435.5 nm 516.0-573.0 nm 610.0-679.0 nm
Resolution	~5 000 ~20 000	~4 000-6 000 (full) ~9 000 (770-900 nm) ~20 000 (1520-1630 nm)	<5 000 (Lrm)	>18 000 (Hrm)
Targets	~1 000 000 stars		stars with $-70 < \mathrm{DEC} < +30$ deg (>25 000 000 spectra)	
Date range	from 2018 onwards	from 2019 onwards	from 2020-2022 onwards (5 years)	

chemical composition and the rate at which new stars were forming at the time we are seeing them. By determining v_{rad} values for serveral million galaxies, Weave will make it possible to look for small deviations from uniformity in the expansion of the Universe, and hence to get some clues about dark energy. First light of Weave is expected in March 2018 and the start of its science observations is currently scheduled for June 2018.

Moons

The Multi-Object Optical and Near-infrared Spectrograph (Moons; `http://www.roe.ac.uk/~ciras/MOONS/VLT-MOONS.html`; [40]) is a third generation instrument for the 8.2-m Vlt in Paranal (Chile) in the southern hemisphere. With its full wavelength range being 640-1800 nm, it will extend the observations to the short-wavelength infrared. Moons will be offered in two resolution modes. In one mode, spectra with R ranging from ~4 000 to ~6 000 will be obtained in the full wavelength range. In the other mode, spectra with a higher R will be gathered in small wavelength regions: the region of the Ca II triplet (770-900 nm) with $R \sim 9000$ (to measure v_{rad}) and part of the H-band in the infrared (1520-1630 nm) with $R \sim 20000$ (for a detailed abundance analysis). It will be a powerful tool to perform pioneering research in a wide range of galactic, extragalactic and cosmological studies. Moreover, it will provide crucial follow-up observations for space missions like Gaia ([6]) and Euclid ([41]), and ground-based facilities like Lsst ([42]). Moons is expected to perform scientific observations from 2019 onwards.

4Most

The 4-metre Multi-Object Spectroscopic Telescope (4Most; `https://www.4most.eu/`; [43]) will use the 4.1-m Vista telescope of Eso in Paranal (Chile) in the southern hemisphere. The hexagonal FoV of 4.1 deg^2 will be covered with 2436 fibers. It will use three spectrographs with 812 fibers each:

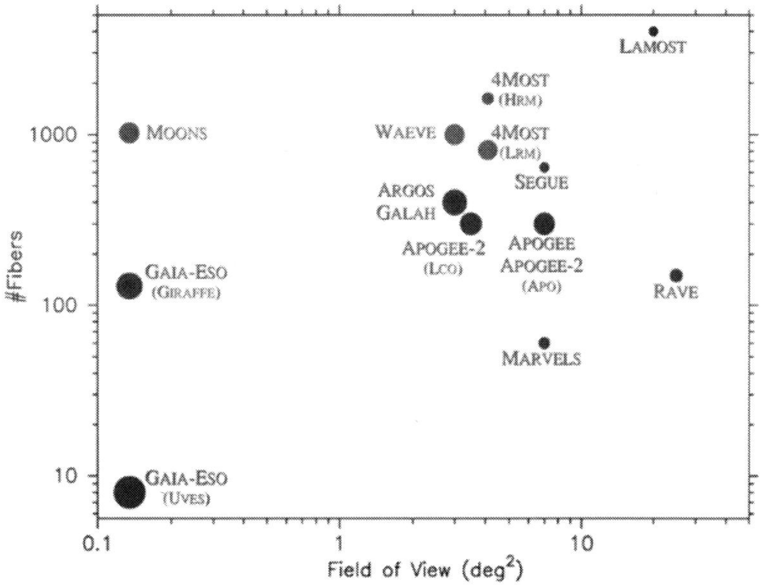

Figure 1. The number of fibers relative to the size of the field of view for the instrument/telescope combinations as used for the large spectroscopic surveys that are discussed in this review. The past/ongoing and future large spectroscopic surveys are given with dots in shades of blue and red, respectively. The size of the dots and their colour brightness are proportional to the (maximum) spectral resolution of the surveys. For the meaning of the abbreviations, we refer to the text and to Tables 1-5.

two to gather spectra in low-resolution mode (LRM) with $R < 5\,000$ in the full optical range (370-950 nm) and one to obtain spectra in high-resolution mode (HRM) with a $R > 18\,000$ in three small wavelength ranges (392.6–435.5 nm, 516.0–573.0 nm, and 610.0–679.0 nm). The main aim of 4MOST is to complement the all-sky space missions GAIA ([6]), EUCLID ([41]), and eROSITA ([44]) by providing v_{rad} values with an accuracy below 2 km s^{-1} for faint stars observed by GAIA ($r < 19$ mag), abundances of up to 15 chemical elements for stars with $V \leq 16$ mag, and redshifts of galaxies and active galactic nuclei with $r \leq 22$ mag. In total, more than 25 000 000 spectra will be collected in a timespan of 5 years for targets spread over a large fraction of the southern sky (declination between −70 and +30 deg). 4MOST will obtain $R > 18\,000$ resolution spectra for nearly all stars with $V < 12$ mag in that area every ~2 years. The observations are expected to start somewhere in 2020-2022.

4 Summary and conclusions

Large spectroscopic surveys are extremely useful for the research of pulsating stars. In general, they make the largest contribution by providing accurate values of stellar parameters and radial velocities. In this review, we have introduced 11 past or ongoing surveys (LAMOST, APOGEE, APOGEE-2, MARVELS, SEGUE-1, SEGUE-2, RAVE, ARGOS, GALAH, GAIA, and GAIA-ESO) and 3 future surveys (WEAVE, MOONS, and 4MOST) that are based in 3 northern observatories (XING, APO, ORM), 3 southern observatories (LCO, SSO, ESO) and one in space (GAIA). All those that are based on Earth use multi-fiber spectrographs to increase the efficiency of the data gathering process. In Figure 1, we show how the number of fibers relates to the FoV of the telescopes. LAMOST is the instrument with the highest number of fibers while RAVE uses the telescope with the largest FoV. MOONS, that will use the 8.2-m VLT@ESO,

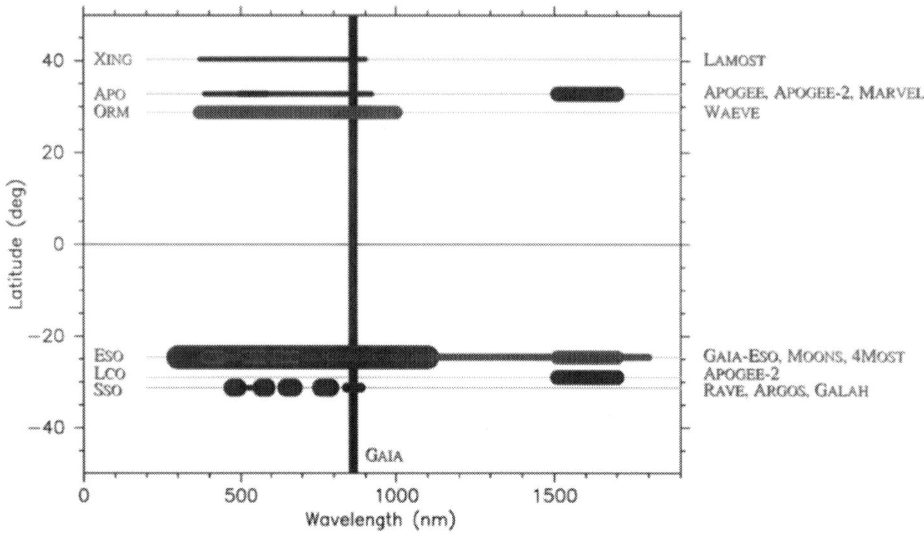

Figure 2. Overview of the spectral ranges covered from the different observatories by the large spectroscopic surveys that are discussed in this review. The space-based mission Gaia is given by the vertical line. The past/ongoing and future large spectroscopic surveys are given with lines in shades of blue and red, respectively. The width of the lines and the brightness of their colour are proportional to the (maximum) spectral resolution of the surveys. For the meaning of the abbreviations, we refer to the text and to Tables 1-5.

will have by far the highest number of fibers per deg^2 on the sky. Figure 2 shows that the large spectroscopic surveys discussed in this review have wavelength ranges starting in the near ultraviolet (300 nm) and ending in the short-wavelength infrared (1.8 μm). When combining all of them, those based in the southern hemisphere cover the full wavelength region. In the northern hemisphere, there is no survey that included/includes/will include the range from 1 μm to 1.51 μm.

A huge amount of spectroscopic data of large spectroscopic surveys is publicly available and this amount is steadily increasing. Therefore, it might be worthwhile to check if the spectroscopic data that you might need for the research of your favorite star(s) is already available in the spectroscopic databases before applying for precious telescope time.

Acknowledgments: The author is very thankful to the organisers for the invitation to present this review.

References

[1] G. Walker, J. Matthews, R. Kuschnig, R. Johnson, S. Rucinski, J. Pazder, G. Burley, A. Walker, K. Skaret, R. Zee et al., PASP **115**, 1023 (2003)

[2] A. Baglin, M. Auvergne, L. Boisnard, T. Lam-Trong, P. Barge, C. Catala, M. Deleuil, E. Michel, W. Weiss, *CoRoT: a high precision photometer for stellar ecolution and exoplanet finding*, in *36th COSPAR Scientific Assembly* (2006), Vol. 36 of *COSPAR Meeting*

[3] W.J. Borucki, D. Koch, G. Basri, N. Batalha, T. Brown, D. Caldwell, J. Caldwell, J. Christensen-Dalsgaard, W.D. Cochran, E. DeVore et al., Science **327**, 977 (2010)

[4] S.B. Howell, C. Sobeck, M. Haas, M. Still, T. Barclay, F. Mullally, J. Troeltzsch, S. Aigrain, S.T. Bryson, D. Caldwell et al., PASP **126**, 398 (2014)

[5] W.W. Weiss, S.M. Rucinski, A.F.J. Moffat, A. Schwarzenberg-Czerny, O.F. Koudelka, C.C. Grant, R.E. Zee, R. Kuschnig, S. Mochnacki, J.M. Matthews et al., PASP **126**, 573 (2014)

[6] M.A.C. Perryman, K.S. de Boer, G. Gilmore, E. Høg, M.G. Lattanzi, L. Lindegren, X. Luri, F. Mignard, O. Pace, P.T. de Zeeuw, A&A **369**, 339 (2001)

[7] G.R. Ricker, J.N. Winn, R. Vanderspek, D.W. Latham, G.Á. Bakos, J.L. Bean, Z.K. Berta-Thompson, T.M. Brown, L. Buchhave, N.R. Butler et al., Journal of Astronomical Telescopes, Instruments, and Systems **1**, 014003 (2015)

[8] H. Rauer, C. Catala, C. Aerts, T. Appourchaux, W. Benz, A. Brandeker, J. Christensen-Dalsgaard, M. Deleuil, L. Gizon, M.J. Goupil et al., Experimental Astronomy **38**, 249 (2014)

[9] T.R. Bedding, H. Kjeldsen, Communications in Asteroseismology **161**, 3 (2010)

[10] T.R. Bedding, S.J. Murphy, I.L. Colman, D.W. Kurtz, *Échelle diagrams and period spacings of g modes in γ Doradus stars from four years of Kepler observations*, in *European Physical Journal Web of Conferences* (2015), Vol. 101 of *European Physical Journal Web of Conferences*, p. 01005

[11] M.A. Dupret, J. De Ridder, P. De Cat, C. Aerts, R. Scuflaire, A. Noels, A. Thoul, A&A **398**, 677 (2003)

[12] M. Briquet, C. Aerts, A&A **398**, 687 (2003)

[13] W. Zima, A&A **455**, 227 (2006)

[14] P. Pietrukowicz, M. Latour, R. Angeloni, F. di Mille, I. Soszyński, A. Udalski, C. Germanà, AcA **65**, 63 (2015)

[15] P.I. Pápics, E. Moravveji, C. Aerts, A. Tkachenko, S.A. Triana, S. Bloemen, J. Southworth, A&A **570**, A8 (2014)

[16] X.Q. Cui, Y.H. Zhao, Y.Q. Chu, G.P. Li, Q. Li, L.P. Zhang, H.J. Su, Z.Q. Yao, Y.N. Wang, X.Z. Xing et al., Research in Astronomy and Astrophysics **12**, 1197 (2012)

[17] A.L. Luo, Y.H. Zhao, G. Zhao, L.C. Deng, X.W. Liu, Y.P. Jing, G. Wang, H.T. Zhang, J.R. Shi, X.Q. Cui et al., Research in Astronomy and Astrophysics **15**, 1095 (2015)

[18] L.C. Deng, H.J. Newberg, C. Liu, J.L. Carlin, T.C. Beers, L. Chen, Y.Q. Chen, N. Christlieb, C.J. Grillmair, P. Guhathakurta et al., Research in Astronomy and Astrophysics **12**, 735 (2012)

[19] P. De Cat, J.N. Fu, A.B. Ren, X.H. Yang, J.R. Shi, A.L. Luo, M. Yang, J.L. Wang, H.T. Zhang, H.M. Shi et al., ApJS **220**, 19 (2015)

[20] A. Ren, J. Fu, P. De Cat, Y. Wu, X. Yang, J. Shi, A. Luo, H. Zhang, S. Dong, R. Zhang et al., ApJS **225**, 28 (2016)

[21] A. Frasca, J. Molenda-Żakowicz, P. De Cat, G. Catanzaro, J.N. Fu, A.B. Ren, A.L. Luo, J.R. Shi, Y. Wu, H.T. Zhang, A&A **594**, A39 (2016)

[22] R.O. Gray, C.J. Corbally, P. De Cat, J.N. Fu, A.B. Ren, J.R. Shi, A.L. Luo, H.T. Zhang, Y. Wu, Z. Cao et al., AJ **151**, 13 (2016)

[23] J. Fu, P. De Cat, M. Smith, *Synergies between spectroscopic and time-series photometric surveys - LAMOST observations for the Kepler field and K2 fields*, in *American Astronomical Society Meeting Abstracts* (2017), Vol. 229 of *American Astronomical Society Meeting Abstracts*, p. 305.02

[24] B. Yanny, C. Rockosi, H.J. Newberg, G.R. Knapp, J.K. Adelman-McCarthy, B. Alcorn, S. Allam, C. Allende Prieto, D. An, K.S.J. Anderson et al., AJ **137**, 4377-4399 (2009)

[25] J. Ge, S. Mahadevan, B. Lee, X. Wan, B. Zhao, J. van Eyken, S. Kane, P. Guo, E. Ford, S. Fleming et al., *The Multi-object APO Radial-Velocity Exoplanet Large-area Survey (MARVELS)*, in *Extreme Solar Systems*, edited by D. Fischer, F.A. Rasio, S.E. Thorsett, A. Wolszczan (2008), Vol. 398 of *Astronomical Society of the Pacific Conference Series*, p. 449

[26] C. Allende Prieto, S.R. Majewski, R. Schiavon, K. Cunha, P. Frinchaboy, J. Holtzman, K. Johnston, M. Shetrone, M. Skrutskie, V. Smith et al., Astronomische Nachrichten **329**, 1018 (2008)

[27] M.H. Pinsonneault, Y. Elsworth, C. Epstein, S. Hekker, S. Mészáros, W.J. Chaplin, J.A. Johnson, R.A. García, J. Holtzman, S. Mathur et al., ApJS **215**, 19 (2014)

[28] S.R. Majewski, APOGEE Team, APOGEE-2 Team, Astronomische Nachrichten **337**, 863 (2016)

[29] M. Steinmetz, T. Zwitter, A. Siebert, F.G. Watson, K.C. Freeman, U. Munari, R. Campbell, M. Williams, G.M. Seabroke, R.F.G. Wyse et al., AJ **132**, 1645 (2006)

[30] ESA, ed., *The HIPPARCOS and TYCHO catalogues. Astrometric and photometric star catalogues derived from the ESA HIPPARCOS Space Astrometry Mission*, Vol. 1200 of *ESA Special Publication* (1997)

[31] K. Freeman, M. Ness, E. Wylie-de-Boer, E. Athanassoula, J. Bland-Hawthorn, M. Asplund, G. Lewis, D. Yong, R. Lane, L. Kiss et al., MNRAS **428**, 3660 (2013)

[32] M. Ness, K. Freeman, E. Athanassoula, E. Wylie-de-Boer, J. Bland-Hawthorn, M. Asplund, G.F. Lewis, D. Yong, R.R. Lane, L.L. Kiss, MNRAS **430**, 836 (2013)

[33] M. Ness, K. Freeman, E. Athanassoula, E. Wylie-de-Boer, J. Bland-Hawthorn, M. Asplund, G.F. Lewis, D. Yong, R.R. Lane, L.L. Kiss et al., MNRAS **432**, 2092 (2013)

[34] G.M. De Silva, K.C. Freeman, J. Bland-Hawthorn, S. Martell, E.W. de Boer, M. Asplund, S. Keller, S. Sharma, D.B. Zucker, T. Zwitter et al., MNRAS **449**, 2604 (2015)

[35] J.D. Simpson, G.M. De Silva, J. Bland-Hawthorn, K.C. Freeman, S.L. Martell, K.J. Schlesinger, S. Sharma, D.B. Zucker, T. Zwitter, J. Kos et al., MNRAS **459**, 1069 (2016)

[36] J. Kos, J. Lin, T. Zwitter, M. Žerjal, S. Sharma, J. Bland-Hawthorn, M. Asplund, A.R. Casey, G.M. De Silva, K.C. Freeman et al., MNRAS **464**, 1259 (2017)

[37] S.L. Martell, S. Sharma, S. Buder, L. Duong, K.J. Schlesinger, J. Simpson, K. Lind, M. Ness, J.P. Marshall, M. Asplund et al., MNRAS **465**, 3203 (2017)

[38] G. Gilmore, S. Randich, M. Asplund, J. Binney, P. Bonifacio, J. Drew, S. Feltzing, A. Ferguson, R. Jeffries, G. Micela et al., The Messenger **147**, 25 (2012)

[39] G. Dalton, S.C. Trager, D.C. Abrams, D. Carter, P. Bonifacio, J.A.L. Aguerri, M. MacIntosh, C. Evans, I. Lewis, R. Navarro et al., *WEAVE: the next generation wide-field spectroscopy facility for the William Herschel Telescope*, in *Ground-based and Airborne Instrumentation for Astronomy IV* (2012), Vol. 8446 of *Proc. SPIE*, p. 84460P

[40] M. Cirasuolo, MOONS Consortium, *MOONS: A New Powerful Multi-Object Spectrograph for the VLT*, in *Multi-Object Spectroscopy in the Next Decade: Big Questions, Large Surveys, and Wide Fields*, edited by I. Skillen, M. Barcells, S. Trager (2016), Vol. 507 of *Astronomical Society of the Pacific Conference Series*, p. 109

[41] L. Amendola, S. Appleby, D. Bacon, T. Baker, M. Baldi, N. Bartolo, A. Blanchard, C. Bonvin, S. Borgani, E. Branchini et al., Living Reviews in Relativity **16**, 6 (2013)

[42] LSST Science Collaboration, P.A. Abell, J. Allison, S.F. Anderson, J.R. Andrew, J.R.P. Angel, L. Armus, D. A(rnett, S.J. Asztalos, T.S. Axelrod et al., ArXiv e-prints (2009), `0912.0201`

[43] R.S. de Jong, 4MOST Consortium, 4MOST Spectroscopy Consortium, Astronomische Nachrichten **337**, 964 (2016)

[44] A. Merloni, P. Predehl, W. Becker, H. Böhringer, T. Boller, H. Brunner, M. Brusa, K. Dennerl, M. Freyberg, P. Friedrich et al., ArXiv e-prints (2012), `1209.3114`

The LAMOST spectroscopic survey of stars in the *Kepler* field of view: Activity indicators and stellar parameters*

Joanna Molenda-Żakowicz[1,**], *Antonio* Frasca[2], *Peter* De Cat[3], and *Giovanni* Catanzaro[2]

[1] *Uniwersytet Wrocławski, Instytut Astronomiczny, ul. Kopernika 11, 51-622 Wrocław, Poland*
[2] *INAF–Osservatorio Astrofisico di Catania, via S. Sofia, 78, 95123 Catania, Italy*
[3] *Royal Observatory of Belgium, Ringlaan 3, 1180 Brussels, Belgium*

Abstract. We summarize the results of the completed first round of the LAMOST–*Kepler* project, and describe the status of its on-going second round. As a result of the first round of this project, the atmospheric parameters ($T_{\rm eff}$, $\log g$, and [Fe/H]), the spectral classification (spectral type and luminosity class), and the radial velocities (*RV*) have been measured for 51,385 stars. For 4031 stars, we were able to measure the projected rotational velocity, while the minimum detectable $v \sin i$ was 120 km s^{-1}. For 8821 stars with more than one observation, we computed the χ-square probability that the detected *RV* variations have a random occurrence. Finally, we classified 442 stars as chromospherically active on the basis of the analysis of their Hα and Ca II-IRT fluxes. All our results have been obtained from the low-resolution ($R \sim 1800$) spectroscopic observations acquired with the LAMOST instrument.

1 LAMOST–*Kepler* project

The LAMOST–*Kepler* project, described in detail by [1], is an activity carried out in the framework of the *Kepler* Asteroseismic Science Consortium (KASC). Started in 2011, this project aims at deriving the effective temperature ($T_{\rm eff}$), the surface gravity ($\log g$), the metallicity ([Fe/H]), the radial velocity (*RV*), the projected equatorial rotational velocity ($v \sin i$), and performing the MK spectral classification for stars in the nominal field of view of the *Kepler* space telescope (hereafter: the *Kepler* field, [2]). The LAMOST–*Kepler* project aims at completing and verification of the atmospheric parameters of stars provided in the *Kepler* Input Catalog (KIC, [3]).

Our activity is crucial for the in-depth studies of stars observed with *Kepler* because among the 13,123,717 stars listed in the KIC, only 2,106,821 (i.e., 16% of the total sample) have the atmospheric parameters provided in that catalogue. Moreover, as shown by, e.g., [4, 5], the precision of the determinations of $T_{\rm eff}$, $\log g$, and [Fe/H] of solar-type stars in the KIC is too low for asteroseismic modelling. For stars of spectral type earlier than A, the situation is even worse since for them the discrepancies between $T_{\rm eff}$ in the KIC and the values derived from spectroscopic observations amount to several thousand Kelvin (see [6–9]).

*Based on observations collected with the Large Sky Area Multi-Object Fiber Spectroscopic Telescope (LAMOST) located at the Xinglong Observatory, China.

**molenda@astro.uni.wroc.pl

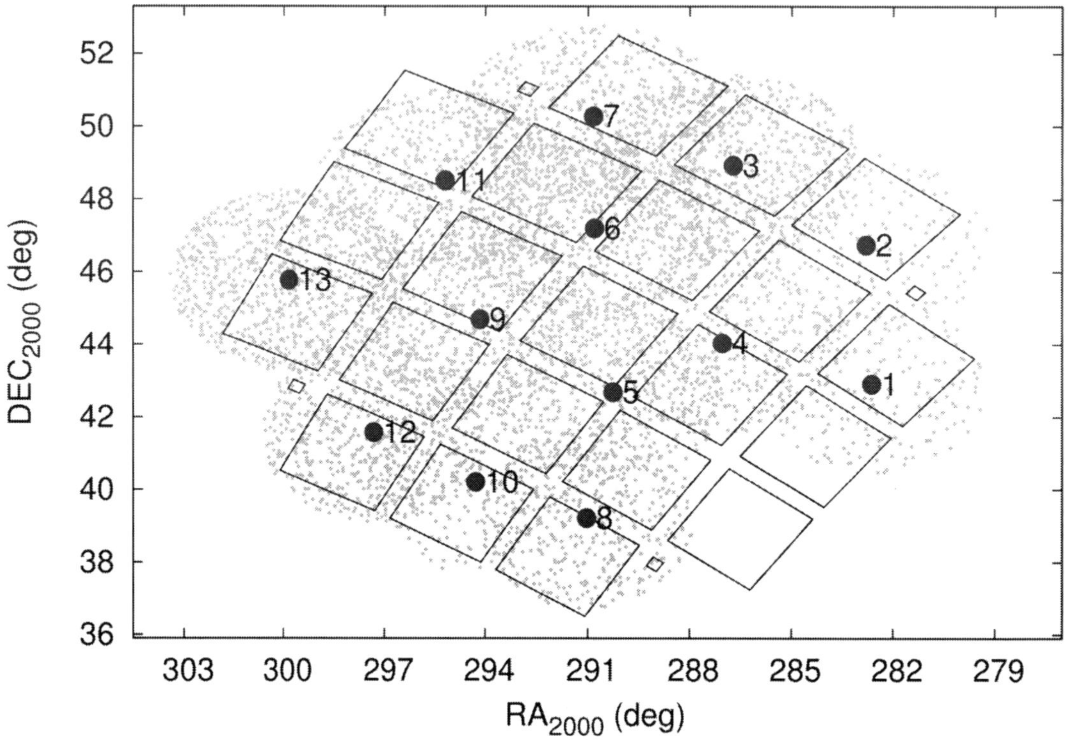

Figure 1. The nominal field of view of the *Kepler* space telescope covered with the LAMOST observations acquired in the second round of the LAMOST–*Kepler* project (in order to reduce the size of the figure, only every 15th target has been plotted). The lines indicate the location of the *Kepler* CCD modules. The gray dots indicate the location of the 86,378 observed stars. The large black dots indicate the centres of the 13 observed fields whose coordinates are provided in Table 1 .

In order to achieve the scientific goals of our project, we have been acquiring low-resolution ($R \sim 1800$) spectroscopic observations using the Large Sky Area Multi-Object Fiber Spectroscopic Telescope (LAMOST) located at the Xinglong Observatory in China. We selected that instrument because it combines a large aperture (3.6–4.9 m) with a large, circular field of view of five degrees in diameter. This allows to cover the whole *Kepler* field with only 14 settings (see Fig. 2 in [1]). The focal surface of the LAMOST instrument is covered with 4000 optical fibers. Since some of them must be used for acquiring sky spectra and for calibration purposes, each exposure allows to acquire spectra of around 3000 stars (cf. Table 1). This gave us an opportunity to obtain spectroscopic observations for a significant number of stars from the *Kepler* field in a relatively short time.

1.1 First round

The first round of the LAMOST–*Kepler* project started in 2011 and ended in 2014. In that round we focused on stars which were included on the list of possible targets of the KASC. Based on those observations, in paper [10] we measured the atmospheric parameters and radial velocities of 51,385 stars. The average accuracy of these parameters is equal to 14 km s^{-1} in RV, 3.5% in $T_{\rm eff}$, 0.3 dex in $\log g$, and 0.2 dex in [Fe/H]. In paper [10], we used the LAMOST data also to measure $v \sin i$ of

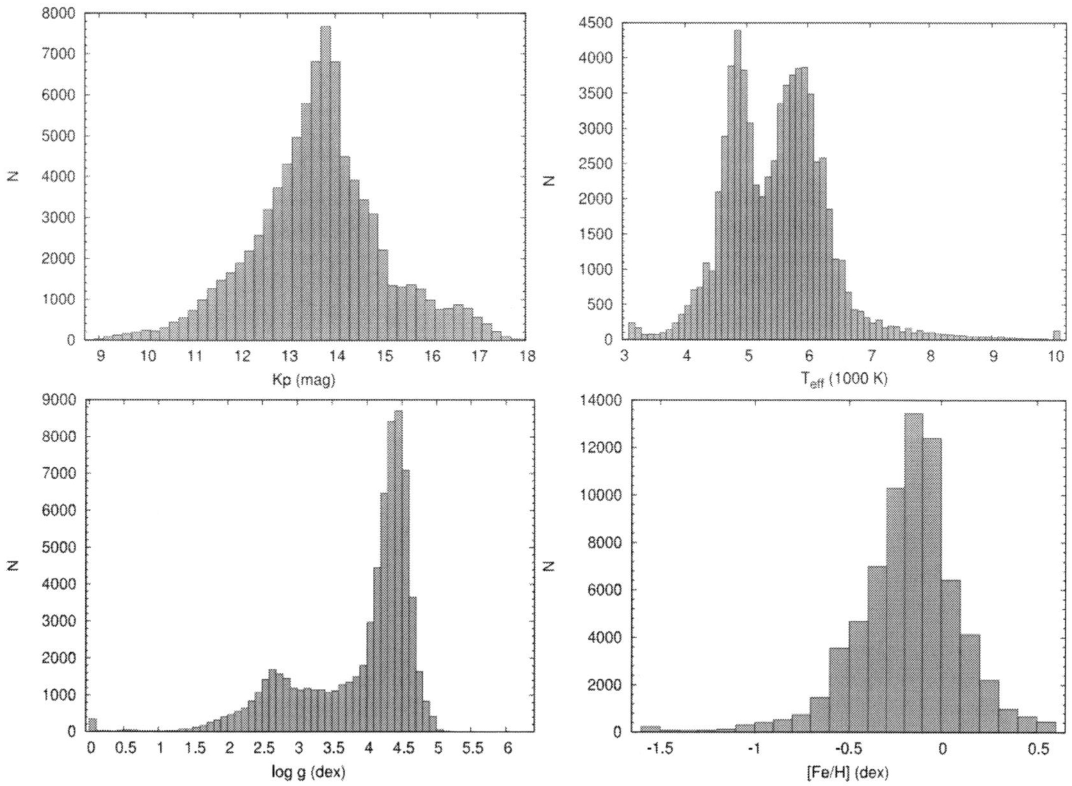

Figure 2. The global characteristics of the targets observed in the framework of the second round of the LAMOST–*Kepler* project. The histogram of the *Kepler* Kp magnitudes has been constructed for 86,284 stars. The histograms of $T_{\rm eff}$, $\log g$, and [Fe/H] have been constructed for 70,298 stars. The last bin of the histogram of Kp includes 34 stars fainter than 18.1 mag. The last bin of the histogram of $T_{\rm eff}$ includes 128 stars hotter than 10,000 K. The first bin of the histogram of $\log g$ includes eight stars with $\log g < 0$ dex. The first bin of the histogram of [Fe/H] includes 134 stars with [Fe/H]< -1.6 dex. The Kp, $T_{\rm eff}$, $\log g$, and [Fe/H] values have been adopted from the *Kepler* Input Catalog.

stars that rotate with $v \sin i > 120\,{\rm km\,s^{-1}}$. The latter value was found as the minimum detectable $v \sin i$ in our data from Monte Carlo simulations. For 547 stars which showed emission or filling-in the Hα line, we measured the equivalent width (EW) of that line. The EW of the Ca II 8498 Å line has been measured for 348 stars, the EW of the Ca II 8542 Å line for 360 stars, and the EW of the Ca II 8542 Å line for 368 stars. The [NII] nebular lines have been detected in 55 stars. Based on the analysis of the Hα and Ca II-IRT fluxes of those 547 stars, we selected 442 stars that we classified as chromospherically active. We showed that one of them, KIC 8749284, is a likely accreting object.

1.2 Second round

The second round of the LAMOST–*Kepler* project started in 2015. In that round, we focused on stars which have already been observed with the *Kepler* telescope. In 2015, we acquired 97,641 spectra of 86,378 individual stars. 77,577 stars from that sample were observed only once, 7266 stars twice, 1195 stars three times, 254 stars four times, 66 stars five times, 14 stars six times, seven stars seven

Table 1. The 13 settings of the LAMOST instrument used in the second round of the LAMOST–*Kepler* project. For each setting, the right ascension and declination of the centre of the respective field of view is provided along with the date(s) of observation, the number of exposures acquired for that field, and the total number of spectra that have been reduced from those data.

Field	R.A. (2000)	DEC (2000)	Date	Number of exposures	Number of spectra
1	18:50:31	42:54:43	2015-10-04	1	3090
2	18:51:11	46:44:17	2015-10-06	1	2947
3	19:06:51	48:55:31	2015-05-30	3	7853
"	"	"	2015-09-15	1	3084
4	19:08:08	44:02:10	2015-09-25	2	6153
5	19:21:02	42:41:13	2015-10-11	2	6223
6	19:23:14	47:11:44	2015-09-14	1	3140
"	"	"	2015-10-02	2	6251
7	19:23:23	50:16:16	2015-05-29	2	5517
"	"	"	2015-09-13	1	3169
8	19:24:09	39:12:42	2015-10-12	2	6303
9	19:36:37	44:41:41	2015-09-21	3	9163
10	19:37:08	40:12:49	2015-10-18	1	3120
11	19:40:45	48:30:45	2015-09-16	1	3135
"	"	"	2015-10-01	2	6314
12	19:49:18	41:34:56	2015-10-08	3	9454
13	19:59:20	45:46:21	2015-09-18	2	6350
"	"	"	2015-10-03	2	6375
					Total: 97,641

times, and one star eight times. The location of our targets in the *Kepler* field is shown in Figure 1. The numbers at the large black dots in Figure 1 refer to the centres of the 13 settings of the LAMOST instrument which were used for those observations. In Table 1, we provide the equatorial coordinates of those settings, the dates of observation, the number of exposures acquired in each setting, and the total number of spectra reduced from each setting.

Figure 1 shows that, in 2015, we covered almost the whole *Kepler* field with dense observations. The missing part and the sparsely sampled areas were scheduled for observing in 2016. However, bad weather conditions at the Xinglong Observatory in that year did not allow us to obtain the requested data. We expect to finish the second round of the LAMOST–*Kepler* project in 2017.

In Figure 2, we illustrate the global characteristics of the sample observed in 2015, adopting a similar approach to that used by us in paper [1] (see Fig. 6 therein). A visual comparison of the statistical properties of the stars observed in the first and the second round of the LAMOST–*Kepler* project shows that in the second round of the project, we observed stars falling in the same range of magnitude as in the first round, however, we included noticeably more stars fainter than 14 mag. The histograms of $T_{\rm eff}$, $\log g$, and [Fe/H] show that the properties of both samples, i.e. stars observed in the first and the second round, are similar. The quantitative analysis of the statistical properties of the sample observed in the second round will be carried out when the round is completed. By that time, we will also start the scientific analysis of the observations acquired in the second round of the LAMOST–*Kepler* project, which is planned to be a continuation of the work done in paper [10].

Acknowledgments: J.M.-Ż. acknowledges grant number NCN 2014/13/B/ST9/00902 and the Wrocław Centre for Networking and Supercomputing grant number 224.

References

[1] De Cat, P., Fu, J. N., Ren, A. B., et al., ApJS, **220**, 19 (2015)
[2] Borucki, W., Koch, D., Boss, A., et al., in *Second Eddington Workshop: Stellar structure and habitable planet finding*, ed. F. Favata, S. Aigrain, & A. Wilson, ESA SP-**538**, 117 (2004)
[3] Brown, T. M., Latham, D. W., Everett, M. E., & Esquerdo, G. A., AJ, **142**, 112 (2011)
[4] Creevey, O. L., Doğan, G.; Frasca, A., et al., A&A, **537**, 111 (2012)
[5] Stello, D., Chaplin, W. J., Bruntt, H., et al., ApJ, **700**, 1589 (2009)
[6] Catanzaro, G., Frasca, A., Molenda-Żakowicz, J., & Marilli, A&A, **517**, 3 (2010)
[7] Lehmann, H., Tkachenko, A., Semaan, T., et al., A&A, **526**, 124 (2011)
[8] McNamara, B. J., Jackiewicz, J., & McKeever, J., AJ, **143**, 101 (2012)
[9] Molenda-Żakowicz, J., Jerzykiewicz, M., Frasca, A., Catanzaro, G., Kopacki, G., & Latham, D. W., preprint (arXiv:1005.0985) (2010)
[10] Frasca, A., Molenda-Żakowicz, J., De Cat, P., et al., A&A, **594**, 39 (2016)

DOI: 10.1051/epjconf/201715204003

Obtaining accurate radial velocities for Cepheid companions using the STIS echelles

C. R. Proffitt[1], *N. R.* Evans[2], *E. M.* Winston[2], *A.* Gallenne[3], and *P.* Kervella[4,5]

[1]*Space Telescope Science Institute, Baltimore, MD 21218, USA*
[2]*Smithsonian Astrophysical Observatory, MS 4, 60 Garden Street, Cambridge, MA 02138, USA*
[3]*European Southern Observatory, Alonso de Cordova 3107, Casilla 19001, Santiago 19, Chile*
[4]*LESIA, Obs. de Paris, CNRS UMR 8109, UPMC,Univ. Paris 7, 5 Pl. Jules Janssen, 92195 Meudon, France*
[5]*CNRS UMI 3386, Departamento de Astronomía, Universidad de Chile, Las Condes, Chile*

Abstract. We discuss the high dispersion echelle observations of the hot binary companions of six Cepheids with known radial-velocity orbits that we have obtained with the STIS FUV E140H mode on board the *Hubble Space Telescope*, with the goal of determining the masses of these Cepheids. We discuss the stability and repeatability of the STIS echelle wavelength scale and other issues that may limit the final accuracy of our mass determinations.

1 Introduction

Over the course of a number of recent *Hubble Space Telescope* (*HST*) GO programs we have obtained FUV echelle spectral observations with the Space Telescope Imaging Spectrograph (STIS) on board *HST* for a number of hot companions of binary Cepheid stars. Our goal is to determine the radial velocity amplitude of each companion's orbit and ultimately the mass of each Cepheid. Complementary observations to measure the astrometric orbit of each Cepheid are also underway, as detailed by the contributions of P. Kervella et al. & A. Gallenne et al. in these proceedings.

2 Wavelength stability and properties of individual companions

We have used the ISM lines superimposed on our spectra to demonstrate the stability of the wavelength scale over time (Fig. 1). These show that the wavelength scale is repeatable to about 0.25 km/s.

In Table 1 we list the approximate $T_{\rm eff}$, projected rotational velocity, $v \sin i$, and FUV flux estimated from our spectra, together with the orbital periods ([1]). For three stars $v \sin i$ is large and the broad lines may limit the obtainable RV accuracy. For two other systems the hot companions are also close binaries. This was known for SU Cyg ([2]) and the sharp lines of this star and the 6 epochs for which we have data may make it possible to solve for both the close and the wide orbit. The binary velocity of the AW Per companion (Fig. 2) confirms its previously suspected binary nature, but additional data are needed to constrain its short period orbit. AX Cir is a triple system, and the previous IUE observations were dominated by the brighter long period companion; this is the first spectrum obtained of the close 18 yr SB companion. Work is ongoing to derive the absolute velocities and stellar parameters, and these results will be used to determine the masses and evolutionary state for each system.

DOI: 10.1051/epjconf/201715204003

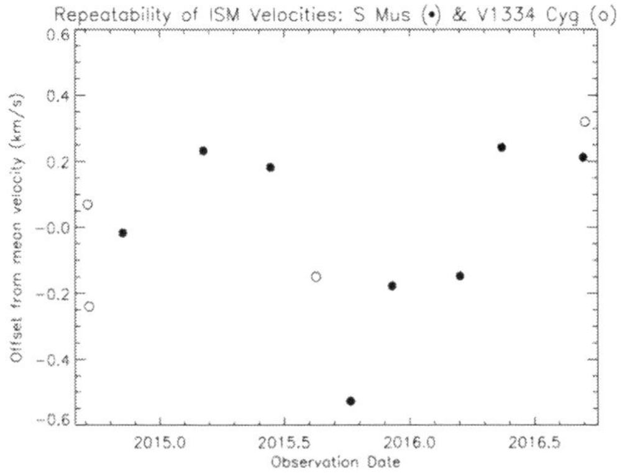

Figure 1. The measured offsets of the O I ISM line wavelength about the mean value measured in each star are shown for our observations of S Mus and V1334 Cyg. The rms scatter about the mean is equivalent to only 0.24 km/s. This demonstrates that the STIS echelle wavelength scale as determined using internal wavelength calibration observations is very stable and can be used to compare STIS observations taken months or years apart.

Table 1. Close Cepheid companions with STIS E140H observations.

Star	P_{orb} (yr)	T_{eff} (K)	$v \sin i$ (km/s)	$F_{1322} \times 10^{-14}$ (ergs cm^{-2}Å^{-1}s^{-1})	# epochs	Comments
V350 Sgr	4.0	10000	180	4.1	2	
SU Cyg	1.5	13250	< 3	160	8	close triple
S Mus	1.4	15000	20	550	8	
V1334 Cyg	5.3	13500	180	1100	3	
AX Cir	17.9	11000	180	88	1	wide triple
AW Per	36	15750	15	9.5	1	close triple

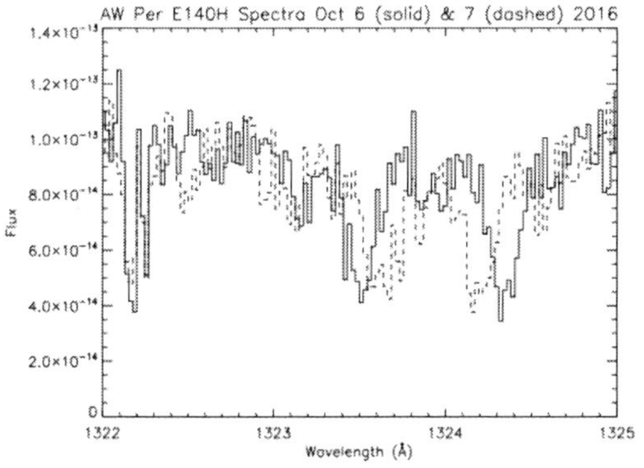

Figure 2. The C II multiplet near 1324 Å is doubled in the spectrum of AW Per and changes significantly in a single day, while the ISM CO line near 1322.2 Å is unchanged. This demonstrates that the hot companion is itself a close short period double-lined binary.

Acknowledgments: Based on observations with the NASA/ESA Hubble Space Telescope which is operated by the Associations of Universities for Research in Astronomy, Incorporated, under NASA contract NAS5-26555. These observations are associated with programs 13368, 14332, 13841, 14340, 13454, 13849, 14339, and 14194. Support for these programs was provided by NASA through grants from the Space Telescope Science Institute, operated by the Associations of Universities for Research in Astronomy, Incorporated, under NASA contract NAS5-26555. PK and AG acknowledge support from the French Agence Nationale de la Recherche (ANR), under grant ANR-15-CE31-0012-01 and the French-Chilean exchange program ECOS-Sud/CONICYT (C13U01).

References

[1] Evans, N. R., & Bolton, C. T., 1990 ApJ, **356**, 630

[2] Evans, N. R., Bond, H. E., Schaefer, G. H., Mason, B. D., Karovska, M., & Tingle, E., 2013 AJ, **146**, 93

Photometric and spectroscopic variability of the B5IIIe star HD 171219

Laerte Andrade[1,*], *Eduardo* Janot-Pacheco[2], *Marcelo* Emilio[1], and *Yves* Frémat[3]

[1] *Universidade Estadual de Ponta Grossa, Av. Carlos Cavalcanti, 4748, 84030-900, Ponta Grossa, Brazil*
[2] *Universidade de São Paulo/IAG, rua do Matão, 1226 - Cidade Universitária, 05508-900, São Paulo, Brazil*
[3] *Royal Observatory of Belgium, 3 avenue circulaire, B1180 Brussels, Belgium*

Abstract. We analyzed the Be star HD 171219, observed with the CoRoT satellite during a 77.56 day run, in order to determine its physical and pulsation characteristics. High-resolution ground-based spectroscopic observations with HARPS and SOPHIE were also obtained during the month preceding the CoRoT observations. Twenty frequencies have been detected in the object, some of them compatible with non-radial g-modes, including a quintuplet centered around 1.113 c/d (12.88 μHz) identified as a pulsation of degree $\ell \sim 2$. The star underwent at least 6 small outbursts during the CoRoT observations. The relative intensity of the main frequencies varied after each outburst suggesting that the stability of the star and the feeding of the envelope are linked to the pulsation regime ([1]).

1 Photometric analysis of CoRoT data

The Be Star HD 171219 was continually observed by the CoRoT satellite during 77.56 days from July to September 2010, with a cadence of 32 seconds. Its light curve (Fig. 1a) contains 209,418 photometric observations. Close examination shows the occurrence of a medium-size outburst around CJD 3876 and four minor outbursts close to CJD = 3847, 3853, 3894 and 3905.

We applied the CLEANEST algorithm ([2, 3]) to look for pulsation frequencies (Fig. 1b). We have twenty periodicities in the light curve, some of them compatible with a g-mode quintuplet (f_1 = 1.072 c d^{-1}, f_2 = 1.089 c d^{-1}, f_3 = 1.113 c d^{-1}, f_4 = 1.130 c d^{-1} and f_5 = 1.146 c d^{-1}).

2 Ground-based spectroscopic analysis

High-resolution spectra were obtained during a 20-day period with two echelle spectrographs, HARPS and SOPHIE. With 53 good quality spectra available, we rederived and improved the stellar parameters by directly comparing 2D spectra that account for first order rapid rotation effects (stellar flattening and gravitational darkening) to observations made of a combination of the best available spectra ([4]).

We also applied the CLEANEST method to the time series of the spectroscopic data. Because of the scarcity of spectra, there was less information on non-radial pulsations from line profile variations to be extracted ([5]), but peaks around the multiplet frequencies and their harmonics appear clearly.

* laerteandrade@uepg.br

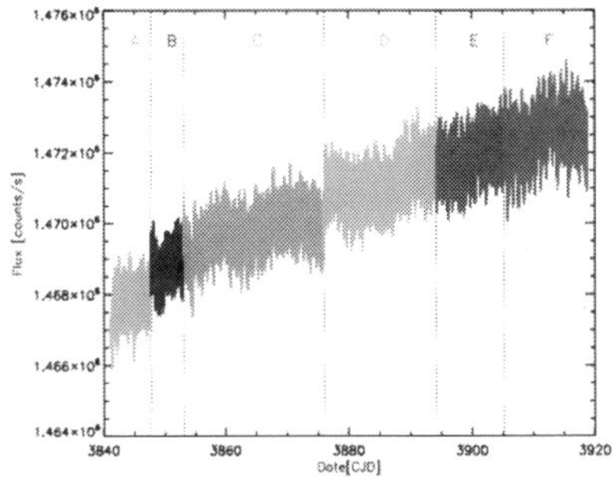

(a) CoRoT light curve of HD 171219. Colored regions (A) through (F) are divided by the ocurrences of minor or medium outbursts.

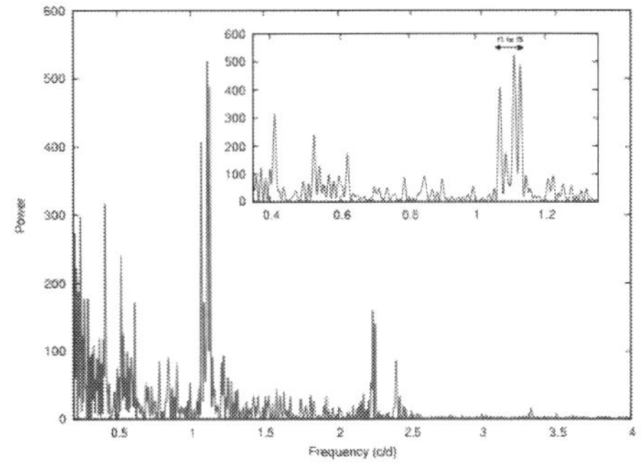

(b) Power spectrum from light curve on the left. Inset: Zoom of the region containing the g-mode $\ell \sim 2$ quintuplet (f_1 to f_5) centered around 1.113 c/d.

Figure 1: Photometric analysis of CoRoT data.

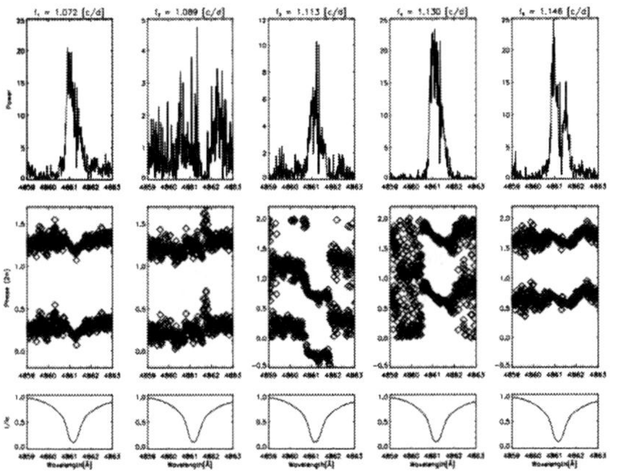

Figure 2. Power (top panel) and phase (middle panel) variation of quintuplet frequencies along the mean HARPS and SOPHIE Hβ line profile (bottom panel).

In Figure 2 we plotted the amplitudes and phases of the variations of the multiplet frequencies as a function of the position in the Hβ Balmer line profile, calculated by CLEANEST along the mean SOPHIE + HARPS observations. Following [6] we estimated the value of the NRP degree $\ell = 2 \pm 1$.

Acknowledgments: Work supported in Brazil by FAPESP, CNPq and CAPES (PNPD). Partially based on observations made with the 3.6-m telescope at La Silla Observatory under the ESO Large Programme LP185.D-0056.

References

[1] Andrade, L., Janot-Pacheco, E., Emilio, M., et al., A&A, **603**, A41 (2017)
[2] Emilio, M., Andrade, L., Janot-Pacheco, E., et al., A&A, **522**, A43 (2010)
[3] Foster, G., ApJ, **109**, 1889 (1995)
[4] Frémat, Y., Zorec, J., Hubert, A.-M., et al., A&A, **440**, 305 (2005)
[5] Neiner, C., Floquet, M., Samadi, R., et al., A&A, **546**, 47 (2012)
[6] Telting, J. H., & Schrijvers, C., A&A, **317**, 723 (1997)

Spectroscopic mode identification of γ Doradus stars

E. Brunsden[1,2,*], K. R. Pollard[2], P. L. Cottrell[2], D. J. Wright[3], and P. De Cat[4]

[1]*Department of Physics, University of York, York, UK*
[2]*Department of Physics and Astronomy, University of Canterbury, Christchurch, New Zealand*
[3]*Department of Astrophysics, University of New South Wales, Sydney, Australia*
[4]*Royal Observatory of Belgium, Brussels, Belgium*

Abstract. The g-mode pulsations in γ Doradus stars are identified using time-series colour photometry and high-resolution spectroscopy. For 22 class members the pulsational frequencies and modes are compared. Ground-based spectroscopic and photometric results show good agreement. The prevalence of $(1, |1|)$ modes is noted and examined.

1 Introduction

The γ Doradus class of stars is of particular interest in asteroseismology. They are defined as A-F type main sequence (or close to main sequence) stars that display pulsational gravity-modes (g-modes) of high radial order. These g-modes probe the deep interior of the stars and thus provide us with a tool to determine the structure and evolutionary changes that occur in regions as deep as the core.

For a successful asteroseismic study, two conditions have to be satisfied: a sufficient number of pulsation modes should be observed and the modes should be well identified. This information feeds into and refines theoretical stellar models as well as revealing new insights into the class. We present a summary of mode identifications found in ground-based spectroscopic and photometric studies of class members.

2 Spectroscopic methods

The MUSICIAN program uses observations that were primarily taken from the 1-metre McLellan telescope at University of Canterbury Mt John Observatory (UCMJO), New Zealand. Spectra were obtained on the High Efficiency and Resolution Canterbury University Large Echelle Spectrograph. Multi-site data were also used where possible. Frequencies were found by a Fourier analysis of the moments of the line profile. A two-dimensional analysis of the line profile was done using the pixel-by-pixel method. Mode identification was done by the Fourier Parameter Fit method in the software package FAMIAS ([1]) which models frequencies, amplitudes and phases of the representative line profile to determine a mode for each frequency. Performing the fit with several frequencies and modes simultaneously gives some basic stellar parameters.

*emily.brunsden@gmail.com

3 Summary of results

A total of 22 stars have been analysed for modes[1], of which two are binary stars and one is a hybrid δ Scuti / γ Doradus star. Nine stars have photometric identifications of l and 16 have full (l, m) identifications. Of the 22 stars, 12 have had full mode identifications as part of the MUSICIAN programme using spectra from UCMJO. Despite the small numbers of classifications, there are enough now known to identify some global features and compare these to other identifications.

In colour photometry 33 modes have been identified, split almost equally between $l = 1$ modes (16) and $l = 2$ modes (17). The spectroscopic identifications (48) have 16 $(1, |1|)$ modes and 9 $l = 2$ modes, which are almost exclusively $(2, -2)$. There is no apparent correlation between the presence of retrograde modes and $v \sin i$ of a star. Stars with low to mid observed rotation speeds ($15\ \mathrm{km\,s^{-1}} < v \sin i < 80\ \mathrm{km\,s^{-1}}$) are preferred for spectroscopic mode identification due to their broadened line profiles but minimum rotational distortion.

Only three stars have both photometric and spectroscopic mode identifications and for two stars these agree for a total of six frequencies, suggesting good agreement between the methods. There is not presently good agreement between ground-based and space-based observations (e.g., [2]), although only a few stars have such observations.

Observations to date suggest a prevalence of $(1, |1|)$ modes in γ Doradus stars ([3–5]). This appears to also extend to colour photometry methods. The 15 $(1, 1)$ modes identified here appear to be clustered with an average of around 1.4 $\mathrm{d^{-1}}$. Retrograde $(1, -1)$ modes are split between higher and lower frequencies. Other modes identified have a higher average of approximately 2.5 $\mathrm{d^{-1}}$ and retrograde modes are all much lower than this. A $(1, 1)$ mode may arise from observational bias due to larger surface deformation. It is also possible the frequency is linked to the proximity of the rotational frequency, constraining the light maxima to once per rotation cycle. This is supported by the frequency distinction. In addition, stars with high inclinations will have higher observed amplitudes for sectoral modes and those with low inclinations, higher-amplitude tesseral and zonal modes ([6]). For rotating stars, the presence of an equatorial wave-guide ([7]) would suppress the tesseral modes and thus only the stars with sectoral modes would be classified as γ Doradus stars.

The analysis of these stars is ongoing as we look for observational patterns in the modes from both ground-based and space-based observations to improve our understanding of these complex stars.

Acknowledgments: This work was supported by the Marsden Fund through the Royal Society of New Zealand. Mode identification results obtained with the software package FAMIAS developed in the framework of the FP6 European Coordination action HELAS (`http://www.helas-eu.org/`).

References

[1] W. Zima, Communications in Asteroseismology **157**, 387 (2008)

[2] E. Brunsden, K.R. Pollard, P.L. Cottrell, K. Uytterhoeven, D.J. Wright, P. De Cat, MNRAS **447**, 2970 (2015)

[3] L.A. Balona, J.A. Guzik, K. Uytterhoeven, J.C. Smith, P. Tenenbaum, J.D. Twicken, MNRAS **415**, 3531 (2011)

[4] E. Brunsden, K.R. Pollard, P.L. Cottrell, D.J. Wright, P. De Cat, MNRAS **427**, 2512 (2012)

[5] T. Van Reeth, A. Tkachenko, C. Aerts, A&A **593**, A120 (2016)

[6] D.R. Reese, V. Prat, C. Barban, C. Van't Veer-Menneret, K.B. MacGregor, A&A **550**, A77 (2013)

[7] R.H.D. Townsend, MNRAS **343**, 125 (2003)

[1] A full table of pulsational frequencies and mode identifications is available on request or can be found in Brunsden et al. (in preparation).

Whiting 1: Confirmation of its accretion by the Milky Way

Julio A. Carballo-Bello[1,2,*], *Jesús M.* Corral-Santana[3,1], *David* Martínez-Delgado[4], *Antonio* Sollima[5], *Ricardo R.* Muñoz[6], *Sonia* Duffau[2,1], and *Márcio* Catelan[1,2]

[1] *Pontificia Universidad Católica de Chile, Instituto de Astrofísica, Av. Vicuña Mackenna 4860, 782-0436 Macul, Santiago, Chile*
[2] *Millennium Institute of Astrophysics, Santiago, Chile*
[3] *European Southern Observatory, Alonso de Córdova 3107, Casilla 19001, Santiago, Chile*
[4] *Astronomisches Rechen-Institut, Zentrum für Astronomie der Universität Heidelberg, Mönchhofstr. 12-14, D-69120 Heidelberg, Germany*
[5] *INAF Osservatorio Astronomico di Bologna, via Ranzani 1, I-40127 Bologna, Italy*
[6] *Departamento de Astronomía, Universidad de Chile, Camino El Observatorio 1515, Las Condes, Santiago, Chile*

Abstract. We investigate the association of Whiting 1 with the Sagittarius tidal stream by obtaining radial velocities for a sample of 101 stars observed with VIMOS. Our results reveal the presence of a component of the Sagittarius tidal stream with a radial velocity – and distance – compatible with that of the globular cluster. Therefore, we conclude that Whiting1 was formed in the interior of the Sagittarius dwarf spheroidal galaxy and later accreted by the Milky Way. In addition, our data also reveal the detection for the first time of an ancient wrap of the Sagittarius tidal stream along the same line-of-sight and at the same heliocentric distance.

1 Introduction

The accretion of massive satellites is believed to have contributed to the formation of a significant fraction of galaxy haloes by transferring a large amount of gas, stars and globular clusters (GC). Compelling evidence for the latter comes from the GC systems of nearby galaxies (e.g. M31), where a large fraction of their globulars seems to align with tidal streams (e.g., [8]). As for the Milky Way, the accretion of GCs in the hierarchical formation scheme has been long proposed because of the existence of at least two distinct subgroups in the Galactic halo (e.g., [6]).

The most pronounced accretion event in the Galaxy is the one corresponding to the Sagittarius (Sgr; [3]) dwarf spheroidal and its associated stellar structure, the so-called Sgr tidal stream (e.g., [4]). Given the spatial extent of that system, it seems likely that some of the Galactic GCs might have formed within Sgr and been later accreted by the Milky Way.

2 Observations and results

In order to confirm the association of Whiting1 with the Sgr tidal stream, unveiled around the cluster by [1], we have observed 101 stars from our photometric catalogs with VLT+VIMOS. We have

* jcarballo@astro.puc.cl

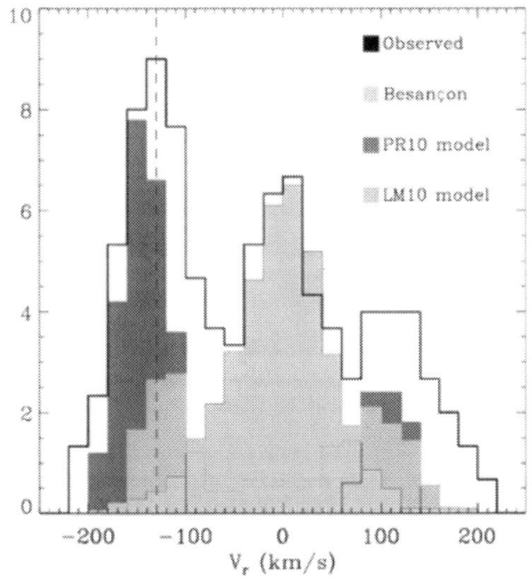

Figure 1. Figure taken from [2]. Radial velocity distribution obtained for the sample of stars around Whiting 1 (black solid line). The grey area indicates the velocity distribution as predicted by the Besançon model for the same line-of-sight, while the blue and yellow areas represent the predicted velocities by [7] and [5] models, respectively. The vertical line indicates the radial velocity of the GC

derived radial velocities for the targets by cross-correlating their spectra with a sample of templates corresponding to all spectral types and the obtained radial velocity distribution is shown in Figure 1. Three different components are clearly observed with approximate radial velocities $v_r \approx -130$, 0 and +150 km s^{-1}. The expected Galactic kinematics for that region of the sky according to the synthetic Milky Way model of Besançon suggests that the central peak is generated by Galaxy stars. On the other hand, two additional components are found and the most prominent, located at negative velocities, is compatible with the radial velocity measured for the globular Whiting 1. Therefore, our results confirm that Whiting1 is coincident in phase-space with the trailing arm of the Sagittarius tidal stream, being the first photometric+spectroscopic evidence of this association. The additional peak with positive radial velocities might correspond to the first detection of the predicted "old" leading arm of Sagittarius along that line-of-sight.

References

[1] Carballo-Bello, Sollima, A., Martínez-Delgado, D., et al., MNRAS, **445**, 2971 (2014)
[2] Carballo-Bello, J.A., Corral-Santana, J. M., Martínez-Delgado, D., Sollima, A., Muñoz, R. R., Duffau, S., Catelan, M., & Grebel, E. K. , MNRAS, **467**, L91 (2017)
[3] Ibata, R. A., Gilmore, G., & Irwin, M. J., Nature, **370**, 194 (1994)
[4] Koposov, S. E., Belokurov, V., Evans, N. W., et al., ApJ, **750**, 80 (2012)
[5] Law, D. R., & Majewski, S. R., ApJ, **714**, 229 (2010)
[6] Leaman, R., VandenBerg, D. A., & Mendel J. T., MNRAS, **436**, 122 (2013)
[7] Peñarrubia, J., Belokurov, V., Evans, N. W., et al., MNRAS, **408**, 26 (2010)
[8] Veljanoski, J., Mackey, A. D., Ferguson, A. M. N., et al., MNRAS, **442**, 2929 (2014)

The impact of Gaia DR1 on asteroseismic inferences from *Kepler*

Travis Metcalfe[1], *Orlagh* Creevey[2], and *Jennifer* van Saders[3]

[1]*Space Science Institute, 4750 Walnut Street, Suite 205, Boulder CO 80301, USA*
[2]*Université Côte d'Azur, Observatoire de la Côte d'Azur, CNRS, Laboratoire Lagrange, France*
[3]*Carnegie Observatories, 813 Santa Barbara Street, Pasadena CA 91101, USA*

Abstract. The *Kepler* mission has been fantastic for asteroseismology of solar-type stars, but the targets are typically quite distant. As a consequence, the reliability of asteroseismic modeling has been limited by the precision of additional constraints from high-resolution spectroscopy and parallax measurements. A precise luminosity is particularly useful to minimize potential biases due to the intrinsic correlation between stellar mass and initial helium abundance. We have applied the latest version of the Asteroseismic Modeling Portal (AMP) to the complete *Kepler* data sets for 30 stars with known rotation rates and chromospheric activity levels. We compare the stellar properties derived with and without the measured parallaxes from the first data release of Gaia. We find that in most cases the masses and ages inferred from asteroseismology shift within their uncertainties. For a few targets that show larger shifts, the updated stellar properties only strengthen previous conclusions about anomalous rotation in middle-aged stars.

1 Motivation

The correlation between stellar mass and initial helium abundance is a long-standing problem in modeling solar-type stars. Increasing either the mass or the initial helium yields a model with a higher luminosity, so we can trade off one parameter for the other while still satisfying the observational constraints ([1]). Asteroseismic observations can reduce the severity of this problem by providing a strong constraint on the stellar radius, but the issue cannot be avoided entirely without a more direct constraint on the stellar luminosity. Without such a constraint, the correlation can lead to systematic biases in the stellar properties inferred from asteroseismology. In this paper, we examine the impact of including luminosity constraints derived from Gaia DR1 parallaxes ([2]) for a sample of 30 *Kepler* targets with known rotation rates ([3]) and chromospheric activity levels ([4]). Our goal is to evaluate the possibility of systematic biases in the asteroseismic masses and ages for stars with anomalous rotation compared to empirical gyrochronology relations ([5]), with implications for a new theory of magnetic evolution beyond middle age ([6]).

Until recently, it was presumed that rotation and magnetism decay together throughout the lives of solar-type stars ([7, 8]). Although stars are formed with a range of initial rotation rates, the stellar winds entrained in their magnetic fields lead to angular momentum loss from magnetic braking. This forces convergence to a single rotation rate at a given mass after roughly 500 Myr. The *Kepler* mission allowed the measurement of rotation periods in old field stars whose masses and ages could be determined from asteroseismology. These new data revealed a population of field stars rotating

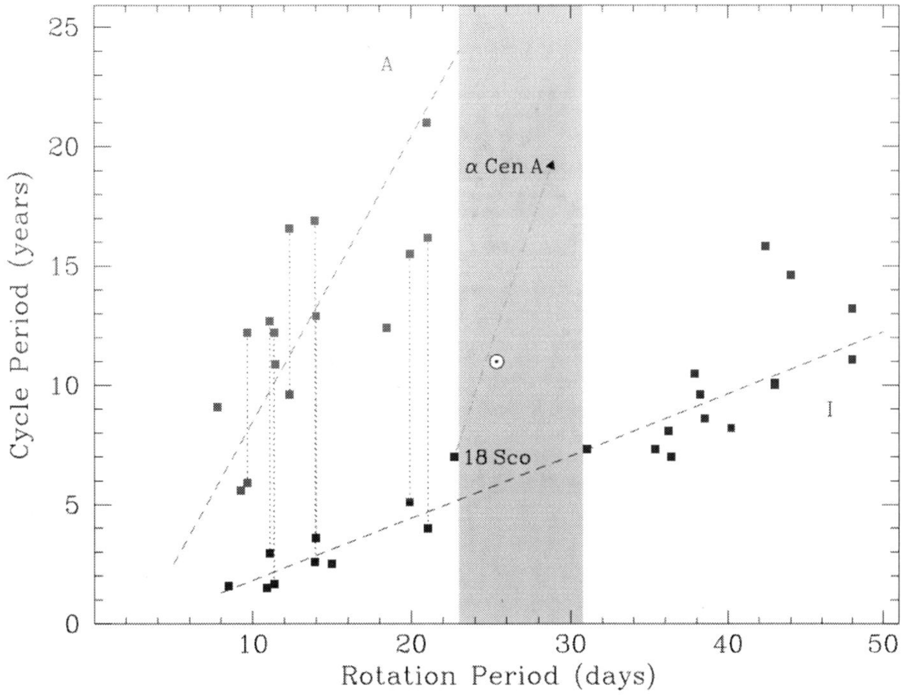

Figure 1. Updated version of a diagram published in [12] using data from [13], showing the active (A) and inactive (I) sequences. Multiple cycles observed in the same star are connected with vertical dotted lines. The shaded region indicates the rotation periods around the Sun (⊙) where Mount Wilson stars do not show cycles.

more quickly than expected from gyrochronology ([9]), suggesting that magnetic braking may operate with a dramatically reduced efficiency beyond a critical Rossby number, Ro ~ 2 ([5]). A magnetic counterpart to this rotational transition was recently identified by [6]. They proposed that a change in the character of differential rotation is the underlying mechanism that ultimately disrupts the large-scale organization of magnetic fields in solar-type stars. The process begins at Ro ~ 1, where many global convection simulations exhibit a transition from solar-like to anti-solar differential rotation ([10]). This change in the character of differential rotation triggers a phase of rapid magnetic evolution leading to a reduction in the efficiency of magnetic braking, probably due to a shift in magnetic topology ([11]).

The new picture of rotational and magnetic evolution provides a framework for understanding some observational features of stellar activity cycles that have until now been mysterious. An updated version of a diagram published in [12] is shown in Figure 1, using data from [13]. More recent data have been added from [14–20]. All of the slower rotators ($P_{\rm rot}$ > 30 days) are K-type stars, which is now understandable – magnetic braking ceases in more massive main-sequence stars before they reach these long rotation periods. The transition across Ro ~ 2 for G-type stars occurs at rotation periods comparable to the Sun ($P_{\rm rot}$ ~ 23–30 days). If we consider the evolutionary sequence defined by 18 Sco, the Sun, and α Cen A, the data suggest that a normal cycle on the inactive sequence may grow longer across this transition (dot-dash line) before disappearing entirely.

DOI: 10.1051/epjconf/201715205001

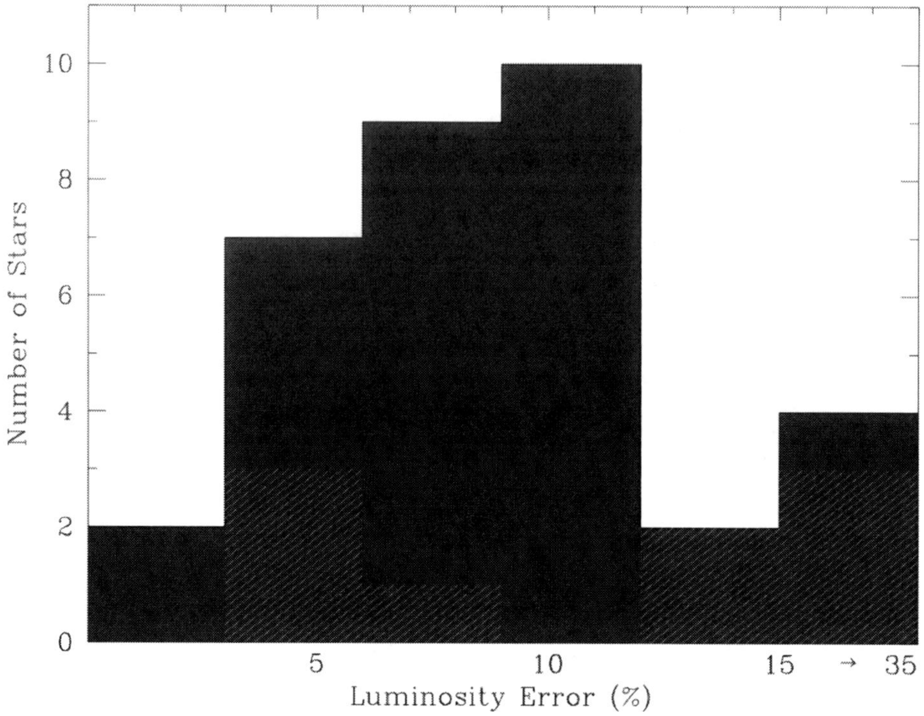

Figure 2. Histogram of the available luminosity constraints for our sample from the Tycho catalog (red) and from Gaia DR1 (blue). All of the stars in our sample now have luminosity constraints, and most have a precision better than 10%. In late 2017, Gaia DR2 promises to improve the precision by nearly an order of magnitude.

2 Asteroseismology with *Kepler*

The *Kepler* space telescope yielded unprecedented data for the study of solar-like oscillations in other stars. Initial indications of anomalous rotation and magnetic activity in old field stars relied on asteroseismic properties from [21], which were based on an analysis of only 9 months of data ([22]). *Kepler* completed its primary mission in 2013, but the large samples of multi-year observations posed an enormous data analysis challenge that has only recently been surmounted ([23–25]). The longer data sets improved the signal-to-noise ratio (S/N) of the power spectrum for fainter stars with previously marginal detections, and yielded additional oscillation frequencies for the brighter targets. This expanded the asteroseismic sample of stars with known rotation rates ([3]) and chromospheric activity levels ([4]), from 21 targets ([26]) to more than 30.

We have used the latest version of the Asteroseismic Modeling Portal (AMP, [27]) to obtain reliable stellar properties from the complete *Kepler* data sets ([28]). For each star in the sample, we repeated the modeling with and without a luminosity constraint derived from Gaia DR1 parallaxes ([2]), allowing us to isolate the impact of this constraint on the inferred stellar properties. Parallaxes from the Tycho catalog were previously available for a few of the targets, but all of the stars in our sample now have luminosity constraints – and most have a precision better than 10% (see Fig. 2).

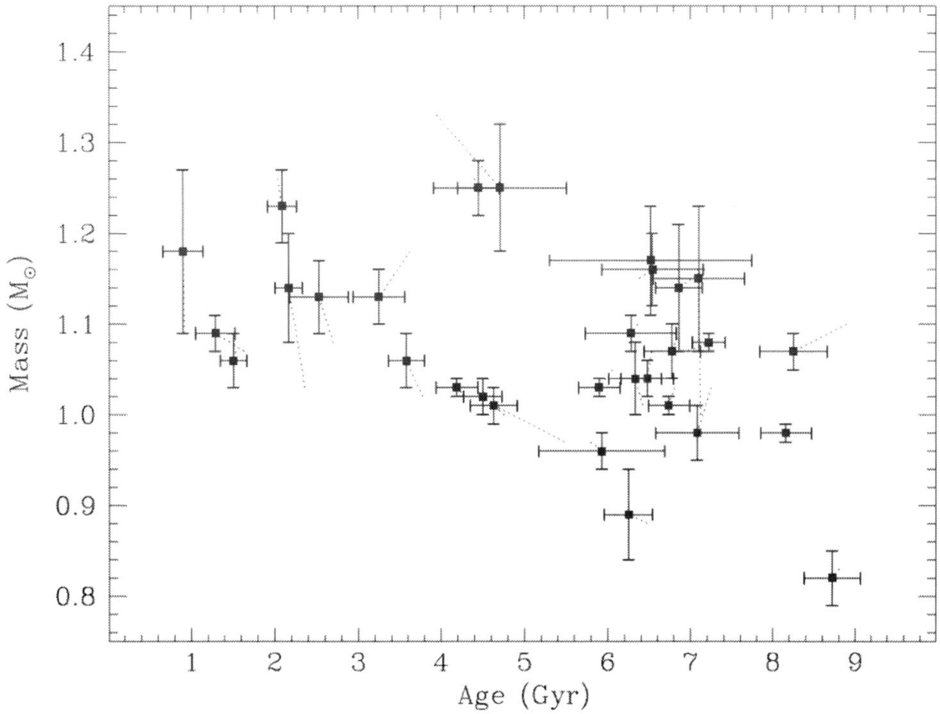

Figure 3. Stellar masses and ages inferred from asteroseismology before and after including luminosity constraints derived from Gaia DR1 parallaxes. Blue points with uncertainties show the values obtained when including a luminosity constraint, and dotted red lines connect to the values inferred without a luminosity constraint.

3 Impact of Gaia parallaxes

Claims of anomalous rotation in old field stars depend most directly on the inferred asteroseismic ages. The masses are also important because magnetic braking depends on the depth of the surface convection zone, and angular momentum evolution ultimately follows changes in the moment of inertia as a star expands into a subgiant. In Figure 3 we illustrate the impact of the available luminosity constraints on the inferred asteroseismic masses and ages. Most of the values shift within their respective uncertainties. For a few targets that show larger shifts, the updated stellar properties only strengthen previous conclusions about anomalous rotation in middle-aged stars.

In late 2017, Gaia DR2 promises to improve the parallax precision for all of our targets by nearly an order of magnitude. This will allow us to eliminate any remaining biases in the inferred stellar properties for our sample. It may also yield the asteroseismic composition with sufficient accuracy to distinguish between competing Galactic enrichment models.

Acknowledgments: This work was supported in part by NASA grants NNX15AF13G and NNX16AB97G, and by White Dwarf Research Corporation through the Non-profit Adopt a Star program. Computational time at the Texas Advanced Computing Center was provided through XSEDE allocation TG-AST090107.

References

[1] Y. Lebreton, M.J. Goupil, A&A, **569**, A21 (2014)
[2] Gaia Collaboration, A.G.A. Brown, A. Vallenari, T. Prusti, J.H.J. de Bruijne, F. Mignard, R. Drimmel, C. Babusiaux, C.A.L. Bailer-Jones, U. Bastian et al., A&A, **595**, A2 (2016)
[3] R.A. García, T. Ceillier, D. Salabert, S. Mathur, J.L. van Saders, M. Pinsonneault, J. Ballot, P.G. Beck, S. Bloemen, T.L. Campante et al., A&A, **572**, A34 (2014)
[4] C. Karoff, T.S. Metcalfe, W.J. Chaplin, S. Frandsen, F. Grundahl, H. Kjeldsen, J. Christensen-Dalsgaard, M.B. Nielsen, S. Frimann, A.O. Thygesen et al., MNRAS, **433**, 3227 (2013)
[5] J.L. van Saders, T. Ceillier, T.S. Metcalfe, V. Silva Aguirre, M.H. Pinsonneault, R.A. García, S. Mathur, G.R. Davies, Nature, **529**, 181 (2016)
[6] T.S. Metcalfe, R. Egeland, J. van Saders, ApJL, **826**, L2 (2016)
[7] A. Skumanich, ApJ, **171**, 565 (1972)
[8] S.A. Barnes, ApJ, **669**, 1167 (2007)
[9] R. Angus, S. Aigrain, D. Foreman-Mackey, A. McQuillan, MNRAS, **450**, 1787 (2015)
[10] T. Gastine, R.K. Yadav, J. Morin, A. Reiners, J. Wicht, MNRAS, **438**, L76 (2014)
[11] V. Réville, A.S. Brun, S.P. Matt, A. Strugarek, R.F. Pinto, ApJ, **798**, 116 (2015)
[12] E. Böhm-Vitense, ApJ, **657**, 486 (2007)
[13] S.H. Saar, A. Brandenburg, ApJ, **524**, 295 (1999)
[14] J.C. Hall, G.W. Henry, G.W. Lockwood, AJ, **133**, 2206 (2007)
[15] L.E. DeWarf, K.M. Datin, E.F. Guinan, ApJ, **722**, 343 (2010)
[16] T.S. Metcalfe, S. Basu, T.J. Henry, D.R. Soderblom, P.G. Judge, M. Knölker, S. Mathur, M. Rempel, ApJL, **723**, L213 (2010)
[17] T.S. Metcalfe, A.P. Buccino, B.P. Brown, S. Mathur, D.R. Soderblom, T.J. Henry, P.J.D. Mauas, R. Petrucci, J.C. Hall, S. Basu, ApJL, **763**, L26 (2013)
[18] T.R. Ayres, AJ, **147**, 59 (2014)
[19] R. Egeland, T.S. Metcalfe, J.C. Hall, G.W. Henry, ApJ, **812**, 12 (2015)
[20] D. Salabert, C. Régulo, R.A. García, P.G. Beck, J. Ballot, O.L. Creevey, F. Pérez Hernández, J.D. do Nascimento, Jr., E. Corsaro, R. Egeland et al., A&A, **589**, A118 (2016)
[21] T.S. Metcalfe, O.L. Creevey, G. Doğan, S. Mathur, H. Xu, T.R. Bedding, W.J. Chaplin, J. Christensen-Dalsgaard, C. Karoff, R. Trampedach et al., ApJS, **214**, 27 (2014)
[22] T. Appourchaux, W.J. Chaplin, R.A. García, M. Gruberbauer, G.A. Verner, H.M. Antia, O. Benomar, T.L. Campante, G.R. Davies, S. Deheuvels et al., A&A, **543**, A54 (2012)
[23] G.R. Davies, W.J. Chaplin, W.M. Farr, R.A. García, M.N. Lund, S. Mathis, T.S. Metcalfe, T. Appourchaux, S. Basu, O. Benomar et al., MNRAS, **446**, 2959 (2015)
[24] G.R. Davies, V. Silva Aguirre, T.R. Bedding, R. Handberg, M.N. Lund, W.J. Chaplin, D. Huber, T.R. White, O. Benomar, S. Hekker et al., MNRAS, **456**, 2183 (2016)
[25] M.N. Lund, V. Silva Aguirre, G.R. Davies, W.J. Chaplin, J. Christensen-Dalsgaard, G. Houdek, T.R. White, T.R. Bedding, W.H. Ball, D. Huber et al., ApJ, **835**, 172 (2017)
[26] T. Ceillier, J. van Saders, R.A. García, T.S. Metcalfe, O. Creevey, S. Mathis, S. Mathur, M.H. Pinsonneault, D. Salabert, J. Tayar, MNRAS, **456**, 119 (2016)
[27] T.S. Metcalfe, O.L. Creevey, J. Christensen-Dalsgaard, ApJ, **699**, 373 (2009)
[28] O.L. Creevey, T.S. Metcalfe, M. Schultheis, D. Salabert, M. Bazot, F. Thévenin, S. Mathur, H. Xu, R.A. García, A&A, **601**, A67 (2017), 1612.08990

The γ Dor stars as revealed by *Kepler*: A key to reveal deep-layer rotation in A and F stars

S. J. A. J. Salmon[1,*], *R.-M.* Ouazzani[2], *V.* Antoci[2], *T. R.* Bedding[3], and *S. J.* Murphy[3,2]

[1] *STAR Institute, Université de Liège, Allée du Six Août 19C, B-4000 Liège, Belgium*
[2] *Stellar Astrophysics Centre, Department of Physics and Astronomy, Aarhus University, Ny Munkegade 120, DK-8000 Aarhus C, Denmark*
[3] *Sydney Institute for Astronomy (SIfA), School of Physics, University of Sydney, Sydney, NSW 2006, Australia*

Abstract. The γ Dor pulsating stars present high-order gravity modes, which make them important targets in the intermediate- and low-mass main-sequence region of the Hertzsprung-Russell diagram. Whilst we have only access to rotation in the envelope of the Sun, the g modes of γ Dor stars can in principle deliver us constraints on the inner layers. With the puzzling discovery of unexpectedly low rotation rates in the core of red giants, the γ Dor stars appear now as unique targets to explore internal angular momentum transport in the progenitors of red giants. Yet, the γ Dor pulsations remain hard to detect from the ground for their periods are close to 1 day. While the CoRoT space mission first revealed intriguing frequency spectra, the almost uninterrupted 4-year photometry from the *Kepler* mission eventually shed a new light on them. It revealed regularities in the spectra, expected to bear signature of physical processes, including rotation, in the shear layers close to the convective core. We present here the first results of our effort to derive exploitable seismic diagnosis for mid- to fast rotators among γ Dor stars. We confirm their potential to explore the rotation history of this early phase of stellar evolution.

1 Introduction

In the last decade, asteroseismology has gathered in a wealth of results thanks, among others, to the CoRoT and *Kepler* space missions. In particular, with the detection of pulsations in large samples of red giant stars, important progress has been made in our exploration of this more advanced phase of stellar evolution (e.g., [2, 3, 21, 22]). The asteroseismic determination of the core rotation rate in red giants ([4, 10, 13, 23]) has revealed values significantly lower than theoretical predictions (e.g., [12, 30]). It has pinpointed the need of an efficient mechanism to extract angular momentum out of the core layers ([9, 14, 15, 20]). Several candidates, as e.g. angular momentum transport by internal gravity waves ([16]) or mixed modes ([5]), were suggested but appear insufficient to explain the low contrast of rotation between core and surface observed in red giants. These new results push for exploring the preceding phases of the evolution to bring additional constraints on the unknown mechanism.

*sebastien.salmon@ulg.ac.be

Among stars evolving to the giant branch, solar analogues would be the natural targets to explore the internal rotation during the main sequence. The inversion of the solar rotation profile is indeed one of the main achievements of helioseismology (e.g., [8, 18]). But the nature of its oscillations, acoustic modes, hampers the knowledge of the rotation profile near the core. Inferring the internal rotation in other pulsating solar-like stars ([6, 25]) deals with the same limitation. Here we explore the pulsations of slightly more massive, A- and F-type, stars known as γ Dor stars. Their gravity modes of pulsation propagate in the deepest layers of the star, and bear a signature of the rotation in these layers ([7]). We present and summarize in the next section the first results obtained on the internal rotation in γ Dor stars observed by *Kepler*. We then conclude on how further investigation of these stars may pave the way to retrieve the evolution of rotation during the main sequence of intermediate- and low-mass stars.

2 Revealing the internal rotation of γ Dor stars

Remarkably, with the almost 4 years of uninterrupted data gathered by the *Kepler* satellite, the resolution has now become sufficient to unravel the individual modes in the time-series analysis of γ Dor stars. The first observational constraints on the internal rotation profile were obtained for γ Dor-δ Scuti hybrid stars ([17, 19, 24, 28, 29]. All of them are slow rotators with projected surface velocities of about 10 km/s. Thanks to the detection of rotational splitting in both their g (γ Dor) and p mode (δ Scuti) regimes, these authors unveiled that they present all a nearly rigid rotation profile. The values of their deep-layer rotation frequencies are reported in Figure 1, along with the core rotation from samples of subgiants ([11]) and red giants ([22]). This first sample of main-sequence objects presents rather low values of rotation frequency in their deep layers but show us it is possible to complete the picture of the rotation history from the main sequence to the helium-burning phase of intermediate-mass stars.

For γ Dor rotating faster, [7] illustrated theoretically the importance of analysing the period spacing of g modes, on which rotation imprints its signatures under the form of regular patterns in the spacings. Recently, such features were indeed detected in the period spectra of *Kepler* stars ([1, 31]). Motivated by these findings, we have developed in [27] practical tools to infer the rotation in the deep layers (close to the convective core) from the observed period spacings. In particular, we have demonstrated the existence of a relation between the slope of the linear trend in the period spacings and the rotation frequency, almost independent of the metallicity or the mass of the pulsators.

Exploiting this relation resulted in the determination of the deep-layer rotation frequencies of 4 stars, represented in Figure 1. Based on a similar approach, [32] determined rotation frequencies for another sample of *Kepler* γ Dor stars, also depicted in Figure 1. Both studies reveals the existence of fast rotation in central layers of intermediate-mass stars, which was absent from the first results from the study of hybrid γ Dor-δ Scuti stars.

From a first and qualitative comparison (P. Eggenberger, private communication), the values we found seem compatible with prediction from stellar models including rotation. The next step is now to refine the estimated rotation frequencies with detailed modelling of the observed stars, and proper comparison with theoretical predictions.

3 Conclusion

With the unprecedented quality and length of the *Kepler* data, we have now access to the rotation of the deep layers in γ Dor stars. These intermediate- and low-mass stars are fundamental to trace back the rotation history during the main sequence. As a part to the puzzling question of the spin down of

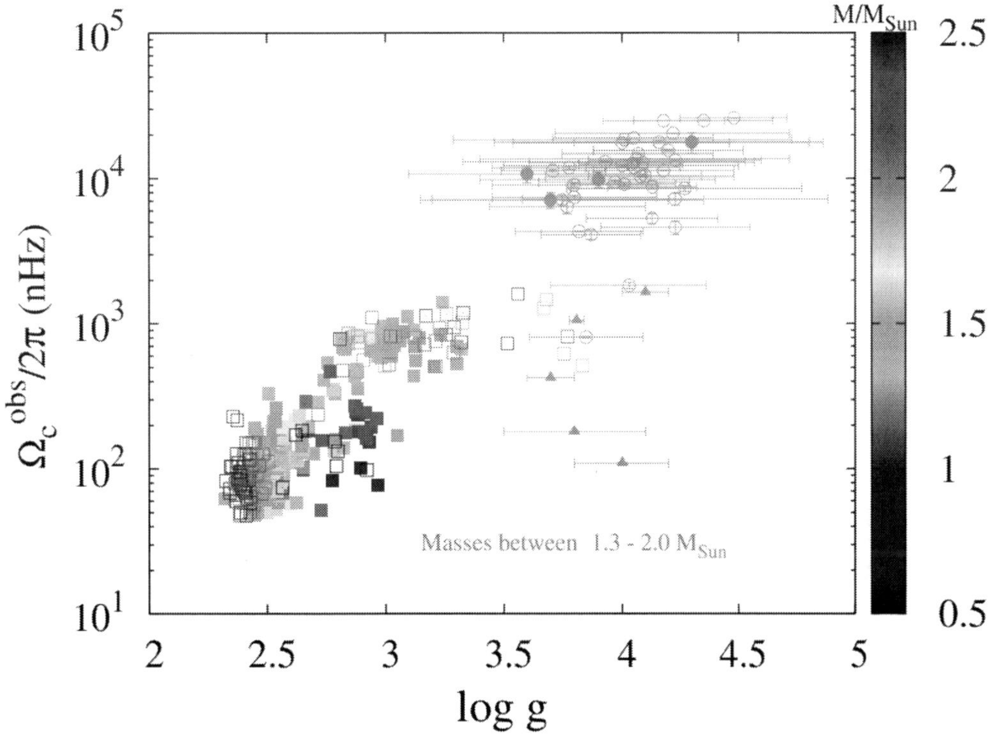

Figure 1. Seismicly determined core rotation frequency as a function of the surface gravity of red giants and γ Dor stars observed by *Kepler*. Color-coded filled squares (1.3 – 2.5 M$_\odot$) and empty squares (< 1.3 M$_\odot$) represent the subgiants and red giants analyzed in [11] and [22]. Grey triangles show the slowly rotating hybrid γ Dor-δ Scuti stars from [17, 19, 24, 28, 29]. Results from the γ Dor studied by [32] and [27] are represented by grey open and filled circles, respectively. Masses of stars in grey are expected to be between ~ 1.3 – 2 M$_\odot$.

core rotation in red giants, we have access to an earlier phase of evolution to determine then what are the angular momentum transport mechanisms at work.

To get firm constraints, we need better determination of the fundamental parameters, such as the surface gravity, of the γ Dor stars observed by *Kepler*. A better precision on these parameters is first necessary to evaluate their evolutionary stage on the main sequence. An ideal objective would be to build eventually the complete evolution of the rotation profile during the core hydrogen burning phase. Secondly, detailed modelling of the stars is requested to get precise profile of rotation with help of rotational kernels from a pulsation code including a complete treatment of the rotation effects (e.g., [26]). However, such a code requests an intense computing effort. Thus it is crucial to reduce the parameter space of the stellar models used for the modelling of observed pulsators.

Finally, to have a global and unbiased picture of the angular momentum transport mechanisms at work during the main sequence, we have to extend the sample of hybrid or γ Dor stars from a thorough analysis of the *Kepler* archives. We will then be able to make comprehensive comparison with stellar evolutionary codes and help determining whether their physical description of the rotational processes are adequate.

References

[1] Bedding, T. R., in *The Space Photometry Revolution – CoRoT Symposium 3, Kepler KASC-7 Joint Meeting*, ed. R. A. García & J. Ballot, EPJWC, **101**, 01005 (2015)
[2] Bedding, T. R., Mosser, B., Huber, D., et al., Nature, **471**, 608 (2011)
[3] Beck, P. G., Bedding, T. R., Mosser, B., et al., Science, **332**, 205 (2011)
[4] Beck, P., Montalbán, J., Kallinger, T., et al., Nature, **481**, 55 (2012)
[5] Belkacem, K., Marques, J. P., Goupil, M. J., et al., A&A, **579**, A31 (2015)
[6] Benomar, O., Takata, M., Shibahashi, H., Ceillier, T., & García, R. A., MNRAS, **452**, 2654 (2015)
[7] Bouabid, M. P., Dupret, M. A., Salmon, S., Montalbán, J., Miglio, A., & Noels, A., MNRAS, **429**, 2500 (2013)
[8] Brown, T. M., Christensen-Dalsgaard, J., Dziembowski, W. A., Goode, P., Gough, D. O., & Morrow, C. A., ApJ, **343**, 526 (1989)
[9] Ceillier, T., Eggenberger, P., García, R. A., & Mathis, S., A&A, **555**, A54 (2013)
[10] Deheuvels, S., Ballot, J., Beck, P., Mosser, B., Ostensen, R., García, R. A., & Goupil, M. J., A&A, **580**, A96 (2015)
[11] Deheuvels, S., Dogan, G., Goupil, M. J., et al., A&A, **564**, A27 (2014)
[12] Denissenkov, P. A., & Tout, C. A., MNRAS, **316**, 395 (2011)
[13] Di Mauro, M. P., Ventura, R., Cardini, D., et al., ApJ, **817**, 65 (2016)
[14] Eggenberger, P., Lagarde, N., Miglio, A., et al., A&A, **599**, 18 (2017)
[15] Eggenberger, P., Montalbán, J., & Miglio, A., A&A, **544**, L4 (2012)
[16] Fuller, J., Lecoanet, D., Cantiello, M., & Brown, B., ApJ, **796**, 17 (2014)
[17] Keen, M. A., Bedding, T. R., Murphy, S. J., Schmid, V. S., Aerts, C., Tkachenko, A., Ouazzani, R. M., & Kurtz, D. W., MNRAS, **454**, 1792 (2015)
[18] Kosovichev, A. G., Schou, J., Scherrer, P. H., et al., SoPh, **170**, 43 (1997)
[19] Kurtz, D. W., Saio, H., Takata, M., Shibahashi, H., Murphy, S. J., & Sekii, T., MNRAS, **444**, 102 (2014)
[20] Marques, J. P., Goupil, M. J., Lebreton, Y., et al., A&A, **549**, A74 (2013)
[21] Mosser, B., Barban, C., Montalbán, J., et al., A&A, **532**, A86 (2011)
[22] Mosser, B., Goupil, M. J., Belkacem, K., et al., A&A, **540**, A143 (2012)
[23] Mosser, B., Goupil, M. J., Belkacem, K., et al., A&A, **548**, A10 (2012)
[24] Murphy, S. J., Fossati, L., Bedding, T. R., Saio, H., Kurtz, D. W., Grassitelli, L., & Wang, E. S., MNRAS, **459**, 1201 (2016)
[25] Nielsen, M. B., Schunker, H., Gizon, L., & Ball, W. H., A&A, **582**, 10 (2015)
[26] Ouazzani, R. M., Dupret, M. A., & Reese, D. R., A&A, **547**, 75 (2012)
[27] Ouazzani, R. M., Salmon, S. J. A. J., Antoci, V., Bedding, T. R., Murphy, S. J., & Roxburgh, I. W., MNRAS, **465**, 2294 (2017)
[28] Saio, H., Kurtz, D. W., Takata, M., Shibahashi, H., Murphy, S. J., Sekii, T., & Bedding, T. R., MNRAS, **447**, 3264 (2015)
[29] Schmid, V. S., Tkachenko, A., Aerts, C., et al., A&A, **584**, A35 (2015)
[30] Sills, A., & Pinsonneault, M. H., ApJ, **540**, 489 (2011)
[31] Van Reeth, T., Tkachenko, A., Aerts, C., et al., ApJS, **218**, 27 (2015)
[32] Van Reeth, T., Tkachenko, A., & Aerts, C., A&A, **593**, 120 (2016)

Pulsating star research from Antarctica

Merieme Chadid[1,*]

[1]*Université Nice Sophia–Antipolis, Observatoire Côte d'Azur, UMR 7293, Parc Valrose, F-06108, Nice Cedex 02, France*

Abstract. This invited talk discusses the pulsating star research from the heart of Antarctica and the scientific polar challenges in the extreme environment of Antarctica, and how the new polar technology could cope with unresolved stellar pulsation enigmas and evolutionary properties challenges towards an understanding of the mysteries of the Universe.
PAIX, the first robotic photometer Antarctica program, has been successfully launched during the polar night 2007. This ongoing program gives a new insight to cope with unresolved stellar enigmas and stellar oscillation challenges with a great opportunity to benefit from an access to the best astronomical site on Earth, Dome C. PAIX achieves astrophysical measurement time-series of stellar fields, challenging photometry from space. A continuous and an uninterrupted series of multi-color photometric observations has been collected each polar night – 150 days – without regular interruption, Earth's rotation effect. PAIX shows the first light curve from Antarctica and first step for the astronomy in Antarctica giving new insights in remote polar observing runs and robotic instruments towards a new technology.

1 Introduction

Stellar pulsations and asteroseismogy are currently among the fundamental techniques to improve our understanding of the internal structure of stars and the hydrodynamics of their atmosphere. On the observational side, progress is limited by the data accuracy needed to detect numerous modes of oscillations with small amplitudes and by the discontinuous nature of typical ground–based data strings which often introduce ambiguities in the determination of oscillation frequencies. Space missions such as MOST ([12]), CoRoT ([1]) and *Kepler* ([2]) enable to overcome both difficulties, and indeed have considerably enhanced the scope of pulsation and asteroseismology methods. However, the outcome of the space missions on the stellar oscillation fields shows large gaps in terms of the flexibility during the observing runs, the choice of targets, the repair of failures and the inexorable high costs. Now the time has come to implement a new way to study the stellar oscillations with long uninterrupted and continuous observations over 150 days from the ground. PAIX – Photometer AntarctIca eXtinction – is a polar program ([6]) made of low cost commercial components, and achieves astrophysical measurement time-series of stellar fields, challenging photometry from space. and even has more advantages than the space telescopes in observing in $UBVRI$ bands photometric colors and then collecting simultaneously multicolor light curves of several targets within the same 12.4×8.3 arcmin

*chadid@unice.fr

field of view. PAIX has been "antarctized" to run under extreme weather conditions with temperatures as low as −80°C, and has been robotized, designed and built by PAIXTeam.

In what follows we summarize a brief review of the PAIX polar mission, the concepts and polar-craft. We discuss the technical and financial gaps between space telescopes and PAIX Polar Mission and then we present the first PAIX outcomes on the stellar pulsation fields.

2 PAIX first robotic Antarctic photometer

Gaps in the time variability of stellar brightness or radial velocity are a major problem in pulsating and oscillating study. It is mandatory to measure the frequency spectrum with the narrowest bandwidth $\Delta\nu$ possible, which means that the observing time T must be as long as possible following the relationship $\Delta\nu = 1/T$. In classical ground observatories, stars are observed only during few consecutive weeks, so a low duty cycle, which is the percentages of time observed over the span of the data set, brings confusions in determining frequencies caused by the day–night interruptions such as one–day aliases appearing around $\nu = 11.6\,\mu$Hz.

2.1 Site testing

Several site testing studies from the heart of Antarctica has been done and show that the best polar astronomical site of observation is the Antarctic plateau Dome C (DMS $75^{deg}\,6^{mn}\,0^{s}$ S, $123^{deg}\,21^{mn}\,0^{s}$ E, Decimal −75.1, 123.35). A comprehensive set of site testing experiments has been deployed on the Dome C base, dedicated to astronomical characterization of the optical turbulence and atmosphere transparency such as Differential Image Motion Monitor ([14]), Single Star Scidar ([13]), Scintillation Detection and Ranging ([15]), and PAIX ([3]). These deep investigations showed that Dome C benefits from exceptional astronomical conditions within a large wavelength range, from blue remote millimeter. Several factors allow such good astronomical conditions such as the high altitude 3300 m a.s.l., the extremely low temperature oscillating between −50 and −83° Celsius, the low cloudiness and air-glow emission, the extremely low light pollution and the very pure air. We clearly demonstrated that most of the optical turbulence is concentrated within the first 30 m of the atmosphere ([10]), and the rest of the atmosphere is very quiet with a seeing of about 0.3 – 0.4 arcsec, as early noticed by [11]. Dome C lies within the best astronomical places in the world.

We show in Figure 1 that the sky is dark when the Sun elevation is less than −8 and not −18 degrees as generally accepted in classical observatories. With such threshold, stars are tracked without any interruption during several months, and the overall observing duration is 135 days. At classical observatories there is no way to achieve continuous and uninterrupted observation over 24 hours.

2.2 How does polar pulsation research happen

PAIX, the first Robotic Antarctica Photometer, launched in 2007 at Dome C, is made of low cost commercial components, and achieves astrophysical measurement multi-color time-series of stellar pulsation fields. The PAIX CCD camera is a SBIG ST10-XME connected to a filter wheel, attached to an Astro-Physics AP3600 equatorial mount which supports a 0.4 m Ritchey-Chretien telescope (Fig. 2). All telescope systems, motors, encoders, main CCD camera and CCD finder are under a thermal stabilization control. Two remote cameras are placed each side of the telescope, as well as a switchable light, in order to check for whole instrument safeness during the dark winter.

A remote command of the whole instrument, Paix Automatic Control Software (PACS, Fig. 4) allows, from any area in the world equipped with an internet link, to control the status and to pilot the

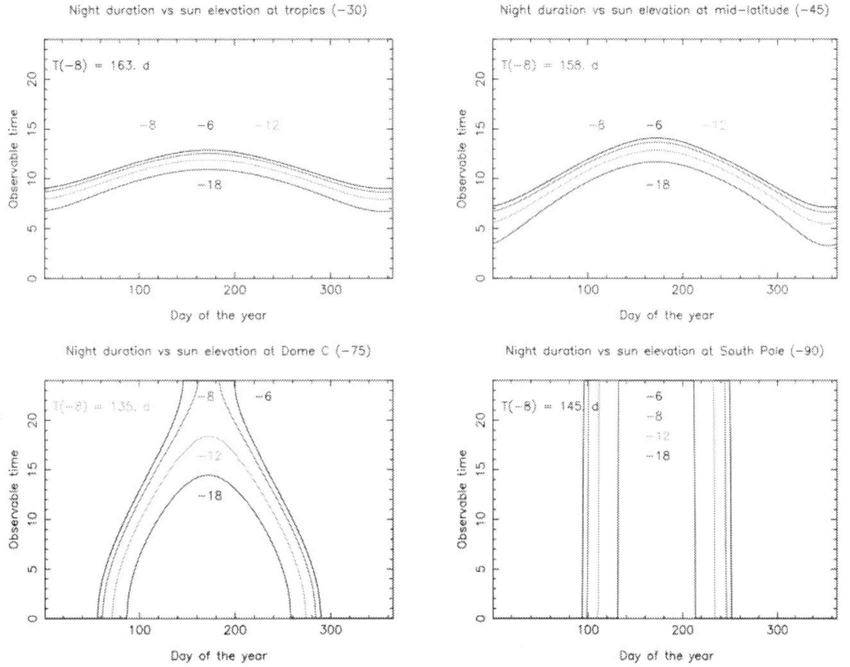

Figure 1. Night duration as a function of the Sun threshold, −6, −8, −12 and −18 degrees below horizon, and depending upon the latitude λ = −30, −45, −75 and −90 degrees. Of course, this night duration is equal to the star observing time if the star is circum-polar and is the time duration where observation is possible when the Sun is −8 degrees below horizon.

Figure 2. PAIX at Dome C during a quick check before starting the polar observing run

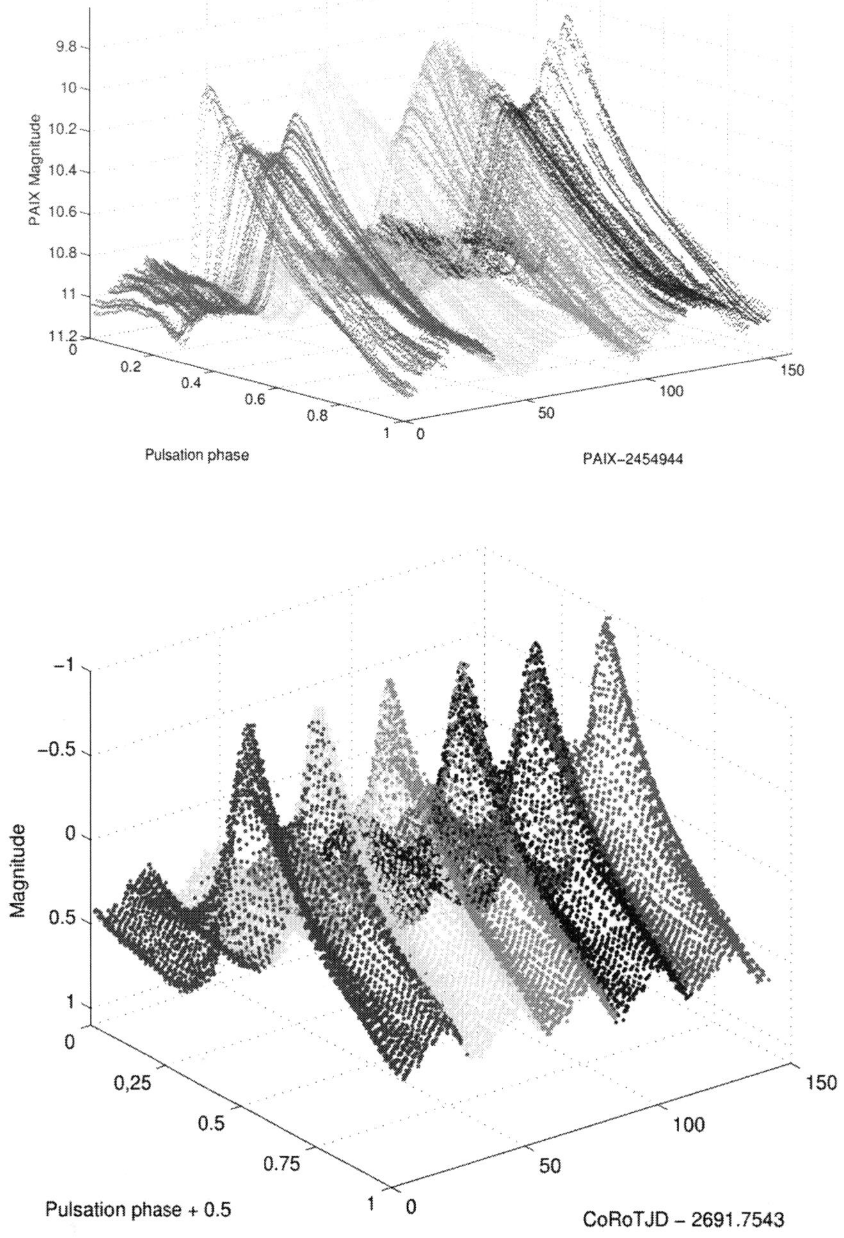

Figure 3. Top: two–dimensional PAIX light curve of the RR Lyrae star S Arae folded with the pulsation period (0.452 d) over 3 Blazhko cycles, 150 days, and showing a strong Blazhko strength. Bottom: two–dimensional CoRoT light curve of RR Lyrae star V1127 Aql folded with the pulsation period (0.356 d) over five Blazhko cycles, 150 days. The variations in the light curve due to the Blazhko effect are clearly visible.

Figure 4. PACS Control Window. All PAIX components and devices are visible and remotely ordered.

instrument. The PAIX Pipeline package – PP – provides work flows for processing PAIX observed photometric data. The PAIX pipeline is dedicated to reduce the raw images and to measure both magnitude and extinction extractions towards light-curves and Bouger's lines simultaneously ([7]). This completely makes PAIX a robotic telescope, and an automatic experiment towards an executing of the polar night observing runs without any local human intervention. Thus, this new experience is a great step forward for the stellar oscillation in Antarctica and gives, for the first time, new insights in remote polar observing runs and robotic polar instruments towards new polar technology developments.

3 Scientific impact and results

3.1 Star–hunting Antarctica

The fields of stellar evolution and pulsation, have been flourished during the last few years. The new generation of largest optical and infrared ground–based telescopes with 8-10 m class, VLT, Keck, LBT and Subaru, are providing major constraints on the pulsation and evolutionary properties, the surface abundances, the internal stellar structure and the lifetimes of stellar populations in the nearby Universe. These observational data have been thoroughly complemented with high resolution images and time-series photometric data from space telescopes, *HST*, *Chandra*, *Spitzer*, MOST, CoRoT and *Kepler*. Despite this incredible observational progress, we are facing big challenges and many old and new astrophysical issues for which we still lack a robust explanation. The new circumpolar technology PAIX opens wide eyes to the Universe in helping to deal with challenges in the stellar pulsation and evolution, specially (1) the period search, search for secondary modulations and their origin, and mysterious temporal hydrodynamical phenomena in RR Lyrae stars which are still not understood a century after their discovery, with large extension to several type of pulsating stars across the Hertzsprung-Russell diagram, and (2) the time-series multi-color and wide-field photometric observations of globular clusters provide the opportunity to investigate the cosmic distance scale and to study the evolutionary properties of stellar populations.

 DOI: 10.1051/epjconf/201715205003

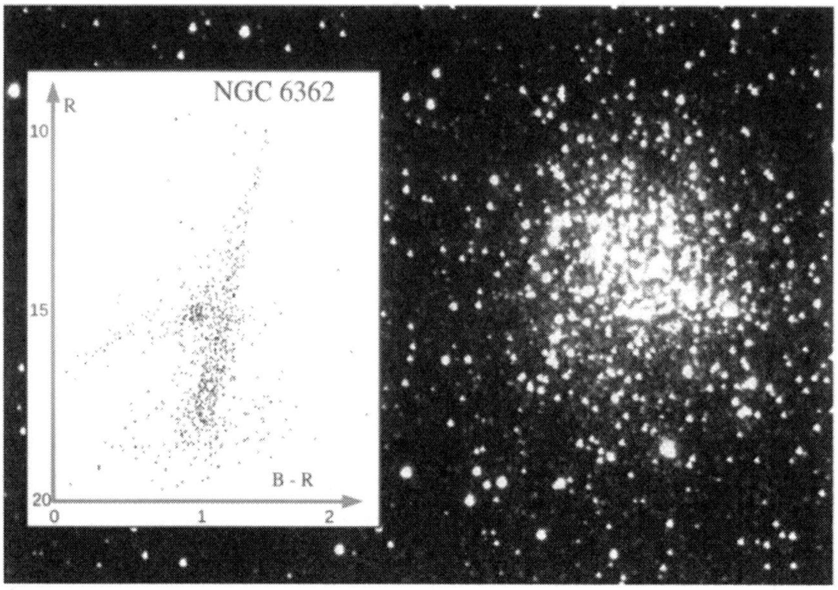

Figure 5. Globular cluster NGC 6362 and color-magnitude diagram of the observed fields based on simultaneous multi color time-series data from PAIX.

3.2 Large photometric survey

The PAIX polar mission is a great opportunity for monitoring stars with excellent time-sampling and unprecedented multicolor photometric precision for up to 150 days. As an important benefit, high-quality RR Lyrae light curves are obtained with a quasi-uninterrupted coverage over several pulsation and Blazhko cycles. Due to the proper data sampling, our first analysis shows new results on the frequency content, the amplitude behavior, new light curve properties, physical characteristics derived from Fourier parameters, namely metal abundances, mass, temperature, surface gravities and distances. The details and strategies of the frequency analysis studies can be found in [3] and [6]. Furthermore, we described the high-temperature hydrodynamic phenomena challenges on a higher level revealing more details to constrain the theoretical models. As shown in Figure 6, we demonstrated that the atmosphere of RR Lyrae stars is crossed by a multiple shock structure intimately connected to the Blazhko modulation. Figure 7 shows a systematic change of the shock intensity with the period of the Blazhko effect. We suggested then that the Blazhko enigma in RRab stars may be explained as a dynamical coupling interplay between a multi–shock structure and an outflowing wind in a coronal structure ([6]). On the other hand, monitoring the globular clusters without the regular interruptions imposed by the earth rotation and with long-term multicolor photometric observations is an outstanding opportunity of determining accurate ages and distances of globular clusters. In the framework of an ongoing high resolution spectroscopic survey aimed at to study southern RR Lyrae stars ([5, 9]), [8] performed a spectroscopic comparison of metal-rich and metal-poor RR Lyrae stars and showed that the period distribution is different between the metal-rich Galactic field and globular cluster RR Lyrae stars. The reason of the different period distributions of these two metal-rich populations is still unknown. NGC 6362 was selected as one of the targets of our ongoing polar photometric survey aimed at identification and pulsating star study towards an understanding of the evolutionary issues of stellar population in the nearby Universe. The field of the southern hemisphere globular cluster NGC 6362

DOI: 10.1051/epjconf/201715205003

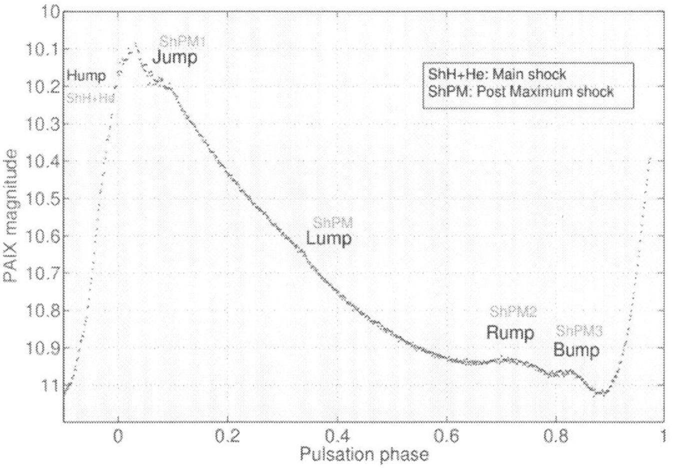

Figure 6. PAIX light curve of S Arae showing the hump, jump, lump, rump, and bump induced by shock waves.

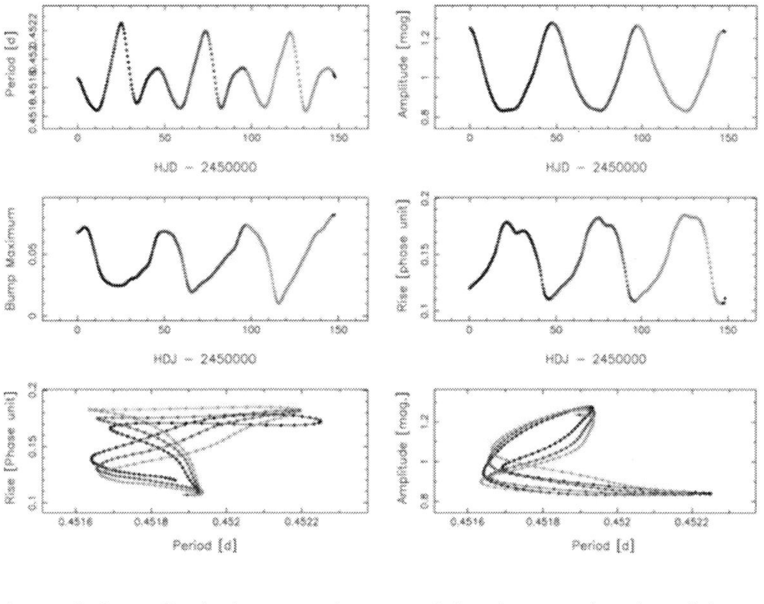

Figure 7. Pulsation period, amplitude, bump maximum, and rise time as a function of time. The rise-period diagram and the amplitude-period diagram are shown at the first (black), second (red), and third (green) Blazhko cycles of S Arae.

has been monitored over one polar night, 150 days, simultaneous uninterrupted *UBVRI*-band photometric data were collected and Fourier decomposition of the light curves is used to estimate the basic properties of cluster variables. Figure 5 presents a color-magnitude diagram (R magnitudes versus $B - V$ colors) from our ongoing NGC 6362 data analysis.

4 Science for less money

PAIX results demonstrate that Antarctic observations are scientifically and financially competitive with space missions. In fact, in some area, Antarctic Sciences compete with spatial missions notwithstanding development costs which are not comparable. A brief comparison between CoRoT ([3]) and PAIX ([6]) shows that PAIX scientifically challenges the CoRoT space mission (Fig. 3) and even exceeds the space mission with more advantages in multi-color time-series performances. The total PAIX cost was around 150 k€, including various costs related to the facilities in terms of Dome C base building, maintenance and polar expeditions. In comparison, the CoRoT satellite mission cost 50 M€, without including the related price of Baïkonour Cosmodrome base, ESA and CNES agencies and Thales Aliena company. Besides the financial side, technical gaps arise between space and polar missions as evidenced by the fact that the polar missions show a large flexibility during the observing runs in choosing the targets, in changing the observation set-up and the repair of any failures that happen during the observing runs.

Acknowledgments: I warmly thank Hiromoto Shibahashi for having given this invited talk on behalf of me during my endurance in Antarctica. I would also to acknowledge Márcio Catelan and Wolfgang Gieren for the invitation to give this talk and the considerable effort required to organize the San Pedro LASPCS 2016 meeting.

References

[1] Baglin, A., Auvergne, M., Catala, C., Michel, E., Goupil, M. J., Samadi, R., Popielsky, B., & COROT Team, in IAU Colloq. 185: *Radial & Nonradial Pulsations as Probes of Stellar Physics*, ed. C. Aerts, T. R. Bedding, & J. Christensen- Dalsgaard, Astronomical Society of the Pacific Conference Series, **259**, 626 (2002)

[2] Borucki, W. J., Koch, D. G., Lissauer, J., et al., in *Transiting Extrapolar Planets Workshop*, ed. C. Afonso, D. Weldrake, & T. Henning, Astronomical Society of the Pacific Conference Series, **366**, 309 (2007)

[3] Chadid, M., Vernin, J., Trinquet, H., & Bono, G., A&A, **516**, L15 (2010)

[4] Chadid, M., et al., A&A, **510**, 39 (2010)

[5] Chadid, M., & Preston, G. W., MNRAS, **434**, 552 (2013)

[6] Chadid, M., Vernin, J., Preston, G., Zalian, C., Pouzenc, C., Aristidi, E., Liu, L. Y., & Trinquet, H., AJ, **148**, 88 (2014)

[7] Chadid, M., Vernin, J., Abe, L., et al., Proceedings SPIE, Volume **9908** (2016)

[8] Chadid, M., Sneden, C., & Preston, G. W., ApJ, **835**, 187 (2017)

[9] For, B.-Q., Preston, G. W., & Sneden, C., ApJS, **194**, 38 (2011)

[10] Giordano, C., Vernin, J., Chadid, M., Aristidi, E., Agabi, A., & Trinquet, H., PASP, **124**, 494 (2012)

[11] Lawrence, J. S., Ashley, M. C. B., Tokovinin, A., & Travouillon, T., Nature, **431**, 278 (2004)

[12] Matthews, J., J. of the Roy. Astron. Soc. of Canada, **92**, 314 (1998)

[13] Vernin, J., Chadid, M., Aristidi, E., Agabi, A., Trinquet, H., & van der Swaelmen, M., A&A, **500**, 1271 (2009)

[14] Vernin, J., Chadid, M., Aristidi, E., Azouit, M., Sadibekova, T., & Trinquet, H., Highlights of Astronomy, **14**, 695 (2007)

[15] Trinquet, H., Agabi, A., Vernin, J., Azouit, M., Aristidi, E., & Fossat, E., PASP, **120**, 203 (2008)

Asteroseismology of the *Kepler* target KIC 9204718

C. Ulusoy[1,2,⋆], I. Stateva[3], I. Kh. Iliev[3], B.Ulaş[4], and M. Napetova[3]

[1]*School of Aviation, Girne American University, University Drive, P.O. Box 5, 99428 Karmi Campus, Karaoğlanoğlu, Kyrenia, Turkey*
[2]*College of Graduate Studies, University of South Africa, PO Box 392, Unisa, 0003 Pretoria, South Africa*
[3]*Institute of Astronomy with NAO, Bulgarian Academy of Sciences, blvd.Tsarigradsko chaussee 72, Sofia 1784, Bulgaria*
[4]*Izmir Turk College Planetarium, 8019/21 sok., No: 22, İzmir, Turkey*

Abstract. The high precision data obtained by the *Kepler* satellite allows us to detect hybrid-type pulsator candidates more accurately than the data obtained by ground–based observations. In this study, we present preliminary results on the new analysis of the *Kepler* light curve and high-resolution spectroscopic observations of the pulsating Am star KIC 9204718. Our tentative analysis shows that the star has hybrid-type pulsational characteristics.

1 Introduction

New generation space missions provide accurate detections of hybrid-type pulsations in a variety of different stars. Recently, using the data of A stars from the *Kepler* Telescope, [1] reported that six Am stars have been known in the *Kepler* field showing δ Sct type pulsations. The star KIC 9204718 (HD 176843, A3mF0) was first classified as an Am star by [2]. KIC 9204718 was also listed as a marginal Am type star showing δ Sct type pulsations and a possible member of a contact binary system ([3]). Since Am stars were known as a class of non-pulsating variables, this new result makes them very important to test the new diffusion scenario regarding their location in the HR diagram ([4]).

2 *Kepler* photometry

The *Kepler* time-series of KIC 9204718 was used to derive pulsational content. The frequencies were extracted from the Long Cadence data by using the software package `SigSpeC` ([5]). The light curve of KIC 9204718 is therefore dominated by two peaks with frequencies $f_1 = 0.11421\,\mathrm{d}^{-1}$ and $f_2 = 0.024384\,\mathrm{d}^{-1}$.

3 Spectroscopic observations

Spectroscopic observations have been carried out with ESPERO, a newly commissioned fiber-fed echelle spectrograph attached to the 2-m RCC telescope of Rozhen National Astronomical Observatory. High dispersion spectroscopic data were used to extract the projected rotational velocity and atmospheric parameters.

⋆`cerenulusoy@gau.edu.tr`

Table 1. Radial velocity measurements of KIC 9204718.

Date	HJD2457+	RV [km/s]	RVerr [km/s]
21.07.2016	591.44155	−14.94	± 1.88
22.07.2016	592.40175	−15.05	± 1.82
23.07.2016	593.44898	−14.85	± 1.80
17.08.2016	618.41679	−16.16	± 1.98
20.08.2016	621.36269	−16.43	± 2.29
15.09.2016	647.48922	−19.71	± 1.94
16.09.2016	648.32971	−19.65	± 2.03
17.09.2016	649.38379	−19.61	± 2.11

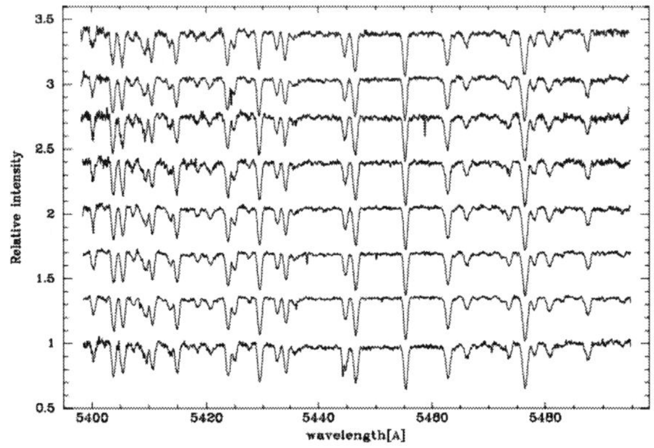

Figure 1. Normalized spectra of KIC 9204718

4 Conclusion

We have presented preliminary results of both frequency and spectrum analysis of marginal Am type star KIC 9204718 observed by the *Kepler* satellite.The radial velocity measurements reveal not very significant variations (Table 1). The radial velocity is changed in the framework of 5 km/s. A careful check of the obtained spectra does not show any signs of binarity (Figure 1). In conclusion, more observations may enable us to identify at least some of the modes and our present work will support in future asteroseismic studies of this star.

Acknowledgments: The authors acknowledge the whole *Kepler* team for providing the unprecedented data sets that makes these results possible. CU wishes to thank Miss Cemile Gür for her valuable support. MN acknowledges the support of the project DFNP-1 03/11.05.2016 "Support for young scientists" funded by the Bulgarian Academy of Sciences.

References

[1] Balona, L. A., Catanzaro G., Abedigamba O. P., Ripepi V., & Smalley, B., MNRAS, **448**, 1378 (2015)
[2] Floquet, M., A&AS, **21**, 25 (1975)
[3] Uytterhoeven, K., Moya, A., Grigahcène, A., et al., A&A, **534**, A125 (2011)
[4] Turcotte, S., Richer, J., Michaud, G., & Christensen-Dalsgaard, J., A&A, **360**, 603, (2000)
[5] Reegen, P., A&A, **467**, 1353 (2007)

Determining the core stratification in white dwarfs with asteroseismology

S. Charpinet[1,2,*], N. Giammichele[1,2,**], P. Brassard[3,***], and G. Fontaine[3,****]

[1] *Université de Toulouse, UPS-OMP, IRAP, Toulouse, F-31400,France*
[2] *CNRS, IRAP, 14 avenue Edouard Belin, F-31400 Toulouse, France*
[3] *Département de Physique, Université de Montréal, CP 6128, Succursale Centre-Ville, Montréal, QC H3C 3J7, Canada*

Abstract. Using the forward modeling approach and a new parameterization for the core chemical stratification in ZZ Ceti stars, we test several situations typical of the usually limited constraints available, such as small numbers of observed independent modes, to carry out asteroseismology of these stars. We find that, even with a limited number of modes, the core chemical stratification (in particular, the location of the steep chemical transitions expected in the oxygen profile) can be determined quite precisely due to the significant sensitivity of some confined modes to partial reflexion (trapping) effects. These effects are similar to the well known trapping induced by the shallower chemical transitions at the edge of the core and at the bottom of the H-rich envelope. We also find that success to unravel the core structure depends on the information content of the available seismic data. In some cases, it may not be possible to isolate a unique, well-defined seismic solution and the problem remains degenerate. Our results establish that constraining the core chemical stratification in white dwarf stars based solely on asteroseismology is possible, an opportunity that we have started to exploit.

1 Introduction

In Giammichele et al. (2017a, [1]), we introduced a new generation of hydrostatic equilibrium white dwarf models designed specifically for in-depth asteroseismic probing of the stellar structure. These new models most notably provide a new parameterization for the C/O stratification allowing us to explore a vast range of chemical configurations in the core. We hereafter use these tools to investigate the core seismic signatures. To estimate the potential of asteroseismology for probing the core, we also propose a series of three "hare and hounds" (blind) tests.

*stephane.charpinet@irap.omp.eu
**ngiammichele@irap.omp.eu
***brassard@astro.umontreal.ca
****fontaine@astro.umontreal.ca

DOI: 10.1051/epjconf/201715205005

2 Core seismic signature and blind tests

The impact of the core structure on the oscillation periods of white dwarf pulsators and the series of tests presented at this conference are now discussed extensively in Giammichele et al. (2017b, [2]). We refer the reader to this article.

3 Conclusion

We conclude that determining the core stratification in white dwarf stars with asteroseismology is possible even with a limited number of observed periods. Success or failure to do so depends on the star analyzed and its ability to show modes that are partially confined in its core (see also Giammichele et al. 2016, [3], for a discussion on such modes).

Acknowledgments: S. Charpinet acknowledges financial support from "Programme National de Physique Stellaire" (PNPS) of CNRS/INSU, France, and from the Centre National d'Études Spatiales (CNES, France). This work was granted access to the HPC resources of CALMIP under allocation number 2016-p0205. This work was supported by the FQRNT (Québec) through a postdoctoral fellowship awarded to N. Giammichele. G. Fontaine also acknowledges the contribution of the Canada Research Chair Program.

References

[1] Giammichele, N., Charpinet, S., Fontaine, G., & Brassard, P., ApJ, **834**, 136 (2017a)
[2] Giammichele, N., Charpinet, S., Brassard, P., & Fontaine, G., A&A, **598**, A109 (2017b)
[3] Giammichele, N., Fontaine, G., Brassard, P., & Charpinet, S., ApJS, **223**, 10 (2016)

EPJ Web of Conferences **152**, 05006 (2017) DOI: 10.1051/epjconf/201715205006
Wide-Field Variability Surveys: A 21st Century Perspective

Asteroseismology of the ZZ Ceti and DAZ GD 133

J.-N. Fu[1], *G. Vauclair*[2,3], and *J. Su*[1]

[1] *Department of Astronomy, Beijing Normal University, 100875 Beijing, China*
[2] *Université de Toulouse, UPS-OMP, IRAP, Toulouse, France*
[3] *CNRS, IRAP, 14 avenue Edouard Belin, F-31400 Toulouse, France*

Abstract. GD 133 is a DAZ white dwarf with an atmosphere polluted by heavy elements accreted from a debris disk, which is formed by the disruption of rocky planetesimals with orbits bringing them at the white dwarf tidal radius. To reach such orbits implies the potential presence of a perturbing planet. GD 133 is a ZZ Ceti pulsator close to the blue edge of the instability strip. The presence of a planet could be revealed by the periodical variation of the observed pulsation periods induced by the orbital motion of the white dwarf. We started a multi-site photometric follow-up aimed at detecting the signature of this potential planet. As a partial result of this work in progress, we give the parameters of a preliminary best-fit model derived from asteroseismology.

1 Context: GD 133, an accreting DA white dwarf and ZZ Ceti pulsator

GD 133 is one of the increasing number of white dwarf stars surrounded by debris disks resulting from the tidal disruption of rocky planetesimals ([2–4]). Part of the matter of the disk falls onto the white dwarf surface, polluting the chemical composition of its outer layers. The spectrum of GD 133 shows the lines of a number of heavy elements (Ca, O, Mg, Si). The derived abundance ratios are quite similar to the bulk Earth composition ([10]). GD 133 is a ZZ Ceti pulsator ([9]) close to the blue edge of the instability strip.

The scenarii for the formation of the debris disks by the tidal disruption of rocky planetesimals suggest that one (or more) undetected planet(s) could be orbiting the white dwarfs. Through gravitational interactions the planet(s) perturbs their orbit so that they may come close enough to the white dwarf tidal radius.

GD 133 is a good candidate to search for the signature of such a potential planet through the effect of the orbital motion on the observed pulsation periods.

2 Observations

We started a photometric follow-up of GD 133 in 2011. A multisite campaign including the 1.5 m telescope of the San Pedro Martir observatory (Mexico), the 2.4 m telescope of the Lijiang observatory (Yunnan, China), and the 1 m telescope of Pic-du-Midi observatory (France) was organized in March 2011 during which we obtained 106 hours of photometric data. Table 1 gives the list of the identified frequencies and their amplitudes. As a preliminary result of this work in progress we used these data to determine the parameters of GD 133 from asteroseismology.

Table 1. Frequencies of GD 133 during the 2011 campaign. The frequencies and their uncertainties are given in units of μHz; the amplitude and their uncertainties are given in mmag.

$\nu(\mu\text{Hz})$	$\sigma(\nu)(\mu\text{Hz})$	A (mmag)	$\sigma(A)$(mmag)
2821.74	0.34	0.47	0.15
5106.04	0.31	0.55	0.20
6818.48	0.14	1.33	0.51
6821.89	0.14	1.24	0.38
8308.81	0.16	1.95	0.43
8310.28	0.12	2.24	0.58
8625.25	0.15	0.72	0.25
8628.57	0.15	0.62	0.19

3 Modelling GD 133

We use MESA ([6–8]) to obtain a set of white dwarf models in the range $0.45 - 0.9\,M_\odot$ from initial main-sequence stellar models with masses between 0.8 and $8\,M_\odot$. DA white dwarf models were then obtained with an updated version of WDEC ([1, 5]) by adjusting and relaxing the C-O core models prepared with MESA. The spectroscopic analysis by [10] leads to the atmospheric parameters of GD 133 as $T_{\rm eff}$ = 12600 K (200 K) and $\log(g)$ = 8.10 (0.10) from which they derive a mass of 0.65 (0.05) $M_\odot$. We built a grid of models exploring the space of stellar parameters: $0.60 \le M/M_\odot \le 0.70$; $12400 \le T_{\rm eff}(\rm K) \le 12800$; $-4 \le \log(M_{\rm He}/M) \le -2$; $-10 \le \log(M_{\rm H}/M) \le -4$.

We computed the adiabatic ℓ=1 and ℓ=2 non-radial oscillation modes for all these models. By comparing their periods with the observed periods through a χ^2 method, we derived a preliminary best-fit model for GD 133 with parameters: $M/M_\odot$ = 0.63; $T_{\rm eff}$ = 12400 K; $\log(M_{\rm He}/M)$ = −2; $\log(M_{\rm H}/M)$ = −4.5.

4 Conclusions

We have undertaken a photometric follow-up of the ZZ Ceti and DAZ GD 133, which we consider as a promising candidate for the search of a potential planet. As a preliminary result we used the data obtained during the 2011 first multisite campaign to derive preliminary values of the star main parameters. This is a work in progress and the follow-up of GD 133 is going on.

References

[1] Bradley, P. A., 1993, Ph.D. Thesis, University of Texas at Austin
[2] Farihi, J., Jura, M., & Zuckerman, B., 2009, ApJ, **694**, 805
[3] Jura, M., Farihi, J., & Zuckerman, B., 2007, ApJ, **663**, 1285
[4] Jura, M., Farihi, J., & Zuckerman, B., 2009, AJ, **137**, 3191
[5] Montgomery, M., 1998, Ph.D. Thesis, University of Texas at Austin
[6] Paxton, B., Bildsten, L., Dotter, A., Herwig, F., Lesaffre, P., & Timmes, F., 2011, ApJS, **192**, 3
[7] Paxton, B., Cantiello, M., Arras, P., et al., 2013, ApJS, **208**, 4
[8] Paxton, B., Marchant, P., Schwab, J., et al., 2015, ApJS, **220**, 15
[9] Silvotti, R., Pavlov, M., Fontaine, G., Marsh, T., & Dhillon, V., 2006, MmSAI, **77**, 486
[10] Xu, S., Jura, M., Koester, D., Klein, B., & Zuckerman, B., 2014, ApJ, **783**, 79

Asteroseismological analysis of the GW Virginis stars SDSS J0349-0059 and VV 47

Leila M. Calcaferro[1,2,★], *Alejandro H.* Córsico[1,2,★★], and *Leandro G.* Althaus[1,2,★★★]

[1] *Facultad de Ciencias Astronómicas y Geofísicas, Universidad Nacional de La Plata, Paseo del Bosque s/n, (1900) La Plata, Argentina*
[2] *Instituto de Astrofísica La Plata, CONICET-UNLP, Paseo del Bosque s/n, (1900) La Plata, Argentina*

Abstract. We present an asteroseismological analysis of the GW Vir stars SDSS J0349−0059 and VV 47. We found good agreement between our mass determinations and previous results. For SDSS J0349−0059, we found a seismological model that provides us with additional information on the star.

1 Introduction

GW Vir stars (or pulsating PG 1159 stars) are very hot H-deficient post-Asymptotic Giant Branch (AGB) stars, that show surface layers rich in He, C and O ([7]). These stars exhibit multiperiodic luminosity variations with periods ranging from ~ 300 to 3000 s due to nonradial pulsation g modes.

In this work, we focus on two GW Vir stars: SDSS J0349-0059 (hereafter J0349), characterized by $T_{\rm eff} = 90\,000 \pm 900$ K and $\log g = 7.5 \pm 0.01$ (cgs, [4]); and VV 47, with $T_{\rm eff} = 130\,000 \pm 13\,000$ K and $\log g = 7 \pm 0.5$ (cgs, [7]). Given the large uncertainty in the $T_{\rm eff}$, this star may be evolving either before or after the "evolutionary knee". We computed adiabatic nonradial g-mode pulsation periods on PG 1159 evolutionary models with stellar masses between 0.515 and 0.741 $M_\odot$ that take into account the complete evolution of the progenitor stars through the thermally pulsing AGB phase and born-again episode (that explains their H deficiency; [1, 2, 5]).

2 Analysis and results

The pulsations exhibited by GW Vir stars are produced by g modes with low harmonic degree (ℓ) and high radial order (k). The difference between the periods of modes with consecutive values of k reaches a constant value, known as the asymptotic period spacing ([6]). Given that in GW Vir stars the period spacing mainly depends on the stellar mass, it is possible to constrain the mass of the star by comparing the observed period spacing with the asymptotic period spacing, or with the average of the period spacings computed on a grid of models with different masses and effective temperatures. So, the first step is to search for a constant period spacing (if it exists) in the pulsation spectrum of the star. According to [8], J0349 shows periods in the range 300.93 – 963.48 s. According to [3],

★lcalcaferro@fcaglp.unlp.edu.ar
★★acorsico@fcaglp.unlp.edu.ar
★★★althaus@fcaglp.unlp.edu.ar

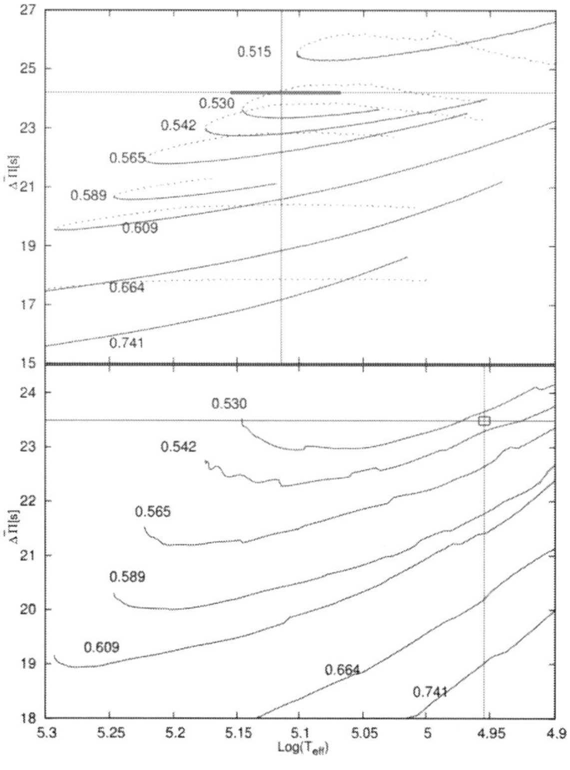

Figure 1. Average of the computed $\ell = 1$ period spacings for VV 47 (upper panel) and J0349 (lower panel) corresponding to our PG 1159 evolutionary sequences in terms of $\log(T_{\rm eff})$. Dashed (solid) lines correspond to evolutionary stages before (after) the "evolutionary knee". Also shown are the observed period spacings and the $T_{\rm eff}$, along with their uncertainties.

VV 47 shows a set of periods in the range 131 − 5682 s. By employing three different and independent significance tests to the periods observed for each star, namely the Kolmogorov-Smirnov, the Inverse Variance and a Fourier Transform based method, we determined a mean period spacing of 23.49 s for J0349 and 24.2 s for VV 47. When we compared these values with the asymptotic period spacing for our models, we obtained $M_* = 0.569\,M_\odot$ for J0349 and $M_* = 0.523\,M_\odot$ for VV 47, which is an average value considering both the cases the star is evolving before and after the "evolutionary knee". When we compared the observed period spacing for each star with the average of the computed period spacings, as shown in Figure 1, we obtained $M_* = 0.535\,M_\odot$ for J0349 and $M_* = 0.528\,M_\odot$ for VV 47 (once again, this is an average value). Next, we carried out period-to-period fits, consisting in search for models that best reproduce the individual observed periods of each star. This is achieved by assessing a quality function that measures the difference between the observed individual periods and the theoretical pulsation periods for our grid of models. We display in Figures 2 and 3 the inverse of the quality function in terms of $T_{\rm eff}$ for J0349 and VV 47, respectively. The model that shows the greatest value of the inverse of the quality function, if it exists, is adopted as the best-fit model. If there is not a unique maximum, we need to employ some external constraint, like the uncertainty in $T_{\rm eff}$. Following this procedure, we found for J0349 a possible solution (within the range of allowed $T_{\rm eff}$) for the model with $M_* = 0.542\,M_\odot$ and $T_{\rm eff} = 91\,255$ K. Once we adopt a model, we have access to theoretical information, such as $\log(g) = 7.488$ (cgs), $\log(R_*/R_\odot) = -1.658$ and $\log(L_*/L_\odot) = 1.475$. For the case of VV 47, as seen in Figure 3, there is no unambiguous asteroseismological solution, not even within the range of allowed $T_{\rm eff}$. Finally, there is another mass determination for the two stars under analysis we obtained using the spectroscopic data given by [4] and [7] combined with our grid of models ([5]). It results in $M_* = 0.543M_\odot$ for J0349 and $M_* = 0.529M_\odot$ for VV 47 (average value).

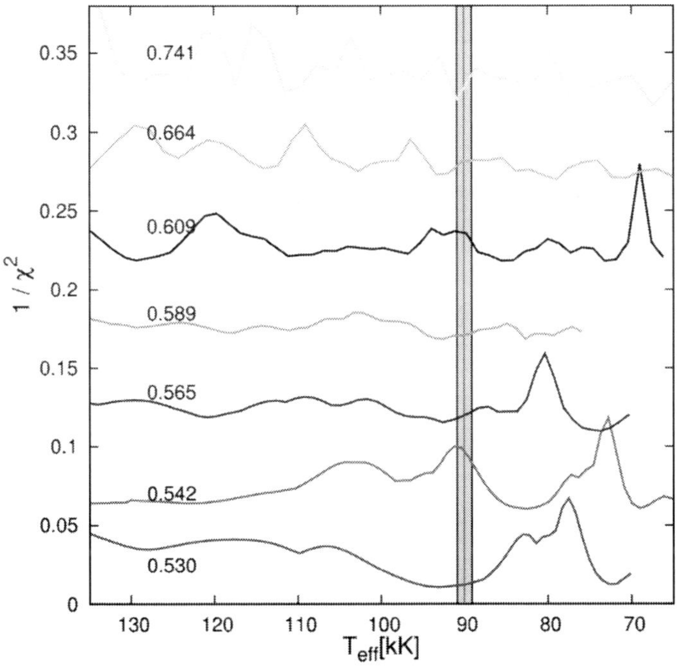

Figure 2. Inverse of the quality function of the period fit considering $\ell = 1$ g modes vs $T_{\rm eff}$ for J0349. The curves have been arbitrarily shifted upward for clarity (with a step of 0.05).

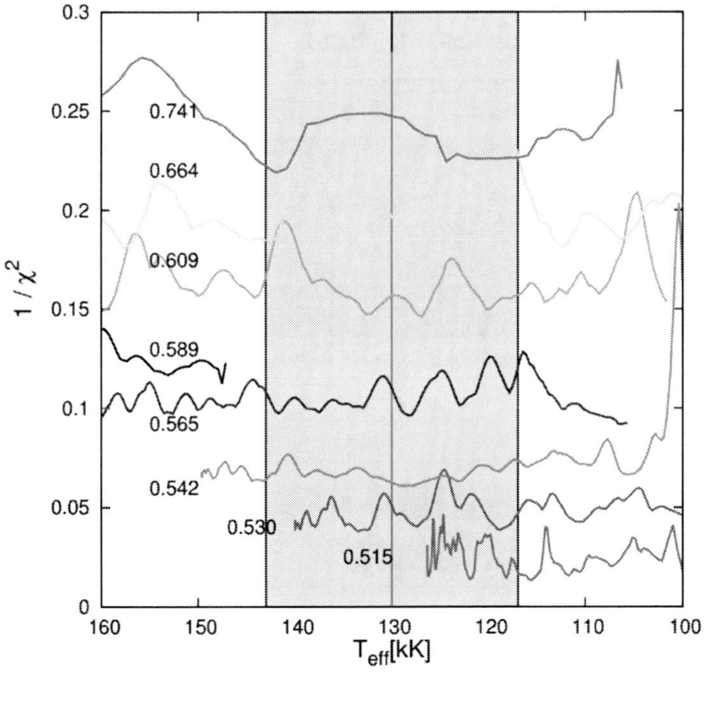

Figure 3. Same as Figure 2 but for VV 47, for the case "before the knee". The curves have been arbitrarily shifted upward for clarity (with a step of 0.025).

DOI: 10.1051/epjconf/201715205007

3 Conclusions

In this work we employed the photometric and spectroscopic data given for J0349 and VV 47 ([3, 4, 7, 8]), and our computations of adiabatic nonradial g-mode pulsation periods on PG 1159 evolutionary models ([1, 2, 5]), in order to determine the mass and the internal structure of these stars. We found that all the mass estimates we obtained through the different methods employed are in good agreement with each other and also in line with the spectroscopic mass. As we were able to find an asteroseismological model for J0349 that fits the observed periods, we have access to theoretical information that otherwise is not possible to infer by any other methods. Unfortunately, we could not find an unambiguous asteroseismological model for VV 47. It would be really important to have available more accurate spectroscopic determinations for VV 47, in order to be able to choose a seismological model and hence, know more about this star.

Acknowledgments: L.M.C. warmly thanks the LOC of the 22nd Los Alamos Stellar Pulsation Conference Series Meeting for support that allowed her to attend this conference.

References

[1] Althaus, L. G., Serenelli, A. M., Panei, J. A., Córsico, A. H., García Berro, E., & Scóccola, C. G., A&A, **435**, 631 (2005)
[2] Córsico, A. H., Althaus, L. G., & Miller Bertolami, M. M., A&A, **458**, 259 (2006)
[3] González Pérez, J. M., Solheim, J. E., & Kamben, R., A&A, **454**, 527 (2006)
[4] Hügelmeyer, S. D., Dreizler, S., Homeier, D., Krzesinski, J., Werner, K., Nitta, A., & Kleinman, S. J., A&A, **454**, 617 (2006)
[5] Miller Bertolami, M. M., & Althaus, L. G., A&A, **454**, 845 (2006)
[6] Tassoul, M., ApJS, **43**, 469 (1980)
[7] Werner, K., & Herwig, F., PASP, **118**, 183 (2006)
[8] Woudt, P. A., Warner, B., & Zietsman, E., MNRAS, **426**, 2137 (2012)

Comparing the asteroseismic properties of pulsating extremely low-mass pre-white dwarf stars and δ Scuti stars

J. P. Sánchez Arias[1,*], *A. H. Córsico*[1], *A. D. Romero*[2], and *L. G. Althaus*[1]

[1]*Facultad de Ciencias Astronómicas y Geofísicas (UNLP), La Plata, Buenos Aires, Argentina*
[2]*Instituto de Física, Universidade Federal do Rio Grande do Sul, Brazil*

Abstract. We present the first results of a detailed comparison between the pulsation properties of pulsating Extremely Low-Mass pre-white dwarf stars (the pre-ELMV variable stars) and δ Scuti stars. The instability domains of these very different kinds of stars nearly overlap in the $\log T_{\rm eff}$ vs. $\log g$ diagram, leading to a degeneracy in the classification of the stars. Our aim is to provide asteroseismic tools for their correct classification.

1 Introduction

Main sequence δ Scuti stars, with masses between 1.2 and 2.5 $M_{\odot}$, pulsate in low-order p and g modes in the range 700 – 36300 sec approximately, driven by the κ mechanism operating in the He II ionization zone. They are variables of spectral type between A0 and F5, lying on the extension of the Cepheid instability strip towards low luminosities, at effective temperatures between 6000 K and 9000 K and $3.25 \lesssim \log g \lesssim 4.75$. On the other hand, pulsating pre-Extremely Low-Mass (ELM) white dwarfs (WD), with masses between $\sim 0.15\,M_{\odot}$ and $\sim 0.30\,M_{\odot}$, constitute a class of pulsating stars called pre-ELMV stars which exhibit radial-, p- and possibly g-mode pulsations with periods between 380 and 3500 sec, driven by the κ mechanism operating in the zone of the second ionization of He. The pre-ELMV WDs and the δ Scuti instability domains nearly overlap in the $\log T_{\rm eff}$ vs. $\log g$ diagram (see Fig. 1), and their period ranges also overlaps in the range 700 – 3500 sec approximately, leading to a degeneracy in the classification of the stars, since their spectra could exhibit similar atmosphere parameters. Asteroseismology can be employed to distinguish these very different kinds of stars. Motivated by this fact, we addressed an asteroseismic comparison between these two families of pulsating stars, with the aim of providing tools for their correct classification.

2 Models and results

The pulsation analysis of this work makes use of full stellar evolution models of δ Sct stars and pre-ELM WDs generated with the LPCODE stellar evolution code ([1]) and the LP-PUL pulsation code ([2]). δ Sct and pre-ELM WD stellar models were calculated as in [3] and [4], respectively. We choose two template models with similar surface gravity and effective temperature, one of them representative of δ Scuti stars and the other one representative of pulsating pre-ELM WDs, both with $T_{\rm eff} \approx 7660$ K

* `jsanchez@fcaglp.unlp.edu.ar`

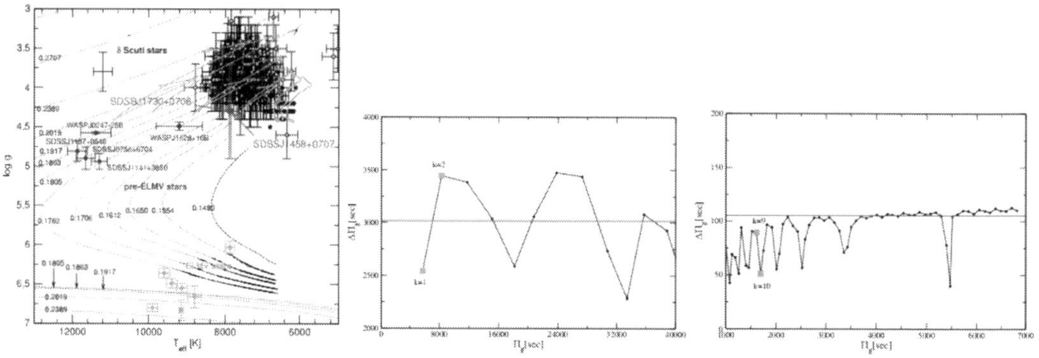

Figure 1. Left: $T_{\rm eff} - \log g$ diagram with the location of ELM WDs and evolutionary tracks of low-mass He-core WDs ([5], black curves), and the location of a δ Sct stars sample (black dots) along with some evolutionary tracks (cyan curves). Also included is the location of the stars reported in [6] (red dots), along with the pre-ELMV WDs ([7, 8]). Centre: period spacing vs. period for g modes corresponding to the δ Sct template model. Right: same as in the centre panel, but for the pre-ELM WD template model. We include the asymptotic period spacing (red line).

and $\log g \approx 4.078$. The δ Sct model is characterized by a H/He core surrounded by an H-dominated envelope. The pre-ELM WD model has a pure He core, with surface layers composed by H and He. The H/He chemical transition leaves notorious signatures in the Brunt Väisälä (B-V) frequency. In the δ Sct model, this frequency becomes null in the convective core and in an outer convective layer. Also, there exists a "bump" in the chemical gradient region at the boundary of the convective core. In the pre-ELM WD model, the B-V frequency becomes null only in a convective outer layer, and the bump is located at more internal layers than in the δ Sct model. These differences lead to notorious differences in p- and g- mode behavior. Among other results, we found that the g-mode period spacing for the δ Sct model is much larger than for the pre-ELM WD model (Fig. 1); in addition, the pre-ELM WD model shows a frequency spacing somewhat larger than the δ Sct model for p modes.

3 Conclusion

The very different g-mode period spacing ($\Delta\Pi_g$) of the two classes of stars indicates that it would be possible to distinguish δ Sct stars from pre-ELMVs using asteroseismology. A typical value of $\Delta\Pi_g$ for a δ Sct is $\approx$ 2500 sec, while for a pre-ELM WD it is $\approx$ 90 sec. Then, the detection of g-modes with consecutive radial orders could shed light on the nature of these stars.

References

[1] Althaus, L. G., Serenelli, A. M., Panei, J. A., et al., A&A, **435**, 631 (2005)
[2] Córsico, A. H., & Althaus, L. G., A&A, **454**, 863 (2006)
[3] Sánchez Arias, J. P., Córsico A. H., & Althaus, L. G., A&A, **597**, A29 (2017)
[4] Córsico, A. H., Althaus, L. G., Serenelli, A. M., et al., A&A, **588**, A74 (2016)
[5] Althaus, L. G., Miller Bertolami, M. M., & Córsico, A. H., A&A, **577**, A19 (2013)
[6] Corti, M. A., Kanaan, A., Córsico, A. H., et al., A&A, **587**, L5 (2016)
[7] Gianninas, A., Curd, B., Fontaine, G., Brown, W. R., & Kilic, M., ApJL, **822**, L27 (2016)
[8] Maxted, P. F. L., Serenelli, A. M., Miglio, A., et al., Nature, **498**, 463 (2013)

Multi-site, multi-colour and space photometry of a southern δ Scuti star, 38 Eri

Margit Paparó[1,*], *József* Benkö[1], and ground-based and MOST photometry teams

[1] *Konkoly Observatory, MTA CSFK, Konkoly Thege u 15-17, H-1125 Budapest, Hungary*

Abstract. 38 Eri is a δ Scuti star in the shell hydrogen burning stage of evolution, with a 0.04 mag peak-to-peak amplitude and a very complex light variation. It was discovered in 1971. In the subsequent investigation the star always showed a different frequency content. A ground-based multi-site campaign and a MOST observation were devoted to resolve the contradicting pulsational behaviour. We present the preliminary result of our investigation obtained on Strömgren b and MOST data. The complexity of the pulsation is confirmed, but the comprehensive analysis helps to resolve the frequencies of the fast rotating star.

1 Introduction

All the previous ground-based observations, both photometric and spectroscopic, revealed a very complex frequency content of 38 Eri. None of them confirmed the previous set of frequencies, some frequencies were missing or new ones appeared. It was questionable that the excited amplitudes of the frequencies were changing or the frequency content was not properly resolved. A ground-based multi-site campaign and the MOST space data were devoted to answer the question.

2 Single site results: effect of different sampling

The preliminary analyses revealed how serious a task it is to find the frequencies using single site data. A basic problem is the daily alias structure. Additionally, the rotational splitting due to the fast rotation plays an important role. The different subsets of the ground-based campaign show the serious effect of the different sampling. The left and right panels of Figure 1 show the variety of the frequencies that we obtained for two different segments. The dominant frequencies, around 13.3 d^{-1}, are the same in both datasets, but the subsequent frequencies in the prewhitening process are different. We can recognize a doublet (left panel) or triplet (right panel) at around 10.47 d^{-1}, but the peaks around 5.2 d^{-1} appear only in the dataset with better sampling. An example for the possible confusion between the 1 d^{-1} alias and the independent frequency is the two peaks at 9.47 and 10.47 d^{-1}. A peak at 12.62 d^{-1} appears only in the subset with better sampling. The preliminary investigation confirms that the single site observation is not enough for resolving the complexity of the pulsation in 38 Eri.

*paparo@konkoly.hu

DOI: 10.1051/epjconf/201715205009

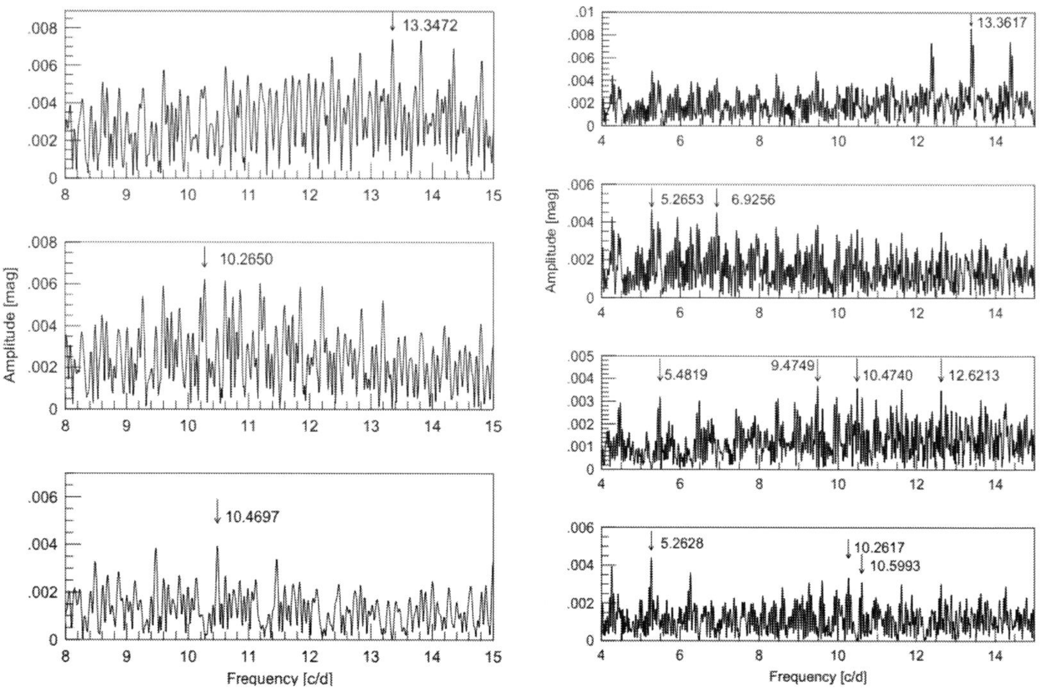

Figure 1. Left panel: the original spectrum and two pre-whitened spectra of one segment of the ground-based campaign are shown. Right panel: Another set of the ground-based campaign represents a different sampling. Some distinguished spectra are shown.

3 MOST results

The δ Scuti star 38 Eri was a switch target with two other targets, so only part of each MOST 101-minute orbit was devoted to it. This is a regular sampling along the orbital period, but it results in an irregular sampling along the light curve. The effect of the different sampling may be followed in the MOST data, too. Due to the larger timebase we can resolve the frequencies with even lower separation. Dividing the whole dataset into two subsets, different frequencies dominate the frequency spectrum of the subsets. It is suggested that the sampling influences not only the frequency content, but the amplitudes, too.

4 Conclusion

Each previous investigation of the δ Scuti star 38 Eri could catch part of the complexity of the pulsation. The amplitude of the excited modes seems to vary from epoch to epoch resulting in different modes being dominant at a certain epoch. The final analysis may find the whole set of frequencies and lead to the localization of the rotational triplets of the non-radial modes.

Acknowledgments: This work was supported by the grant: ESA PECS No 4000103541/11/NL/KML.

The evolved slowly pulsating B star 18 Peg

A testbed for upper main sequence stellar evolution

Andreas Irrgang[1,*], *Peter* De Cat[2,**], and *Andrew* Tkachenko[3]

[1]*Dr. Karl Remeis-Observatory & ECAP, Astronomical Institute, Friedrich-Alexander University Erlangen-Nürnberg (FAU), Sternwartstr. 7, 96049 Bamberg, Germany*
[2]*Royal Observatory of Belgium, Ringlaan 3, 1180 Brussels, Belgium*
[3]*Instituut voor Sterrenkunde, KULeuven, Celestijnenlaan 200D, B-3001 Leuven, Belgium*

Abstract. The predicted width of the upper main sequence in stellar evolution models depends on the empirical calibration of the convective overshooting parameter. Despite decades of discussions, its precise value is still unknown and further observational constraints are required to gauge it. Irrgang et al. ([1]) showed that the B3 III giant 18 Peg is one of the most evolved members of the class of slowly pulsating B (SPB) stars and, thus, bears tremendous potential to derive a tight lower limit for the width of the upper main sequence. In addition, 18 Peg turns out to be part of a single-lined spectroscopic binary system with an eccentric, more than 6-year orbit. The orbital solution, in combination with the absence of additional signatures of the secondary component in the spectroscopic data and the spectral energy distribution, lead to the conclusion that all the observations of 18 Peg are fully compatible with the assumption that the secondary component is either a main-sequence star with a mass of 1-4 $M_{\odot}$ or a neutron star.

1 18 Peg as an asteroseismic target in a binary

To exploit the full potential of 18 Peg (RA(2000) = 22:00:07.93; DEC(2000) = +06:43:02.78; V = 6 mag; $v \sin i = 15(3)\,\mathrm{kms}^{-1}$; [2]) as benchmark object for stellar evolution theory, additional observations are required. In 2016, 18 Peg is approaching periastron (see Fig. 1) and is hence in the orbital phase were the orbital velocity is changing fast. We therefore organized a follow-up campaign with the HERMES spectrograph attached to the 1.2m Mercator telescope (Roque de Los Muchachos Observatory, La Palma, Canary Islands; [3]) to gather high-quality, high-resolution spectra appropriate for the fine-tuning of the orbit and for a detailed study of the main pulsation mode (see gray area on Fig. 1). The new orbit found after inclusion of the new spectroscopic data (open diamonds in Fig. 1) is drawn with a solid curve in Figure 1 and the corresponding orbital parameters are given in the right column of Table 1. Given that only few of the revised values of the orbital parameters agree within the 1 σ errors with those published by Irrgang et al. ([1]), more data is needed to fine-tune the orbit.

So far, only one pulsation mode with a period of ~1.4 days has been detected for 18 Peg, which is insufficient to derive an accurate asteroseismic position in the Hertzsprung-Russell diagram. 18 Peg is (1) being observed in the Strömgren u- and y-band with the 0.75m T6 APT at Fairborn Observatory

*andreas.irrgang@fau.de
**Peter.DeCat@oma.be

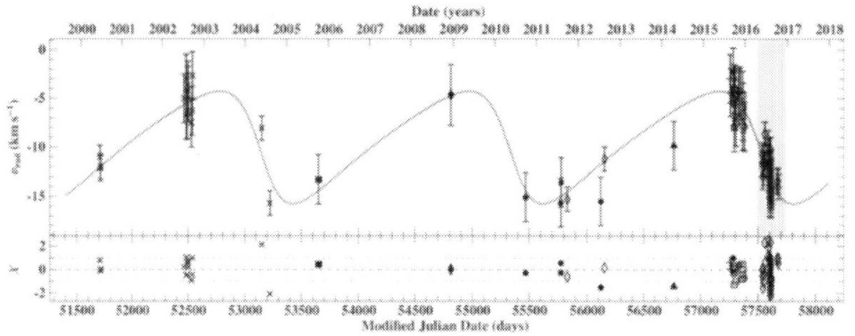

Figure 1. Radial velocity curve of 18 Peg. The radial velocities were derived from spectra taken with multiple instruments represented by different symbols. The HERMES spectra (open diamonds) in the grey zone were not available yet at the time of publication of Irrgang et al. ([1]).

Table 1. Comparison of the orbit published by Irrgang et al. ([1], left) to the one found after inclusion of HERMES spectra obtained up to November 8, 2016 (right). The orbital and derived parameters are given in the top and bottom part, respectively. Incompatible values are given in italics.

Parameter	Published	This work
Period P (in days)	*2245^{+25}_{-30}*	*2191 ± 14*
Epoch of periastron $T_{\rm periastron}$ (in MJD)	*57730^{+40}_{-60}*	*57550^{+60}_{-70}*
Eccentricity e	*$0.60^{+0.07}_{-0.08}$*	*0.33 ± 0.06*
Longitude of periastron ω (in deg)	123^{+12}_{-7}	103 ± 14
Velocity semiamplitude K_1 (in km s^{-1})	*$7.74^{+1.9}_{-1.1}$*	*5.7 ± 0.5*
Systemic velocity γ (in km s^{-1})	-9.9 ± 0.4	-9.6 ± 0.4
Mass function $f(M)$ (in $M_\odot$)	$0.054^{+0.035}_{-0.017}$	$0.036^{+0.011}_{-0.009}$
Projected semimajor axis $a_1\sin(i)$ (in AU)	$1.27^{+0.23}_{-0.15}$	$1.09^{+0.10}_{-0.09}$
Projected periastron distance $r_{\rm p}\sin(i)$ (in $R_\odot$)	*108^{+21}_{-17}*	*156 ± 23*

(Arizona, USA; kindly offered by G. Handler), (2) has been accepted as a target for the BRITE-constellation and (3) will (probably) be observed for 27 days with the space mission TESS (to be launched in March 2018) to gather complementary time-series of accurate photometry with a sufficiently long time-base to be able to study the expected low amplitude modes that are unobserved so far.

Acknowledgments: Based on new observations obtained with the HERMES spectrograph that is attached to the Mercator Telescope, operated on the island of La Palma by the Flemish Community, at the Spanish Observatorio del Roque de los Muchachos of the Instituto de Astrofísica de Canarias. The research leading to these results has partially received funding from The Research Foundation – Flanders (FWO), Belgium, from the European Research Council (ERC) under the European Union's Horizon 2020 research and innovation programme (grant agreement N°670519: MAMSIE)

References

[1] A. Irrgang, A. Desphande, S. Moehler, M. Mugrauer, D. Janousch, A&A **591**, L6 (2016)
[2] M.F. Nieva, N. Przybilla, A&A **539**, A143 (2012)
[3] G. Raskin, H. van Winckel, H. Hensberge, A. Jorissen, H. Lehmann, C. Waelkens, G. Avila, J.P. de Cuyper, P. Degroote, R. Dubosson et al., A&A **526**, A69 (2011)

Rotation and magnetic activity of oscillating solar-like stars with the *Kepler* mission

Savita Mathur[1,⋆], *Lauren* Balliet[1,2], *Rafael A.* García[3], and *David* Salabert[3]

[1]*Space Science Institute, 4750 Walnut street Suite#205, Boulder, CO 80301, USA*
[2]*Clarion University of Pennsylvania, 840 Wood Street, Clarion, PA 16214, USA*
[3]*Laboratoire AIM, CEA/DRF-CNRS-Université Paris Diderot; IRFU/SAp, Centre de Saclay, 91191 Gif-sur-Yvette Cedex, France*

Abstract. For the last few decades the investigation of stellar magnetic activity has been conducted through spectroscopic and spectropolarimetric surveys. This led not only to the detection of magnetic cycles in other stars but also to variable and magnetic activity. For the Sun, the magnetic activity is described as the interaction between convection, rotation, and magnetic field. To study magnetic activity of solar-like stars we need to have the knowledge of the surface rotation period, the properties of magnetic activity, and the structure of the stars. We present the results obtained from the studies of *Kepler* solar-like targets in terms of rotation periods, magnetic activity proxies and magnetic activity cycles detected. We can then combine this information with asteroseismic studies to have a broader picture of stellar magnetic activity.

1 Introduction

The *Kepler* mission has observed more than around 200,000 stars. Recently, a new catalog of the *Kepler* star properties was released for 197,096 targets observed providing for instance surface gravity, effective temperature, radius, but also distance and extinction ([1]). This catalog was based on isochrone fitting using inputs from asteroseismology, spectroscopy, photometry along with a Bayesian method. While all these properties are provided on the exoplanet archive, they have been used by the latest planet transit search pipeline to characterize all the planets and planet candidates discovered by the mission, in particular the ones in the habitability zone of the host star. However, stellar magnetic activity is not taken into account in the habitability zone. Indeed, the magnetic activity of the host stars could wipe away the atmosphere of the planet. The study of magnetic activity of solar-like stars also allows us to have a better understanding of the magnetic activity of the Sun.

2 Analyzing photometric data

1. **Rotation and magnetic activity of solar-like pulsating stars:** Different methods can be used to measure surface rotation periods ($P_{\rm rot}$): periodogram, time-frequency analysis (wavelets; e.g., [2]) as shown in Figure 1, and auto-correlation function (ACF; e.g., [3]).

⋆smathur@spacescience.org

We define a photometric magnetic activity index, <Sph>. We compute the standard deviation of subseries of length $5 \times P_{\rm rot}$ that gives Sph(t) ([4]). Its mean value provides the <Sph> index.

We analyzed 540 stars showing solar-like oscillations observed by *Kepler* ([5]). We obtained a reliable rotation period for 310 stars using the light curves calibrated by [6]. We found that hot stars rotate faster and the photospheric magnetic activity of the pulsating solar-like stars is compatible with the solar magnetic activity during the full cycle, which is also the case for 18 solar analogs in the sample ([7]). This also allowed us to study gyrochronology relationships ([8]).

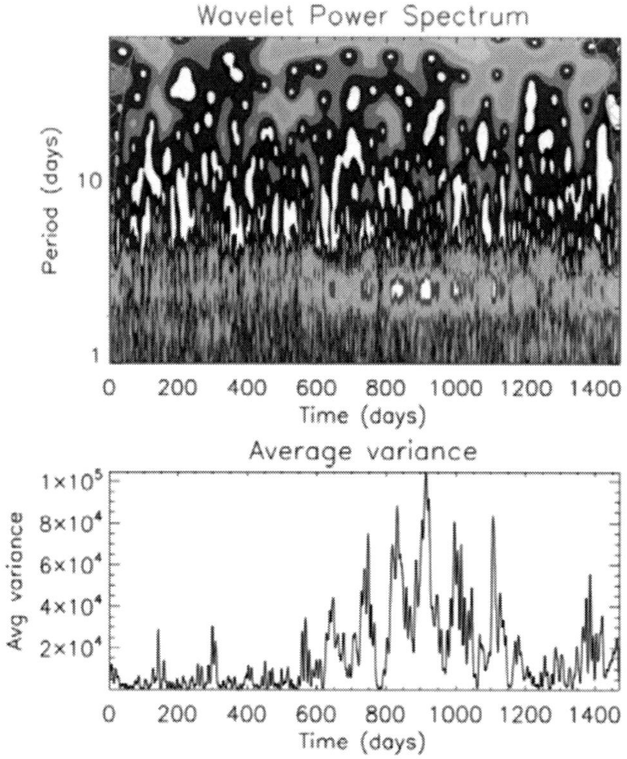

Figure 1. Wavelet analysis of the *Kepler* target KIC 3733735. Upper panel: wavelet power spectrum resulting from the correlation between the time series and a mother wavelet that we slide along time (x-axis) and change the period (y-axis). Red and white areas correspond to higher power. Bottom panel: projection of the wavelet power spectrum on the x-axis showing the change in the magnetic activity with time.

2. **Magnetic Activity Cycles:** Time-frequency analysis allows us to study possible magnetic activity cycles ([9]). Figure 1 shows an example of a cycle observed for an F-type star.

   Acoustic modes are also known to be affected by magnetic activity ([10]). During maximum activity the amplitudes of the modes decrease while the frequencies increase. This has been observed on the Sun and a few solar-like stars targeted by CoRoT and *Kepler* (e.g., [11, 12]).

Acknowledgments: SM would like to acknowledge support from NASA grants NNX12AE17G, NNX14AB92G and NNX15AF13G and NSF grant AST-1411685. LB acknowledges NSF REU grant 1157020. DS and RAG acknowledge the financial support from the CNES GOLF and PLATO grants.

References

[1] Mathur, S., Huber, D., Batalha, N. M., et al., ApJS, **229**, 30 (2017)
[2] Mathur, S., García, R. A., Régulo, C., et al., A&A, **511**, A46 (2010)
[3] Ceillier, T., van Saders, J., García, R. A., et al., MNRAS, **456**, 119 (2016)
[4] Mathur, S., Salabert, D., García, R. A., & Ceillier, T., JSWSC, **4**, A15 (2014)
[5] García, R. A., Ceillier, T., Salabert, D., et al., A&A, **572**, A34 (2014)
[6] García, R. A., Hekker, S., Stello, D., et al., MNRAS, **414**, L6 (2011)
[7] Salabert, D., García, R. A., Beck, P. G., et al., A&A, **596**, A31 (2016)
[8] van Saders, J. L., Ceillier, T., Metcalfe, T. S., et al., Nature, **529**, 181 (2016)
[9] Mathur, S., García, R. A., Ballot, J., et al., A&A, **562**, A124 (2014)
[10] García, R. A., Mathur, S., Salabert, D., et al., Science, **329**, 1032 (2010)
[11] Salabert, D., Régulo, C., García, R. A., et al., A&A, **589**, A118 (2016)
[12] Kiefer, R., Schad, A., Davies, G., & Roth, M., A&A, **598**, A77 (2017)

Expected oscillation parameters for red giants from dynamical masses and radii

Nathalie Themeßl[1,2,3,*] and *Saskia* Hekker[1,2]

[1] *Max Planck Institute for Solar System Research, Justus-von-Liebig-Weg 3, 37077 Göttingen, Germany*
[2] *Stellar Astrophysics Centre, Aarhus University, Ny Munkegade 120, 8000 Aarhus C, Denmark*
[3] *Institute for Astrophysics, Georg-August Univ. Göttingen, Friedrich-Hund-Platz 1, 37077 Göttingen, Germany*

Abstract. Stellar masses and radii of stars with stochastic (solar-like) oscillations can be derived using asteroseismic scaling relations. Here, we predict the asteroseismic observables using the dynamical masses and radii from red-giants in eclipsing binary systems. We show that the predicted frequency of maximum oscillation power ($\nu_{\rm max}$) is generally lower than the observed one, while the predicted large frequency separation ($\Delta\nu$) is in most cases larger than the observed value. This shows that both the scaling relations for $\Delta\nu$ and $\nu_{\rm max}$ used with solar references contribute to the observed differences between dynamical and asteroseismically determined masses and radii.

In the era of space missions, in particular *Kepler* and K2, solar-like oscillations have been observed in a large number of stars. The study of these stochastically excited oscillations provides knowledge about physical properties, such as mean density and surface gravity and thus stellar mass and radius. From the Fourier spectrum of red giants it is possible to directly measure two asteroseismic parameters, the frequency of maximum oscillation power $\nu_{\rm max}$ and the large frequency separation $\Delta\nu$, i.e. the spacing between modes of the same spherical degree and consecutive radial order. These quantities together with the effective temperature ($T_{\rm eff}$) of the star can be used to derive stellar masses (M) and radii (R) through the well-known asteroseismic scaling relations ([1, 2]):

$$\frac{\Delta\nu}{\Delta\nu_{\odot}} \approx \sqrt{\frac{M/\mathrm{M}_{\odot}}{(R/\mathrm{R}_{\odot})^3}} \quad \text{and} \quad \frac{\nu_{\rm max}}{\nu_{\rm max,\odot}} \approx \frac{M/\mathrm{M}_{\odot}}{(R/\mathrm{R}_{\odot})^2\sqrt{T_{\rm eff}/\mathrm{T}_{\rm eff,\odot}}}, \tag{1}$$

which are calibrated to the solar reference values $\Delta\nu_{\odot}$ and $\nu_{\rm max,\odot}$. Here, we use these relations to calculate $\Delta\nu$ and $\nu_{\rm max}$ for known double-lined spectroscopic binaries with red-giant components ([3]). For these systems, dynamical masses and radii can be accurately measured from the binary analysis. We use these measurements in equation 1 and vary the solar reference values as follows: $\Delta\nu_{\odot}$ = [134.9, 135.0, 135.1, 135.2] μHz and $\nu_{\rm max,\odot}$ = [3025, 3050, 3100, 3125, 3150, 3175] μHz. This allows us to predict a range of $\Delta\nu$ and $\nu_{\rm max}$ values for each star (see Fig. 1). The ratios between the asteroseismic parameters determined from the scaling relations and those directly measured from the oscillations are shown in Figure 2. We find that the predicted $\nu_{\rm max}$ ($\Delta\nu$) estimates are lower (higher) than the asteroseismically determined values, with a clear correlation. The right panel of Figure 2 shows $\Delta\nu$ ratio as function of temperature with the metallicity colour-coded. With this small number of stars we are not able to recover the trend in $T_{\rm eff}$ and [Fe/H] that is present in stellar models ([4]).

*themessl@mps.mpg.de

DOI: 10.1051/epjconf/201715205012

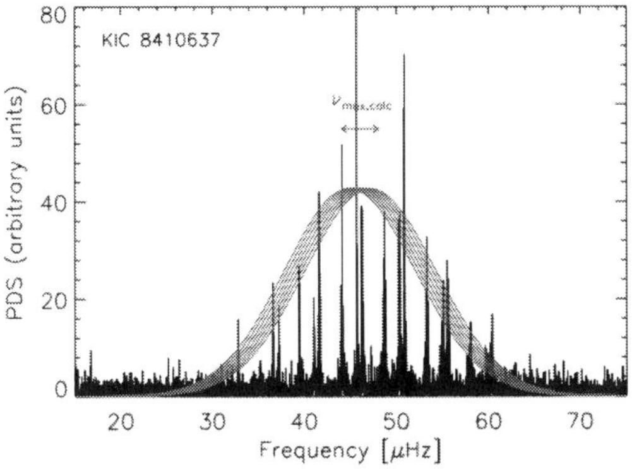

Figure 1. Power density spectrum (in black) of KIC 8410637 centred around $\nu_{\rm max}$. The range of predicted $\nu_{\rm max}$ ($\nu_{\rm max,calc}$) values for this red giant is indicated by the red arrow. The corresponding Gaussian fits (multiplied by a factor for visual purposes) are shown in red. No inconsistencies between the predicted and measured values for $\nu_{\rm max}$ are apparent.

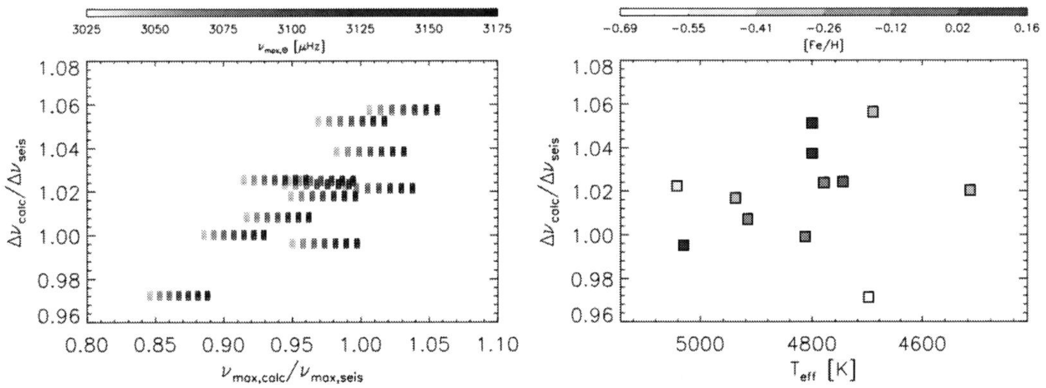

Figure 2. The ratio of predicted $\Delta\nu$ ($\Delta\nu_{\rm calc}$) to the seismically determined $\Delta\nu$ ($\Delta\nu_{\rm seis}$) as a function of the ratio of predicted $\nu_{\rm max}$ ($\nu_{\rm max,calc}$) over observed $\nu_{\rm max}$ ($\nu_{\rm max,seis}$), left, and as a function of effective temperature $T_{\rm eff}$, right. The left panel shows all results for the different solar values (see colour bar) and the right panel shows results using a single solar value ($\Delta\nu_{\odot} = 135.1$ μHz) with the points colour-coded for metallicity (see colour bar).

This pilot study tentatively indicates that for red giants not only the $\Delta\nu$ scaling relation but also the $\nu_{\rm max}$ scaling relation suffers from biases in case the solar reference values are used.

Acknowledgments: NT and SH have received funding from the European Research Council under the European Communitys Seventh Framework Programme (FP7/2007-2013) / ERC grant agreement no 338251 (StellarAges). Funding for the Stellar Astrophysics Centre (SAC) is provided by The Danish National Research Foundation (Grant agreement no.: DNRF106).

References

[1] T.M. Brown, R.L. Gilliland, R.W. Noyes, L.W. Ramsey, ApJ **368**, 599 (1991)
[2] H. Kjeldsen, T.R. Bedding, A&A **293** (1995), astro-ph/9403015
[3] P. Gaulme, J. McKeever, J. Jackiewicz, M.L. Rawls, E. Corsaro, B. Mosser, J. Southworth, S. Mahadevan, C. Bender, R. Deshpande, ApJ **832**, 121 (2016), 1609.06645
[4] E. Guggenberger, S. Hekker, S. Basu, E. Bellinger, MNRAS **460**, 4277 (2016), 1606.01917

Recent progress in the theoretical modelling of Cepheids and RR Lyrae stars

Marcella Marconi[1,*]

[1] *INAF-Osservatorio Astronomico di Capodimonte, Via Moiariello 16 – 80131 Napoli, Italy*

Abstract. Cepheids and RR Lyrae are among the most important primary distance indicators to calibrate the extragalactic distance ladder and excellent stellar population tracers, for Population I and Population II, respectively. In this paper I first mention some recent theoretical studies of Cepheids and RR Lyrae obtained with different theoretical tools. Then I focus the attention on new results based on nonlinear convective pulsation models in the context of some international projects, including VMC@VISTA and the Gaia collaboration. The open problems for both Cepheids and RR Lyrae are briefly discussed together with some challenging future application.

1 Introduction

An important challenge for next decade astrophysics is the determination of stellar distances with a precision better than 5% from the Local Group to about 200 Mpc as the crucial ingredient not only to constrain the Universe geometry but also to characterize several physical properties of stars and galaxies (including dynamical time scales, masses, ages, linear sizes, etc.). In the Milky Way and in photometrically resolved galaxies, the study of the different types of hosted pulsating stars (in different evolutionary phases) allows us to trace the intrinsic properties (age and chemical composition) and, through individual distance estimates, the spatial distribution of the associated stellar populations. This approach also permits to point out the possible presence of radial trends, haloes and/or streams, and in turn to reconstruct the star formation history of the investigated galaxy and to obtain information on galactic formation and evolution mechanisms. Pulsating stars such as Classical Cepheids and RR Lyrae are excellent primary distance indicators and stellar population tracers ([5, 10, 21, 22, 26–28]). This twofold role is particularly challenging in the era of present and future space astrometric missions. Through a multi-epoch monitoring of the full sky, Gaia ([11, 12]) will get the position, the parallax, the proper motion and the time series photometry of thousands of pulsating stars in the Milky Way and its surroundings, down to a faint magnitude limit of $G \sim 20.7$ mag. In particular, Gaia's complete census of the Galactic Cepheids and RR Lyrae will allow a breakthrough in our understanding of the Galactic structure by tracing variable stars of various ages in the Galactic bulge, disc, halo, likely revealing new streams and faint satellites as the signatures of the Milky Way's hierarchical build-up. Moreover Gaia parallaxes will provide an unprecedented absolute calibration of the extragalactic distance scale ([11, 12]). Similarly, in the near future, missions such as LSST will allow

*marcella.marconi@oacn.inaf.it

us to extend Gaia capabilities to Local Group and external galaxies. Finally, using JWST and subsequently MICADO@E-ELT (with spatial resolution similar to the one of the *Hubble Space Telescope*) it will be possible to extend Classical Cepheid observations up to the crowded fields of galaxy clusters well beyond Virgo and RR Lyrae observations in elliptical and giant spiral galaxies up to 6 Mpc. In order to properly exploit the results of these crucial missions, accurate hydrodynamical models are needed in order to constrain the relation between the observed pulsation properties and the intrinsic stellar parameters, that is the physical basis of the cosmic distance scale calibration through pulsating stars ([2]). On the other hand, the comparison between theory and observation is crucial to constrain the physical and numerical assumptions adopted in both evolutionary and pulsational models and, in turn, to improve our knowledge of stellar physics ([14, 19, 20, 22]).

2 Some recent theoretical studies of Cepheids and RR Lyrae

To properly exploit the importance of pulsating stars as property tracers and standard candles of different stellar populations, accurate hydrodynamical models are needed. During the last few decades there has been an increasing interest in predicting the observed properties of pulsating stars by adopting different theoretical approaches: linear non adiabatic hydrodynamical models, non linear convective hydrodynamical 1D models, multidimensional hydrodynamical simulations. Among the most recent results: [1] adopted linear non adiabatic hydrodynamical models to investigate the effect of rotation on Cepheid observables as a function of metallicity (see also Anderson et al., this volume); [29] used non linear convective hydrodynamical 1D models to investigate Type II Cepheid dynamical instability and period doubling as a function of stellar parameters (see also Smolec et al., this volume); assuming a similar theoretical scenario our team developed powerful theoretical tools to use RR Lyrae and Classical Cepheids as distance indicators and stellar population tracers (see, e.g., [5, 10, 20, 21], and references therein); finally, with pioneering multidimensional hydrodynamical simulations, [23, 24] produced realistic simulations of the convection-pulsation interaction in Cepheids, as powerful guidelines for improving the description of convection in 1D modelling. In the following we will focus on some results of recent nonlinear convective 1D pulsation models.

3 The model fitting of light and radial velocity curves

One of the advantages of nonlinear convective models is the possibility to directly compare observed and predicted light, radial velocity, radius variations. The first application of such a comparison was performed by [30] who compared the predicted and the observed light curve of a Magellanic Cepheid. The method was then extended to a Galactic field RR Lyrae by [3]. Since then, many applications to Cepheids and RR Lyrae followed in the literature (see, e.g., [16, 18–20, 22, 25], and references therein). In particular in [22] we performed the model fitting of light and radial velocity curves for a sample of 12 Small Magellanic Cloud (SMC) Cepheids with optical data from the OGLE collaboration and near-infrared data from the ESO Public Survey VMC@VISTA (see Cioni's paper in this volume and [22] for details). Some interesting implications of this analysis were: i) the deviation from a canonical mass-luminosity relation for SMC Cepheids possibly due to a combination of mass-loss, overshooting, rotation; ii) a period-radius relation in agreement with previous radius measurements in the SMC; iii) a mean value of the SMC distance modulus in agreement with the literature; iv) period-luminosity and period-Wesenheit relations in agreement with empirical OGLE and VMC relations. A similar analysis but for Large Magellanic Cloud (LMC) Cepheids is in progress (Ragosta et al., in preparation). Preliminary results are shown in Figure 1 for the inferred mass-luminosity relation. We note that, in agreement with previous results in the Milky Way ([8]) and the SMC ([22]), the predicted

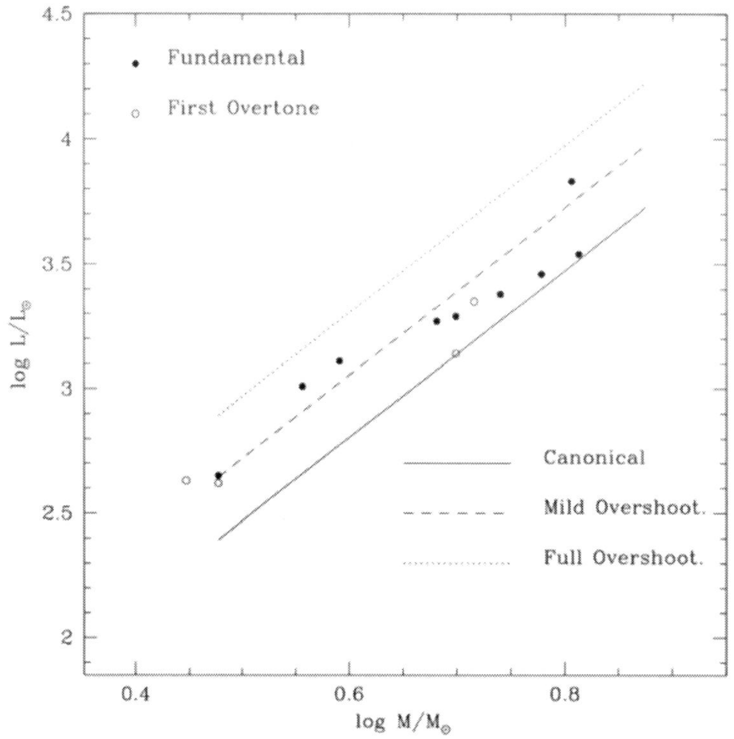

Figure 1. The inferred mass-luminosity relation from the model fitting of a sample of LMC Cepheids with OGLE optical and VMC@VISTA near-infrared photometry (Ragosta et al., in prep).

mass-luminosity relation is somewhat intermediate between the canonical and the full overshooting evolutionary relations. We also took into account a number of Galactic Cepheids for which the Gaia first data release ([11, 12]) provided new trigonometric parallaxes and optical and near-infrared data were available from the literature. We applied the multi-filter light curve model fitting to them and compared the inferred model parallaxes with Gaia values, finding a satisfactory agreement. An example is shown in Figure 2 for the Galactic Cepheid RS Pup. For this star, characterized by a pulsation period of 41.5 d, an independent geometric parallax has been provided on the basis of the circumstellar nebula reflecting the light from the central star ([15]). This value corresponds to $\Pi_{K08} = 0.502 \pm 0.007$ mas and is consistent with the Gaia first data release value $\Pi_{TGAS} = 0.63 \pm 0.26$ mas. On the other hand, the model fitting technique illustrated in Figure 2 provides an intrinsic distance modulus of 11.1 ± 0.1 mag, that implies $\Pi_{FIT} = 0.58 \pm 0.03$ mas, in excellent agreement with both the previous estimates.

4 The role of the helium content in Cepheid models

Cepheid pulsation properties are affected not only by metallicity but also by the helium content ([7, 17]). This implies that once the metallicity of a Cepheid sample is known, e.g. from spectroscopic measurements, the investigation of Cepheid pulsation properties can be used to constrain the helium content and in turn the helium to metal enrichment ratio. An example could be the well known

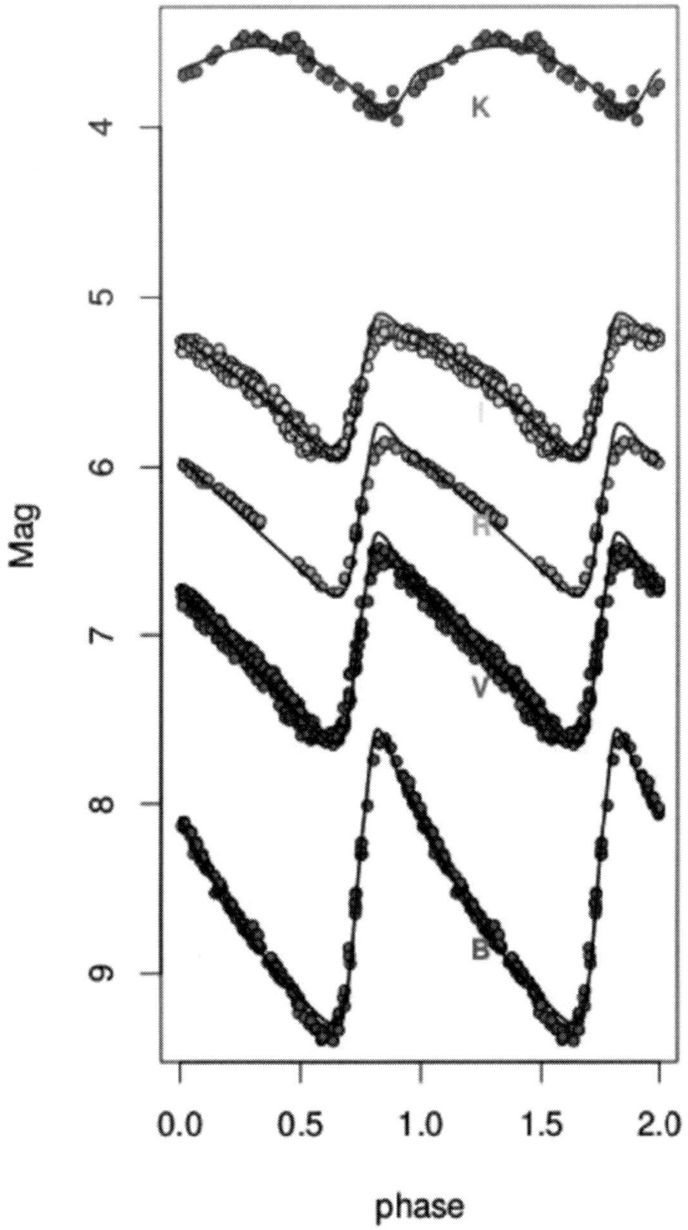

Figure 2. Model fitting of the multi-filter light curve of Gaia Galactic Cepheid RS Pup.

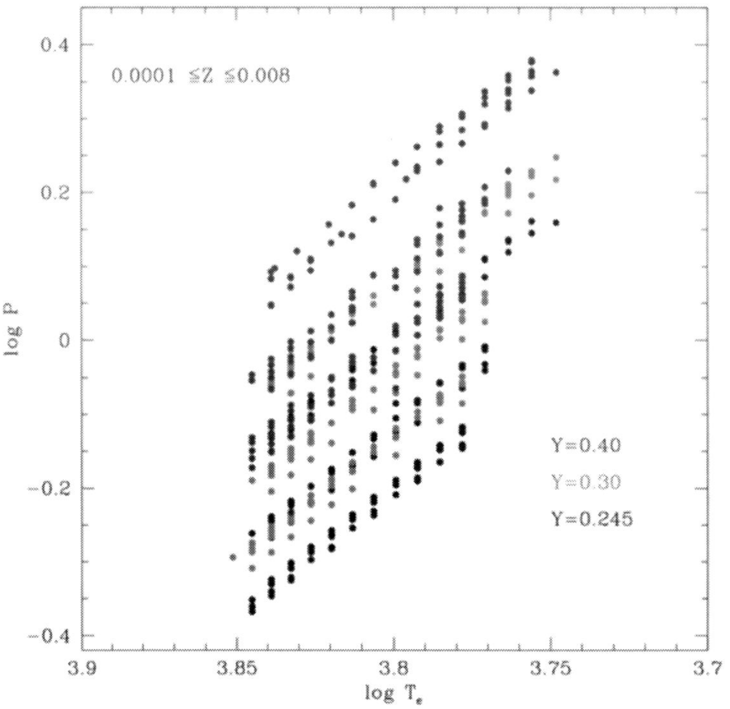

Figure 3. Predicted period distribution of fundamental RR Lyrae as a function of the assumed helium content.

Hertzsprung progression ([13]) of Cepheid light curves with periods roughly between 6 and 16 d. The pulsators in this period range show a bump along both the light and the velocity curves. The phase of this secondary feature varies with the period: it is on the curve descending branch for periods up to about 9 days, then it appears close to maximum light up to about 12 d, to move to the rising branch for longer periods (see, e.g., [4], and references therein). Current non linear convective models support the observational evidence that the Hertzsprung progression central period (P_{HP}) moves to longer values as the metallicity decreases. On the other hand, fixing the metallicity and increasing the HSelium content moves P_{HP} to longer values. For example assuming the LMC metallicity Z=0.008 and increasing the He content from the standard Y = 0.25 value to Y = 0.35, one again obtains $P_{HP} \approx$ 13 d (see [7]). This implies that once known the metallicity, the comparison between observed and predicted light curves, for bump Cepheids, might help constrain the helium content and in turn the helium to metal enrichment ratio of the associated stellar population.

5 A new theoretical scenario for RR Lyrae stars: The dependence on chemical composition

During the last few years we have updated the theoretical scenario for RR Lyrae stars as based on a new set of metal-dependent non linear convective pulsation models (see [21] for details). One of the most important results of this investigation was the predicted universality of the period-Wesenheit

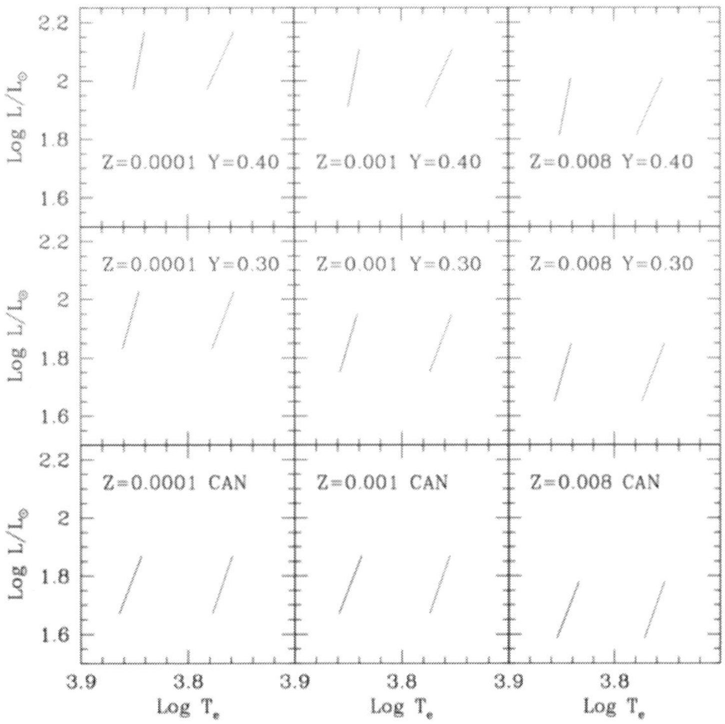

Figure 4. Predicted instability strip extreme boundaries (first overtone blue edge and fundamental red edge) as a function of both metallicity and helium content (see labels).

relation in V, $(B-V)$ for RR Lyrae stars. Indeed this relation was found to show a negligible dependence on the assumed model metal content, whereas for other band combinations a metallicity effect is expected. This theoretical framework was applied with success to observational samples, e.g. in the Galactic globular cluster M4 (see [6]) and in the dwarf spheroidal galaxy Carina (see [9]). An important new prediction of these models was an analytical relation to infer double mode RR Lyrae masses as a function of the period ratio and the metallicity. This relation is

$$\log(M/M_{\odot}) = -0.85(\pm0.05) - 2.8(\pm0.3)\log(P_{FO}/P_F) - 0.097(\pm0.003)\log Z,$$

where P_{FO} and P_F are the first overtone and fundamental pulsation period, respectively (see [21] for details).

There are also some new results on the dependence of RR Lyrae evolutionary and pulsation properties on the He abundance (Marconi et al., in preparation). The main effect of changing the helium content from the canonical value, for a given metal abundance, to values as high as Y = 0.30 or Y = 0.40, is that the evolutionary zero age horizontal branch luminosity significantly increases, with the consequent shift of predicted pulsation periods towards longer values. In Figure 3 we show the dependence of the predicted period distribution of fundamental RR Lyrae on the assumed helium content. The period distribution is shifted to longer values as Y increases. An interesting implication of the behaviour shown in Figure 3 is that also the minimum RR_{ab} period is predicted to increase with the helium abundance, thus representing a powerful observable to constrain the helium to metal enrichment ratio of a stellar population.

On the other hand, we expect a quite small effect on the predicted instability strip. As shown in Figure 4, the location of the extreme boundaries of the RR Lyrae instability strip, namely the first overtone blue edge and the fundamental red edge, as a function of the chemical composition, shows that the width of the instability strip is almost unaffected by abundance variations, with the main effect related to the increase of the corresponding luminosities.

6 Some (not all) open problems and possible future developments

Even if the presented theoretical scenario is able to predict all the relevant pulsation observables and has been successfully applied to both Galactic and extragalactic samples, there are a number of open problems that deserve further theoretical investigation. Just to quote a few of them:

- Current non linear convective 1D models are not completely able to reproduce the light curve morphology of the reddest Cepheids and RR Lyrae due to extreme sensitivity to details of the convective turbulent model;
- It is very difficult to disentangle mass loss and core overshooting (and rotation) effects on the mass-luminosity relation of Classical Cepheids;
- Metallicity and helium variations simultaneously affect Cepheid and RR Lyrae light curves, amplitudes, instability strips, period-luminosity and period-luminosity-colour relations and it is hard to disentangle the two contributions.

In order to clarify these important issues we need to improve the treatment of turbulent convection in pulsation models and to provide accurate independent constraints on physical and numerical assumptions. In this context we expect that Gaia and future LSST measurements will have a crucial impact. For example if accurate distances are available from these missions and metallicities are known from complementary spectroscopic surveys, we will be able to constrain the helium content and the stellar mass.

7 Conclusions

Classical Cepheids and RR Lyrae continue to be of great interest for theoreticians. The model fitting of light and radial velocity curves of Magellanic Classical Cepheids with OGLE optical and VMC@VISTA NIR data provides interesting results for the mass-luminosity relation, the period-radius, the period-luminosity and the Wesenheit relations. The application to Galactic Cepheids with parallaxes obtained from the Gaia first data release also gives promising results. The investigation of the dependence of RR Lyrae nonlinear convective models on metallicity has predicted the universality of the period-Wesenheit relation, a 3D version of the Petersen diagram and distances in agreement with the literature. The He content also affects the pulsation properties of both Cepheids and RR Lyrae. In spite of the power of the current theoretical scenario in predicting all the relevant pulsation observables, there are a number of open problems that require further theoretical work. On the other hand, we expect that both Gaia and LSST results will represent a benchmark for pulsation theories and a challenging opportunity to constrain numerical and physical assumptions in both evolutionary and pulsation models.

Acknowledgments: I deeply thank the organizers of the meeting "Wide-field variability surveys: a 21st-century perspective" for inviting me.

References

[1] Anderson, R. I., Saio, H., Ekström, S., Georgy, C., & Meynet, G., A&A, **591**, A8 (2016)
[2] Bono, G., Caputo, F., Castellani, V., & Marconi, M., ApJ, **512**, 711 (1999)
[3] Bono, G., Castellani, V., & Marconi, M., ApJL, **532**, L129 (2000a)
[4] Bono, G., Marconi, M., & Stellingwerf, R. F., A&A, **360**, 245 (2000b)
[5] Bono, G., Caputo, F., Marconi, M., & Musella, I., ApJ, **715**, 277 (2010)
[6] Braga, V. F., Dall'Ora, M., Bono, G., et al., ApJ, **799**, 165 (2015)
[7] Carini, R., Brocato, E., Marconi, M., & Raimondo, G., A&A, **561**, A110 (2014)
[8] Caputo, F., Bono, G., Fiorentino, G., Marconi, M., & Musella, I., ApJ, **629**, 1021 (2005)
[9] Coppola, G., Marconi, M., Stetson, P. B., et al., ApJ, **814**, 71 (2015)
[10] Fiorentino, G., Musella, I., & Marconi, M., MNRAS, **434**, 2866 (2013)
[11] Gaia Collaboration, Brown, A. G. A., Vallenari, A., et al., A&A, **595**, A2 (2016a)
[12] Gaia Collaboration, Prusti, T., de Bruijne, J. H. J., et al., A&A, **595**, A1 (2016b)
[13] Hertzsprung, E., Bulletin of the Astronomical Institutes of the Netherlands, **3**, 115 (1926)
[14] Keller, S. C., & Wood, P. R., ApJ, **642**, 834 (2006)
[15] Kervella, P., Mérand, A., Szabados, L., et al., A&A, **480**, 167 (2008)
[16] Marconi, M., & Clementini, G., AJ, **129**, 2257 (2005)
[17] Marconi, M., Musella, I., & Fiorentino, G., ApJ, **632**, 590 (2005)
[18] Marconi, M., & Degl'Innocenti, S., A&A, **474**, 557 (2007)
[19] Marconi, M., Molinaro, R., Ripepi, V., Musella, I., & Brocato, E., MNRAS, **428**, 2185 (2013a)
[20] Marconi, M., Molinaro, R., Bono, G., et al., ApJL, **768**, L6 (2013b)
[21] Marconi, M., Coppola, G., Bono, G., et al., ApJ, **808**, 50 (2015)
[22] Marconi, M., Molinaro, R., Ripepi, V., et al., MNRAS, **466**, 3206 (2017)
[23] Mundprecht, E., Muthsam, H. J., & Kupka, F., MNRAS, **435**, 3191 (2013)
[24] Mundprecht, E., Muthsam, H. J., & Kupka, F., MNRAS, **449**, 2539 (2015)
[25] Natale, G., Marconi, M., & Bono, G., ApJL, **674**, L93 (2008)
[26] Riess, A. G., Macri, L. M., Hoffmann, S. L., et al., ApJ, **826**, 56 (2016)
[27] Ripepi, V., Marconi, M., Moretti, M. I., et al., ApJS, **224**, 21 (2016)
[28] Saha, A., Sandage, A., Tammann, G. A., et al., ApJ, **562**, 314 (2001)
[29] Smolec, R., MNRAS, **456**, 3475 (2016)
[30] Wood, P. R., Arnold, A. S., & Sebo, K. M., ApJL, **485**, L25 (1997)

How rotation affects masses and ages of classical Cepheids

Richard I. Anderson[1,⋆], *Sylvia* Ekström[2], *Cyril* Georgy[2], *Georges* Meynet[2], and *Hideyuki* Saio[3]

[1] *Department of Physics and Astronomy, The Johns Hopkins University, Baltimore, MD 21218, USA*
[2] *Département d'Astronomie, Université de Genève, 1290 Sauverny, Switzerland*
[3] *Astronomical Institute, Graduate School of Science, Tohoku University, Sendai, 980-8578 Miyagi, Japan*

Abstract. Classical Cepheid variable stars are both sensitive astrophysical laboratories and accurate cosmic distance tracers. We have recently investigated how the evolutionary effects of rotation impact the properties of these important stars and here provide an accessible overview of some key elements as well as two important consequences. Firstly, rotation resolves the long-standing Cepheid mass discrepancy problem. Second, rotation increases main sequence lifetimes, i.e, Cepheids are approximately twice as old as previously thought. Finally, we highlight the importance of the short-period ends of Cepheid period distributions as indicators for model adequacy.

1 Introduction

Classical Cepheid variable stars (henceforth: Cepheids) are yellow (super-)giant stars of intermediate mass ($3 - 10\,M_{\odot}$) well-suited for investigating how rotation affects stellar evolution in general. The evolutionary status of Cepheids can be accurately inferred, since these stars occupy a well-defined region in the Hertzsprung-Russell diagram, the classical instability strip (IS). Additionally, Cepheid variability is very noticeable and straightforward to identify. Due to the timescales involved, nearly all Cepheids are observed during the blue loop evolutionary phase, i.e., during second or third IS crossings, which are inferred using observed rates of period change ([1]). Since blue loop evolution is highly sensitive to model assumptions and input physics ([2]), the same holds for Cepheid properties.

The impact of rotation on stellar evolution increases with mass ([3]), and Cepheids occupy somewhat of a "sweet spot" in this regard. They are massive enough for rotational mixing to be relevant, but not as dominated by binary interactions and radiatively-driven winds as more massive stars ([4, 5]). That said, many Cepheids are part of binary systems ([6–8]) and interactions may occur during various stages of a star's evolution ([9, 10]).

Geneva stellar evolution models provide insights into the effects of rotation for a wide range of masses and metallicities ([11, 12]). Capitalizing on their sensitivity to stellar physics, we have carried out the first detailed investigations of how rotation impacts the evolution and pulsation properties of Cepheids ([13, 14]). In this contribution, we provide an accessible overview of some key notions regarding rotation and argue that rotation is required to correctly predict Cepheid properties. Specifically, we focus on the Cepheid mass discrepancy (§2) and period-age relations (§3). We close by highlighting the importance of correctly reproducing the short-period end of the Cepheid period distribution as well as the need to test predictions that are unmistakably linked to rotation (§4).

⋆ria@jhu.edu

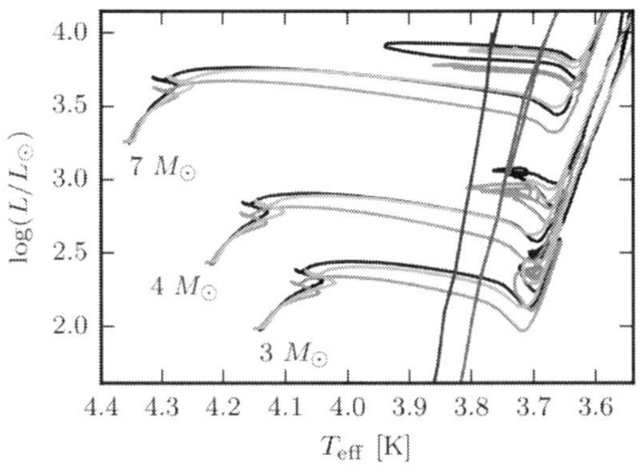

Figure 1. Evolutionary tracks for 3, 4, and 7 $M_\odot$ models with LMC metallicity ($Z = 0.006$). Cyan and black lines: Geneva models without and with average initial (ZAMS) rotation ($\omega = \Omega/\Omega_{\rm crit} = 0.5$). Yellow lines: MESA models ([15]) with $v/v_{\rm crit} = 0.4$, i.e., $\omega = 0.586$. 5 $M_\odot$ models not shown for clarity. Blue and red IS boundaries from [14]. Cepheids are highly sensitive to model assumptions and input physics, being observed almost exclusively during blue loop evolution, i.e., 2nd and 3rd IS crossings. Rotation modifies the mass-luminosity relation and resolves the long-standing Cepheid mass discrepancy (§2).

2 How rotation resolves the Cepheid mass discrepancy problem

Much work has been done to resolve the long-standing Cepheid mass discrepancy problem (cf. [16, 17], and references therein), which manifests in a 10 – 20% overestimate of Cepheid masses inferred from stellar evolution models compared to other (pulsational, or model-independent) mass estimates. Conceptually, there are two possibilities of resolving the mass discrepancy: a) lowering the mass of a given model at fixed luminosity, e.g. via enhanced mass-loss ([18]), and b) increasing luminosity at fixed mass, e.g. by increasing effective stellar core sizes via enhanced convective core overshooting (e.g., [19]) or rotational mixing ([13]). Despite growing observational evidence for circumstellar material in the vicinity of Cepheids ([20–22]), pulsationally-enhanced mass-loss rates ([18], $\dot{M} \sim 10^{-10} - 10^{-8}\,M_\odot{\rm yr}^{-1}$) do not appear sufficient to resolve the mass discrepancy given that the Cepheid evolutionary phase is relatively short-lived (10^4 – 10^6 yr). Rotation, however, provides a sufficient increase in luminosity to resolve the mass discrepancy, cf. Figures 1 and 2.

Geneva stellar evolution models include convective core overshooting ($d_{\rm over}/H_P = 0.1$) to match the observed width of the main sequence over the mass range $1.35 - 9\,M_\odot$ ([11]) and have not been adjusted to reproduce properties of evolved stars. Notably, Geneva models include a consistent treatment of (radially) differential rotation that induces mixing throughout a star at all times during its evolution. Rotational mixing occurs even in radiative layers, notably replenishing the core's hydrogen supply while bringing processed materials to the surface during main sequence evolution. Thus, rotational mixing increases the core's effective mass, although hydrostatic effects (centrifugal forces) counteract this increase for very fast rotation ([13]). Due to this competition, luminosity does not increase

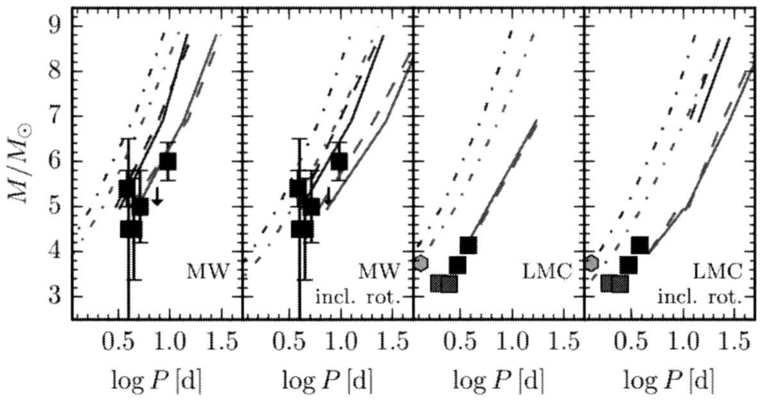

Figure 2. Fundamental mode mass-period relations for Solar ($Z = 0.014$, MW) and LMC ($Z = 0.006$) models with and without rotation compared to model-independent mass estimates for fundamental mode (black squares), first overtone (green squares), and one anomalous Cepheid (yellow hexagon); cf. [14], and references therein.

monotonously with increasing rotation. Rather, luminosity reaches a maximum near typical initial angular rotation rates, varies relatively slowly over the range of most probable $\omega = \Omega/\Omega_{\rm crit} \sim 0.5$, and falls off towards both extremes. However, even the models with the fastest rotation feature higher luminosities than models with no rotation.

Rotation thus modifies the mass-luminosity relation (MLR). Since initial angular velocities (ω) differ from star to star, it follows that MLRs have intrinsic scatter. Of course, MLRs also change during the evolution of a star, e.g. between 2nd and 3rd IS crossings (Fig. 1). Hence, to resolve the mass discrepancy, both effects (rotation and crossing) must be considered simultaneously ([13]). Mass-period relations provide more direct observational tests, since pulsation periods are easily measured and since luminosity differences between 2nd and 3rd crossings imply changes in stellar structure that affect the pulsation equation ($P = Q/\sqrt{\bar{\rho}}$), where $\bar{\rho}$ denotes average density and Q is the pulsation "constant" ([23]). Figure 2 shows that observed and predicted mass-period relations in the Milky Way (MW) and Large Magellanic Cloud (LMC) agree for models that include average levels of rotation.

The intrinsic scatter of MLRs implied by rotation is further relevant for predicting the period-luminosity-color (PLC) relation, which follows from inserting an *assumed* MLR into the pulsation equation combined with the Stefan-Boltzmann law. Given that the MLR itself is not unique, the PLC-relation cannot be unique, either. It would be interesting to test the impact of any associated effects (e.g., changes in He abundance) on Cepheid light and radial velocity variations ([24]).

3 Rotational rejuvenation: Cepheids older than previously thought

Period-age relations ([25]) provide useful age estimates of stellar populations based on readily-observed pulsation periods. The age of a star in the Cepheid stage is dominated by its main sequence lifetime, followed by the duration of core He burning before the onset of blue loop evolution and the evolution along it. Rotation significantly prolongs the duration over which a star can produce energy via fusion in its core by effectively increasing core size, leading to increased lifetimes for increasing levels of rotation. Note that period-age relations for models with average and fast rotation are comparable, in contrast to the effects of rotation on MLRs (§2, [13]). Using Geneva models ([14]), we have quantified this effect and find that typical levels of rotation imply Cepheids to be older by approximately 60 – 100% than if rotation is not taken into account (cf. Fig. 3).

This change in Cepheid age scale is particularly important if Cepheid ages are compared with other age indicators. To illustrate the impact, ages for three $\sim$ 20 d Cepheids located in the nuclear bulge of the Galaxy ([26]) have been estimated to be 25 ± 5 Myr using solar-metallicity period-age relations that do not account for rotation ([27]). Yet, period-age relations that take into account typical

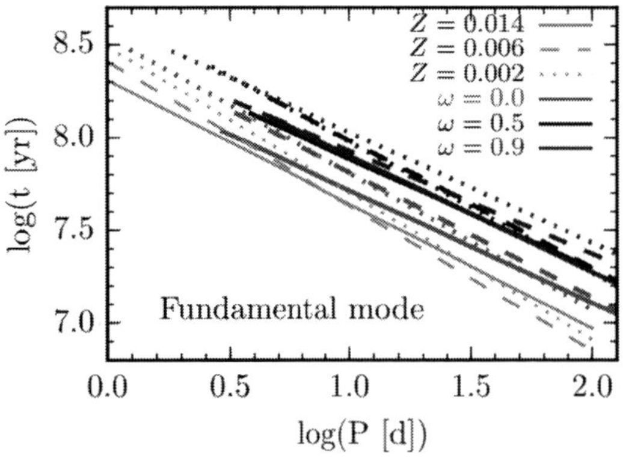

Figure 3. Period-age relations for fundamental mode Cepheids [14]. Line styles differentiate Solar, LMC, and SMC metallicity and colors indicate ZAMS rotation rate (ω) as indicated. Cyan lines are literature period-age relations that do not take into account rotation and are on a slightly different metallicity scale ($Z_\odot = 0.02$, $Z_{\rm LMC} = 0.008$, $Z_{\rm SMC} = 0.004$, [27]). Rotation extends main sequence lifetimes by mixing unprocessed material into the core during H-burning, resulting in significantly ($\Delta \log t = 0.2 - 0.3$ dex) older Cepheids.

initial rotation rates imply typical ages of 50 – 52 Myr, depending on the IS crossing. In the future, it will be important to cross-check the Cepheid age scale with other (hopefully independent) age indicators, such as asteroseismic studies of lower-mass stars in which rotational mixing is less efficient. In particular, Cepheids located in open clusters ([28]) are important test-beds for models, providing a means to compare period-age relations to cluster isochrone ages or potentially asteroseismic ages of cluster members. Further tests of the rotational mixing efficiency, especially via CNO surface abundances, will provide important cross-checks to evaluate the accuracy of the predictions.

4 Conclusions

Rotation has significant implications for Cepheids, resolving the mass discrepancy problem and implying significantly older ages than previously thought. Rotation leads to additional testable predictions, since rotation impacts stellar structure, evolution, mass-loss, and surface abundances ([3]). To this end, we have carried out extensive comparisons with different kinds of observational data ([13, 14]) in the past and will continue to confront model predictions to observations.

A related key observable is *the distribution of Cepheid pulsation periods*, which is sensitive to the blueward extent of blue loops: the lower the minimum mass of a model entering the IS, the lower the predicted minimum period. Short-period Cepheids are by far the most numerous in a population due to long IS crossing timescales and the shape of the initial mass function. Given that blue loop evolution is suppressed in models with high values of convective core overshooting, the short-period peak sets a stringent constraint on this important parameter. Figure 4 shows that Geneva models correctly predict the peak in period for fundamental mode (FU) Cepheids[1]. Intriguingly, the predicted minimum period for first overtone (FO) Cepheids is longer than observed. These points will be investigated further using synthetic populations ([29]).

CNO surface abundances probe the efficiency of main sequence rotational mixing, which is directly related to the Cepheid age scale. In a preliminary comparison using literature abundances, it appeared that CNO surface abundance enhancement is over-predicted by Geneva models ([30]). However, rotation does appear to be required to correctly reproduce the extended main sequence turnoffs of intermediate-age open clusters ([31]). Further observational tests of these effects are required to improve the models and ensure the accuracy and consistency of stellar age scales.

The ESA space mission *Gaia* will provide crucial information in this regard, allowing to firmly establish the often uncertain cluster membership of Cepheids ([28]) and thus significantly improving

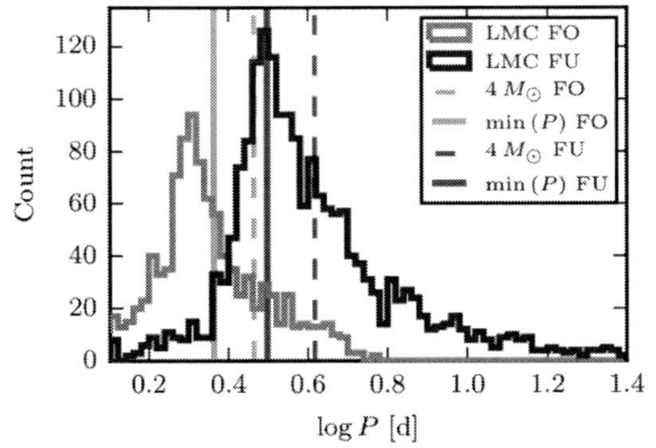

Figure 4. Period distribution of OGLE FU and FO Cepheids ([32]) compared to periods predicted by Geneva models with $Z = 0.006$ and average rotation ([14]). Dashed vertical lines: lowest-mass ($4\,M_\odot$) computed model to enter IS. Solid vertical lines: grid-interpolated $3.7\,M_\odot$ model whose blue loop just barely reaches the red edge of the IS. Pulsation periods for interpolated models are based on a linear extrapolation of $\log(\sqrt{\bar{\rho}}) - \log(P)$ relations ([14]). The FU peak is well-matched by predictions. The puzzling mismatch for the FO peak requires further investigation.

[1] Minimum mass of models entering IS during blue loop evolution determined by interpolation

the ability to use cluster populations as test-benches for different assumptions, input physics, and numerical implementations required to model stellar evolution.

References

[1] Turner, D. G., Abdel-Sabour Abdel-Latif, M., & Berdnikov, L. N., PASP, **118**, 410 (2006)
[2] Kippenhahn, R. & Weigert, A., *Stellar Structure and Evolution* (Springer-Verlag, Berlin–Heidelberg–New York; also Astronomy and Astrophysics Library XVI, 1994)
[3] Maeder, A., & Meynet, G., ARA&A, **38**, 143 (2000)
[4] Puls, J., Vink, J. S., & Najarro, F., A&ARv, **16**, 209 (2008)
[5] Sana, H., de Mink, S. E., de Koter, A., et al., Science, **337**, 444 (2012)
[6] Szabados, L., IBVS, **5394**, 1 (2003)
[7] Evans, N. R., Berdnikov, L., Lauer, J., et al., AJ, **150**, 13 (2015)
[8] Anderson, R. I., Casertano, S., Riess, A. G., et al., ApJS, **226**, 18 (2016)
[9] Anderson, R. I., Sahlmann, J., Holl, B., Eyer, L., Palaversa, L., Mowlavi, N., Süveges, M., & Roelens, M., ApJ, **804**, 144 (2015)
[10] Neilson, H. R., Schneider, F. R. N., Izzard, R. G., Evans, N. R., & Langer, N., A&A, **574**, A2 (2015)
[11] Ekström, S., Georgy, C., Eggenberger, P., et al., A&A, **537**, A146 (2012)
[12] Georgy, C., Ekström, S., Granada, A., Meynet, G., Mowlavi, N., Eggenberger, P., & Maeder, A., A&A, **553**, A24 (2013)
[13] Anderson, R. I., Ekström, S., Georgy, C., Meynet, G., Mowlavi, N., & Eyer, L., A&A, **564**, A100 (2014)
[14] Anderson, R. I., Saio, H., Ekström, S., Georgy, C., & Meynet, G., A&A, **591**, A8 (2016)
[15] Paxton, B., Cantiello, M., Arras, P., et al., ApJS, **208**, 4 (2013)
[16] Bono, G., Caputo, F., & Castellani, V., MmSAI, **77**, 207 (2006)
[17] Keller, S. C., ApJ, **677**, 483 (2008)
[18] Neilson, H. R., & Lester, J. B., ApJ, **684**, 569 (2008)
[19] Prada Moroni, P. G., Gennaro, M., Bono, G., Pietrzyński, G., Gieren, W., Pilecki, B., Graczyk, D., & Thompson, I. B., ApJ, **749**, 108 (2012)
[20] Kervella, P., Mérand, A., & Gallenne, A., A&A, **498**, 425 (2009)
[21] Marengo, M., Evans, N. R., Barmby, P., et al., ApJ, **725**, 2392 (2010)
[22] Nardetto, N., Poretti, E., Rainer, M., et al., A&A, **597**, A73 (2017)
[23] Saio, H., & Gautschy, A., ApJ, **498**, 360 (1998)
[24] Marconi, M., Molinaro, R., Bono, G., et al., ApJ, **768**, L6 (2013)
[25] Kippenhahn, R., & Smith, L., A&A, **1**, 142 (1969)
[26] Matsunaga, N., Kawadu, T., Nishiyama, S., et al., Nature, **477**, 188 (2011)
[27] Bono, G., Marconi, M., Cassisi, S., Caputo, F., Gieren, W., & Pietrzynski, G., ApJ, **621**, 966 (2005)
[28] Anderson, R. I., Eyer, L., & Mowlavi, N., MNRAS, **434**, 2238 (2013)
[29] Georgy, C., Granada, A., Ekström, S., Meynet, G., Anderson, R. I., Wyttenbach, A., Eggenberger, P., & Maeder, A., A&A, **566**, A21 (2014)
[30] Anderson, R. I., PhD thesis, Université de Genève (2013)
[31] Brandt, T. D., & Huang, C. X., ApJ, **807**, 25 (2015)
[32] Soszynski, I., Poleski, R., Udalski, A., et al., AcA, **58**, 163 (2008)

Petersen diagram revolution

Radoslaw Smolec[1,*], *Wojciech* Dziembowski[1,2], *Pawel* Moskalik[1], *Henryka* Netzel[2,1], *Zdenek* Prudil[3], *Marek* Skarka[4], and *Igor* Soszynski[2]

[1]*Nicolaus Copernicus Astronomical Center, Polish Academy of Sciences, ul. Bartycka 18, 00-716 Warszawa, Poland*
[2]*Warsaw University Observatory, Al. Ujazdowskie 4, 00-478 Warszawa, Poland*
[3]*Astronomisches Rechen-Institut, Zentrum für Astronomie der Universität Heidelberg, Mönchhofstr. 12-14, 69120 Heidelberg, Germany*
[4]*Konkoly Observatory, Research Centre for Astronomy and Earth Sciences, Hungarian Academy of Sciences, Konkoly Thege Miklós út 15-17, H-1121 Budapest, Hungary*

Abstract. Over the recent years, the Petersen diagram for classical pulsators, Cepheids and RR Lyr stars, populated with a few hundreds of new multiperiodic variables. We review our analyses of the OGLE data, which resulted in a significant extension of the known, and in the discovery of a few new and distinct forms of multiperiodic pulsation. The showcase includes not only radial mode pulsators, but also radial-non-radial pulsators and stars with significant modulation observed on top of the beat pulsation. First theoretical models explaining the new forms of stellar variability are briefly discussed.

1 Introduction

Classical pulsators, Cepheids and RR Lyrae stars, are among the most important variables. As excellent standard candles, they are invaluable for distance determination and for studying the structure and evolution of nearby stellar systems. For years they were regarded as simple, typically single-periodic, radially pulsating stars. The only serious, unsolved problem was the Blazhko effect, quasi-periodic modulation of pulsation amplitude and/or phase, observed in a significant fraction of RR Lyr stars. In the recent years, thanks to precise observations from space and to the huge amount of top-quality data gathered by the ground-based photometric sky surveys, many new dynamical phenomena were discovered in classical pulsators. In these proceedings, we focus on the new forms of multi-periodic pulsation. Most of the discoveries were done in the data collected by the Optical Gravitational Lensing Experiment (OGLE, [19]).

2 Classical Cepheids

Several new classes of multi-mode radial pulsation in classical Cepheids were reported in the OGLE data (see [17], and references therein). These include triple-mode radial pulsators and rare double-mode radial pulsators with higher order overtones excited. In the most interesting group, in which

*smolec@camk.edu.pl

the dominant variability is due to the radial first overtone, and additional variability falls in the $P_x = (0.60, 0.65)\,P_{1O}$ range, non-radial pulsation must be involved. More than 200 such stars were reported e.g. in [4, 15, 17]. The first detailed analysis of the largest sample of 138 SMC pulsators of the new type was presented in [11]. In the Petersen diagram in Figure 1 these stars are marked with filled squares and form three tight and close sequences, which immediately rules out the possibility that additional periodicities are due to consecutive radial overtones. Additional periodicities are always of low amplitude, in the mmag range. In the frequency spectrum, these periodicities manifest as a relatively broad power excess rather than a single and coherent peak. Most interestingly, in many stars, significant signal centered at a sub-harmonic frequency, $1/2\nu_x$, was detected. These signals were detected in 74 per cent of stars forming the middle sequence in the Petersen diagram and in 31 per cent of stars forming the top sequence. The signal was not detected in any star from the bottom sequence. This observation is crucial for the model explaining the nature of the additional variability, which we describe below, after discussing an analogous form of pulsation in RR Lyr stars.

3 RR Lyr stars with the dominant radial first overtone

$RR_{0.61}$ stars. AQ Leo was the first RR Lyr star in which additional variability with P_x/P_{1O} close to 0.61 was detected ([2]). Now we know more than three hundred RR Lyr stars in which the first overtone is excited (either RRc or RRd) and in which additional variability in the $(0.60, 0.64)\,P_{1O}$ range was detected. The majority of these stars were discovered in the OGLE Galactic bulge collection ([8, 9]). This form of pulsation must be common, as 14 out of 15 RRc/RRd stars observed at utmost precision from space show the additional variability (see [9], and references therein). We denote these stars as $RR_{0.61}$. In the Petersen diagram in Figure 1 they are marked with small filled squares. Three sequences are apparent, but not as distinct as in the case of Cepheids. The most populated sequence is centered at $P_x/P_{1O} \approx 0.613$. Amplitude of the additional signal is low, typically 2 per cent of the first overtone amplitude. Just as in the case of Cepheids, in the frequency spectra of these stars, significant power at sub-harmonic frequency, $1/2\nu_x$, is often detected. Again, stars with this signal are not distributed randomly in the Petersen diagram: nearly all of them form the top sequence and only a few are among the stars forming the most numerous bottom sequence. Signal at $1/2\nu_x$ was not detected in stars forming the middle sequence. This least populated sequence arises because of stars in which three additional signals are simultaneously observed in the frequency spectrum. Interestingly, the middle signal is centered at the arithmetic average of the frequencies of the two remaining signals.

Recently, Dziembowski proposed a model explaining the nature of additional periodicity observed at $(0.60, 0.65)\,P_{1O}$ in Cepheids and in RR Lyr stars ([1]). In this model, additional variability is due to non-radial pulsation modes with frequencies equal to $1/2\nu_x$ – it is the signal at sub-harmonic frequency (which is not always detected) that corresponds to non-radial mode. These are acoustic, strongly trapped unstable modes of moderate degrees. Their degrees are $\ell = 7$, 8 and 9 in classical Cepheids (top, middle and bottom sequence, respectively) and $\ell = 8$ and 9 in RR Lyr stars (top and bottom sequence, respectively). The middle sequence in the case of RR Lyr variables is due to a combination frequency, $\nu_8 + \nu_9$. Observed amplitude of a non-radial mode is reduced by geometric cancellation, which increases with increasing ℓ, and, for moderate and high degrees, is stronger for odd ℓ. Hence, there are higher chances to detect non-radial modes of $\ell = 8$ than of $\ell = 7$, and then, of $\ell = 9$. This explains why signal at the sub-harmonic frequency is predominantly detected for the top sequence in the case of RR Lyr stars and for the middle sequence in the case of Cepheids, as these sequences correspond to $\ell = 8$. On the other hand, geometric and non-linear effects lead to high amplitude of the harmonic of the non-radial mode – the signal observed at $P_x = (0.60, 0.65)\,P_{1O}$.

Stars first reported by Netzel, Smolec & Dziembowski ([7]). This group was discovered in the OGLE data for Galactic bulge RR Lyr stars. In eleven first overtone pulsators additional low-

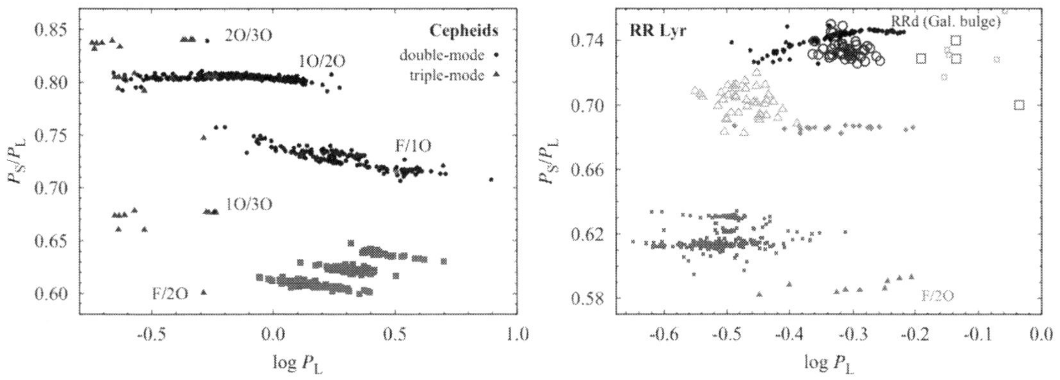

Figure 1. Petersen diagram (shorter-to-longer period ratio vs. log of longer period) for classical Cepheids (left panel) and RR Lyr stars (right panel) with various classes of multi-mode pulsators plotted with different symbols.

amplitude periodicity of longer period was discovered. Literature search revealed one additional star of the same type, KIC9453114, observed by *Kepler* and reported in [5]. An additional 8 stars were reported in [6]. In Figure 1 these stars are marked with filled diamonds. Period ratios tightly cluster at $P_{1O}/P_x \approx 0.686$, while first overtone periods cover a wide range from ≈ 0.22 to ≈ 0.43 d. In all stars additional variability is coherent and of low amplitude. As this group is located well below the RRd sequence, the period of the additional variability must be longer than the expected period of the fundamental mode in these stars. Hence, not only the additional variability cannot correspond to radial mode, but its interpretation in terms of non-radial modes is also difficult. As the frequency is below the radial fundamental mode frequency, the supposed non-radial mode must be of gravity or of mixed character. Gravity modes are not expected to be excited in giant type stars. The nature of the additional variability in these stars remains unexplained at the moment.

4 RR Lyr stars with the dominant radial fundamental mode

Anomalous RRd stars. Till recently, Blazhko effect was reported exclusively in single-periodic RR Lyr stars. First detection of Blazhko effect in RRd pulsators was reported in the OGLE Galactic bulge data ([13, 16]) and in M3 observations ([3]). Later, Soszyński et al. ([18]) reported similar stars in the OGLE Magellanic Cloud observations and defined a new class of *anomalous* RRd stars, which encompass the Blazhko RRd stars reported before. In the Petersen diagram in Figure 1, these stars are plotted with blue open circles. Their anomalous nature is revealed when the stars are compared to *classical* RRd stars, the double-mode pulsators plotted with black solid dots in Figure 1. The latter stars form a well defined and relatively tight progression in the Petersen diagram. Its short period, low period ratio tail is formed by metal-rich Galactic bulge RRd stars. There are three anomalous properties of the discussed stars. First, the period ratios are either too low, or, in a few cases, too high as compared to classical RRd stars of the same fundamental mode period. Still, these period ratios are easily explained as due to simultaneous pulsation in the radial fundamental and in radial first overtone modes. Second, unlike in classical RRd stars, in the majority of anomalous RRd stars the amplitude of the fundamental mode is higher than the amplitude of the first overtone. Third, in the majority of anomalous RRd stars, periodic modulation of the pulsation modes is observed.

The cause of the anomalous period ratios and of the modulation remains to be understood. In [18] we suggested that the parametric resonance between the three radial modes, $2\nu_{1O} = \nu_F + \nu_{2O}$, might be

DOI: 10.1051/epjconf/201715206003

essential in driving the pulsation of anomalous RRd stars, but dedicated study of this mechanism is needed.

Stars reported by Smolec et al. ([14]). In a few long-period fundamental mode RR Lyr stars of the OGLE Galactic bulge collection, additional variability of low amplitude and with a period shorter than the period of the radial fundamental mode was found. In the Petersen diagram, these stars are marked with open squares. The stars do not form any tight sequence in the Petersen diagram. The light curves corresponding to the dominant variability are among the typical for long-period fundamental mode RRab stars (Fig. 2, top left), which is confirmed by the analysis of their Fourier parameters. Interestingly, the amplitudes and amplitude ratios are among the highest, while Fourier phases are among the lowest observed in RRab stars of similar period (Fig. 2, right panels).

Linear pulsation calculations indicate that stars in this group might be extreme RRd stars, with the longest fundamental mode periods known. The only exception is OGLE-BLG-RRLYR-07283. Its low period ratio, nearly exactly equal to 0.7 ($P_{\rm X}/P_{\rm F}$ = 0.700008), cannot be fit with the models. This star has the largest amplitude of the additional variability, 8.5 per cent of the radial mode amplitude.

Stars reported by Prudil et al. ([10]). Another group of double-periodic pulsators is located at the short period extension of the classical RRd sequence. In 42 RR Lyr stars of the OGLE Galactic bulge collection, of which 38 were originally classified as RRab and 4 as RRc, additional variability was detected. Its period is always shorter than the period of the dominant variability. Period ratios are in the (0.68, 0.72) range. In the Petersen diagram in Figure 1 these stars are marked with green open triangles. Additional variability is of relatively high amplitude, typically 20 per cent of the amplitude of the dominant variability. In 10 stars (24 per cent) Blazhko-type modulation was detected. The most interesting and distinct feature of the discussed group is the shape of the light curve of the dominant variability – in all stars the light curve is a smooth triangular shape, void of any sharp features, as illustrated in Figure 2 (bottom left). The corresponding Fourier parameters are in between those characteristic for single-periodic RRab stars and single-periodic RRc stars (Fig. 2, right panels).

The nature of this group remains unknown. The characteristic triangular shape of the light curve suggests that dominant variability is due to radial fundamental mode. The additional variability cannot correspond to the radial first overtone then, at least assuming that physical parameters of these variables are similar to those of RR Lyr stars or of RR Lyr impostors (Binary Evolution Pulsators; see, e.g., [12]).

5 Concluding remarks

Over the recent years we witness the revival of classical pulsators. New dynamical effects, low-amplitude instabilities or new forms of multi-periodic pulsation were discovered. In the Petersen diagram the revolution took place – ten years ago, the only group in the case of RR Lyr stars were RRd variables. The radial mode paradigm for classical pulsators no longer holds – non-radial modes are excited in at least one of the just described groups, with additional periodicities in the (0.60, 0.65) $P_{\rm 1O}$ range. Their identification as $\ell = 7$, 8 and 9 modes might open the new window of asteroseismic investigations of classical pulsators. For a few other groups, the nature of additional periodicities remains unknown. Fortunately, the additional periodicities are either rare or of low amplitude. Hence, they have no practical impact on the use of classical pulsators for distance determination. It is disturbing however that we are far from a complete understanding of the most useful astrophysical tools.

Acknowledgments: This research is supported by the Polish National Science Centre through grant DEC-2015/17/B/ST9/03421. HN is supported by the Polish Ministry of Science and Higher Education through grant no. 0192/DIA/2016/45. MS acknowledges financial support from the Hungarian NKFIH Grant K-115709.

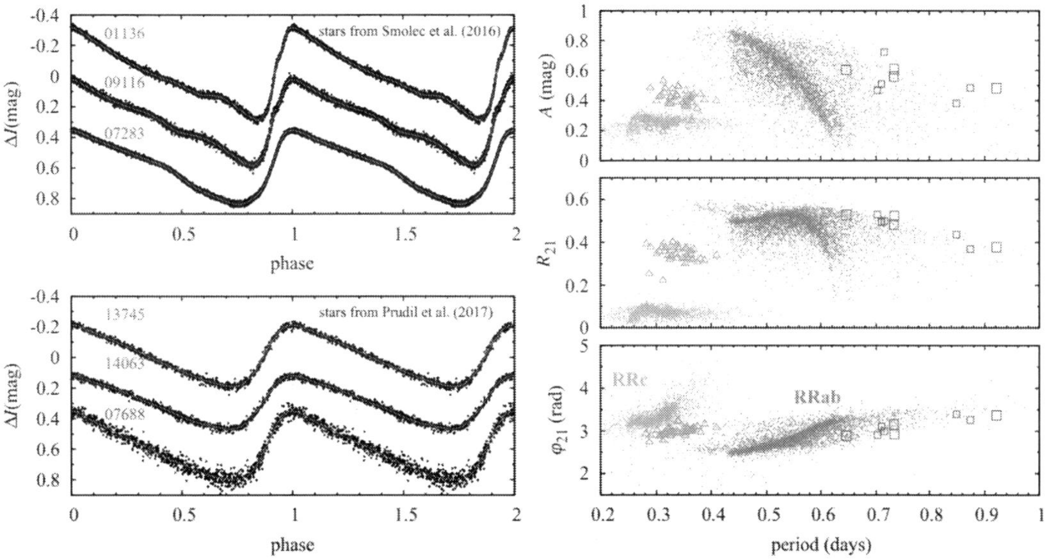

Figure 2. Left panels: exemplary light curves corresponding to the dominant periodicity in stars from Smolec et al. ([14], top) and Prudil et al. ([10], bottom). Star's ids (OGLE-BLG-RRLYR-xxxxx) are given along each light curve. Right panel: peak-to-peak amplitudes and low-order Fourier parameters (R_{21} and φ_{21}) for all stars of these two groups. Symbols are the same as in Figure 1.

References

[1] Dziembowski, W., in *RRL2016 – High-Precision Studies of RR Lyr Stars*, ed. L. Szabados, R. Szabo, & K. Kinemuchi, Comm. Konkoly Obs., **105**, 23 (2016)
[2] Gruberbauer, M., Kolenberg, K., Rowe, J. F., et al., MNRAS, **379**, 1498 (2007)
[3] Jurcsik, J., Smitola, P., Hajdu, G., Nuspl, J., ApJL, **797**, L3 (2014)
[4] Moskalik, P., & Kolaczkowski, Z., MNRAS, **394**, 1649 (2009)
[5] Moskalik, P., Smolec, R., Kolenberg, K., et al., MNRAS, **447**, 2348 (2015)
[6] Netzel, H., Smolec, R., in *XXXVII Meeting of the Polish Astronomical Society*, ed. A. Rozanska, Proc. Polish Astron. Soc., **3**, 36 (2016)
[7] Netzel, H., Smolec, R., & Dziembowski, W., MNRAS, **451**, L25 (2015)
[8] Netzel, H., Smolec, R., & Moskalik, P., MNRAS, **447**, 1173 (2015)
[9] Netzel, H., Smolec, R., & Moskalik, P., MNRAS, **453**, 2022 (2015)
[10] Prudil, Z., Smolec, R., Skarka, M., & Netzel, H., MNRAS, **465**, 4074 (2017)
[11] Smolec, R., & Śniegowska, M., MNRAS, **458**, 3561 (2016)
[12] Smolec, R., Pietrzyński, G., Graczyk, D., et al., MNRAS, **428**, 3034 (2013)
[13] Smolec, R., Soszyński, I., Udalski, A., et al., MNRAS, **447**, 3756 (2015)
[14] Smolec, R., Prudil, Z., Skarka, M., & Bąkowska, K., MNRAS, **461**, 2934 (2016)
[15] Soszyński, I., Poleski, R., Udalski A., et al., Acta Astron., **60**, 17 (2010)
[16] Soszyński, I., Udalski A., Szymański, M. K., et al., Acta Astron., **64**, 177 (2014)
[17] Soszyński, I., Udalski A., Szymański, M. K., et al., Acta Astron., **65**, 329 (2015)
[18] Soszyński, I., Smolec, R., Dziembowski, W., et al., MNRAS, **463**, 1332 (2016)
[19] Udalski, A., Szymański, M. K., & Szymański, G., Acta Astron., **65**, 1 (2015)

EPJ Web of Conferences **152**, 06004 (2017) DOI: 10.1051/epjconf/201715206004
Wide-Field Variability Surveys: A 21st Century Perspective

Faint variable stars observed with *Kepler*

Catherine Lovekin[1,*] and *Jasmin* Tompkins[1]

[1] *Physics Department, Mount Allison University, Sackville, NB, E4L 1C6, Canada*

Abstract. We present preliminary analysis of approximately 10 variable stars observed with *Kepler*. The sample stars are faint, and have temperatures greater than 8000 K. The stars were observed for up to three quarters (Q14-Q16) in long cadence mode. Frequencies were extracted with Period04, and 1-21 frequencies were detected in each quarter, with an average of 8 frequencies per quarter. Some variability is detected from quarter to quarter, while the dominant frequencies remain unchanged. We fit the frequencies using MESA models between 1.5 and 3 solar masses, and varied the core overshoot. Best fitting properties of each of these stars will be discussed.

1 Introduction

δ Scuti stars are typically between 1.4 and 2.5 $M_{\odot}$, and are found in the HR diagram where the classical instability strip intersects the main sequence. As such, they are expected to be found in the transition region between stars with convective envelopes and radiative cores (< 1.5 $M_{\odot}$) and stars with radiative envelopes and convective cores (> 1.5 $M_{\odot}$). At the same time, δ Scuti stars are often rapidly rotating ([1]). This makes δ Scuti stars ideal for constraining the effects of rotation and convective overshoot.

2 Frequency analysis

We selected ten stars with listed temperatures greater than 8000 K in the *Kepler* input catalogue. These stars were previously known to be variable ([2]), but are quite faint, from 14th to 16th magnitude. These stars were observed from Q14 to Q16, and most of the stars have continuous coverage across all three quarters. We used Period04 ([3]) to analyse the light curves and extract the frequencies. Since these stars are quite faint, and the noise level is correspondingly high, only a few frequencies were detected in each star with SNR > 4. Across all quarters, we were able to extract between 1 and 21 frequencies per star. We detected some quarter-to-quarter variability, but the dominant frequency remains unchanged in all stars. A sample Fourier spectrum is shown for KIC 9818160 in Figure 1.

The observed frequencies are primarily greater than 5 c/d, which suggests these stars are primarily δ Scuti stars. Uncertainties in the frequencies, amplitudes, and phases were calculated in Period04 using 500 iterations of a Monte Carlo simulation. At the present time, we have not made any attempt to search for patterns in the observed frequencies. With as few as 3 or 4 observed modes in some of the stars, it may not be possible to measure large frequency separations.

*clovekin@mta.ca

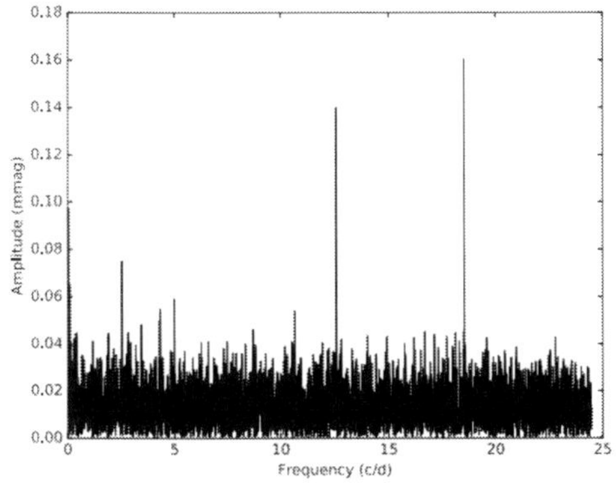

Figure 1. The raw Fourier spectrum for KIC 9818160. Four frequencies with SNR > 4 were detected in this star.

3 Models

Preliminary fits to the observed frequencies were calculated using models calculated with MESA ([4]) and GYRE ([5]). The model grid extended from 2.0 to 2.8 $M_{\odot}$ in steps of 0.05 $M_{\odot}$. We also calculated a few models outside the main grid at 1.5 and 3.2 $M_{\odot}$. We also varied the convective core overshoot of these models, varying f_{ov} from 0.1 to 0.2 in steps of 0.02. The observed frequencies were directly compared to the model frequencies via a χ^2 algorithm. The best fitting models cover the full range of our model grid. No obvious trends were discovered in either mass or convective core overshoot. However, the best fitting models were universally young, with core hydrogen fractions from 0.69 to 0.59.

4 Conclusion

We present preliminary modelling of 10 faint variable stars observed with *Kepler*. The observed frequencies are consistent with δ Scuti pulsations, as expected based on the parameters of these stars in the KIC. More detailed fitting of these stars, including a larger model grid is in progress. We also plan to expand our grid by including the effects of rotation as well as convective overshoot.

Acknowledgments: This paper includes data collected by the *Kepler* mission as part of the Guest Observer program. Funding for the *Kepler* mission is provided by the NASA Science Mission Directorate. This research has made use of NASA's Astrophysics Data System and Astropy, a community developed core Python package for astronomy.

References

[1] Breger, M., Communications in Asteroseismology, **150**, 25 (2007)
[2] Kinemuchi, K., Still, M., Fanelli, M., & Kepler Science, Bulletin of the American Astronomical Society, **43**, 201.06 (2011)
[3] Lenz, P., & Breger, M., Communications in Asteroseismology, **146**, 53 (2005)
[4] Paxton, B., Bildsten, L., Dotter, A., et al., ApJS, **192**, 3 (2011)
[5] Townsend, R. H. D., & Teitler, S. A., MNRAS, **435**, 3406 (2013)

Inference for stellar opacities from seismic studies of the hybrid β Cep/SPB pulsators

Przemysław Walczak[1,⋆], *Jadwiga* Daszyńska-Daszkiewicz[1], and *Alexey* Pamyatnykh[2]

[1]*Astronomical Institute Wrocław University, ul. Kopernika 11, 51-622 Wrocław, Poland*
[2]*Nicolaus Copernicus Astronomical Center Polish Academy of Sciences, ul. Bartycka 18, 00-716 Warsaw, Poland*

Abstract. We present a comprehensive seismic study of the three pulsating stars of β Cep/SPB type: ν Eridani, 12 Lacertae and γ Pegasi. Models with the modified mean opacity profile are constructed in order to account for both the observed frequency range and the values of some individual frequencies. To decrease the number of possible solutions, we make use of the non-adiabatic parameter f, whose value is very sensitive to subphotospheric layers where pulsations are driven. This complex seismic modelling shows the need for a significant modification of the opacity profile.

1 Introduction

The hybrid pulsators, exhibiting oscillations both in low order p/g modes and high order g modes, pose a challenge for pulsation theory because the range of their observed frequencies cannot be explained by standard models. To overcome this problem an increase of the opacity at the certain depths was postulated by many authors (i.e., [1–3]). This approach is justified by recent experiments ([4]), where higher than predicted opacities of iron and nickel were measured.

Here we analyse the three well known hybrid β Cep/SPB pulsators: ν Eri (HD 29248), γ Peg (HD 886) and 12 Lac (HD 214993). The stars pulsate in many modes of different types, i.e. in low frequency high order g modes (SPB-type) and high frequency low order p/g-modes (β Cep-type).

In Section 2, we summarize the basic information about the stars. In Section 3, we search the pulsational models with the standard and modified opacity profile. Conclusions end the paper.

2 The three analysed pulsators

All the three pulsators are massive stars of the B2 spectral type. The effective temperature of ν Eri amounts to $\log T_{\rm eff} = 4.345 \pm 0.014$, as determined by [5]. γ Peg is the coolest star in our sample with $\log T_{\rm eff} = 4.316 \pm 0.017$ ([6]). For 12 Lac we used $\log T_{\rm eff} = 4.375 \pm 0.018$ ([7]). With the parallaxes by [8] we derived luminosities $\log L/L_{\odot} = 3.807 \pm 0.045$, $\log L/L_{\odot} = 3.710 \pm 0.065$ and $\log L/L_{\odot} = 4.077 \pm 0.11$ for ν Eri, γ Peg and 12 Lac, respectively.

The position of the three target stars on the Hertzsprung-Russell diagram is shown in Figure 1. The evolutionary tracks were calculated with the metallicity parameter $Z = 0.015$, initial hydrogen abundance $X = 0.7$ and with no overshooting from the convective core. We assumed the solar chemical

⋆walczak@astro.uni.wroc.pl

composition by [9] and the new opacities from Los Alamos National Laboratory, OPLIB ([10, 11]). Evolutionary calculations were performed using the Warsaw-New Jersey code ([12]). All stars seems to be well on the main-sequence evolution phase. The models marked with squares are explained in Section 3.

ν Eri was identified as the first early B-type hybrid pulsator. The analysis of the data gathered during extensive 2003-2005 world-wide observational campaigns ([13–16]) revealed 14 pulsational frequencies. Two of them were low frequency modes, and 12 were high frequency modes. Recently, the star has been observed by the BRITE-Constellation ([17]). [18] derived 17 pulsational frequencies including 7 low frequency modes and 10 high frequency modes. γ Peg has been observed by the MOST satellite ([19]). The analysis of the data revealed 14 pulsational frequencies: 8 in the high frequency range and 6 in the low frequency range ([20]). 12 Lac was the aim of the multisite observational campaign carried out in the years 2003-2004. [7] reported 11 pulsational frequencies derived form the light curves (one of them was a low frequency mode and 10 were high frequency modes). The oscillation spectra of the three pulsators are shown in Figure 2.

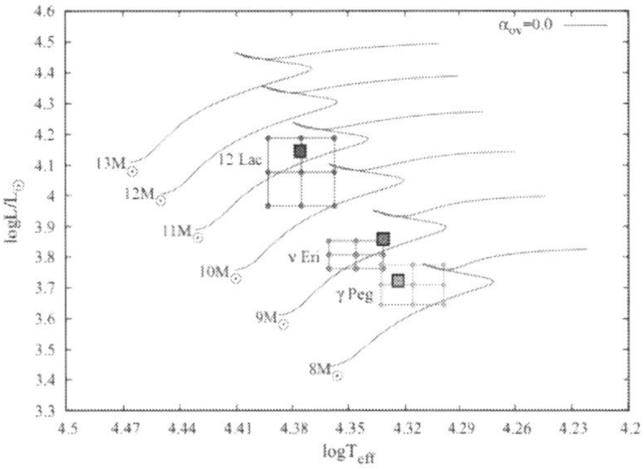

Figure 1. The observational error boxes of ν Eri, γ Peg and 12 Lac in the Hertzsprung-Russell diagram. The evolutionary tracks were computed for metallicity Z = 0.015 and the overshooting parameter $\alpha_{\rm ov} = 0$. Models marked as squares are described in the text.

3 Pulsational models

In order to perform instability analysis, we calculated non-adiabatic pulsational models using the code written by Dziembowski ([21]). The computations were based on the three opacity sources: OPAL ([22]), OP ([23]) and OPLIB.

In the case of ν Eri, we selected models that fit the three frequencies, the radial mode (5.76 d^{-1}) and two centroids of the dipole modes (5.63 d^{-1} and 6.24 d^{-1}). An example of such models calculated with the three opacity data are plotted in the upper left panel of Figure 2, where we show the instability parameter, η, as a function of the frequency, ν. A mode is excited if $\eta > 0$. We marked the observed frequencies of ν Eri with vertical lines. Their height indicates the amplitude in the BRITE Blue filter. We assumed metallicity $Z = 0.015$, mass $M = 9.5\,M_\odot$, and the overshooting parameter, $\alpha_{\rm ov}$, from 0.07 up to 0.09. The value of $\alpha_{\rm ov}$ was changed to fit the frequencies of the dipole mode centroids.

We see that neither OPAL nor OP nor OPLIB models can explain the whole observed frequency range. The OPLIB model has unstable high frequency modes, but low frequency modes are not excited. The OPAL model is very similar to the OPLIB model with slightly smaller instabilities, especially for high frequencies. The OP model has the highest instability in the low frequency region but the instability of p modes at the highest frequencies ($\nu \gtrsim 6.5$ d^{-1}) is much reduced. Changing the model parameters in a reasonable range do not solve the problem. The results for ν Eri have been already published in [24] and here we give only a summary and brief outline of the approach.

In order to increase the excitation of modes, we changed artificially the standard mean opacity profile according to the formula:

$$\kappa(T) = \kappa_0(T)\left\{1 + \sum_{i=1}^{N} b_i \cdot \exp\left[-\frac{(\log T - \log T_{0,i})^2}{a_i^2}\right]\right\}, \tag{1}$$

 DOI: 10.1051/epjconf/201715206005

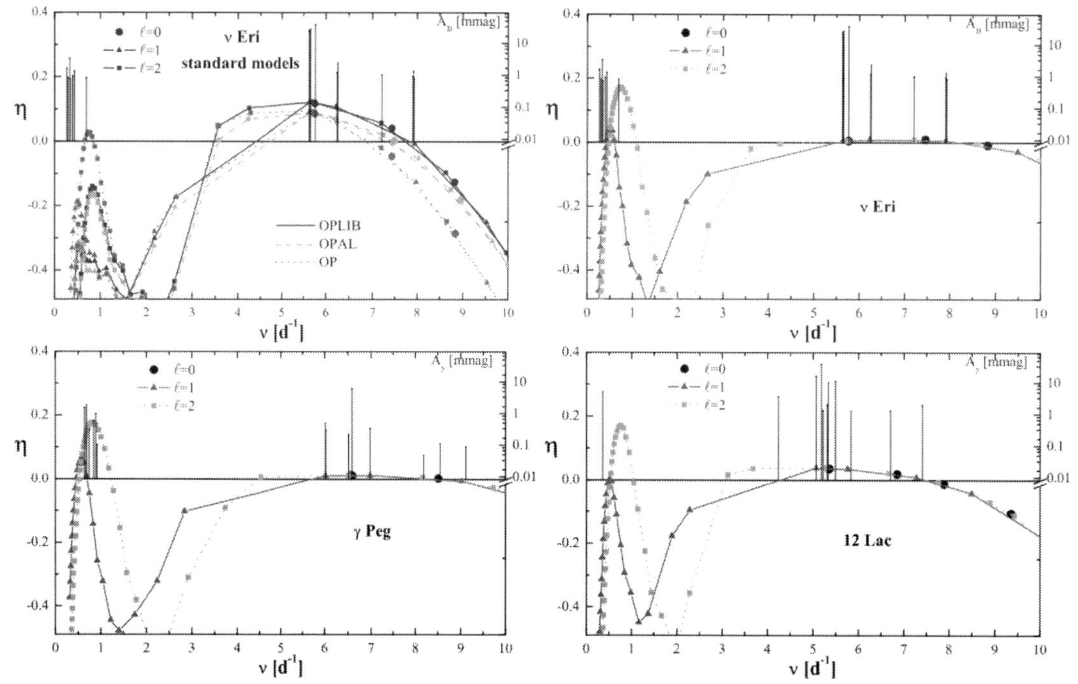

Figure 2. The instability parameter, η, as a function of frequency for modes with degree $\ell = 0-2$. In the upper-left panel, we show the standard models of ν Eri calculated with three opacity data. In the upper-right panel, the best model calculated with the modified OPLIB opacities is shown. In the bottom panels, we plot the modified OPLIB models of γ Peg and 12 Lac, respectively. The opacity modification parameters are given in Table 1. The vertical lines indicate the observed frequencies.

where $\kappa_0(T)$ is the original opacity profile and $(a,\ b,\ T_0)$ are parameters of the Gaussian describing the width, height and position of the maximum, respectively.

By changing the value of the parameters a, b and T_0, we can find a lot of seismic models that are unstable in the regions of interest. To constrain the number of combinations of a, b and T_0, we included the nonadiabatic parameter f in our fitting. The parameter f is defined as the ratio of the bolometric flux perturbation to the radial displacement at the level of the photosphere ([5, 25]) and its value is very sensitive to the opacity of the outer layers. The empirical values of f can be derived from multicolor photometry and radial velocity measurements ([5, 25]). They depend also on atmosphere models, therefore we used both LTE and non-LTE atmosphere models by [26] and [27], respectively. Since the non-adiabatic effects of pulsation cause the phase shift of the flux changes, we need to consider both the amplitude, $|f|$, and the phase lag, $\Psi = \arg(f) - 180°$. The value of Ψ describes the phase shift between the maximum of the flux and the minimum of the radius.

Finding the models that fit simultaneously the instability range, the values of the frequencies and the value of f for the radial mode was a challenging task. We searched an extensive grid of models changing parameters of equation (1) with steps: $\Delta a = 0.001$, $\Delta b = 0.05$ and $\Delta T_0 = 0.005$. The $\log T_0$ parameter was chosen from the range 5.0 – 5.5. It turned out that only models with quite complicated opacity modification meet all requirements.

Our best model calculated with the OPLIB opacity modification is shown in the upper right panel of Figure 2, where we plotted the instability parameter as a function of the mode frequency. The

Table 1. Parameters of the opacity modification for models described in the text.

star	$\log T_{0,1}$	a_1	b_1	$\log T_{0,2}$	a_2	b_2	$\log T_{0,3}$	a_3	b_3
ν Eri	5.06	0.500	−0.54	5.22	0.088	+0.34	5.46	0.061	+1.80
γ Peg	5.06	0.447	−0.60	5.22	0.067	+0.50	5.46	0.063	+2.10
12 Lac	5.06	0.447	−0.25	5.22	0.082	+0.50	5.46	0.077	+2.00

position of the model on the HR diagram is marked as a big square nearly inside the ν Eri error box in Figure 1. The model parameters are as follows: $M = 9.0\,M_\odot$, $\log T_{\rm eff} = 4.332$, $\log L/L_\odot = 3.858$, $Z = 0.015$ and $\alpha_{\rm ov} = 0.163$. The parameters of the opacity modifications are given in the first line of Table 1. The mean opacity has been changed at the three depths, $\log T_{0,1} = 5.06$ (corresponding to the opacity bump identified by [28, 29]), $\log T_{0,2} = 5.22$ (corresponding to the maximum contribution of chromium and manganese, [2]) and $\log T_{0,3} = 5.46$ (corresponding to the maximum contribution of nickel, [2]). This $\kappa-$modified model can account for all observed high frequency modes. The lowest observed frequency modes are not excited, but the instability parameter in this frequency region is significantly higher than in standard models. Moreover, most of the peaks can be explained by the rotational splitting.

We performed the same analysis for γ Peg and 12 Lac, and found that in these cases a very similar opacity modification works as well. In case of γ Peg we calculated models that fit two frequencies: the radial mode, $\nu = 6.59$ d^{-1}, and the dipole mode, $\nu = 6.02$ d^{-1}. Also models fitting the two frequencies were constructed for 12 Lac: the radial mode, $\nu = 5.33$ d^{-1}, and the dipole mode, $\nu = 5.07$ d^{-1}. The instability parameters of the best models of γ Peg and 12 Lac are shown in the bottom left and right panels of Figure 2, respectively. The vertical lines correspond to the observed frequencies and their height indicates the amplitude in the Strömgrem y filter. For γ Peg, the model parameters are: $M = 8.06\,M_\odot$, $\log T_{\rm eff} = 4.324$, $\log L/L_\odot = 3.723$, $Z = 0.012$ and $\alpha_{\rm ov} = 0.262$. The model of 12 Lac has the parameters: $M = 11.152\,M_\odot$, $\log T_{\rm eff} = 4.376$, $\log L/L_\odot = 4.145$, $Z = 0.015$ and $\alpha_{\rm ov} = 0.1$. The parameters of the opacity modifications are given in the second and third lines of Table 1.

The model of γ Peg has a high instability in the low frequency range and all observed SPB-type modes are explained by this model. In the case of high frequencies only one mode is slightly stable, i.e. that with the highest frequency. For 12 Lac, the modified model has unstable high frequency modes, but the only low frequency mode is stable. Nevertheless, the instability parameter is quite high in the SPB-type mode region for all three stars.

The empirical values of f for ν Eri and 12 Lac are inside of the empirical uncertainties. In the case of γ Peg, the amplitude $|f|$ is reproduced within 1 σ errors, while the phase lag, Ψ, within 2 σ errors.

4 Conclusions

The standard asteroseismic models of the hybrid β Cep/SPB pulsators cannot explain the observed properties. The biggest problem is the theoretical instability range, which is usually too small in comparison with the observations. As a possible solution, we proposed a modification of stellar opacities at certain depths. It goes in line with the recent experimental results which give higher opacities for iron and nickel. The aim was to reproduce the observed frequency range, the values of some individual frequencies and the parameter f which is very sensitive to the opacity changes in the subphotospheric layers. Such complex seismic modelling reduces greatly the number of possible opacity modifications and make our modelling more justifiable.

The most interesting and striking result is that the modification of the opacity profile of our best model is very similar for all the three analysed stars. In each case, the opacity modifications are necessary at the three depths: an increase by about 200% at $\log T_{0,3} = 5.470$, an increase by about 30-50% at $\log T_{0,2} = 5.220$ and a decrease by about 50% in the uppermost layers at $\log T_{0,1} = 5.065$.

The necessity of decreasing opacity near $\log T = 5.065$, where a new maximum was identified by [28, 29], is quite bothersome and we do not have an explanation for that. Also a huge increase of κ at $\log T = 5.470$ seems to be very surprising. This may result from both the inhomogeneous chemical composition and inaccuracies in the opacity calculations. Intensive further studies are needed to resolve this issue.

Acknowledgments: This work was financially supported by the Polish NCN grant 2015/17/B/ST9/02082. Calculations have been partly carried out using resources provided by Wrocław Centre for Networking and Supercomputing (`http://www.wcss.pl`), grant No. 265.

References

[1] A.A. Pamyatnykh, G. Handler, W.A. Dziembowski, MNRAS **350**, 1022 (2004)
[2] S. Salmon, J. Montalbán, T. Morel, et al., MNRAS **422**, 3460 (2012)
[3] E. Moravveji, MNRAS **455**, L67 (2016)
[4] J.E. Bailey, T. Nagayama, G.P. Loisel, et al., Nature **517**, 56 (2015)
[5] J. Daszyńska-Daszkiewicz, W.A. Dziembowski, A.A. Pamyatnykh, A&A **441**, 641 (2005)
[6] M. Koleva, A. Vazdekis, A&A **538**, A143 (2012)
[7] G. Handler, M. Jerzykiewicz, E. Rodríguez, et al., MNRAS **365**, 327 (2006)
[8] F. van Leeuwen, A&A **474**, 653 (2007)
[9] M. Asplund, N. Grevesse, A.J. Sauval, P. Scott, ARAA **47**, 481 (2009)
[10] J. Colgan, D.P. Kilcrease, N.H. Magee, et al., High Energy Density Physics **14**, 33 (2015)
[11] J. Colgan, D.P. Kilcrease, N.H. Magee, et al., ApJ **817**, 116 (2016)
[12] A.A. Pamyatnykh, W.A. Dziembowski, G. Handler, H. Pikall, A&A **333**, 141 (1998)
[13] G. Handler, R.R. Shobbrook, M. Jerzykiewicz, et al., MNRAS **347**, 454 (2004)
[14] M. Jerzykiewicz, G. Handler, R.R. Shobbrook, et al., MNRAS **360**, 619 (2005)
[15] C. Aerts, P. De Cat, G. Handler, et al., MNRAS **347**, 463 (2004)
[16] J. De Ridder, J.H. Telting, L.A. Balona, et al., MNRAS **351**, 324 (2004)
[17] W.W. Weiss, S.M. Ruciński, A.F.J. Moffat, et al., PASP **126**, 573 (2014)
[18] G. Handler, M. Rybicka, A. Popowicz, et al., MNRAS **464**, 2249 (2017)
[19] G. Walker, J. Matthews, R. Kuschnig, et al., PASP **115**, 1023 (2003)
[20] G. Handler, J.M. Matthews, J.A. Eaton, et al., ApJ **698**, L56 (2009)
[21] W. Dziembowski, AcA **27**, 95 (1977)
[22] C.A. Iglesias, F.J. Rogers, ApJ **464**, 943 (1996)
[23] M.J. Seaton, MNRAS **362**, L1 (2005)
[24] J. Daszyńska-Daszkiewicz, A.A. Pamyatnykh, P. Walczak, et al., MNRAS **466**, 2284 (2017)
[25] J. Daszyńska-Daszkiewicz, W.A. Dziembowski, A.A. Pamyatnykh, A&A **407**, 999 (2003)
[26] R. Kurucz, http:// kurucz.harvard.edu (2004)
[27] T. Lanz, I. Hubeny, ApJ **169**, 83 (2007)
[28] H. Cugier, A&A **547**, 42 (2012)
[29] H. Cugier, A&A **565**, 76 (2014)

Constraints on rotation of B-type stars from the nitrogen abundance

Jakub Ostrowski[1,*], *Jadwiga* Daszyńska-Daszkiewicz[1,**], and *Henryk* Cugier[1,***]

[1] *Instytut Astronomiczny, Uniwersytet Wrocławski, ul. Kopernika 11, 51-622 Wrocław, Poland*

Abstract.
We study the effects of rotation on surface abundances of CNO elements in massive stars. Nitrogen enrichment caused by rotationally induced mixing can help to constrain parameters of observed stars, especially their rotational velocities. Here we present theoretical results and apply them to the star HD 163899.

1 Introduction

Rotation is one of the most important parameters that determine the evolution and ultimate fate of a star. Rotationally induced mixing strongly modifies the internal structure of a star. Besides the difficulty in modelling the effects of rotation, there is also the problem of determining the rotation velocity from observations, and usually only its projected value is known. In the presence of moderate-to-fast rotation, the abundance of surface nitrogen should increase during the course of evolution as an effect of rotational mixing processes (e.g., [2]). Abundances of carbon and oxygen are affected as well. Here, we study this effect and use the observed abundance of nitrogen of the star HD 163899 (B2 II) to constrain its rotational velocity.

2 Results

Evolutionary models were calculated with the MESA code ([4–6]) assuming a solar metallicity and chemical composition ([1]), initial masses $M \in [16, 26]\ M_\odot$ and initial rotation rates $\Omega < 0.5\ \Omega_{\rm crit}$, where Ω and $\Omega_{\rm crit}$ are the angular rotational velocity and its critical value, respectively. The considered evolution covers stages from the zero-age main sequence (ZAMS) up to the point where $\log T_{\rm eff} \approx 4.2$ during a supergiant phase. An increase of surface nitrogen abundance with evolution is clearly visible in the left panel of Figure 1, where we show ratios of mass abundances, $X(^{14}{\rm N})/X(^{12}{\rm C})$ vs. $X(^{14}{\rm N})/X(^{16}{\rm O})$, for representative models. In the right panel we show similar diagrams but for models contained in the error box of HD 163899. Theoretical abundances are compared to the values derived from the spectra obtained for the star. Such comparison allows to constrain stellar parameters, especially the rotation rate. We derived an initial rotation rate in range of $0.26 - 0.30\ \Omega_{\rm crit}$. All details of our analysis are provided in [3].

*ostrowski@astro.uni.wroc.pl
**daszynska@astro.uni.wroc.pl
***cugier@astro.uni.wroc.pl

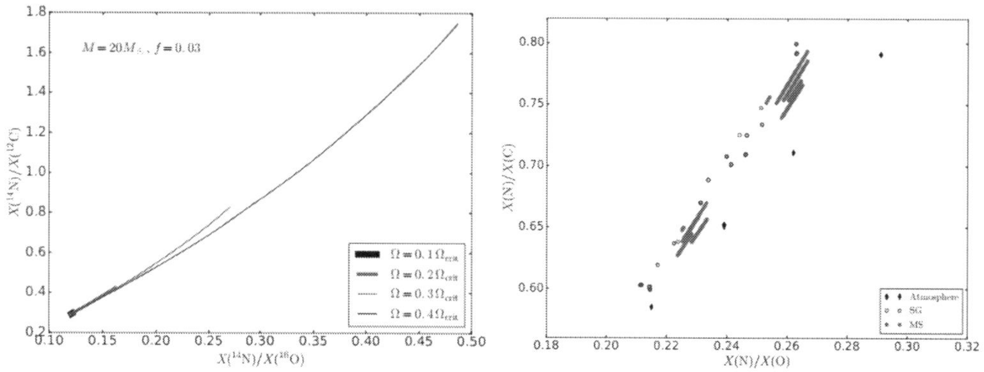

Figure 1. Left panel: the ratios of mass abundances, $X(^{14}N)/X(^{12}C)$ vs. $X(^{14}N)/X(^{16}O)$, for models with initial mass 20 $M_\odot$, overshooting parameter $f = 0.03$ and the four values of the rotation rate at ZAMS. Right panel: the ratios of $X(^{14}N)/X(^{12}C)$ vs. $X(^{14}N)/X(^{16}O)$, showing values derived from atmosphere models with nitrogen enrichment calculated for HD 163899 (black diamonds) and for evolutionary models in the error box of HD 163899 on the main sequence (MS, red dots) and supergiant phase (SG, blue open circles).

3 Conclusions

Enrichment of stellar atmospheres in CNO elements due to rotationally induced mixing can be successfully applied to the observed nitrogen-rich stars, as shown for HD 163899. The method helps to obtain the intrinsic rotational velocity disentangled from the inclination angle. More details can be found in [3].

Acknowledgments: Based on the archive "HARPS spectra of CoRoT targets", prepared in the framework of the FP7 project no. 312844 *SpaceInn – Exploitation of Space Data for Innovative Helio- and Asteroseismology*. This work was financially supported by the Polish NCN grants 2015/17/B/ST9/02082 and 2013/09/N/ST9/00611. Calculations have been carried out using resources provided by Wroclaw Centre for Networking and Supercomputing (`http://wcss.pl`), grant No. 265.

References

[1] Asplund, M., Grevesse, N., Sauval, A. J., & Scott, P., ARA&A, **47**, 481 (2009)
[2] Maeder, A., Przybilla, N., Nieva, M.-F., Georgy, C., Meynet, G., Ekström, S., & Eggenberger, P., A&A, **565**, A39 (2014)
[3] Ostrowski, J., Daszyńska-Daszkiewicz, J., & Cugier, H., ApJ, **835**, 290 (2017)
[4] Paxton, B., Bildsten, L., Dotter, A., Herwig, F., Lesaffre, P., & Timmes, F., ApJS, **192**, 3 (2011)
[5] Paxton, B., Cantiello, M., Arras, P., et al., ApJS, **208**, 4 (2013)
[6] Paxton, B., Marchant, P., Schwab, J., et al., ApJS, **220**, 15 (2015)

DOI: 10.1051/epjconf/201715206007

Important consequences of atomic diffusion inside main-sequence stars: opacities, extra-mixing, oscillations

M. Deal[1,2,3], *S.* Vauclair[2,3], and *O.* Richard[1]

[1] *LUPM, Université de Montpellier, France*
[2] *Université de Toulouse, UPS-OMP, IRAP, Toulouse, France*
[3] *CNRS, IRAP, 14 avenue Edouard Belin, F-31400 Toulouse, France*

Abstract. Atomic diffusion, including the effects of radiative accelerations on individual elements, leads to important variations of the chemical composition inside stars. The accumulation of important elements in specific layers leads to a local increase of the average opacity and to hydrodynamic instabilities that modify the internal stellar structure. This can also have important consequences for asteroseismology.

1 Introduction

Atomic diffusion is a fundamental process which occurs in all stars, except when prevented by large mixing processes. Due to selective effects, each ion undergoes individual motions during small timescales, before sharing its momentum through collisions. These short motions, statistically added, lead to atomic diffusion processes, which may have important consequences for the global stellar structure. Most stellar evolution codes include helium and sometimes heavy elements gravitational settling. Very few include the computations of the selective radiative accelerations of individual species in a complete way. Without these computations, the results are not correct. In particular, the resulting accumulation and depletion of individual elements strongly influence the local opacities. We have computed the consequences of introducing these effects in stellar models, especially for A-type stars.

2 Numerical computations

The stellar models were computed using the Toulouse-Geneva Evolution Code (TGEC), which includes atomic diffusion with radiative accelerations calculated for 21 species, namely 12 elements and their main isotopes: H, ^{3}He, ^{4}He, ^{6}Li, ^{7}Li, ^{9}Be, ^{10}B, ^{12}C, ^{13}C, ^{14}N, ^{15}N, ^{16}O, ^{17}O, ^{18}O, ^{20}Ne, ^{22}Ne, ^{24}Mg, ^{25}Mg, ^{26}Mg, ^{40}Ca, and ^{56}Fe ([5]). The diffusion computations were based on the Boltzmann equation for a dilute collision dominated plasma. The details of the physical parameters and computational methods included in the TGEC are described in several papers, including [5]. OPCD v3.3 codes and data ([4]) are used to compute Rosseland opacity at each time step to take the variations of the abundances of each element with depth and time into account. The opacity data are used to compute individual radiative accelerations on C, N, O, Ne, Mg, Ca, and Fe. This is done by using the improved semi-analytical prescription proposed by [1] and [3]. The models were evolved from pre-main sequence up to hydrogen core exhaustion. Atomic diffusion was introduced at the beginning of the main sequence.

Fingering (thermohaline) convection was also computed. This instability occurs in stars in case of inverse μ-gradients. This may be due to several effects, including accretion of heavy matter onto stars, or accumulation of heavy elements in specific stellar layers due to atomic diffusion. The first treatments of fingering convection in stars were purely analytical. Recently, 2D and 3D numerical simulations were performed (see references in [2]). In the present work, the prescriptions obtained by the most recent 3D simulations ([6]) have been used. They show that the accumulation of heavy elements in specific layers inside A stars leads to important extra mixed zones.

3 Results

We present important consequences of atomic diffusion obtained in 1.7 $M_\odot$ models. Iron accumulates at the place where it is the main contributor of the overall opacity, which creates an important local opacity increase. This opacity increase may lead to a dynamical convective zone induced by iron κ mechanism. When fingering convection is taken into account, the dynamical convective zone is less efficient, but in this case new mixed zones due to the fingering (thermohaline) effect do appear. One of them is due to the local calcium accumulation which occurs around $\log(\Delta M/M) = 7.5$, as the result of the rapid upward decrease of the radiative acceleration on this element. We computed the surface abundances obtained in these models, to compare them with those observed in several Am stars. When the specific fingering mixed zones are taken into account, the abundances remain somewhat different from the observed ones. However, if we assume that these mixed zones are connected, we find abundances very close to the observed ones (see [2] for more detailed results)

4 Conclusion

Atomic diffusion, which is generally considered as a second order process inside stars, may have important macroscopic consequences on the stellar structure, on the hydrodynamic instabilities as well as on the surface abundances. It leads to a local increase of the opacity which must be taken into account when computing the stellar internal structure, the triggering of stellar oscillations and the oscillation frequencies. This needs stellar evolution codes including these effects, which are not trivial. Fingering convection, the only macroscopic process directly induced by the local accumulation of heavy elements in specific stellar layers, has to be taken into account in the evolution of A-type stars. It is an unavoidable consequence of the diffusion-induced inverse μ-gradient. We show that, if this process is the only hydrodynamic instability taken into account in the computations, the stellar gas below the outer convective zones is locally mixed in several layers, which move with time during stellar evolution. If we assume that these zones are dynamically connected, we find a very good fit between the computed and observed abundances.

References

[1] Alecian, G., & Leblanc, F., IAU Symp., **224**, 587 (2004)
[2] Deal, M., Richard, O., & Vauclair, S., A&A, **589**, 140 (2016)
[3] Leblanc, F., & Alecian, G., MNRAS, **352**, 1329 (2004)
[4] Seaton, M. J., MNRAS, **362**, L1 (2005)
[5] Théado, S., Alecian, G., LeBlanc, F., & Vauclair, S., A&A, **546**, 100 (2012)
[6] Zemskova, B., Garaud, P., Deal, M., & Vauclair, S., ApJ, **795**, 118 (2014)

DOI: 10.1051/epjconf/201715206008

Non-radial modes in AGB stars

Josefina Montalbán[1,3,★], *Michele* Trabucchi[1], *Paola* Marigo[1], *Peter R.* Wood[2], and *Giada* Pastorelli[1]

[1]*Department of Physics and Astronomy "Galileo Galilei", University of Padua, Vicolo dell'Osservatorio 3, I-35122 Padova, Italy*
[2]*Research School of Astronomy and Astrophysics, Australian National University, Canberra, ACT2611, Australia*
[3]*Observatory of Padua-INAF, Vicolo dell'Osservatorio 5, I-35122 Padova, Italy*

Abstract. The success of asteroseismology in characterising G-K giants has motivated the extension of the same techniques to stars after the central He-burning and M-giants. The latter have been usually studied only as radial pulsators; the presence, however, of fine-structure in the period-luminosity diagram of red variables in the Magellanic Clouds could result from the presence of non-radial oscillations, offering the potential of observational indexes based on non-radial oscillations also for luminous red giants. We present here the results of a first approach aiming to identify the origin of the sub-ridges in the sequence A of the LMC red variables.

1 Method and results

After the central H- and He-burning, the density contrast between the core and the envelope increases with luminosity, and these stellar regions become dynamically decoupled. Radial and non-radial modes are then strongly trapped in the envelope, and their oscillation frequencies are determined only by the physical properties of that region. This hypothesis is generally accepted for radial and quadrupole modes, but it does not always hold for dipole ones. We have verified its validity for dipole modes ($\ell = 1$) in 1.5 $M_\odot$ AGB models, by computing $\ell = 1$ frequency oscillations for complete and envelope models. These results are consistent with those by [3] for luminous RGB stars.

Using the adiabatic oscillation code LOSC ([2]), we have computed radial and non-radial ($\ell = 1, 2$) oscillation frequencies for AGB envelope models ([1]) for three masses (0.8, 1.5 and 2.0 $M_\odot$) with a significant representation in the LMC population synthesis. The left panel of Figure 1 shows the period-luminosity (PL) diagram for a luminosity sequence of a 1.5 $M_\odot$ envelope model. Adiabatic computations do not allow to identify the dominant mode (P1), hence, we have considered all the possible theoretical period-pairs from the 6 modes closest to $P_{\max}$ (period at maximum emission in the framework of stochastic oscillations). The results are shown in the right panel of Figure 1, a Petersen diagram where we plot the period ratios $\log(P2/P1)$ and $\log(P3/P1)$ ($P1$, $P2$ and $P3$ are the three observed periods for each star – black dots – in the LMC, with P1 being the dominant one) against $\log P1[W_{JK} = 12]$ (the value of $\log P1$ projected along sequence A, see [4] for details).

★josefa.montalbaniglesias@unipd.it

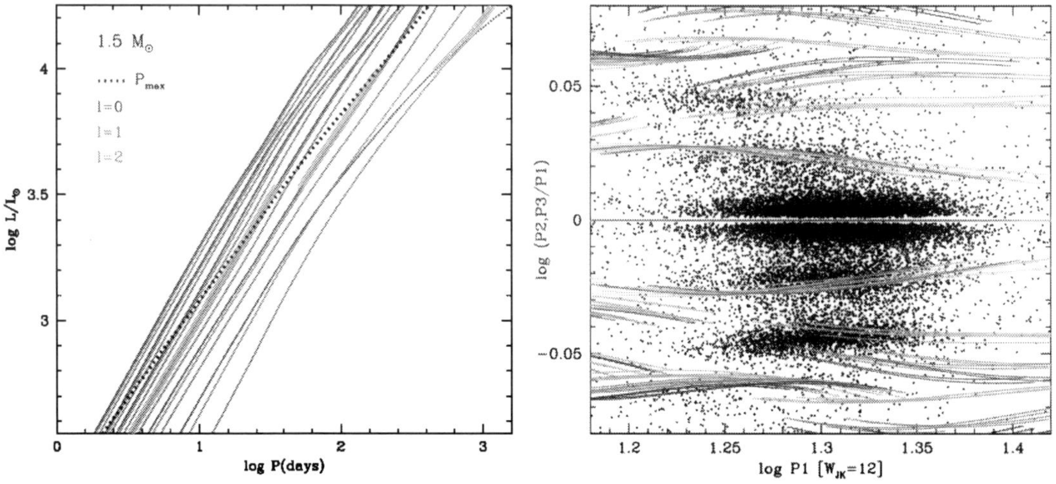

Figure 1. Left panel: Luminosity *versus* period for radial and non-radial modes (dipole and quadrupole) of 1.5 M$_\odot$ envelope models along a luminosity sequence. Different lines correspond to fundamental and five consecutive overtones. Dotted black line is the PL relation corresponding to $\nu_{\rm max}$ from scaling relation ([6]). Orange symbols indicate the dominant modes as those with the largest growth rate given by the non-adiabatic radial computation. Right panel: Petersen diagram as that in Figure 3 of [4]. Black dots correspond to period ratios from LMC observations ([5]), and coloured ones to theoretical values obtained from luminosity sequences for 0.8, 1.5 and 2.0 M$_\odot$, LMC metallicity, O- and C-rich chemical composition. Different colours corresponds to different pairs of angular degrees involved in the period ratio: $\ell = 0$ and $\ell = 1$ modes, blue dots; $\ell = 1$ and $\ell = 2$, red-orange ones; $\ell = 0$ and $\ell = 2$, green ones. The lightest colours are for the highest mass, and the darkest ones for lowest mass.

While the sub-ridges at $\log(P2/P1) \sim \pm 0.05$ seem defined by the periods of $\ell = 1$ and $\ell = 2$ modes and their characterisation should require more detailed modelling, the sub-ridges at $\log(P2/P1) \sim \pm 0.02$ are completely determined by the period difference between $\ell = 0$ and $\ell = 2$ modes, and their location does not depend on total stellar mass, He-core mass, or chemical composition (O- or C-rich models).

Acknowledgments: We acknowledge the support from the ERC Consolidator Grant funding scheme (project STARKEY, G.A. n. 615604).

References

[1] Fox, M., & Wood, P. R., ApJ, **259**,198 (1982)
[2] Scuflaire, R., Montalbán, J., Théado, S., Bourge, P.-O., Miglio, A., Godart, M., Thoul, A., & Noels, A., Ap&SS, **316**, 149 (2008)
[3] Dziembowski, W. A., A&A, **539**, A83 (2012)
[4] Wood, P. R., MNRAS, **448**, 3829 (2015)
[5] Soszyński, I., et al., Acta Astron., **59**, 239 (2009)
[6] Kjeldsen, H., & Bedding, T. R., A&A, **293**, 87 (1995)

Pulsation models of O-rich and C-rich long period variables

Michele Trabucchi[1,*], *Paola* Marigo[1], *Josefina* Montalbán[1], *Peter R.* Wood[2], and *Léo* Girardi[3]

[1]*Dipartimento di Fisica e Astronomia "Galileo Galilei", Università di Padova, Vicolo dell'Osservatorio 3, I-35122 Padova, Italy*
[2]*Research School of Astronomy and Astrophysics, Australian National University, Canberra, ACT2611, Australia*
[3]*Astronomical Observatory of Padova – INAF, Vicolo dell'Osservatorio 3, I-35122 Padova, Italy*

Abstract. The intrinsic variability of luminous red giants is an observable providing important information about stellar structure and evolution. Our aim is to provide a solid theoretical background to the wide amount of data coming from present and future variability surveys. We present here the first comparison between long period variables in the LMC (from OGLE-III) and our grid of pulsation models computed for a wide range of stellar parameters and input physics.

1 Introduction

Low to intermediate mass stars experience the Asymptotic Giant Branch (AGB) during their late evolution, undergoing a number of complex and strongly interacting phenomena such as stellar pulsations, mass loss, dust formation, thermal pulses and dredge-up events. Because of their high luminosity they are often used to characterise stellar populations of external galaxies. Their observed variability is currently attributed to pulsation in modes of low radial orders ([6]). Their periods, that strongly depend on the stellar parameters, provide an additional and powerful observational constraint to be matched by theoretical models, allowing to refine our understanding of late stellar evolution. Existing and future large surveys require large grids of pulsation models and their interpretation as a function of stellar properties. We are currently computing such a grid, planned to be made public (Trabucchi et al. 2017, in preparation), and present here the first comparison with observation.

2 Methods

We compute a large grid of pulsation models covering stellar parameters characteristic of the thermal pulsing AGB (TP-AGB) using the code of [1]. For each model, the pulsation period of the fundamental mode and four overtone modes are provided, as well as their growth rates (stability criterion).

In addition to the large coverage of stellar parameters, the main improvement with respect to previous work is the update of opacity, that now includes low temperature atomic and molecular data for the specific chemical composition of oxygen- and carbon-rich AGBs. We use combined data from the ÆSOPUS code ([2]) and the Opacity Project ([4]).

*michele.trabucchi@studenti.unipd.it

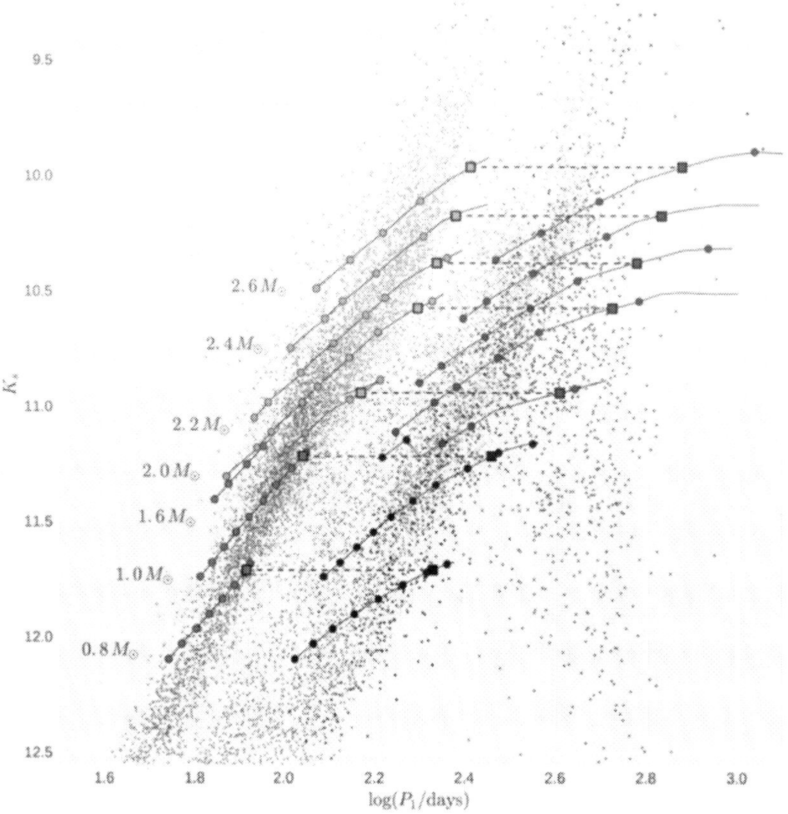

Figure 1. Radial first overtone and fundamental periods along TP-AGB tracks (left and right groups of solid lines, respectively, and corresponding large symbols) crossing sequences C' and C (small dots in backgrounds) of long period variables in the LMC (OGLE-III Collection of Variable Stars, [5]). Observations are colour-coded according to their OGLE classification: O- and C-rich semi-regular variables (green and orange), O- and C-rich Miras (blue and red).

3 Results

Here we present the results of pulsation models along TP-AGB evolutionary tracks ([3]) for masses between 0.8 and 2.6 $M_\odot$ and metallicity suitable for the LMC, and compare the periods of excited fundamental (F) and first overtone (1O) modes with the period-luminosity (P-L) sequences in the LMC ([5]). In Figure 1 we show the results. We note that: 1) the theoretical and observed properties are in good agreement, including the transition from O- to C-rich and 2) during TP-AGB evolution, the tracks cross the P-L sequences. Models also predict the luminosity at which the radial order of dominant mode shift downwards, from 1O to F (square symbols in Fig. 1).

Acknowledgments: We acknowledge the support from the ERC Consolidator Grant funding scheme (*project STARKEY*, G.A. n. 615604).

References

[1] Fox, M. W., & Wood, P. R., ApJ, **259**, 198 (1982)
[2] Marigo, P., & Aringer, B., A&A, **508**, 1539 (2009)
[3] Marigo, P., et al., MNRAS, **434**, 488 (2013)
[4] Seaton, M. J., MNRAS, **362**, L1 (2005)
[5] Soszyński, I., et al., Acta Astron., **59**, 239 (2009)
[6] Wood, P. R., MNRAS, **448**, 3829 (2015)

Pulsational instabilities in hot pre-horizontal branch stars

Tiara Battich[1,2,⋆], *Marcelo M.* Miller Bertolami[2], *Alejandro H.* Córsico[1,2], and *Leandro G.* Althaus[1,2]

[1] *Facultad de Ciencias Astronómicas y Geofísicas, Universidad Nacional de La Plata, Argentina*
[2] *Instituto de Astrofísica de La Plata, CONICET-UNLP, Argentina*

Abstract. The ϵ-mechanism is a self-excitation mechanism of pulsations which acts on the regions where nuclear burning takes place. It has been shown that the ϵ-mechanism can excite pulsations in models of hot helium-core flash, and that the pulsations of LS IV-14°116, a He-enriched hot subdwarf star, could be explained that way. We aim to study the ϵ-mechanism effects on models of hot pre-horizontal branch stars and determine, if possible, a domain of instability in the $\log g - \log T_{\rm eff}$ plane. We compute non-adiabatic non-radial pulsations on such stellar models, adopting different values of initial chemical abundances and mass of the hydrogen envelope at the time of the main helium flash. We find an instability domain of long-period ($400\,{\rm s} \lesssim P \lesssim 2500\,{\rm s}$) g-modes for models with $22000\,{\rm K} \lesssim T_{\rm eff} \lesssim 50000\,{\rm K}$ and $4.67 \lesssim \log g \lesssim 6.15$.

1 Introduction

Hot horizontal branch (HB) stars are core helium-burning low-mass stars ($\sim 0.5\,M_\odot$). Before settling in this phase, low-mass stars undergo a helium-core flash and subsequent subflashes. The blue horizontal branch (BHB; $7000\,{\rm K} \lesssim T_{\rm eff} \lesssim 21000\,{\rm K}$) is the hot part of the HB to the left of the RR Lyrae gap in the Hertzsprung-Russell diagram. The extreme horizontal branch (EHB) forms an extension of the HB at higher temperatures. The position of stars in the BHB or the EHB depends mainly on their hydrogen-rich envelope mass. EHB stars have hydrogen envelopes too thin to sustain hydrogen burning ([1]). The pre-EHB and pre-BHB stars are going through He-subflashes. EHB stars are identified with the hot subdwarf B stars in the field (sdB). The atmospheres of sdB stars are mostly hydrogen pure due to the action of diffusion. But there are some sdBs with mild helium enrichment. These stars could be at the pre-EHB phase where the ongoing diffusion has had no time to turn the envelope hydrogen pure ([2]). It has been shown that the ϵ-mechanism can excite pulsations in models of hot helium-core flash, and in the subsequent subflashes ([2]). However, a complete exploration of the properties of ϵ-driven pulsations in pre-EHB and pre-BHB stars is lacking. Here, we aim to extend the study of the ϵ-mechanism effects on stellar models appropriate for pre-BHB and pre-EHB stars.

2 Metodology and results

We constructed pre-EHB and pre-BHB models in the hot-flasher scenario, i.e., removing different amounts of the hydrogen-rich envelope by an artificially enhanced mass loss at the red giant branch

⋆tbattich@fcaglp.unlp.edu.ar

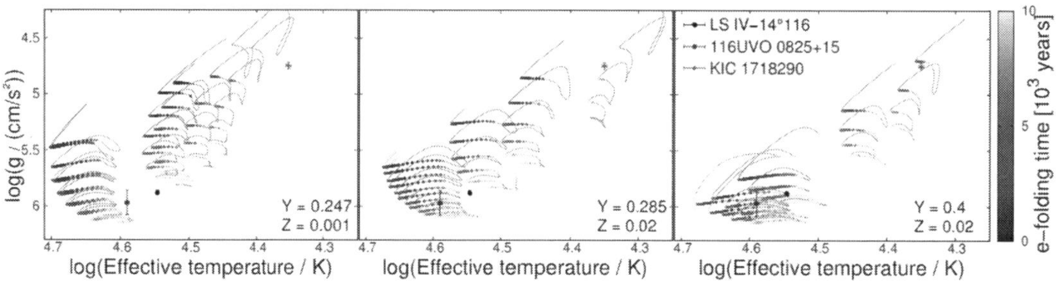

Figure 1. He-subflashes phase (grey lines) along with the location of the known He-rich hot subdwarf pulsators ([5–7]). Colored points are models with excited modes and the color coding shows the minimum e-folding time.

tip. We computed adiabatic and non-adiabatic non-radial pulsations in stellar models going through He-subflashes. The calculations were carried out with LPCODE and LP-PUL codes ([3, 4]). We found in all cases that the periods excited are in the range $400\,\mathrm{s} \lesssim P \lesssim 2500\,\mathrm{s}$. The longest and most unstable periods correspond to the first subflashes due to the larger nuclear energy release. We found an instability region at about $4.67 \lesssim \log g \lesssim 6.15$ and $4.3 \lesssim \log T_{\mathrm{eff}} \lesssim 4.7$ (Fig. 1). In models with lower surface gravities the modes are stabilized by radiative damping in the outer layers. Within the hot-flasher scenario there are no stellar models with greater temperatures and gravities. The characteristic times in which modes are excited are shorter than the typical timescales of subflashes. Therefore, they could grow to observable amplitudes. We found that ~3 out of 100 pre-EHB stars should be pulsating.

3 Conclusion

We found an instability domain of long-period g-modes driven by the ϵ-mechanism associated to pre-EHB stellar models, but on the other hand, we did not obtain pulsational instability in pre-BHB stellar models. The location of known He-rich hot subdwarf pulsators (Fig. 1) is well reproduced by our theoretical predictions, although the observed periods in these stars are systematically longer than the predicted ones. Our results could constitute a theoretical basis for future searches of pulsators in the Galactic field and stellar clusters.

Acknowledgments: TB thanks the LOC for the financial support for her assistance to the conference.

References

[1] Heber, U., PASP, **128**, 082001 (2016)
[2] Miller Bertolami, M. M., Córsico, A. H., & Althaus, L. G., ApJL, **741**, L3 (2011)
[3] Althaus, L. G., Serenelli, A. M., Panei, J. A., Córsico, A. H., García-Berro, E., & Scóccola, C. G., A&A, **435**, 631 (2005)
[4] Córsico, A. H., Althaus, L. G., & Miller Bertolami, M. M., A&A, **458**, 259 (2006)
[5] Randall, S. K., Bagnulo, S., Ziegerer, E., Geier, S., & Fontaine, G., A&A, **576**, A65 (2015)
[6] Jeffery, C. S., Baran, A. S., Behara, N. T., et al., MNRAS, **465**, 3101 (2017)
[7] Østensen, R. H., Degroote, P., Telting, J. H., et al., ApJL, **753**, L17 (2012)

The connection between period spectra and constraints in white dwarf asteroseismology

Agnès Bischoff-Kim[1]

[1]*Penn State Worthington Scranton, Dunmore, PA 18512, USA*

Abstract. White dwarfs are the end product of evolution for around 98% of the stars in our Galaxy. Buried in their interiors are the records of physical processes that take place during earlier stages in the life of the star. In recent years, a well-established theory of non-radial oscillations, improved white dwarf models, year of expertise built up in the field of white dwarf asteroseismic fitting, and computing power have culminated in the asteroseismology finally delivering what it promised: a detailed map of the interior structure of white dwarfs. As always in science, new results raise new questions. We perform a number of numerical experiments to better understand the connection between a given set of periods varying in the number of periods and in the set of radial overtones and the quality of the constraints on interior structure one obtains from fitting these periods.

1 Introduction

A few rules of thumb are used in white dwarf asteroseismology: 1) The number of parameters should be of the same order and no greater than the number of periods fitted, and 2) Lower radial overtones (low k modes) carry information about the core while higher k modes do not. The fact that these are only rules of thumb was most recently made clear when two stars each with 6 observed lower k modes (R458 and GD165) were analyzed side by side ([1]). The amount of constraint on core structure and other parameters varied greatly between the two stars, even though their pulsation spectra were similar. This made evident the fact that not all low k modes are created equal. In this work, we investigate the connection between period spectra and constraints.

2 Numerical experimental setup

We picked a fiducial model similar to the best fit model of Giammichele et al. (2016, [1]); that is, a hydrogen layer mass white dwarf of average stellar mass, inside the ZZ Ceti instability strip. The properties of the model are listed in Table 1. For that fiducial model we computed a set of ($\ell = 1$) modes, then went on to sample that list as if we were observing actual stars, which often only show a few of the possible excited modes. We selected 5 modes, using weight functions to guide our choices. Weight functions determine where in the model a mode is strongly trapped ([2]). For our first fit, we specifically selected modes that appeared strongly trapped in the core, while for the second fit, we selected lower k modes that were not strongly trapped in the core.

We fit the chosen periods of the fiducial model to models in an extensive grid of DAV models ([3]). For each fit, we calculate the fitness function

Table 1. Properties of the fiducial model. X_0 is the central oxygen abundance and X_{fm} is the location of the edge of the homogeneous C/O core. For more details on how parameters are defined, see [3].

$T_{\rm eff} = 12{,}000\ K$	$M_{\rm H} = -4.80$
Mass $= 0.595\ M_\odot$	$X_0 = 0.70$
$M_{\rm He} = -2.20$	$X_{\rm fm} = 0.45\ M_\star$

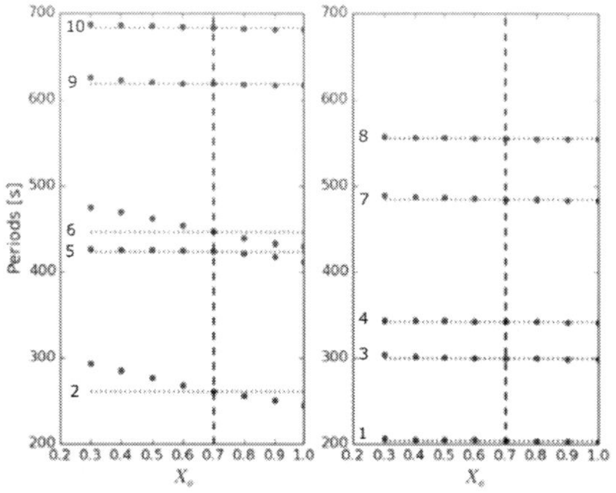

Figure 1. Dependence of each mode on the central oxygen abundance. Each mode is labeled with its radial overtone number. The vertical dashed line marks the value set for the fiducial model. *left*: Fit strongly sensitive to core structure. *right*: Fit not very sensitive to core structure.

$$\sigma_{\rm RMS} = \sqrt{\frac{1}{n_{\rm obs}} \sum_{1}^{n_{\rm obs}} \left(P_i^{\rm calc} - P_i^{\rm obs}\right)^2}, \quad (1)$$

where in this case $P_i^{\rm obs}$ refers to the periods of the fiducial model, while $P_i^{\rm calc}$ refers to the periods of each model in the grid. Evidently, when we match the fiducial model to itself in the grid, we get $\sigma_{\rm RMS} = 0$ s.

3 Results

We find that indeed, when choosing modes that are strongly trapped in the core for our fitting, we obtain tight constraints on core structure, while when we specifically choose modes that show weak trapping in the core, we obtain very little constraints on core structure. While this is a nice demonstration of the usefulness of weight functions, this result also shows strikingly that even very low k modes can be of little help in constraining core structure. This is best illustrated in Figure 1, where we show how sensitive each individual mode is to the central abundance of oxygen in the model.

References

[1] Giammichele, N., Fontaine, G., Brassard, P., & Charpinet, S., ApJS, **223**, 10 (2016)
[2] Montgomery, M. H., Metcalfe, T. S., & Winget, D. E., MNRAS, **344**, 657 (2003)
[3] Bischoff-Kim, A., Montgomery, M. H., Winget, D. E., ApJ, **675**, 1505 (2008)

Pulsational instability of high-luminosity H-rich pre-white dwarf stars

Leila M. Calcaferro[1,2,*], *Alejandro H.* Córsico[1,2,**], *María E.* Camisassa[1,2,***], *Leandro G.* Althaus[1,2,****], and *Hiromoto* Shibahashi[3,†]

[1] *Facultad de Ciencias Astronómicas y Geofísicas, Universidad Nacional de La Plata, Paseo del Bosque s/n, (1900) La Plata, Argentina*
[2] *Instituto de Astrofísica La Plata, CONICET-UNLP, Paseo del Bosque s/n, (1900) La Plata, Argentina*
[3] *Department of Astronomy, The University of Tokyo, Japan*

Abstract. We present a pulsational stability analysis on high-luminosity H-rich (DA) white dwarf models evolved from low-metallicity progenitors. We found that the ε mechanism due to H-shell burning is able to excite low-order g modes.

1 Introduction

The majority of the stars will end their lives as white dwarf (WD) stars. Hence, it is important to study them in order to understand stellar evolution and the history of the stars in our galaxy. There are two main types of WDs: H-rich WD ($\sim$ 80% of the WD stars), which are called "DA" WD stars, and H-deficient WDs, called "non-DA" WD stars. Although their progenitors are different, both of them can undergo pulsational instabilities, making possible an asteroseismological analysis. In general, they are found to pulsate with g modes with periods in the range of $\sim$ 100 – 7000 s. The mechanisms responsible for the pulsations are usually the $\kappa - \gamma$ mechanism acting in the ionization zones of the abundant chemical elements or the convective driving mechanism in cool WDs. There is another excitation mechanism that may produce instabilities of short-period g modes: the ε mechanism, induced by nuclear reactions and which depends strongly on the temperature ([5]).

In this work, we extend the work of [2] who showed that low-luminosity DA WDs exhibit pulsations excited by the ε mechanism, and now, we concentrate on the high-luminosity phase.

2 Analysis

We computed non-adiabatic non-radial g mode pulsation periods, employing the pulsation code described in [3], for DA WD evolutionary models with stellar masses in the range of 0.53 to 0.75 $M_{\odot}$, and progenitor metallicities Z = 0.0001, 0.0005 and 0.0010. In this high-temperature regime, there

* lcalcaferro@fcaglp.unlp.edu.ar
** acorsico@fcaglp.unlp.edu.ar
*** camisassa@fcaglp.unlp.edu.ar
**** althaus@fcaglp.unlp.edu.ar
† shibahashi@astron.s.u-tokyo.ac.jp

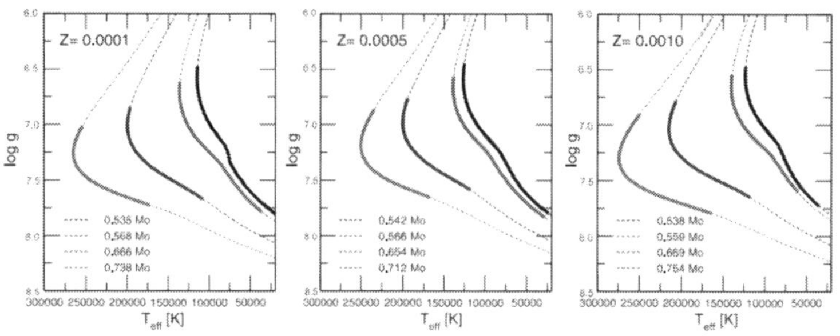

Figure 1. Evolutionary tracks of WD models on the $T_{\rm eff}$-log g plane. Each panel corresponds to a different metallicity value of the WD progenitors. Thick segments on the tracks depict DA WD models with unstable low-order g modes destabilized by the ε mechanism.

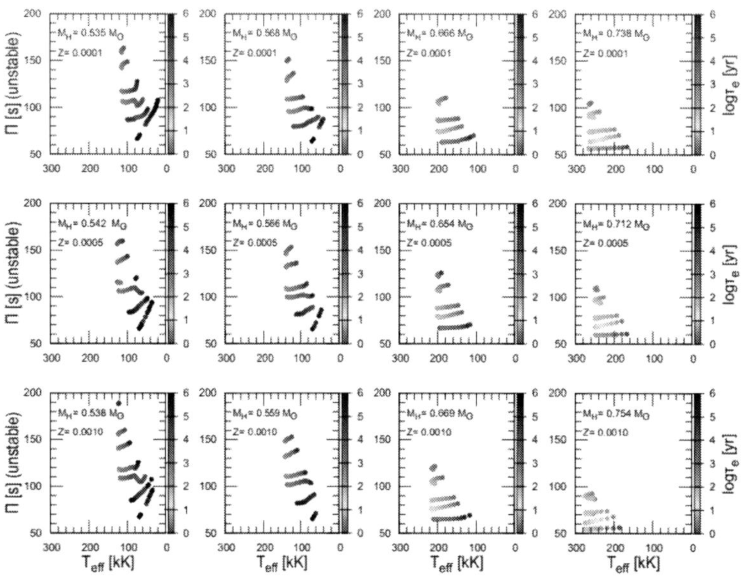

Figure 2. Periods of unstable dipole g modes in terms of the effective temperature for our WD models for different progenitor metallicities. Color coding indicates the e-folding time (τ_e) of each unstable mode (right scale).

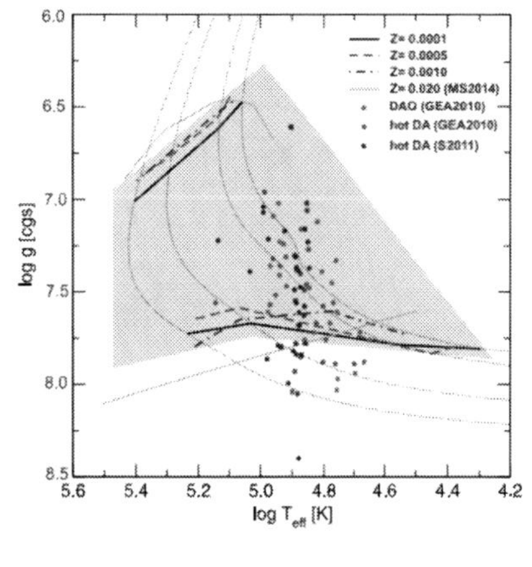

Figure 3. Instability domain (shaded area) of low-order g modes excited by the ε mechanism on the $\log(T_{\rm eff}) - \log(g)$ plane. Thick lines show the edges of the instability region for different progenitor metallicities. Also included are the results of [6] for solar metallicity. Dots of different colors show the DAO and hot DA stars from [4] and hot DA stars from [7]. Dotted lines depict evolutionary tracks corresponding to Z = 0.0001.

are no uncertainties in the pulsation stability computations due to convection. Initial WD models are the result of the complete evolution from the zero-age main sequence, considering H- and He-burning, thermal pulses in the AGB and post-AGB phases, until the pre-WD and WD stages (see [1]). Thus, the initial H content and the residual nuclear burning are the result of the complete previous evolution. During the WD stage, element diffusion due to chemical and thermal diffusion and gravitational settling was allowed to operate.

In Figure 1 we show the evolutionary tracks of the WD models on the $T_{\rm eff}$ – $\log(g)$ plane for different metallicity values (Z) of the WD progenitors. As shown by the thick segments on the tracks, there is a wide region where DA WD models have unstable low-order g modes destabilized by the ε mechanism. It is clear that, for low-metallicity progenitors and for lower mass, the range of instability extends to lower $T_{\rm eff}$. Next, we analyzed the pulsation spectrum in terms of the effective temperature, for different progenitor metallicities and different masses. We display the results in Figure 2, where we included a color coding for the e-folding time (τ_e) of each unstable mode, that represents a measure of the time taken for the perturbation to reach an observable amplitude. We can see from this figure that, for a fixed value of Z, the pulsation spectrum spans a wider range of periods for lower mass, and becomes narrower with increasing mass. Furthermore, for larger mass, the periods are destabilized at higher $T_{\rm eff}$. If we change the value of Z and compare the situation for approximately the same values of the mass, we find that the unstable period domain remains almost unaltered. Thus, the range of periods excited is not sensitive to Z. Finally, in Figure 3 we show the instability domain of low-order g modes excited by the ε mechanism on the $\log(T_{\rm eff})$ – $\log(g)$ diagram. We can see that the instability region extends to significantly lower effective temperature for lower values of Z, in comparison with the results of [6] for solar metallicity.

3 Summary and conclusions

We found that some dipole ($\ell = 1$) low-order ($k = 1-7$) g modes with periods in the range $\sim 50-200$ s are destabilized by nuclear burning through the ε mechanism in DA WD models at high luminosities, in line with previous results ([6]). Our results indicate that the instability domain is sensitive to both the stellar mass and the metallicity of the progenitor stars. In particular, the instability domain extends to lower $T_{\rm eff}$ for lower Z and lower masses. By comparing with the computations of [6], who considered only solar metallicity, we found that the instability given by the ε mechanism extends to substantially lower effective temperatures, particularly for low-mass ($M_* \sim 0.53 - 0.55\,M_\odot$) WD models. We also found that the range of unstable periods is sensitive to the stellar mass, and does not depend on the progenitor metallicity. Specifically, the longest unstable periods are longer for lower mass. The results presented in this work indicate that a search for photometric variations in hot DA WDs is worth doing.

Acknowledgments: L.M.C. warmly thanks the Local Organising Committee of the 22nd Los Alamos Stellar Pulsation Conference Series Meeting for support that allowed her to attend this conference.

References

[1] Althaus, L. G., Camisassa, M. E., Miller Bertolami, M. M., Córsico, A. H., & García-Berro, E., A&A, **576**, A9 (2015)
[2] Camisassa, M. E., Córsico, A. H., Althaus, L. G., & Shibahashi, H., A&A, **595**, A45 (2016)
[3] Córsico, A. H., Althaus, L. G., & Miller Bertolami, M. M., A&A, **458**, 259 (2006)
[4] Gianninas, A., Bergeron, P., Dupuis, J., & Ruiz, M. T., ApJ, **720**, 581 (2010)
[5] Kawaler, S. D., Winget, D. E., Hansen, C. J., & Iben, I., Jr., ApJL, **306**, L41 (1986)
[6] Maeda, K., & Shibahashi, H., PASJ, **66**, 76 (2014)
[7] Sion, E. M., *Hot White Dwarfs* (Wiley-VCH, Weinheim, 2011)

Pulsating stars and the distance scale

Lucas Macri[1,*]

[1] *Mitchell Institute for Fundamental Physics & Astronomy and Department of Physics & Astronomy, Texas A&M University, College Station, Texas, United States of America*

Abstract. I present an overview of the latest results from the SH0ES project, which obtained homogeneous *Hubble Space Telescope* (*HST*) photometry in the optical and near-infrared for ~ 3500 and ~ 2300 Cepheids, respectively, across 19 supernova hosts and 4 calibrators to determine the value of H_0 with a total uncertainty of 2.4%.
I discuss the current 3.4σ "tension" between this local measurement and predictions of H_0 based on observations of the CMB and the assumption of "standard" ΛCDM. I review ongoing efforts to reach $\sigma(H_0) = 1\%$, including recent advances on the absolute calibration of Milky Way Cepheid period-luminosity relations (PLRs) using a novel astrometric technique with *HST*.
Lastly, I highlight recent results from another collaboration on the development of new statistical techniques to detect, classify and phase extragalactic Miras using noisy and sparsely-sampled observations. I present preliminary Mira PLRs at various wavelengths based on the application of these techniques to a survey of M33.

1 Introduction & motivation

Pulsating stars have played a key role in the extragalactic distance scale since the discovery of the Cepheid period-luminosity relation (hereafter, PLR, [1]) and its first use ([2]) to calibrate a secondary distance indicator and estimate the local expansion rate of the Universe (hereafter, H_0). Figure 1 shows a schematic representation of the modern extragalactic distance "ladder" based on Cepheids and type Ia supernovae (hereafter, SNe Ia). The latter are excellent "standardizable candles" as first shown by [3]. Further refinements to the calibration of SNe Ia magnitudes based on their light curve properties ([4, 5]) enabled the discovery of dark energy ([6, 7]).

In the two decades since the discovery of dark energy, a considerable amount of theoretical and observational effort has been devoted to understanding its properties. Given an equation of state of dark energy $w = p/\rho$ that evolves with scale factor a as $w = w_0 + w_a(1 - a)$ ([11]), one possible "figure of merit" for dark energy surveys ([9]) is the inverse of the area of the ellipse that represents the allowed parameter space for w_0 and w_a (see left panel of Fig. 2). The combination of SNe Ia with other observational probes of dark energy (such as baryon acoustic oscillations and the cosmic microwave background) can significantly improve this figure of merit by reducing the impact of degeneracies affecting any given method. More precise and accurate determinations of H_0 play a critical role in this effort, and their impact can be quite significant if $\sigma(H_0) \lesssim 2\%$ ([10]; see right panel of Fig. 2).

*lmacri@tamu.edu

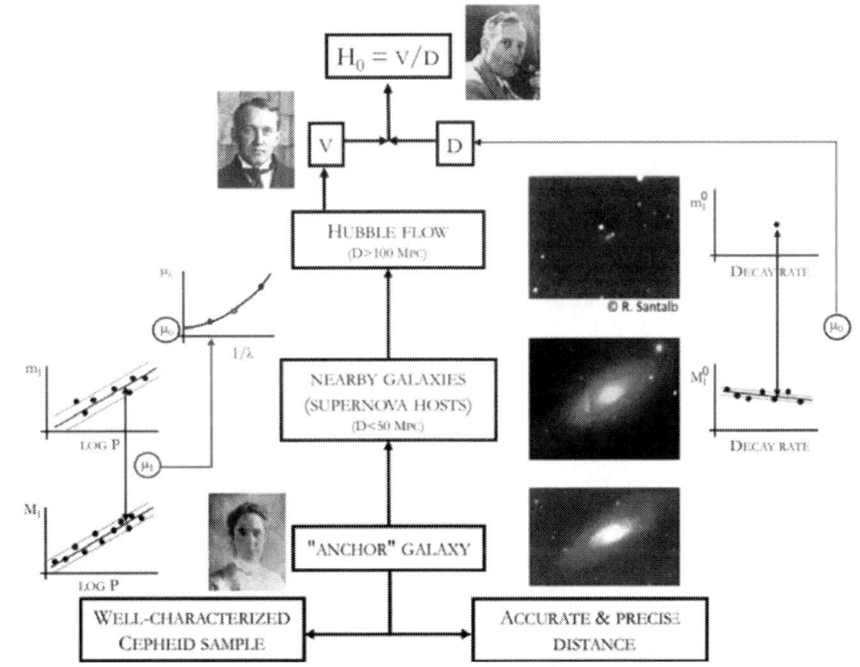

Figure 1. Steps to the Hubble Constant. This distance ladder begins with the absolute calibration of the Cepheid PLR in one or more nearby "anchor" galaxies that have well-characterized samples of these variables as well as precise and accurate distances. The second "rung" in the ladder consists of multi-wavelength observations of Cepheid-bearing galaxies that have hosted modern and "ideal" SNe Ia (low reddening, well-sampled digital light curve). This enables a robust calibration of the SN Ia luminosity-decline rate relation (or a more sophisticated formulation of this secondary distance indicator). The last "rung" uses SNe Ia in the Hubble flow. Pictured are H. Leavitt (discoverer of the Cepheid PLR), V. Slipher (pioneer of galaxy redshifts) and E. Hubble.

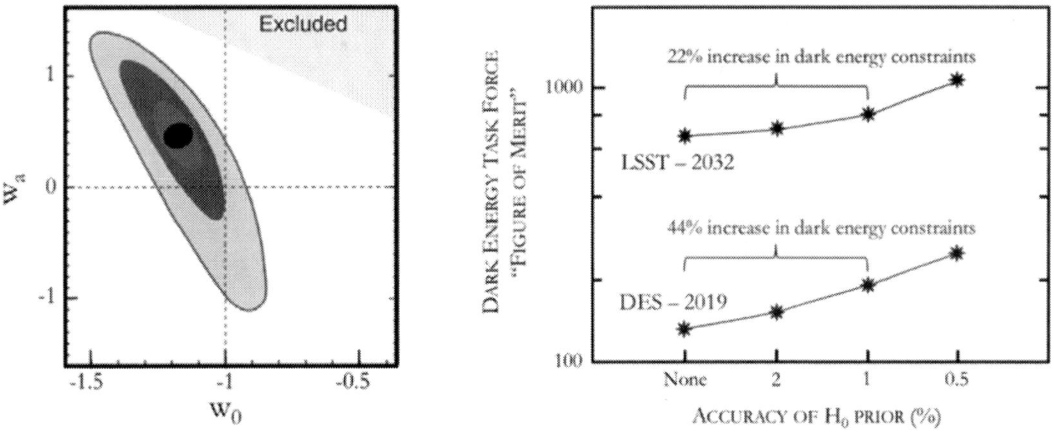

Figure 2. Left: Current and anticipated constraints on the equation of state of dark energy. Red/salmon contours represent $1/2\sigma$ constraints from [8]; blue and black denote expected 2σ constraints by DES and LSST, respectively. Right: Improvement on the Dark Energy Task Force figure of merit ([9]) as a function of $\sigma(H_0)$ ([10]).

2 SH0ES: A decade of efforts leading to $\sigma(H_0) = 2.4\%$

Motivated by the above, the SH0ES team (PI: A. Riess) has worked over the past decade to measure H_0 with increasing accuracy and precision. Our first two publications ([12, 13]) represented improvements of 2× and 3× ($\sigma(H_0) = 4.8\%, 3.3\%$) over previous efforts ([14, 15]), while our most recent work ([16, 17]) halves our initial uncertainty ($\sigma(H_0) = 2.4\%$). Figure 3 shows the Cepheid-bearing galaxies used in our latest analysis, while Figure 4 shows the corresponding *I* and *H* PLRs with ~ 3500 and ~ 2300 variables, respectively.

Figure 5 shows the evolution of the SH0ES error budget over time. Our initial improvement ([12]) over previous work was the result of building a differential distance ladder based exclusively on *HST* photometry of Cepheids (negating systematic uncertainties due to the mix of ground- and space-based measurements) and focused on near-infrared observations (minimizing systematic uncertainties due to extinction and metallicity). Our second publication ([13]) benefited from a substantial increase in the number of near-infrared Cepheid observations, made possible by the installation of WFC3 on *Hubble*. Our third and latest determination ([16]) is based on a significantly larger number of supernova host galaxies, which were more efficiently surveyed for Cepheids by using a long-pass filter ([17]). Equally critical along the way have been the improvements on the distance estimates for our "anchor" galaxies, which are based on independent techniques: maser observations for NGC 4258 ([18, 19]); detached eclipsing binaries for the LMC ([20]); and parallaxes for Milky Way Cepheids ([21–24]).

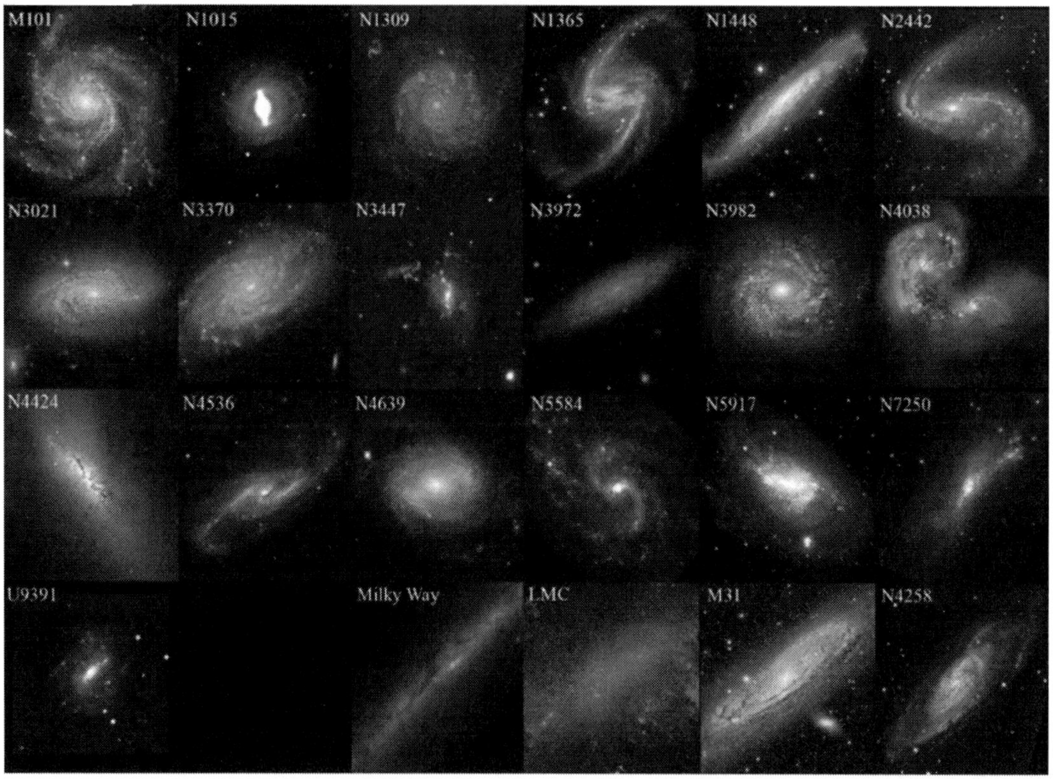

Figure 3. Cepheid-bearing galaxies used in the most recent determination of H_0 by the SH0ES team ([16]). Based on [17].

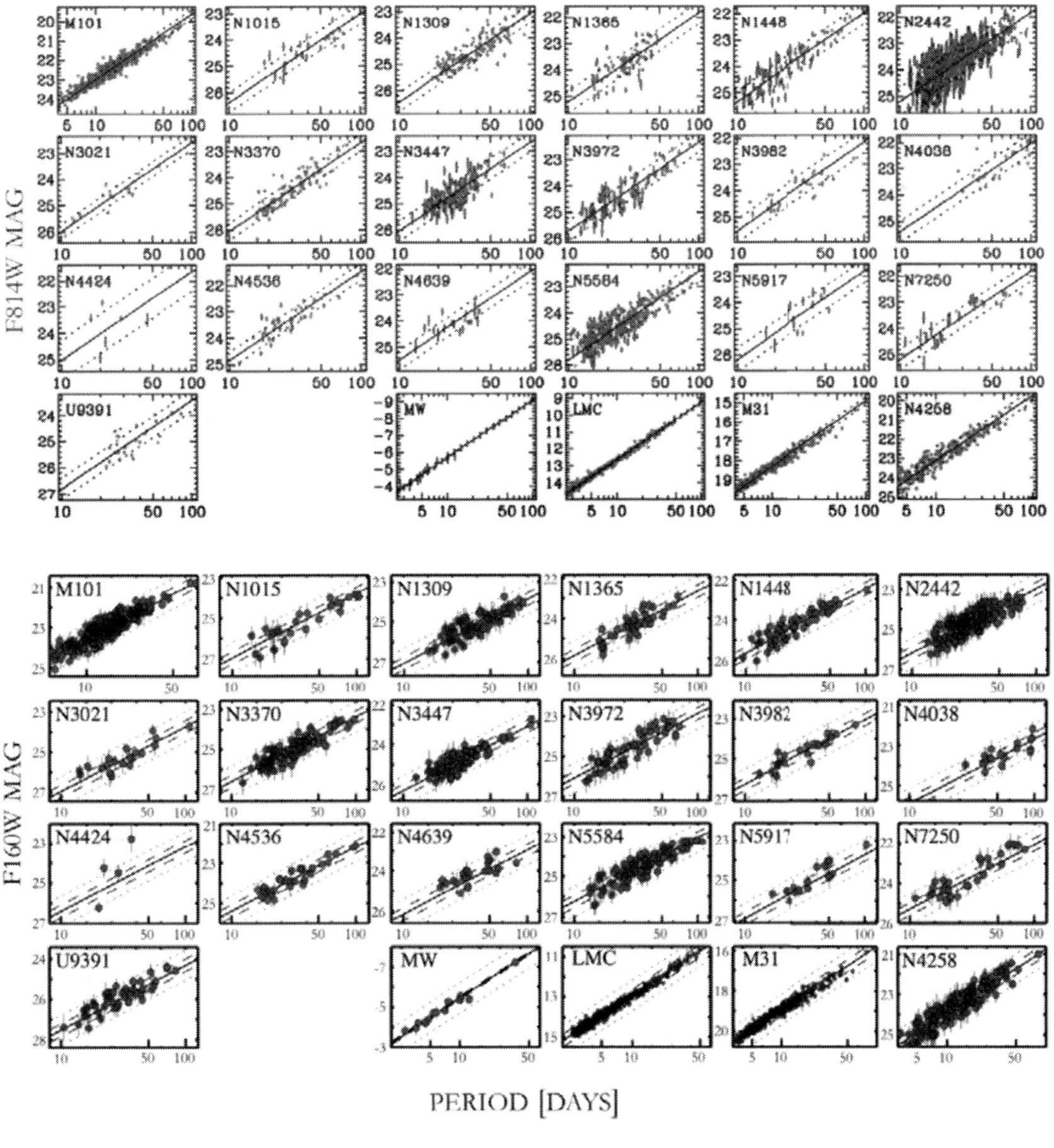

Figure 4. Cepheid period-luminosity relations in *F814W* (top) and *F160W* (bottom) for all galaxies in the most recent study by the SH0ES team. Based on [16, 17].

3 Current and future steps towards $\sigma(\mathrm{H}_0) = 1\%$

The current measurement of H_0 by the SH0ES team ([16]) is larger by $\sim 3.4\sigma$ than the *predicted* value by [25]. The latter is inferred using the Baryon Acoustic Oscillation method, calibrated from observations of the Cosmic Microwave Background and assuming a specific cosmological model (standard ΛCDM). While the discrepancy may be due to systematic effects on one or both sides, an intriguing possibility is that the difference might be due to an additional component in the cosmological model

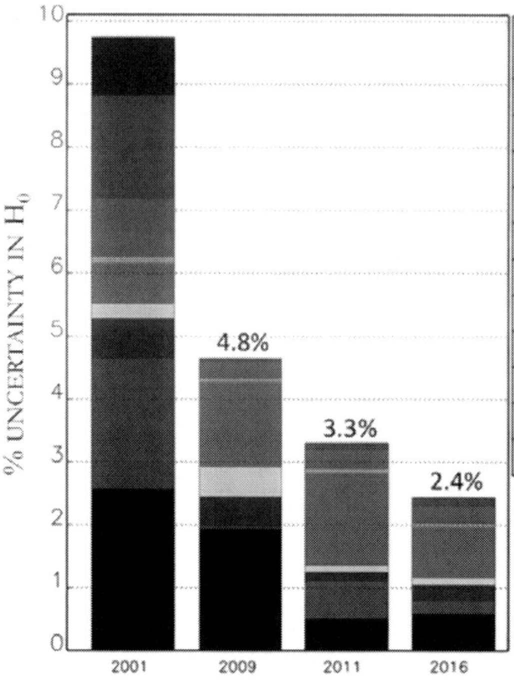

TERM	KP	'09	'11	'16
	%			
ABSOLUTE SCALE	5.0	3.0	1.3	1.3
REDDENING & ZEROPOINTS (ANCHOR-TO-HOSTS)	4.5	0.3	1.4	0.7
P-L SLOPE, D LOG P (ANCHOR-TO-HOSTS)	4.0	0.5	0.6	0.5
METALLICITY DEPENDENCE (ANCHOR-TO-HOSTS)	3.0	1.1	1.0	0.8
GROUND/HST CROSS-CALIBRATION	3.0	--	--	--
MEAN OF SN IA CALIBRATORS	2.5	2.5	1.9	1.3
MEAN OF P-L IN ANCHOR(S)	2.5	1.5	0.8	0.7
MEAN OF P-L IN SN HOSTS	1.5	1.5	0.6	0.4
SN IA M-Z RELATION	1.0	0.5	0.4	0.4
ANALYSIS SYSTEMATICS	??	1.3	1.0	0.9
TOTAL	10	4.8	3.3	2.4

Figure 5. Evolution of the error budget for H_0 in various iterations of the SH0ES project ([12, 13, 16]), compared with the *HST* Key Project on the Distance Scale ([14]). Based on [16].

([26]). Figure 6, adapted from [27], shows that most of the "tension" can be alleviated by including an extra relativistic species in the early Universe.

Given this intriguing discrepancy, further work is warranted to reach $\sigma(H_0) = 1\%$. This will require a better absolute calibration of the Cepheid PLR and a significant increase in the number of SNe Ia hosts with Cepheid distances. Regarding the first issue, the SH0ES team has used a novel observing technique available with WFC3 ("spatial scanning") to measure parallaxes to 19 Milky Way Cepheids located at considerable greater distances ($D \lesssim 4$ kpc) than the previously-studied 10 variables ($D \lesssim 500$ pc). Beyond the significant increase in sample size, our targets span a range in periods that is better matched to the extragalactic sample and helps to minimize any remaining systematics associated with variations in the slope of the PLR (see Fig. 7). The two parallaxes published thus far by this program ([23, 24]) are also significantly more precise than previous determinations despite the considerably smaller parallax signal.

The *Gaia* mission is expected to discover and measure high-quality parallaxes for thousands of Cepheids in the Milky Way ([28]), and to provide much-improved distances for the currently-known variables of this type. This should yield a sub-percent calibration of the Galactic PLR, and it will be critical to place as many of these variables as possible on the same photometric system used to study extragalactic targets to avoid further systematic uncertainties. To this end, the SH0ES team has already obtained *HST* photometry for 50 Cepheids that are expected to have the highest-quality parallaxes from *Gaia*.

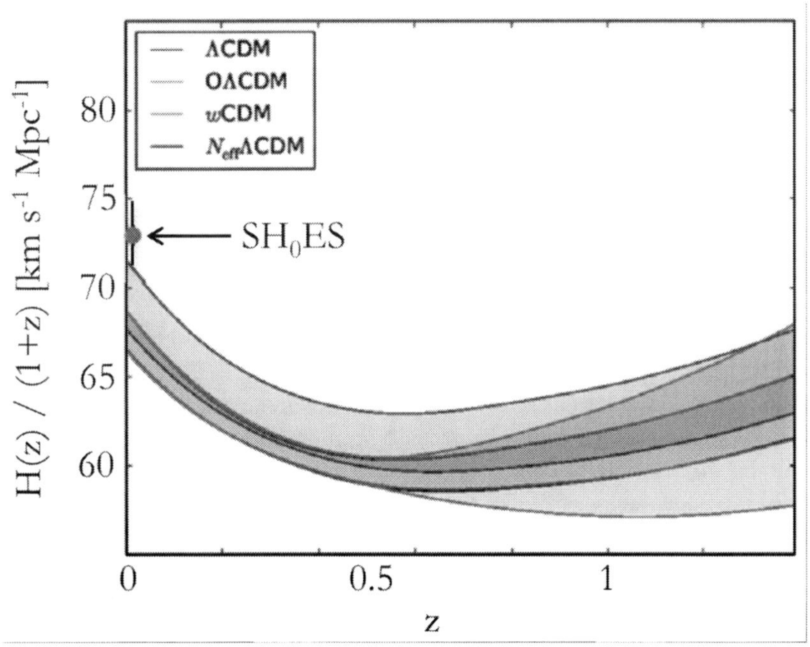

Figure 6. Comparison of *predicted* values of H(*z*) for different cosmological models vs. H_0 measured by SH0ES ([16]). Based on [27].

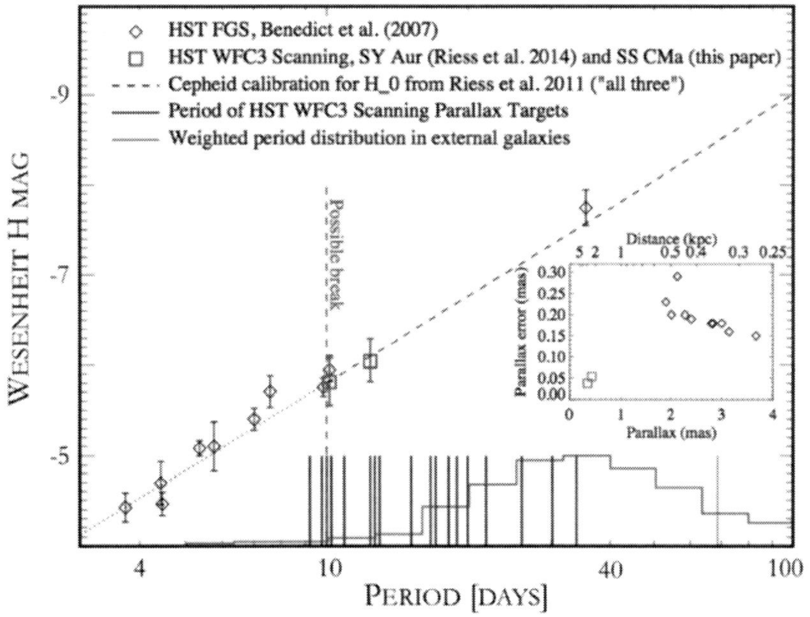

Figure 7. Improved Galactic Cepheid PLR using parallaxes measured with novel *HST* technique, and prospects for future improvement. Based on [24].

DOI: 10.1051/epjconf/201715207001

4 Miras: new techniques and prospects for LSST

Mira variables could play a significant role in the extragalactic distance scale during the next decade, as part of the quest to reach $\sigma(H_0) = 1\%$ with facilities such as *JWST* and *WFIRST*. Given their intermediate-mass progenitors, they are more plentiful than Cepheids and are present in early-type galaxies. As shown in Figure 8, oxygen-rich Miras rival the luminosities of Cepheids at *K* and exhibit a tight PLR at near-infrared wavelengths ([29, 30]). Besides their significantly longer periods relative to Cepheids, another possible roadblock to their wide use is the large variations exhibited by their light curves from cycle to cycle. This complicates the determination of periods using traditional techniques ([31, 32]) on sparsely-sampled light curves.

A collaboration between statisticians and astronomers at Texas A&M University has recently yielded new techniques for determining the periods of Miras in the regime of sparsely-sampled, noisy light curves ([36]). We used the OGLE-III light curves of Miras in the LMC and downsampled them to simulate time-series observations with more typical coverage and photometric precision for extragalactic variables. We developed a novel periodogram based on the Gaussian Process method and tested it using the simulated light curves. We found the new algorithm outperformed the traditional period-finding techniques and was able to successfully recover periods for ~ 70% of the variables (see Fig. 9). We applied the technique to *I*-band observations of M33 ([37, 38]) and discovered over 1800 Mira candidates in this galaxy ([35]). We matched our optical catalog with NIR and MIR observations from [39, 40] and obtained PLRs that compare favorably to their LMC counterparts (see Fig. 10).

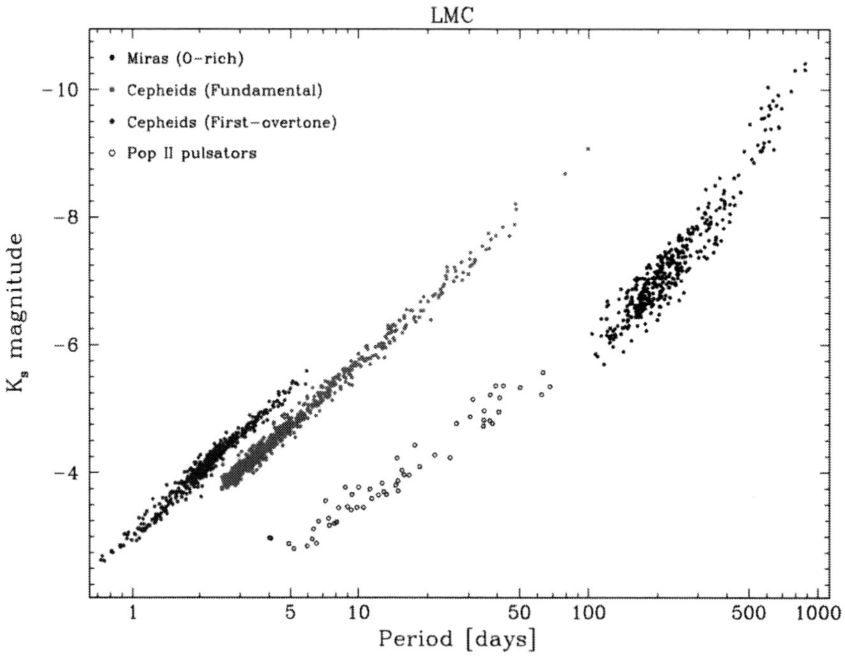

Figure 8. Comparison of *K*-band PLRs for various types of pulsating stars. Based on [33], [34] and [35] for Cepheids, Pop. II pulsators and O-rich Miras, respectively.

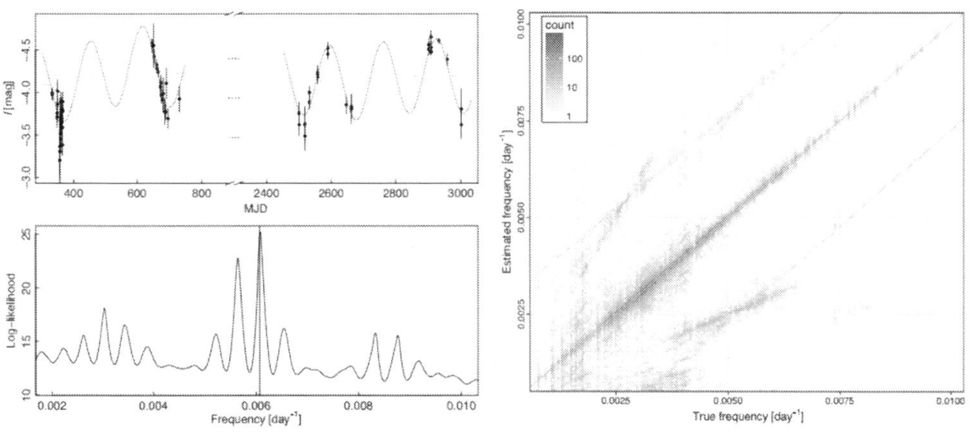

Figure 9. Application of the Gaussian Process to model simulated Mira light curves that are noisy and sparsely sampled (left) and success in recovering the known periods of the variables used to generate the simulations (right). Based on [36].

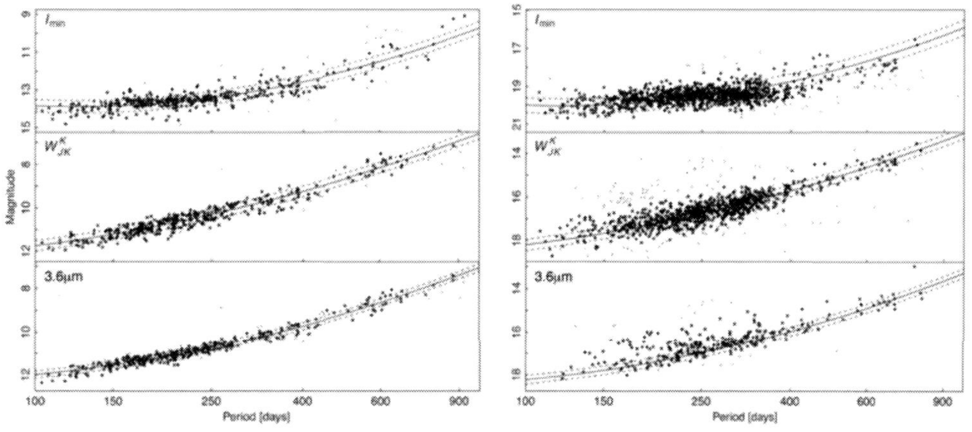

Figure 10. Mira PLRs in various bands for O-rich variables in the LMC (left) and M33 (right). Based on [35].

Motivated by the success of the Gaussian Process periodogram at detecting Mira candidates in M33, we carried out additional simulations to investigate the usefulness of the expected LSST cadence for the "deep-wide-fast" survey for finding O-rich variables in galaxies with $D \lesssim 15$ Mpc ([41]). The preliminary results, shown in Figure 11, are quite encouraging. The largest yield is expected in NGC 5128, given its relatively close distance and large mass; ~ 29000 O-rich Miras could be detected. We expect that LSST will be able to detect 100 or more Miras in ~ 70 galaxies, for a final yield of $\sim 2 \times 10^5$ variables by the end of its initial decade-long survey. Most detections will occur in i, and > 75% of the variables should have $i - z$ colors. Follow-up of selected subsets of these variables with *JWST*, *WFIRST* or other space-based NIR missions would yield accurate and precise distances to the host galaxies.

DOI: 10.1051/epjconf/201715207001

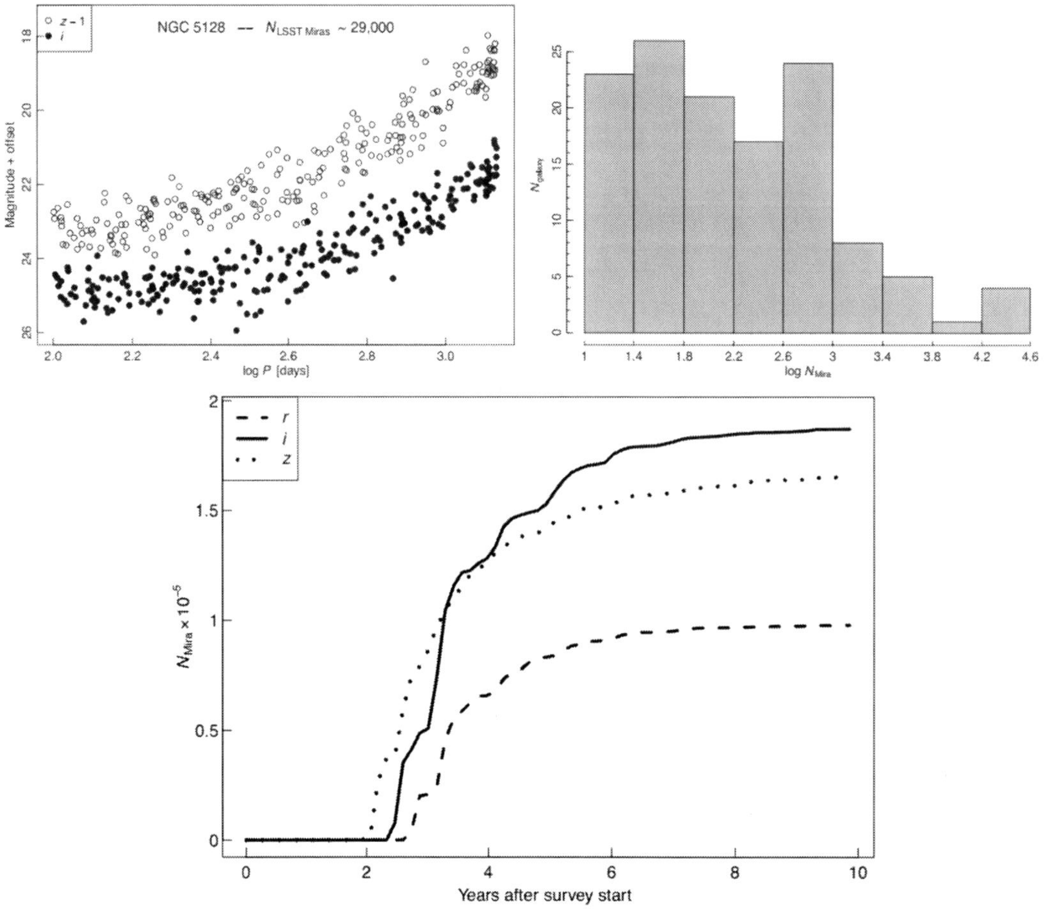

Figure 11. Prospects for detection of O-rich Miras with LSST. Upper left: simulated PLR for Miras in NGC 5128 (only a subset of the ~ 29000 expected variables are plotted). Upper right: Number of galaxies in which LSST should detect a given number of Miras. Bottom: Detectability of Miras with LSST as a function of time since the start of the survey. Based on [41].

Acknowledgments: *HST* observations of extragalactic Cepheids were supported by NASA/STScI through HST-GO awards 9810, 10399, 10802, 11570, 12880 and 13647. *HST* studies of Milky Way Cepheids were supported by NASA/STScI through HST-GO awards 12679, 12879, 13334, 13335, 13344, 13678, 13686, 13928, 14062 and 14648. The ground-based observations of Cepheids and Miras were supported by the US National Science Foundation through grant AST-1211603. All projects received additional support from the Mitchell Institute for Fundamental Physics & Astronomy at Texas A&M University.

References

[1] H.S. Leavitt, E.C. Pickering, Harvard College Observatory Circular **173**, 1 (1912)
[2] E. Hubble, PNAS **15**, 168 (1929)
[3] M.M. Phillips, ApJL **413**, L105 (1993)

[4] M. Hamuy, M.M. Phillips, N.B. Suntzeff, R.A. Schommer, J. Maza, R. Aviles, AJ **112**, 2391 (1996)
[5] A.G. Riess, W.H. Press, R.P. Kirshner, ApJ **473**, 88 (1996)
[6] A.G. Riess, A.V. Filippenko, P. Challis, A. Clocchiatti, A. Diercks, P.M. Garnavich, R.L. Gilliland, C.J. Hogan, S. Jha, R.P. Kirshner et al., AJ **116**, 1009 (1998)
[7] S. Perlmutter, G. Aldering, G. Goldhaber, R.A. Knop, P. Nugent, P.G. Castro, S. Deustua, S. Fabbro, A. Goobar, D.E. Groom et al., ApJ **517**, 565 (1999)
[8] G. Hinshaw, D. Larson, E. Komatsu, D.N. Spergel, C.L. Bennett, J. Dunkley, M.R. Nolta, M. Halpern, R.S. Hill, N. Odegard et al., ApJS **208**, 19 (2013)
[9] A. Albrecht, G. Bernstein, R. Cahn, W.L. Freedman, J. Hewitt, W. Hu, J. Huth, M. Kamionkowski, E.W. Kolb, L. Knox et al., arXiv (2006), `astro-ph/0609591`
[10] D.H. Weinberg, M.J. Mortonson, D.J. Eisenstein, C. Hirata, A.G. Riess, E. Rozo, Phys. Rep. **530**, 87 (2013)
[11] E.V. Linder, PRL **90**, 091301 (2003)
[12] A.G. Riess, L. Macri, S. Casertano, M. Sosey, H. Lampeitl, H.C. Ferguson, A.V. Filippenko, S.W. Jha, W. Li, R. Chornock et al., ApJ **699**, 539 (2009)
[13] A.G. Riess, L. Macri, S. Casertano, H. Lampeitl, H.C. Ferguson, A.V. Filippenko, S.W. Jha, W. Li, R. Chornock, ApJ **730**, 119 (2011)
[14] W.L. Freedman, B.F. Madore, B.K. Gibson, L. Ferrarese, D.D. Kelson, S. Sakai, J.R. Mould, R.C. Kennicutt, Jr., H.C. Ford, J.A. Graham et al., ApJ **553**, 47 (2001)
[15] A. Sandage, G.A. Tammann, A. Saha, B. Reindl, F.D. Macchetto, N. Panagia, ApJ **653**, 843 (2006)
[16] A.G. Riess, L.M. Macri, S.L. Hoffmann, D. Scolnic, S. Casertano, A.V. Filippenko, B.E. Tucker, M.J. Reid, D.O. Jones, J.M. Silverman et al., ApJ **826**, 56 (2016)
[17] S.L. Hoffmann, L.M. Macri, A.G. Riess, W. Yuan, S. Casertano, R.J. Foley, A.V. Filippenko, B.E. Tucker, R. Chornock, J.M. Silverman et al., ApJ **830**, 10 (2016)
[18] J.R. Herrnstein, J.M. Moran, L.J. Greenhill, P.J. Diamond, M. Inoue, N. Nakai, M. Miyoshi, C. Henkel, A. Riess, Nature **400**, 539 (1999)
[19] E.M.L. Humphreys, M.J. Reid, J.M. Moran, L.J. Greenhill, A.L. Argon, ApJ **775**, 13 (2013)
[20] G. Pietrzyński, D. Graczyk, I.B. Gieren, W. Thompson, B. Pilecki, A. Udalski, S. Soszyński, I. Kozłowski, P. Konorski, G. Suchomska, K. Bono, P.G.P. Moroni, S. Villanova et al., Nature **495**, 76 (2013)
[21] G.F. Benedict, B.E. McArthur, M.W. Feast, T.G. Barnes, T.E. Harrison, R.J. Patterson, J.W. Menzies, J.L. Bean, W.L. Freedman, AJ **133**, 1810 (2007)
[22] F. van Leeuwen, A&A **474**, 653 (2007)
[23] A.G. Riess, S. Casertano, J. Anderson, J. MacKenty, A.V. Filippenko, ApJ **785**, 161 (2014)
[24] S. Casertano, A.G. Riess, J. Anderson, R.I. Anderson, J.B. Bowers, K.I. Clubb, A.R. Cukierman, A.V. Filippenko, M.L. Graham, J.W. MacKenty et al., ApJ **825**, 11 (2016)
[25] Planck Collaboration, et al., A&A **596**, A107 (2016)
[26] C. Dvorkin, M. Wyman, D.H. Rudd, W. Hu, Phys Rev D **90**, 083503 (2014)
[27] A.J. Cuesta, L. Verde, A. Riess, R. Jimenez, MNRAS **448**, 3463 (2015)
[28] F. Windmark, L. Lindegren, D. Hobbs, A&A **530**, A76 (2011)
[29] I.S. Glass, T. Lloyd Evans, Nature **291**, 303 (1981)
[30] P.A. Whitelock, M.W. Feast, F. van Leeuwen, MNRAS **386**, 313 (2008)
[31] N.R. Lomb, Ap&SS **39**, 447 (1976)

DOI: 10.1051/epjconf/201715207001

[32] J.D. Scargle, ApJ **263**, 835 (1982)

[33] L.M. Macri, C.C. Ngeow, S.M. Kanbur, S. Mahzooni, M.T. Smitka, AJ **149**, 117 (2015)

[34] A. Bhardwaj, L.M. Macri, M. Rejkuba, S.M. Kanbur, C.C. Ngeow, H.P. Singh, AJ **153**, 154 (2017)

[35] W. Yuan, S. He, L.M. Macri, J. Long, J.Z. Huang, AJ **153**, 170 (2017)

[36] S. He, W. Yuan, J.Z. Huang, J. Long, L.M. Macri, AJ **152**, 164 (2016)

[37] L.M. Macri, K.Z. Stanek, D.D. Sasselov, M. Krockenberger, J. Kaluzny, AJ **121**, 861 (2001)

[38] A. Pellerin, L.M. Macri, ApJS **193**, 26 (2011)

[39] A. Javadi, M. Saberi, J.T. van Loon, H. Khosroshahi, N. Golabatooni, M.T. Mirtorabi, MNRAS **447**, 3973 (2015)

[40] T.A. Thompson, J.L. Prieto, K.Z. Stanek, M.D. Kistler, J.F. Beacom, C.S. Kochanek, ApJ **705**, 1364 (2009)

[41] W. Yuan, Ph.D. thesis, Texas A&M University (2017)

Toward a renewed Galactic Cepheid distance scale from *Gaia* and optical interferometry

Pierre Kervella[1,2,*], *Antoine* Mérand[3], *Alexandre* Gallenne[3], *Boris* Trahin[1,2], *Nicolas* Nardetto[4], *Richard I.* Anderson[5], *Joanne* Breitfelder[2,3], *Laszlo* Szabados[6], *Howard E.* Bond[7], *Simon* Borgniet[1,2], *Wolfgang* Gieren[8], and *Grzegorz* Pietrzyński[9]

[1]*Unidad Mixta Internacional Franco-Chilena de Astronomía (CNRS UMI 3386), Departamento de Astronomía, Universidad de Chile, Las Condes, Chile*
[2]*LESIA (CNRS UMR 8109), Observatoire de Paris, PSL, CNRS, UPMC, Univ. Paris-Diderot, France*
[3]*European Southern Observatory, Alonso de Córdova 3107, Casilla 19001, Santiago 19, Chile*
[4]*Laboratoire Lagrange, UMR 7293, Univ. de Nice Sophia-Antipolis, CNRS, Obs. de la Côte d'Azur, France*
[5]*Physics and Astronomy Department, The Johns Hopkins University, Baltimore, MD 21218, USA*
[6]*Konkoly Observatory, MTA CSFK, Konkoly Thege M. út 15-17, H-1121, Hungary*
[7]*Department of Astronomy & Astrophysics, Pennsylvania State University, University Park, PA 16802, USA*
[8]*Universidad de Concepción, Departamento de Astronomía, Casilla 160-C, Concepción, Chile*
[9]*Nicolaus Copernicus Astronomical Center, Polish Academy of Sciences, Warszawa, Poland*

Abstract. Through an innovative combination of multiple observing techniques and modeling, we are assembling a comprehensive understanding of the pulsation and close environment of Cepheids. We developed the SPIPS modeling tool that combines all observables (radial velocimetry, photometry, angular diameters from interferometry) to derive the relevant physical parameters of the star (effective temperature, infrared excess, reddening, ...) and the ratio of the distance and the projection factor d/p. We present the application of SPIPS to the long-period Cepheid RS Pup, for which we derive $p = 1.25 \pm 0.06$. The addition of this massive Cepheid consolidates the existing sample of p-factor measurements towards long-period pulsators. This allows us to conclude that p is constant or mildly variable around $p = 1.29 \pm 0.04$ ($\pm 3\%$) as a function of the pulsation period. The forthcoming *Gaia* DR2 will provide a considerable improvement in quantity and accuracy of the trigonometric parallaxes of Cepheids. From this sample, the SPIPS modeling tool will enable a robust calibration of the Cepheid distance scale.

1 Introduction

The empirical Leavitt law of Cepheids (the period-luminosity relation; [1]) is a cornerstone of the observational calibration of the cosmological distance scale. The persistent tension between the determinations of the Hubble constant H_0 from the cosmic microwave background modeling and the empirical distance ladder may have far-reaching implications on, for instance, the equation of state of dark energy. The calibration of the empirical ladder relies primarily on Cepheids and Type Ia supernovae ([2], see also [3]), and an improvement of the true calibration accuracy of the Leavitt law

*pierre.kervella@obspm.fr

is required to confirm the significance of the tension. The parallax-of-pulsation method (PoP), also known as the Baade-Wesselink technique, is a powerful way to measure the distances to individual Galactic and Magellanic Cloud Cepheids, and thus calibrate the Leavitt law. It relies on the comparison of the amplitude of the pulsation of the star from (1) the integration of its radial velocity curve measured using spectroscopy and (2) the change in angular diameter of the star estimated from its brightness and color, or measured using interferometry. The major weakness of the PoP technique is that it uses a numerical factor to convert disk-integrated radial velocities into photospheric velocities: the projection factor, or p-factor ([4–6]). The PoP technique provides the ratio d/p, but the distance d and the p-factor are fully degenerate. The calibration of the p-factor of Cepheids therefore requires to know d independently. At the moment, only a dozen reasonably accurate trigonometric parallaxes of nearby Cepheids are available, mainly from the *Hubble Space Telescope*'s Fine Guidance Sensor ([7, 8]) and WFC3 ([9]). But *Gaia* will soon resolve the d/p degeneracy for hundreds of Cepheids, thus removing the major systematics of the PoP technique. The outcome will be a robust calibration of the Cepheid distance scale, based on well-known Galactic and Magellanic Cloud pulsators. For this program, we developed an innovative version of the PoP technique, the *Spectro-Photo-Interferometry of Pulsating Stars* (SPIPS, Sect. 2) that simultaneously integrates multiple observables in a consistent model of the pulsation of the star. We present in Section 3 its application to the Cepheid RS Pup and the dependence of the p-factor on the pulsation period. We discuss in Section 4 the improvement that will come from *Gaia*'s DR2 parallaxes in 2018.

2 The SPIPS algorithm

The general principle of the SPIPS algorithm ([10]) is to fit a parametric model of the star's pulsation simultaneously to the full set of observables: radial velocities from spectroscopy, photometry in all photometric bands and (when available) angular diameter measurements from optical interferometry.

The SPIPS model of the star is built assuming that Cepheids are radially pulsating spheres, for which the pulsational velocity $v_{\text{puls}}(t)$ and the effective temperature $T_{\text{eff}}(t)$ are the basic descriptive parameters (Fig. 1). The cyclic variation of these parameters is represented using classical Fourier series or periodic splines functions. The photometry in all filters is computed using ATLAS9[1] atmosphere models, considering the bandpass and zero points matching the observations (using the SVO[2] or Asiago[3] databases). The interstellar reddening is parametrized using the standard $E(B-V)$ color excess for typical Milky Way dust ($R_V = 3.1$) and computed for the phase-variable T_{eff} of the Cepheid. Angular diameters including limb darkening (LD) are produced by the model to match the interferometric measurements, using SATLAS models ([11]). Finally, circumstellar envelopes are included in the modeling, considering both their photometric contributions in the infrared H and K bands and their influence on interferometric angular diameter measurements.

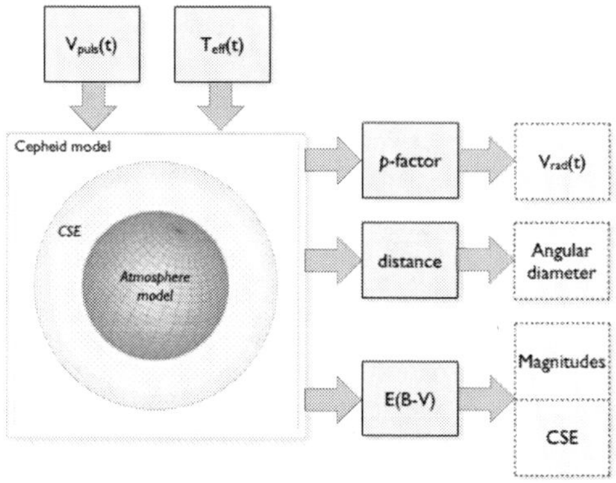

Figure 1. Overview of the principle of the SPIPS modeling. The main model parameters are shown on a light blue background, and the produced observables are shown on the right.

[1] http://wwwuser.oats.inaf.it/castelli/grids.html
[2] http://svo2.cab.inta-csic.es/
[3] http://ulisse.pd.astro.it/Astro/ADPS/Paper/index.html

DOI: 10.1051/epjconf/201715207002

3 The projection factor of Cepheids

Recent applications of the SPIPS parametric modeling to bright Galactic Cepheids are presented in [10] (δ Cep, η Aql), [12] (nine Cepheids with HST/FGS parallaxes) and [13] (Type 2 Cepheid κ Pav).

Figure 2. Color composite image of RS Pup and its nebula (from [19]).

A SPIPS model of the long-period Cepheid RS Pup (P = 41.5 days) has recently been presented by [14]. RS Pup is one of the intrinsically brightest Cepheids in the Galaxy, and it is particularly remarkable due to its large circumstellar nebula (Fig. 2) that reflects the light variations of the Cepheid ([14–19]). From a combination of photometry and polarimetry of the light echoes in the nebula, [19] obtained a distance of $d = 1910 \pm 80$ pc ($\pm 4.2\%$), corresponding to a parallax $\pi = 0.524 \pm 0.022$ mas. Knowing the distance of RS Pup opens the possibility to resolve the distance/p-factor degeneracy and determine the value of p. An overview of the SPIPS model of RS Pup is presented in Figure 3. It was computed using as constraints the photometric and radial velocity measurements available in the literature (e.g., from [20]), completed with a series of VLTI/PIONIER angular diameter measurements ([14]). Interferometric angular diameters are not mandatory for the SPIPS model to converge. But as they are insensitive to reddening, they allow to efficiently constrain both the effective temperature $T_{\rm eff}$ and the color excess $E(B-V)$ of the star, that are otherwise correlated. The overall quality of the fit of the available observables is very good, with a best-fit p-factor of $p = 1.25 \pm 0.06$ ($\pm 5\%$). The color excess $E(B-V)$ is estimated to 0.496 ± 0.006 and a moderate infrared excess of 0.02 to 0.03 mag (± 0.01) is found in the HK bands.

The addition of the p-factor of RS Pup to the limited set of existing p-factor measurements is particularly valuable, as the sample was up to now mostly limited to short and intermediate period Cepheids (Fig. 4). The other exception is ℓ Car (P = 35.5 days), whose parallax was measured by [8], and whose p-factor was derived by [12] and [21]. The p-factors of ℓ Car and RS Pup are statistically identical within their uncertainties. The simple model of a constant $p = 1.29 \pm 0.04$ over the sampled range of Cepheid periods (3.0 to 41.5 days) is consistent with the observations ($\chi^2_{\rm red} = 0.9$, [14]).

4 Perspectives with *Gaia* parallaxes

The precision of the TGAS parallaxes of Galactic Cepheids released in the *Gaia* DR1 ([24]) is insufficient to constrain significantly their p-factors (see, e.g., [25]). The *Gaia* DR2 (currently expected in April 2018) will considerably improve the accuracy of Galactic Cepheid parallaxes. Based on the foreseen accuracy of *Gaia*, 400+ Cepheid parallaxes will be measured with an accuracy better than 3%, out of which approximately 100+ with an accuracy better than 1%. The ongoing observations of a sample of 18 long-period Cepheids using the spatial scanning mode of the HST/WFC3 by [9] will also provide a complementary set of parallaxes for these rare pulsators. The flexibility of the SPIPS modeling enables the integration of all types of new and archival observations in a consistent and robust approach, thus mitigating the influence of systematics of observational origin. Taking into account the contribution of Cepheid envelopes, that can be very significant at infrared wavelengths ([26, 28]), as well as the presence of stellar companions ([27, 29]) will also contribute to mitigate the astrophysical biases on the calibration of the Leavitt law.

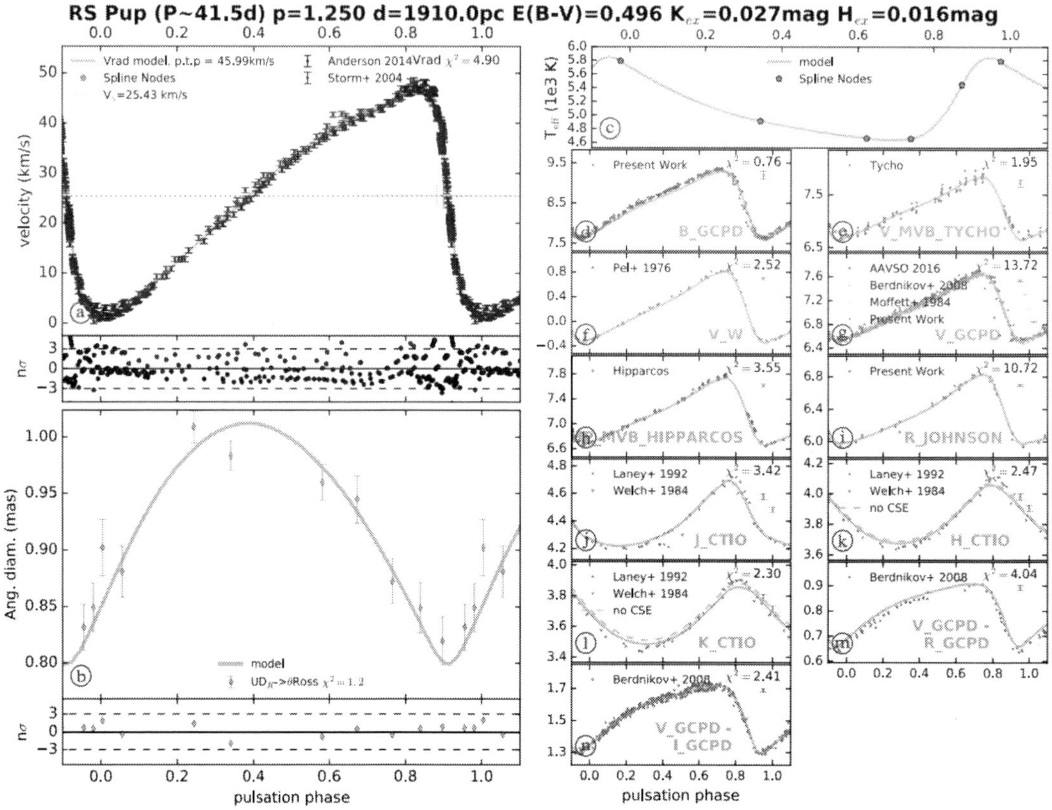

Figure 3. SPIPS combined fit of the available observations of RS Pup (figure from [14]). The radial velocity measurements are presented in the upper left panel, the interferometric angular diameters in the lower left panel, and the different photometric bands in the sub-panels on the right. The effective temperature of the model is shown in the upper right panel. The best-fit model curves are represented using light grey solid lines.

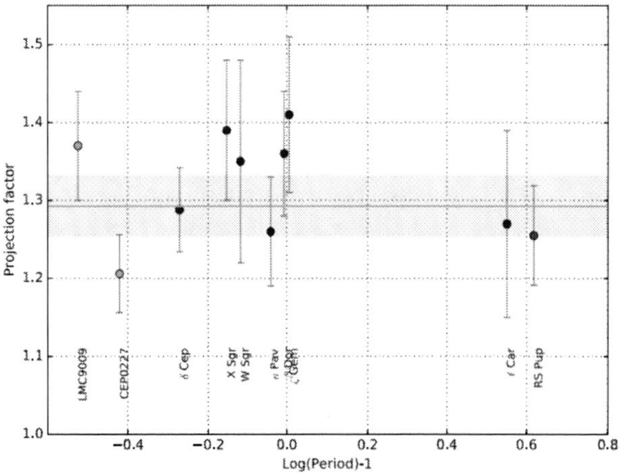

Figure 4. Measured p-factors of Cepheids with better than 10% relative accuracy (figure from [14]). The blue points use HST-FGS distances ([7, 8]), the orange points are the LMC eclipsing Cepheids ([22, 23]) and the red point is RS Pup. The solid line and orange shaded area represent the weighted average of the independent measurements ($p = 1.29 \pm 0.04$).

Acknowledgments: The authors acknowledge the support of the French Agence Nationale de la Recherche (ANR), under grant ANR-15-CE31-0012-01 (project UnlockCepheids). PK, AG, and WG acknowledge support of the French-Chilean exchange program ECOS-Sud/CONICYT (C13U01). WG and GP gratefully acknowledge financial support for this work from the BASAL Centro de Astrofisica y Tecnologias Afines (CATA) PFB-06/2007. WG also acknowledges financial support from the Millenium Institute of Astrophysics (MAS) of the Iniciativa Cientifica Milenio del Ministerio de Economia, Fomento y Turismo de Chile, project IC120009. The research leading to these results has received funding from the European Research Council (ERC) under the European Union's Horizon 2020 research and innovation programme (grant agreement No 695099). HEB acknowledges support from NASA through grant GO-11715 from the Space Telescope Science Institute, which is operated by the Association of Universities for Research in Astronomy, Inc., under NASA contract NAS5-26555

References

[1] Leavitt, H. S., & Pickering, E. C., Harvard College Observatory Circular, **173**, 1 (1912)
[2] Riess, A. G., Macri, L. M., Hoffmann, S. L., et al., ApJ, **826**, 56 (2016)
[3] Bonvin, V., Courbin, F., Suyu, S. H., et al. MNRAS, **465**, 4914 (2017)
[4] Nardetto, N., Mourard, D., Mathias, P., Fokin, A., & Gillet, D., A&A, **471**, 661 (2007)
[5] Barnes, T. G., in AIP Conference Series, Vol. **1170**, ed. J. A. Guzik & P. A. Bradley, p. 3 (2009)
[6] Nardetto, N., Poretti, E., Rainer, M., et al., A&A, **597**, A73 (2017)
[7] Benedict, G. F., McArthur, B. E., Fredrick, L. W., et al., AJ, **124**, 1695 (2002)
[8] Benedict, G. F., McArthur, B. E., Feast, M. W., et al., AJ, **133**, 1810 (2007)
[9] Casertano, S., Riess, A. G., Anderson, J., et al. ApJ, **825**, 11 (2016)
[10] Mérand, A., Kervella, P., Breitfelder, J., et al., A&A, **584**, A80 (2015)
[11] Neilson, H. R., & Lester, J. B., A&A, **554**, A98 (2013)
[12] Breitfelder, J., Mérand, A., Kervella, P., et al., A&A, **587**, A117 (2016)
[13] Breitfelder, J., Kervella, P., Mérand, A., et al., A&A, **576**, A64 (2015)
[14] Kervella, P., Trahin, B., Bond, H. E., et al., A&A, **600**, A127 (2017)
[15] Kervella, P., Mérand, A., Szabados, L., Fouqué, P., Bersier, D., Pompei, E., & Perrin, G., A&A, **480**, 167 (2008)
[16] Kervella, P., Mérand, A., & Gallenne, A., A&A, **498**, 425 (2009)
[17] Kervella, P., Mérand, A., Szabados, L., et al., A&A, **541**, A18 (2012)
[18] Kervella, P., Mérand, A., Szabados, L., et al., The Messenger, **150**, 46 (2012)
[19] Kervella, P., Bond, H. E., Cracraft, M., et al., A&A, **572**, A7 (2014)
[20] Anderson, R. I., A&A, **566**, L10 (2014)
[21] Anderson, R. I., Mérand, A., Kervella, P., et al., MNRAS, **455**, 4231 (2016)
[22] Pilecki, B., Graczyk, D., Pietrzyński, G., et al., MNRAS, **436**, 953 (2013)
[23] Gieren, W., Pilecki, B., Pietrzyński, G., et al., ApJ, **815**, 28 (2015)
[24] Lindegren, L., Lammers, U., Bastian, U., et al., A&A, **595**, A4 (2016)
[25] Casertano, S., Riess, A. G., Bucciarelli, B., & Lattanzi, M. G., A&A, **599**, A67 (2017)
[26] Gallenne, A., Mérand, A., Kervella, P., et al., A&A, **558**, A140 (2013)
[27] Gallenne, A., Mérand, A., Kervella, P., et al., A&A, **579**, A68 (2015)
[28] Gallenne, A., Mérand, A., & Kervella, P., A&A, **538**, A24 (2012)
[29] Gallenne, A., Mérand, A., Kervella, P., Evans, N. R., Proffitt, C., Pietrzyński, G., & Gieren, W., these proceedings

DOI: 10.1051/epjconf/201715207003

The atmosphere, the p-factor and the bright visible circumstellar environment of the prototype of classical Cepheids δ Cep

Nicolas Nardetto[1,*], *Ennio* Poretti[2], *Antoine* Mérand[3], *Richard I.* Anderson[4], *Andrei* Fokin[5], *Pascal* Fouqué[6], *Alexandre* Gallenne[3,7], *Wolfgang* Gieren[7,8], *Dariusz* Graczyk[7,8,9], *Pierre* Kervella[10,11], *Philippe* Mathias[12,13], *Denis* Mourard[1], *Hilding* Neilson[14], *Grzegorz* Pietrzynski[9], *Bogumil* Pilecki[9], *Monica* Rainer[2], and *Jesper* Storm[15]

[1]*Laboratoire Lagrange, UMR7293, Univ. de Nice Sophia-Antipolis, CNRS, Obs.de la Côte d'Azur, France*
[2]*INAF – Osservatorio Astronomico di Brera, Via E. Bianchi 46, 23807 Merate (LC), Italy*
[3]*European Southern Observatory, Alonso de Córdova 3107, Casilla 19001, Santiago 19, Chile*
[4]*Department of Physics and Astronomy, The Johns Hopkins University, Baltimore, MD, 21218, USA*
[5]*Institute of Astronomy of the Russian Academy of Sciences, 48 Pjatnitskaya Str., 109017, Moscow, Russia*
[6]*Observatoire Midi-Pyrénées, UMR 5572, Université Paul Sabatier-Toulouse, France*
[7]*Departamento de Astronomía, Universidad de Concepción, Casilla 160-C, Concepción, Chile*
[8]*Millennium Institute of Astrophysics, Santiago, Chile*
[9]*Nicolaus Copernicus Astronomical Center, Polish Academy of Sciences, Warszawa, Poland*
[10]*LESIA (UMR 8109), Obs. de Paris, PSL, CNRS, UPMC, Univ. Paris-Diderot, Meudon, France*
[11]*Unidad Mixta Internacional Franco-Chilena de Astronomía, CNRS/INSU, France (UMI 3386) and Departamento de Astronomía Universidad de Chile, Camino El Observatorio 1515, Las Condes, Santiago, Chile*
[12]*Univ. de Toulouse, UPS-OMP, Institut de recherche en Astrophysique et Planétologie, Toulouse, France*
[13]*CNRS, UMR5277, Institut de recherche en Astrophysique et Planétologie, Toulouse, France*
[14]*Department of Astronomy & Astrophysics, University of Toronto, Toronto, ON, M5S 3H4, Canada*
[15]*Leibniz Institute for Astrophysics, An der Sternwarte 16, 14482, Potsdam, Germany*

Abstract.
Even ≃16000 cycles after its discovery by John Goodricke in 1783, δ Cep, the prototype of classical Cepheids, is still studied intensively in order to better understand its atmospheric dynamical structure and its environment. Using HARPS-N spectroscopic measurements, we have measured the atmospheric velocity gradient of δ Cep for the first time and we confirm the decomposition of the projection factor, a subtle physical quantity limiting the Baade-Wesselink (BW) method of distance determination. This decomposition clarifies the physics behind the projection factor and will be useful to interpret the hundreds of p-factors that will come out from the next *Gaia* release. Besides, VEGA/CHARA interferometric observations of the star revealed a bright visible circumstellar environment contributing to about 7% to the total flux. Better understanding the physics of the pulsation and the environment of Cepheids is necessary to improve the BW method of distance determination, a robust tool to reach Cepheids in the Milky Way, and beyond, in the Local Group.

*Nicolas.Nardetto@oca.eu

1 Introduction

Classical Cepheids are yellow giants and supergiants pulsating stars used as stellar candles in the universe up to 30 Mpc. The relation between their mean absolute magnitude and the logarithm of their pulsation period, first discovered by [1], is currently used to calibrate the Type 1a supernovae luminosity relation and the expansion rate of the universe (Hubble constant, H_0; [2]). A $3.4\,\sigma$ tension in H_0 is found between the distance scale calibration (Cepheids + SN Ia) and the value derived from the Cosmic Microwave Background (CMB) observed by Planck ([3]). There are two possibilities: uncorrected biases affect one or both approaches, or new physics has to be considered in order to take into account this tension ([2]). With the next *Gaia* release (DR2, expected in April 2018), the distance of about 300 Galactic Cepheids will be derived with a precision of better than 3%. These distances will be used first to constrain the Cepheid period-luminosity relation, but they will also bring strong constrains on the physics of Cepheids, through the so-called inverse Baade-Wesselink (BW) method. The BW method is used to determine the distance of Cepheids in the Milky Way and beyond, in the Magellanic Clouds, and consists in combining the angular size variations of the star with its linear size variation. The angular size variation can be determined using infrared surface-brightness relations ([4, 5]), interferometry ([6]) or even a full set of photometric and interferometric data (SPIPS approach; [7]). The linear size variation is deduced from spectroscopy. The radial velocity curve is first derived from a spectral line profile or a set of spectral line profiles (cross-correlation). The radial velocity curve is then multiplied by a projection factor, which is used to derive the true pulsation velocity curve of the star. Finally this pulsation velocity curve is time-integrated in order to derive the radius variation of the star. In this approach, the projection factor and the distance of the star are fully degenerate. Thus, if the distance is known, the p-factor can be derived. This has been done for several Cepheids already ([8, 9]), but with *Gaia* parallaxes, it should be possible to derive the projection factor of about 300 Cepheids with a 3% precision. In this context, the p-factor decomposition into three sub-concepts proposed by [10] will be useful in order to interpret the *Gaia* p-factors (Sect. 2). This is even more promising since the decomposition of the p-factor has been confirmed observationally in the case of δ Cep thanks to a set of HARPS-N data of exceptional quality (Sect. 3). Besides, a second limiting aspect of the BW method, and even of the period-luminosity calibration, is the circumstellar environment of Cepheids (CSE). These CSEs were already discovered in the infrared, in particular in the case of δ Cep with a K-band flux contribution of 1.5% ([11]). Recently, we discovered a little secret of δ Cep using VEGA/CHARA interferometric observations: the star appears surrounded by a bright visible circumstellar environment (CSE), which was never observed before for any Cepheid. The implication of such visible CSE on the BW method is still under investigation (Sect. 4).

2 The physical concepts behind the p-factor

As already mentioned the projection factor is used to convert the radial velocity into the pulsation velocity of the star. For a Cepheid described simply by a uniform disk pulsating, its value is 1.5 (whatever the pulsation phase). But actually, the radial velocity of each surface element of the star is projected along the light of sight and weighted by the intensity distribution of the Cepheid. The limb-darkening of δ Cep reduces the p-factor significantly, and the so-called geometric projection factor (p_0, **step 1** in Figure 1) is between 1.36 to 1.39, depending on the wavelength in the visible range. The time variation of the p-factor, due mainly to limb-darkening variation, is neglected as it has no impact on the distance ([12]). However, a Cepheid is not simply a limb-darkened pulsating photosphere, it has also an extended atmosphere with various spectral lines (in absorption) forming at different levels from which we derive the radial velocity curve used in the BW method. Moreover, there is a velocity gradient in the atmosphere of the Cepheid, which can be measured from spectroscopic

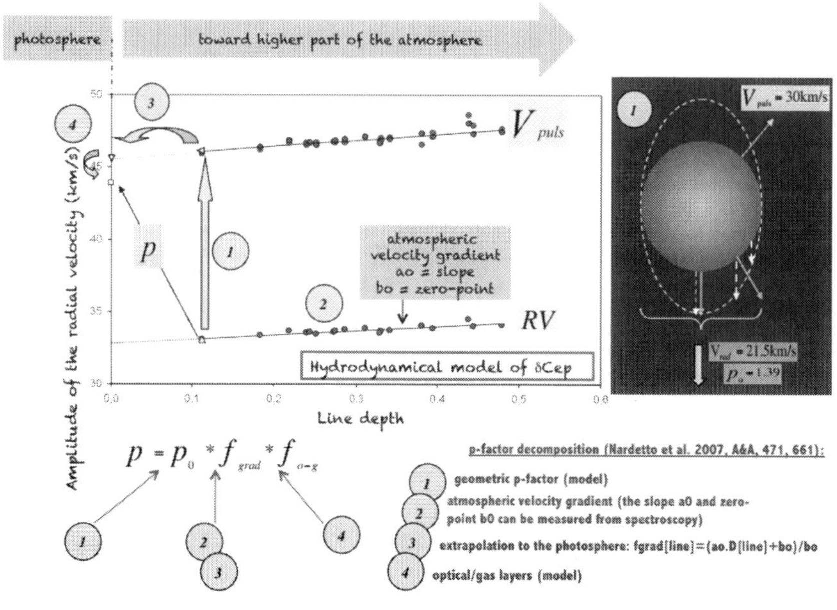

Figure 1. The p-factor decomposition is illustrated based on the hydrodynamical model of δ Cep (see [10]). The different steps are explained in Sect. 2.

observations (**step 2** in Fig. 1). Then, depending on the line considered, the amplitude of the radial velocity curve will not be the same and the resulting projection factor will be different. In Figure 1 ($f_{\rm grad}$, **step 3**), we show the impact of the atmospheric velocity gradient on the p-factor for a line forming rather close to the photosphere (line depth of about 0.1). The higher is the line forming region in the atmosphere, the lower is the projection factor (up to 3% compared to p_0 in the case of δ Cep). The last correction on the projection factor ($f_{\rm o-g}$, **step 4**) is more subtle. In spectroscopy, the radial velocity is actually a velocity associated with the moving *gas* in the line forming region, while in photometry or interferometry, we probe an *optical* layer corresponding to the black body continuum (i.e. the layer from which the photons escape). A correction on the projection factor of several percents (independent of the wavelength or the line considered) has to be considered. A relation between the period of Cepheids and the p-factor has been established using this approach for a specific line ([10]) or using the cross-correlation method ([13]).

3 The case of δ Cep revisited with HARPS-N data

Using HARPS-N data, [14] derived for the first time the atmospheric velocity gradient of δ Cep, i.e. the amplitude of the radial velocity as a function of the line depth (step 1 in Fig. 1). In order to reproduce all observables we had to rescale consistently our hydrodynamical model of δ Cep (already published by [15]), by applying a 7.8% increase on the amplitude of the velocity and radius variation curves, which, by definition, let the projection factor unchanged. The observed and modeled atmospheric velocity gradient are plotted in Figure 2 (left). Two lines in our sample (on the lower part of the figure) do not follow the relation for unknown reasons. Beside the hydrodynamical model reproduces almost perfectly the amplitude and shape of the angular diameter curve of δ Cep obtained from FLUOR/CHARA measurements by [16] (Fig. 2, right). The distance of δ Cep is known to few percents ([17]). By considering FLUOR/CHARA interferometric measurements and this distance, it is possible to inverse the BW approach and derive the projection factor for all lines in our sample (see

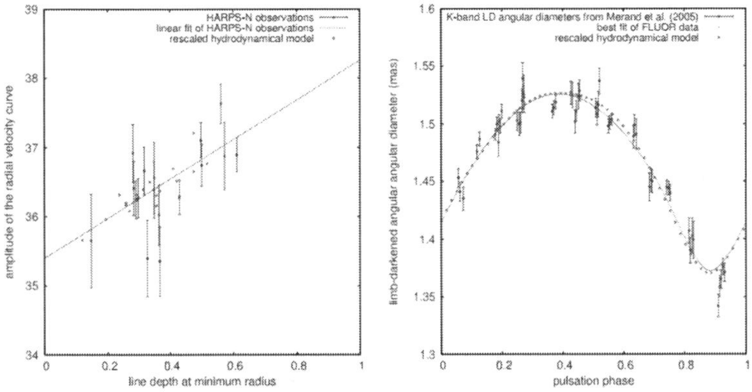

Figure 2. The rescaled hydrodynamical model of δ Cep (Fig. 1) is compared to the atmospheric velocity gradient measured by HARPS-N (left) and the angular diameter curve measured by FLUOR/ CHARA (right).

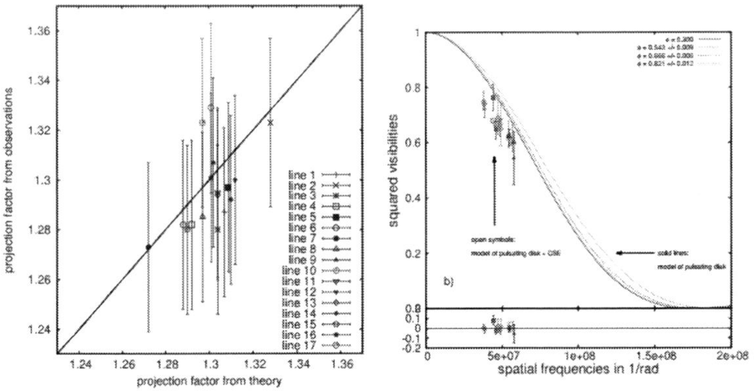

Figure 3. The observed p-factors (derived from HARPS-N, FLUOR/ CHARA and the distance to δ Cep) are compared to the theoretical ones calculated from the hydrodynamical model of δ Cep (left). An visible circumstellar environment is detected by the VEGA/CHARA interferometric measurements (right).

Table 3 in [14]). The comparison between observed and theoretical projection factors (derived from the hydrodynamical code) are plotted in Figure 3 (left). The agreement of the hydrodynamical model with the atmospheric velocity gradient, the angular diameter curve and the individual projection factors associated to specific lines, supports the decomposition of the projection factor as described in Section 2.

4 The bright visible environment of δ Cep revealed by VEGA/CHARA observations

It is well known now that Cepheids, and in particular δ Cep, have infrared circumstellar environments contributing to few percents of the total flux ([11]). Characterizing the CSEs of Cepheids, and in particular their radial brightness profile in several spectral bands, is of high importance in order to unbias the period-luminosity relation. Recently, we observe δ Cep with the VEGA/CHARA instrument ([18]) and we discovered an unexpected visible CSE contributing to 7 ± 1 % to the total flux, with a size of 8.9 ± 3.0 mas (but a visible background filling the field of view of the interferometer is not excluded; see [19]). Figure 3 (right) shows the visibility measurements associated to the shortest baseline of the CHARA interferometer (S1S2). In order to reproduce the squared visibility measurements, one needs to consider the CSE around the pulsating limb-darkened disk of the star, otherwise the model is several σ above the measurements (whatever the observed pulsation phase considered). Furthermore, [20] found the signature of a companion in recent spectroscopic data of δ Cep. Its expected flux contribution is supposed to be lower than 1% in the visible, clearly not detectable by VEGA/CHARA. Interestingly, δ Cep is also found to be a soft X-ray source ([21]).

DOI: 10.1051/epjconf/201715207003

An improved BW method or SPIPS approach (with in particular well calibrated p-factors and CSEs) will provide a robust tool to derive the distance of individual Cepheids up to few Mpc in the Local Group. The problem of the metallicity dependence of distance indicators in the Local Group will be then revisited in the framework of the Araucaria Project ([22]). After a long history, the BW projection factor remains a key quantity in the calibration of the cosmic distance scale, and one century after the discovery of the period-luminosity relation, Cepheid pulsation and environment is still a distinct challenge.

Acknowledgments: The observations leading to these results have received funding from the European Commission's Seventh Framework Programme (FP7/2013-2016) under grant agreement number 312430 (OPTICON). The authors thank the CHARA Array, which is funded by the National Science Foundation through NSF grants AST-0606958 and AST-0908253 and by Georgia State University through the College of Arts and Sciences, as well as the W. M. Keck Foundation. WG gratefully acknowledges financial support for this work from the BASAL Centro de Astrofisica y Tecnologias Afines (CATA) PFB-06/2007, and from the Millenium Institute of Astrophysics (MAS) of the Iniciativa Cientifica Milenio del Ministerio de Economia, Fomento y Turismo de Chile, project IC120009. We acknowledge financial support for this work from ECOS-CONICYT grant C13U01. Support from the Polish National Science Center grant MAESTRO 2012/06/A/ST9/00269 is also acknowledged. EP and MR acknowledge financial support from PRIN INAF-2014. NN, PK, AG, and WG acknowledge the support of the French-Chilean exchange program ECOS-Sud/CONICYT (C13U01). The authors acknowledge the support of the French Agence Nationale de la Recherche (ANR), under grant ANR-15-CE31-0012- 01 (project UnlockCepheids) and the financial support from "Programme National de Physique Stellaire" (PNPS) of CNRS/INSU, France.

References

[1] Leavitt, H. S., & Pickering, E. C., Harvard College Observatory Circular, **173**, 3 (1912)
[2] Riess, A. G., Macri, L. M. , Hoffmann, S. L., et al., ApJ, **826**, 56, (2016)
[3] Planck Collaboration, Ade, P. A. R., Aghanim, N., Arnaud, M., et al., A&A, **594**, 13 (2016)
[4] Storm, J., Gieren, W., Fouqué, P., et al., A&A, **534**, 94, (2011)
[5] Storm, J., Gieren, W., Fouqué, P., et al., ApJ, **534**, 95, (2011)
[6] Kervella, P., Nardetto N., Bersier D., et al., A&A, **416**, 953, (2004)
[7] Mérand, A., Kervella, P., Breitfelder, J., et al., A&A, **584**, A80 (2015)
[8] Breitfelder, J., Mérand, A., Kervella, P., et al., A&A, **587**, 117 (2016)
[9] Kervella, P., Trahin, B., Bond, H. E., et al., A&A, **600**, A127 (2017)
[10] Nardetto, N., Mourard, D., Mathias, P., Fokin, A., & Gillet, D., A&A, **471**, 661 (2007)
[11] Mérand, A., Kervella, P., Coudé du Foresto, V., et al., A&A, **453**, 162, (2006)
[12] Nardetto, N., Fokin, A., Mourard, D., & Mathias, P., A&A, **454**, 332, (2006)
[13] Nardetto, N., Gieren, W., Kervella, P., et al., A&A, **502**, 956, (2009)
[14] Nardetto, N., Poretti, E., Rainer, M., et al., A&A, **597**, 73, (2017)
[15] Nardetto, N., Fokin, A., Mourard, D., Mathias, P., et al., A&A, **428**, 137, (2004)
[16] Mérand, A., Kervella, P., Coudé du Foresto, V., et al., A&A, **438**, 9, (2005)
[17] Majaess, D., Turner, D., & Gieren, W., ApJ, **747**, 145, (2012)
[18] Mourard, D., Clausse, J. M., Marcotto, A., et al., A&A, **508**, 1083, (2009)
[19] Nardetto, N., Mérand, A., Mourard, D., et al., A&A, **593**, 45, (2016)
[20] Anderson, R. I., Sahlmann, J., Holl, B., et al., ApJ, **804**, 144 (2015)
[21] Engle, S. G., Guinan, E. F., Harper, G. M., et al., ApJ, **838**, 67 (2017)
[22] Gieren, W., Pietrzynski, G., Bresolin, F., et al., The Messenger, **121**, 28, (2005)

Observations of field and cluster RR Lyræ with *Spitzer*

Towards high precision distances with Population II stellar tracers

Massimo Marengo[1], *Jillian* Neeley[1], *Giuseppe* Bono[2,3], *Vittorio* Braga[2,3,4], *Massimo* Dall'Ora[5], *Marcella* Marconi[5], *Nicolas* Trueba[1], *Davide* Magurno[2,3], and the CRRP Collaboration

[1]*Department of Physics and Astronomy, Iowa State University, Ames, IA 50010, USA*
[2]*Department of Physics, Università di Roma Tor Vergara, via della Ricerca Scientifica 1, I-00133 Roma, Italy*
[3]*INAF-Osservatorio Astronomico di Roma, via Frascati 33, I-00040 Monte Porzio Catone, Italy*
[4]*Departamento de Física, Facultad de Ciencias Exactas, Universidad Andrés Bello, Santiago, Chile*
[5]*INAF-Osservatorio Astronomico di Capodimonte, Salita Moiarello 16, I-80131 Napoli, Italy*

Abstract. We present our project to calibrate the RR Lyræ period-luminosity-metallicity relation using a sample of Galactic calibrators in the halo and globular clusters.

1 Introduction

The availability of reliable, high-precision, RR Lyræ distances could be a game-changer not just for tracing old stellar populations in the Milky Way and other Local Group galaxies, but also for renewed efforts to establish an independent cosmological distance scale entirely based on Population II indicators ([1]). This will be crucial to resolve the current $\geq 3\sigma$ tension existing between different measurements of the Hubble constant and other cosmological parameters. Measurements obtained with methods based on Population I stellar indicators, anchored on Cepheids, are at odd with determinations based on the Cosmic Background Radiation and Baryonic Acoustic Oscillations ([13]). If confirmed, this tension could be an indication of unknown physics, missing in the current formulation of ΛCDM cosmological models.

Historically, the adoption of RR Lyræ as precision standard candles has been limited by the lack of a proper period-luminosity (PL) relation. Their visible magnitude only shows a shallow dependence from their pulsation period, and the magnitude vs. metallicity relation commonly used to derive their photometric parallaxes is characterized by a large intrinsic scatter ($> 5\%$). The situation however changes at near- and mid-infrared wavelengths (NIR and MIR thereafter), for three main reasons: (i) in the NIR the slope in the RR Lyræ PL relation steadily increases (see Figure 1); (ii) the intrinsic scatter due to evolutionary effects and temperature dependence is greatly reduced in the infrared, because of the narrowing of the instability strip width, and because in the MIR the brightness variations are mainly driven by the radius changes; (iii) in the NIR the effect of reddening is one order of magnitude less than in the visible, with an extra factor of 3 reduction in the MIR ([2]). These predictions were recently confirmed, observationally, both in NIR ([4]) and MIR ([9, 10]) bands.

With the upcoming availability of large aperture infrared space telescopes (JWST in the NIR and MIR, and WFIRST that will operate at wavelengths up to 2 μm), it is time to provide a robust calibration of RR Lyræ PL relations, and characterize their dependence from metallicity. This paper

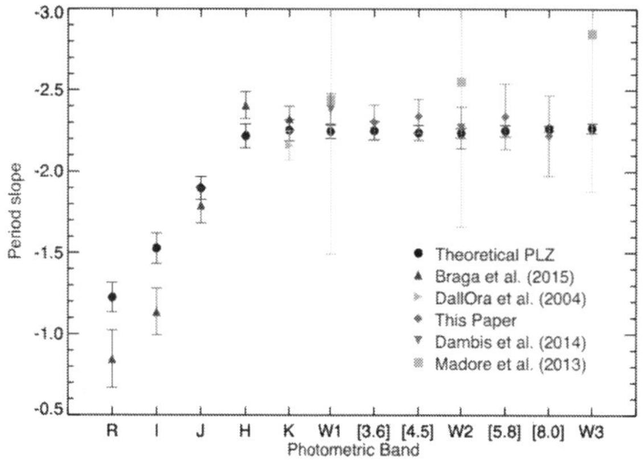

Figure 1. Period slope of RR Lyræ PLZ relations in optical and infrared bands. Different symbols indicate slope observational values from cluster and field RR Lyræ, compared with the theoretical prediction presented in [11] and [12] (black circles). Note that the slope is very shallow in the visible, and then increases in the NIR up to the H band. For wavelengths longer than ~ 2 μm the period slope becomes constant, converging to the value expected for brightness variations entirely driven by the change in radius.

presents a summary of our calibration efforts based on the MIR observation of Galactic RR Lyræ with the *Spitzer Space Telescope* InfraRed Array Camera (IRAC, [5]), part of the Carnegie RR Lyræ Program (CRRP), as well as ground-based observations at visible and NIR wavelengths.

2 Calibrating the period-luminosity-metallicity (PLZ) relations

The absolute brightness of RR Lyræ can be approximated by a linearized expression of their pulsation period P and iron content [Fe/H] (as proxy for their detailed abundances):

$$M_\lambda = a_\lambda + b_\lambda \cdot \log P + c_\lambda \cdot \mathrm{[Fe/H]}, \tag{1}$$

where the dependence on metallicity has been factorized in the c_λ term. Detailed time-dependent hydrodynamic models (such as [11], see also M. Marconi's contribution to this volume) show that the metallicity slope c_λ is only weakly dependent on wavelength, since the metallicity dependence of these stars' bolometric corrections effectively cancels-out the corresponding changes in effective temperature, resulting in a minimal change in color as [Fe/H] is increased. Furthermore, in the MIR the stellar photosphere is well described by the Rayleigh-Jeans approximation, and lacks significant spectral features. For this reasons the PLZ metallicity slope approaches a constant value c. In this formulation, the period slope b_λ is assumed independent of metallicity, a result supported by models; in the MIR this term converges to the value expected for brightness variations driven by changes in radius ($b_\lambda \simeq -2.3$ for fundamental mode pulsators, see Fig. 1).

To be truly useful, RR Lyræ PLZs need however to be calibrated observationally. This can be achieved by observing a number of stars in the field and globular clusters (GCs) with known true distance modulus μ_i, having a broad range of periods P_i and metallicity $\mathrm{[Fe/H]}_i$. The average magnitudes $m_{\lambda,i}$ of each of these calibrators need to be measured at multiple wavelengths, in order to characterize their individual extinction $A_{V,i}$ (after assuming an appropriate reddening law $[A_\lambda/A_V]$), so that:

$$m_{\lambda,i} - \mu_i = a_\lambda + b_\lambda \cdot \log P_i + c_\lambda \cdot \mathrm{[Fe/H]}_i + A_{V,i} \cdot [A_\lambda/A_V]. \tag{2}$$

Equation 2 can then be fit to derive the PLZ coefficients a_λ, b_λ and c_λ (and the nuisance parameter $A_{V,i}$ for each star) for each band in which the calibrators have been observed. To minimize the uncertainty in the best fit PLZ coefficients, the sample of RR Lyræ calibrators in our program has been selected so that they will have Gaia parallax determined better than ~ 1%. For all stars we are also acquiring a homogeneous set of metallicities (with precision better than ~ 0.1 dex), using optical and NIR high resolution spectroscopic observations.

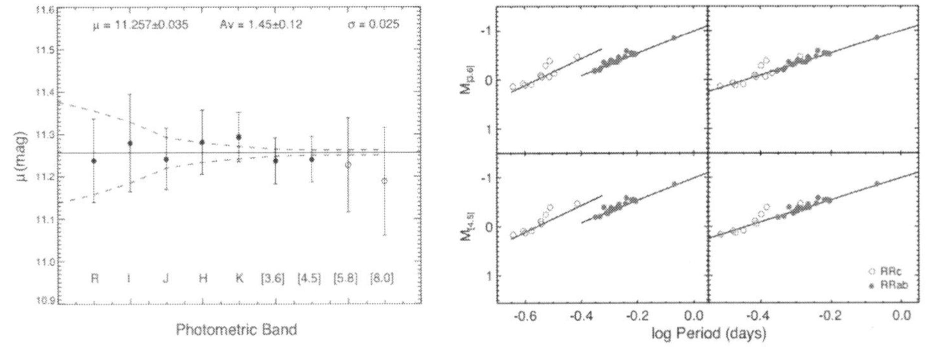

Figure 2. *Left* — Multi-wavelength best fit distance modulus and visual extinction for the Galactic globular cluster M4 (NGC 6121). RR Lyræ stars in the cluster have been fitted using the theoretical PLZ relations in [11], for the [Fe/H] value of the cluster. The two longest wavelength IRAC bands (single-epoch archival observations) have not been included in the fit. The IRAC 3.6 and 4.5 μm bands provide the strongest constraints for the distance modulus, while the optical R and I bands are essential to fit A_V, based on the adopted extinction law. *Right* — PL relations for first overtone (RRc, blue empty circles) and fundamental mode (RRab, red filled circles) M4 RR Lyræ at 3.6 and 4.5 μm (fundamentalized relations on the right). Both figures are adapted from [12].

3 Calibrating the period slope with Galactic GCs

From equation 2 it is clear that even small uncertainties in the observed quantities (especially [Fe/H], rarely known to better than $\sim$ 0.2 dex), and variations in the reddening law, will introduce a large scatter, and possibly a period-dependent bias, in the observed PLZ. This makes it difficult to simultaneously fit all PLZ coefficients, and the extinction A_V, using a heterogeneous sample of calibrator RR Lyræ located at different distances, and having different metallicities. The period slope b_λ, however, can be determined independently from all other parameters by observing a sample of single-metallicity GCs. The advantage of using GCs is of course that the distance of the cluster from Earth is much larger than the relative spatial distribution of the individual stars within the GC, and variable extinction in front of the cluster only contributes to a small unbiased scatter around the overall RR Lyræ PL: for RR Lyræ in a GC the period slope b_λ can be accurately fit as the slope of the PL relation constructed with the reddened apparent magnitudes.

Figure 2 (adapted from [12]) presents an example based on our observations of M4 (NGC 6121). The right panel shows the PLZ for first overtone and fundamental mode RR Lyræ, derived for IRAC 3.6 and 4.5 μm bands. The period slope we measured in the NIR ([3]) and MIR ([12]) are shown in Figure 1, and are in good agreement with their theoretical values. Repeating this measurement for the other single-metallicity clusters observed as part of the CRRP program will allow us to refine the fitted values of b_λ, and check that this term is indeed independent from [Fe/H]. Once the b_λ are determined, we will use our sample of field calibrator RR Lyræ to fit the remaining parameters a_λ and c_λ.

The left panel of Figure 2 assesses how well, if the PLZ is fully calibrated, it can be used to accurately derive the distance and average reddening of GCs. For this purpose we have used the photometric data of all RR Lyræ in M4, at $RIJHK$ and IRAC MIR bands, fitting at each wavelength the cluster reddened distance modulus by assuming the validity of the theoretical PLZ from [11]. We have then derived the average distance modulus and extinction, adopting the peculiar reddening law for M4 derived by [8]. The result is consistent with independent measurements of the cluster's distance and average A_V, and shows that this method is capable of providing accurate distances for individual Galactic GCs with a total uncertainty of just a few percent.

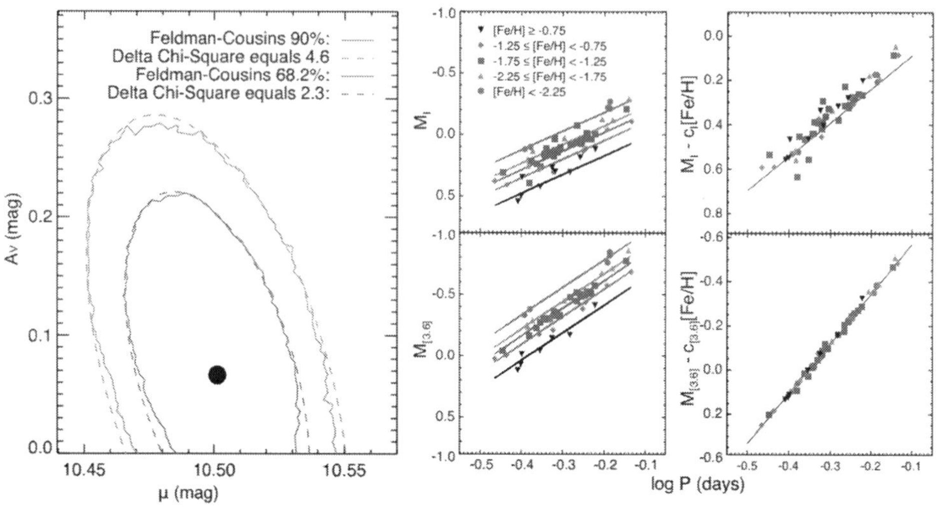

Figure 3. *Left* — Best fit value distance modulus and extinction, 1σ and 90% confidence regions for the field RR Lyræ AP Ser. The fit is based on the theoretical PLZ relations derived in [11]. Confidence regions calculated with the Feldman and Cousins method ([6]) are preferred if one of the fitted parameters is bound by physical boundaries (e.g., $A_V \geq 0$). *Center and Right* — PLZ relations of Galactic field RR Lyræ in the I and IRAC 3.6 μm bands, binned based on their [Fe/H]. The scatter due to the difference in metallicity is ~ 0.13 magnitudes over the metallicity range in the Galactic halo. The two panels on the right show how the scatter is much reduced when the metallicity term is subtracted by the stars' absolute magnitude, especially in the MIR band, where it becomes less than 0.02 magnitudes. Adapted from [12].

4 High precision individual RR Lyræ distances

The theoretical PLZ relations derived by [11] can be used to estimate the effect of metallicity on individual RR Lyræ distances. This is shown in Figure 3, where the distance modulus and extinction of each calibrator in our sample was fit using the theoretical PLZ relations (one example in the left panel, with the best fit confidence level derived using the Feldman and Cousin method ([6]), appropriate when a fitting parameter is bound by a physical boundary, e.g. $A_V \geq 0$). The central panels show the distribution of the calibrators' magnitudes (derived using their best fit distance and extinction) in the I and IRAC 3.6 μm bands, binned based on their [Fe/H]. In both wavelengths the dispersion is of the order of ± 0.13 magnitudes. Once the metallicity term $c_\lambda \cdot$ [Fe/H] is subtracted from the stars' absolute magnitude, the spread is significantly reduced (right). This is especially true in the MIR bands, where temperature and evolutionary effects are vanishing, leading to drastically smaller intrinsic scatter in the resulting PLZ (once metallicity effects are removed, we found $\sigma \simeq 0.02$ mag in the IRAC bands).

To test the validity of the theoretical PLZ derived by [11] and [12] we have compared our best fit distance moduli with the Gaia first data release (DR1, [7]) distances. Figure 4 shows that our results are within 1σ of the Gaia measurements, and that the residuals are not biased in either period, metallicity or the value of extinction derived with our best fitting procedure.

A detailed analysis of these results is provided in [12]. Here we want to highlight how precisely calibrating the metallicity term in the RR Lyræ PLZ relation is essential to have reliable distances for individual stars, as necessary for using them as accurate tracers of old stellar populations. The typical ~ 0.2 dex uncertainty in most spectrophotometric determinations of metallicity, just by itself, would result in a distance uncertainty of almost $\sim 4\%$, stressing the need of accurately measuring the individ-

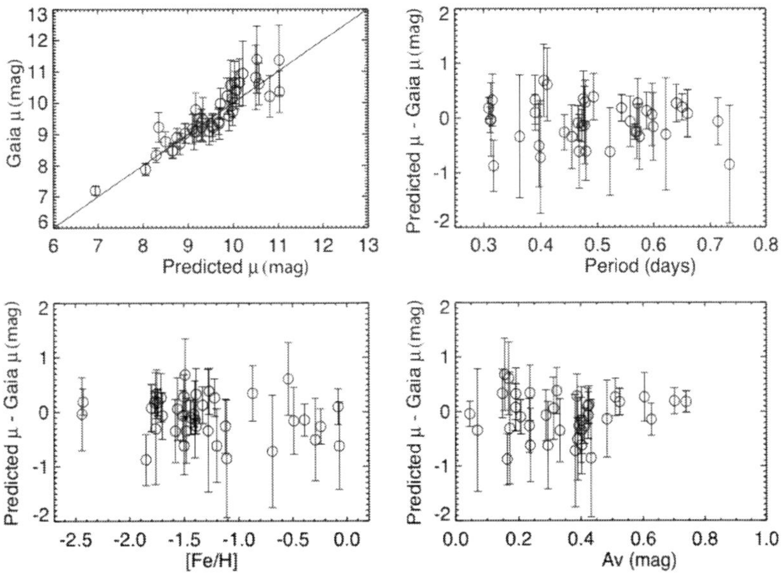

Figure 4. Comparison between the best fit distance moduli of Galactic field RR Lyræ, obtained using the theoretical PLZ relations described in [11], and Gaia DR1 distances. The two sets of values match to within better than 1σ, and do not show any bias due to period, metallicity or best fit extinction. Adapted from [12].

ual star's metallicity to obtain precise distances. Including the metallicity term is also important for adopting RR Lyræ as a new anchor for the distance scale: the reduced scatter in the MIR PLZ relation enables the determination of distances with accuracy comparable to Cepheids, fulfilling the promise of an independent route to the cosmological distance scale based on these Population II pulsators.

Acknowledgments: This work is based in part on observations made with the *Spitzer Space Telescope*, which is operated by the JPL-Caltech under a contract with NASA. This work also made use of data from the ESA mission *Gaia*, processed by the *Gaia* Data Processing and Analysis Consortium (DPAC).

References

[1] Beaton, R. L., Freedman, W. L., Madore, B. F., et al., ApJ, **832**, 210 (2016)
[2] Bono, G., Braga, V. F., Pietrinferni, A., et al., Mem. SAIt, **87**, 358 (2016)
[3] Braga, V. F., Dall'Ora, M., Bono, G., et al., ApJ, **799**, 165 (2015)
[4] Dall'Ora, M., Storm, J., Bono, G., et al., ApJ, **610**, 269 (2004)
[5] Fazio, G. G., Hora, J. L., Allen, L. E., et al., ApJS, **154**, 10 (2004)
[6] Feldman, G. J., & Cousins, R. D., PhRvD, **57**, 3873 (1998)
[7] Gaia Collaboration, Brown, A. G. A., Vallenari, A., et al., A&A, **595**, A2 (2016)
[8] Hendricks, B., Stetson, P. B., VandenBerg, D. A., & Dall'Ora, M., AJ, **144**, 25 (2012)
[9] Klein, C. R., Richards, J. W., Butler, N. R., & Bloom, J. S., ApJ, **738**, 185 (2011)
[10] Madore, B. F., Hoffman, D., Freedman, W. L., et al., ApJ, **776**, 135 (2013)
[11] Marconi, M., Coppola, G., Bono, G., et al., ApJ, **808**, 50 (2015)
[12] Neeley, J. R., Marengo, M., Bono, G., et al., ApJ, **841**, 84 (2017)
[13] Riess, A. G., Macri, L. M., Hoffmann, S. L., et al., ApJ, **826**, 56 (2016)

DOI: 10.1051/epjconf/201715207005

Pulsating stars in ω Centauri

Near-IR properties and period-luminosity relations

Camila Navarrete[1,2,⋆], *Márcio* Catelan[1,2,3,⋆⋆], *Rodrigo* Contreras Ramos[1,2], *Javier* Alonso-García[4,2], *Felipe* Gran[1,2], *István* Dékány[5], and *Dante* Minniti[6,2,7]

[1]*Pontificia Universidad Católica de Chile, Instituto de Astrofísica, Av. Vicuña Mackenna 4860, 782-0436 Macul, Santiago, Chile*
[2]*Millennium Institute of Astrophysics, Santiago, Chile*
[3]*Centro de Astro-Ingeniería, Pontificia Universidad Católica de Chile, Santiago, Chile*
[4]*Unidad de Astronomía, Facultad Cs. Básicas, Universidad de Antofagasta, Av. U. de Antofagasta 02800, Antofagasta, Chile*
[5]*Astronomisches Rechen-Institut, Zentrum für Astronomie der Universität Heidelberg, Mönchhofstr. 12-14, 69120 Heidelberg, Germany*
[6]*Departamento de Fisica, Facultad de Ciencias Exactas, Universidad Andres Bello, Av. Fernandez Concha 700, Las Condes, Santiago, Chile*
[7]*Vatican Observatory, V00120 Vatican City State, Italy*

Abstract. ω Centauri (NGC 5139) contains many variable stars of different types, including the pulsating type II Cepheids, RR Lyrae and SX Phoenicis stars. We carried out a deep, wide-field, near-infrared (IR) variability survey of ω Cen, using the VISTA telescope. We assembled an unprecedented homogeneous and complete J and K_S near-IR catalog of variable stars in the field of ω Cen. In this paper we compare optical and near-IR light curves of RR Lyrae stars, emphasizing the main differences. Moreover, we discuss the ability of near-IR observations to detect SX Phoenicis stars given the fact that the amplitudes are much smaller in these bands compared to the optical. Finally, we consider the case in which all the pulsating stars in the three different variability types follow a single period-luminosity relation in the near-IR bands.

1 Introduction

In the near-infrared (IR) bands, the study of variable stars has some advantages compared to the optical: i) the period-luminosity (PL) relations have lower internal dispersions ([3, 10]), leading to more precise distance determinations; ii) the interstellar extinction is lower (for the K_S band, the extinction is one tenth of the extinction in the V band); and iii) pulsating stars tend to have smaller amplitudes, leading to better constrained mean magnitudes with a smaller number of epochs.

Despite these advantages, only a handful of wide-field time-series surveys are devoted to near-IR bands. Two ongoing near-IR variability surveys are the Vista Variables in the Vía Láctea (VVV) ESO Public Survey, which acquired time-resolved data (in the K_S band) in highly extinct regions of

⋆cnavarre@astro.puc.cl
⋆⋆mcatelan@astro.puc.cl

the Galaxy, at low Galactic latitudes (see, e.g., [4, 15]), and the VISTA survey of the Magellanic Cloud system (VMC; e.g., [6, 19, 20]). A detailed review of the time-series surveys in the near-IR is presented by [14] in these proceedings.

In this present scenario, ω Centauri (NGC 5139) is an excellent laboratory to study pulsation properties in the near-IR bands since it hosts a large and rich variable star content, previously well studied using optical observations. In particular, the cluster has at least 180 RR Lyrae (RRL) stars ([16]); 74 SX Phoenicis (SX Phe), the greatest number found in a globular cluster ([18, 26]); and seven type II Cepheids (T2Cs), among others variability types.

2 Observations

Near-IR images of ω Cen were taken using the VISTA 4.1m telescope, at ESO Paranal Observatory [8]. 42 and 100 images in J and K_S, respectively, were collected. The details of the observations and data reduction have been previously explained in [16]. Point-spread function (PSF) photometry was derived using the DAOPHOT II/ALLFRAME package ([24, 25]) for the innermost ten arcminutes region of the cluster, while the DoPhot package ([1, 22]) was preferred to derive PSF photometry in the outer regions.

From the observations, the intensity-average mean magnitudes for 278 previously known pulsating stars were recovered. Of them, the variability (i.e., sinusoidal light curve for the given period) was recovered for 209 pulsating stars. The 69 stars previously known to be variables ([11, 26]), but without signs of variability in the near-IR, include a subsample of low-amplitude, c-type RRL (RRc) stars, in addition to most of the SX Phe stars that are known to be present in the cluster field.

3 Near-IR and visible light curves of ω Cen RR Lyrae stars

Considering the extensive optical study of variable stars in the cluster by [11], the amplitudes and light curve shapes in V and the near-IR bands J and K_S can be directly compared. Figure 1 shows the light curves in V (top), J (middle) and K_S (bottom) for one ab-type RRL (RRab, left) and one RRc star (right). The light curves in V are from the study of [11]. Each panel shows, at the top right, the number of epochs in each light curve. From the figure, some striking differences can be noticed. First of all, the Blazhko effect that is evident in the V-band light curve of the RRab variable (V5, upper left panel) is not distinguishable in the near-IR light curves. This is the case for all the known Blazhko RRab stars in the cluster ([16]). Even though the resulting amplitude modulation is too small to be recovered in our near-IR light curves, [23] reported increased scatter in their near-IR light curves of RR Lyr, which they attributed to the Blazhko effect. Second, the amplitudes in J and K_S are smaller than in the V band. [16] provide relations to convert between V, J and K_S magnitudes for RRL stars, and the decrease in the amplitudes from the visual to the IR is evident in their Figure 3. Finally, the light curve shape of RRab stars tends to be more sinusoidal as the light curve is observed with redder filters. The same effect is seen for RRc stars but the difference is less prominent, given their already nearly sinusoidal V-band light curves. The different light curve shapes, as one moves from the visual regime to the near-IR, can be understood largely in terms of the reduced dependence of the emerging flux on the star's temperature variations, as compared to the radius ([5], and references therein). The lack of distinctive features in the light curves is one of the difficulties facing the automated classification of variable stars in the near-IR (see, e.g., [2, 7]).

4 Variability amplitudes of SX Phoenicis stars

SX Phe stars typically have smaller variability amplitudes than RRL and T2Cs, and are thus more difficult to detect. ω Cen hosts 74 SX Phe, of which more than 45 have V-band amplitudes ≤ 0.1 mag.

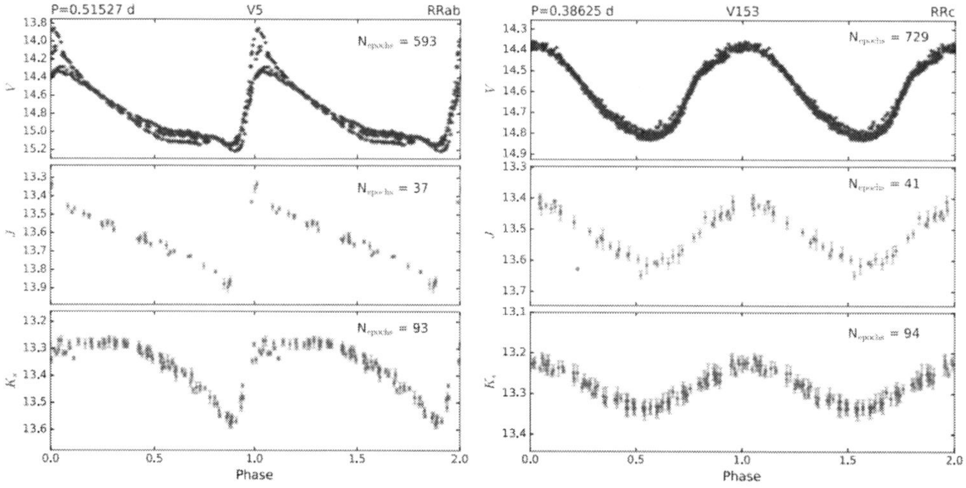

Figure 1. RRL light curves in three different bands. From top to bottom, the light curves are in the V, J and K_S bands, respectively. The left panels correspond to V5, an RRab star, while the right panels show the light curves for V153, an RRc star. The V-band data are taken from [11]. For both stars, the different light curve shapes and amplitudes in different bandpasses are evident.

Using the J and K_S time series, the variability of only 12 of these SX Phe could be recovered. These 12 stars have the highest V-amplitudes among all the SX Phe in the cluster, from 0.13 up to 0.31 mag (based on the minimum and maximum magnitudes reported by [11]). For the remaining SX Phe stars, the variability amplitudes are too small and, therefore, the light curves look indistinguishable from those of non-variable stars.

The amplitude distribution in V, J, and K_S for the SX Phe stars with recovered variability in the near-IR is shown in Figure 2. In this figure, and also in Figure 3, we include only those 7 stars for which we were able to retrieve the corresponding V-band data (from [11]). For these 7 stars, we re-derived the V-band amplitudes in a manner consistent with the technique that was used to obtain the near-IR amplitudes. As can be seen, most of the amplitudes in J tend to be $\lesssim 0.22$ mag, while their respective K_S amplitudes are $\lesssim 0.15$ mag. These small amplitude values make the detection of SX Phe stars on the basis solely of near-IR observations harder. Only high-amplitude pulsators are likely to be found, at least at the magnitude level of $K_S \sim 16$ mag. This group of SX Phe pulsators corresponds to a small fraction of the distribution of amplitudes found for this variability type (see, e.g., [21]). Efforts have been made to collect near-IR light curves for SX Phe stars in the field as well as in some clusters (see, e.g., [2]), again recovering only high-amplitude pulsators.

A comparison between the resulting A_V, A_J, A_{K_S} values is shown in Figure 3. The lines overplotted on this diagram correspond to best-fit relations of the following forms:

$$A_J = (0.29 \pm 0.08)\left[1 - \exp^{(-1.95\pm0.81)A_V}\right], \tag{1a}$$

$$A_{K_S} = (0.20 \pm 0.06)\left[1 - \exp^{(-1.74\pm0.76)A_V}\right], \tag{1b}$$

which are valid for visual amplitudes in the range $A_V < 0.75$ mag.

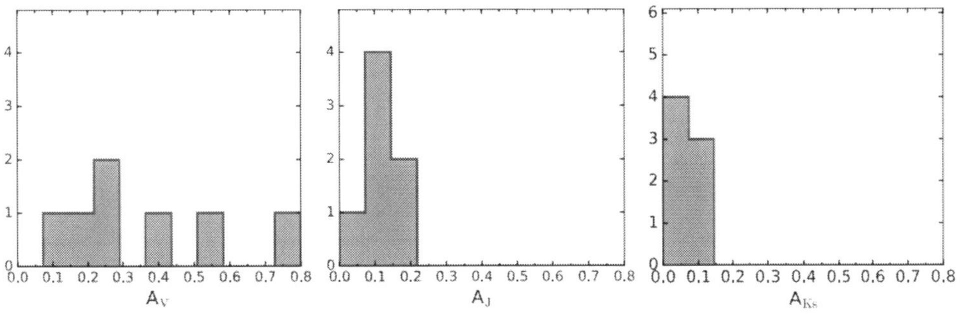

Figure 2. Distribution of amplitudes for those SX Phe stars for which the variability was recovered in the near-IR observations and for which we were able to retrieve V-band light curves (from [11]) and thus measure amplitudes following the same procedure as for the near-IR light curves. From left to right, the amplitudes in V, J and K_S bands are presented. The decrease in amplitude is evident when observing in IR bands.

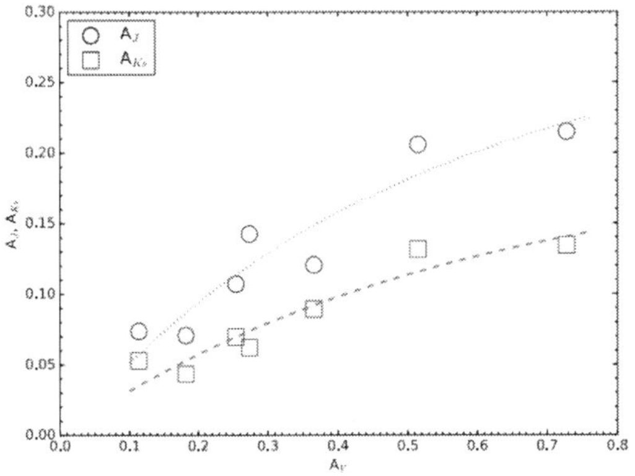

Figure 3. Relationship between visual and near-IR amplitudes for SX Phe stars in ω Cen. The sample is the same as used in Figure 2. J-band values are shown as blue circles, whereas those in the K_S band are displayed as red squares. The corresponding best-fitting relations (Eqs. 1a and 1b) are shown as dotted and dashed lines, respectively.

5 Near-IR period-luminosity relations

It has been suggested that the PL relations for T2Cs and RRL can be seen as a single relation, with nearly the same slope and zero points, at least in the near-IR, extending from periods shorter than 1 day up to $\sim$ 30 days (see, e.g., [9, 13, 17]). At least to first order, this may plausibly also extend to the SX Phe regime (e.g., [5, 12], and references therein). Here we revisit this possibility, using our sample of T2Cs, RRab and (candidate) fundamental-mode SX Phe stars from [17].

The left panels of Figure 4 show the log (P) (in days) and extinction-corrected J (top) and K_S (bottom) magnitudes and the least-squares fit recovered from the data (dashed lines). Despite the fact that the data largely conform to a straight line, resembling a single PL relation for all the aforementioned variability types, small differences are found, mostly in the longest- and shortest-period regimes. In the right panels of the figure, the difference between the observed J and K_S magnitudes (top and bottom panels, respectively) and the best fit to the data is presented. It is clear that the SX Phe stars tend to deviate the most from the relation, suggesting a non-negligible difference in slope between their PL relation and the one for RRL stars. The same is found for the longest-period T2Cs (W Virginis and RV Tauri sub-types). RRab stars present very small deviations from the fit ($\lesssim$ 0.1 mag), in spite of the fact that its well-known metallicity dependence ([17]) was not corrected for when producing this plot.

Table 1. PL relations in different bandpasses.[1]

	a_J	b_J	c_J	Ref.
T2Cs	-2.23 ± 0.05	0	-0.86 ± 0.06	[13]
RRL	-1.77 ± 0.06	0.15 ± 0.03	-0.63 ± 0.08	[17]
SX Phe	-3.04 ± 0.17	0	-1.60 ± 0.22	[17]
T2Cs+RRL[2]	-2.32 ± 0.03	0	-0.77 ± 0.06	This work
All types[2]	-2.46 ± 0.01	0	-0.77 ± 0.01	This work
	a_{K_S}	b_{K_S}	c_{K_S}	Ref.
T2Cs	-2.41 ± 0.05	0	-1.11 ± 0.05	[13]
RRL	-2.23 ± 0.04	0.141 ± 0.02	-0.96 ± 0.05	[17]
SX Phe	-3.39 ± 0.24	0	-2.19 ± 0.30	[17]
T2Cs+RRL[2]	-2.47 ± 0.03	0	-1.11 ± 0.01	This work
All types[2]	-2.53 ± 0.02	0	-1.12 ± 0.01	This work

[1] In the form $m_X = a_X \log(P) + b_X\,[\mathrm{Fe/H}] + c_X$, where X corresponds to the bandpass.
[2] Adopting $(m - M)_0 = 13.708$ mag for ω Cen (from [17]).

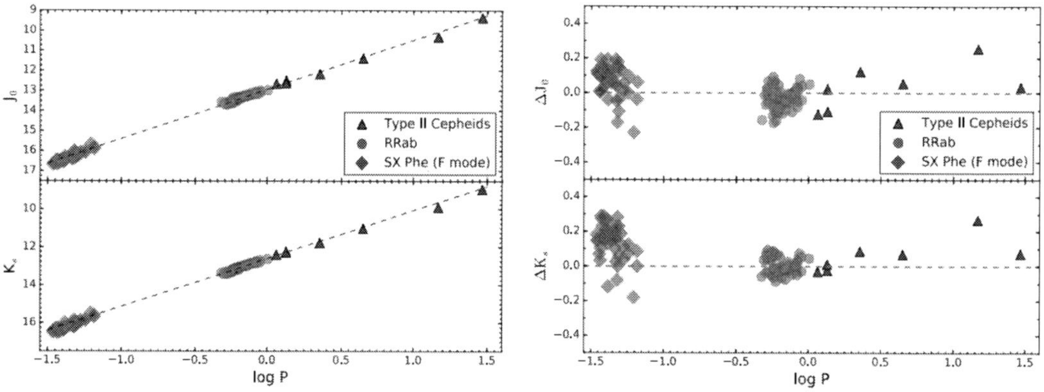

Figure 4. Left: Extinction-corrected magnitudes in J (top) and K_S (bottom) versus $\log(P)$, in days, for T2Cs, RRab and (candidate) fundamental-mode SX Phe stars. The dashed lines correspond to the linear least-squares fit to the data. The three variability types seem to follow, roughly, a single PL relation. Right: Difference between the observed extinction-corrected magnitudes and the derived fits, as a function of the logarithm of the period (in days). The horizontal dashed lines mark a zero-magnitude difference.

The slopes of the fits can be compared with the recently derived PL and PL$-Z$ relations for SX Phe and RRL stars, respectively, as obtained by [17] using the same data. For T2Cs, no such relations were derived, since the cluster hosts only seven such stars. Table 1 shows the coefficients of the PL relations for the different variability types (from [13] and [17]), as well as the slopes derived in this paper simultaneously using the three variability types and when only T2Cs and RRL are considered. There is much better agreement among the different slope values when SX Phe stars are excluded from the fits. This was already evident from Figure 4 in which SX Phe stars are those with the highest deviations from the fit. Reassuringly, the slope values are closer when the K_S band is considered, suggesting that in fact T2Cs and RRL could indeed share a common PL relation, at least in this bandpass.

Acknowledgments: Support for this project is provided by the Ministry for the Economy, Development, and Tourism's Millennium Science Initiative through grant IC 120009, awarded to the Millennium Institute of Astrophysics (MAS); by the Basal Center for Astrophysics and Associated Technologies (CATA) through grant PFB-06/2007; by CONICYT's PCI program through grant DPI20140066; and by FONDECYT grants #1141141 and #1171273 (C.N., M.C.), #1130196 (D.M.), #1150345 (F.G.), #3130320 (R.C.R.), and #11150916 (J.A.-G.). C.N. acknowledges additional support from CONICYT-PCHA/Doctorado Nacional 2015-21151643.

References

[1] Alonso-García, J., Mateo, M., Sen, B., et al. AJ, **143**, 70 (2012)
[2] Angeloni, R., Contreras Ramos, R., Catelan, M., et al. A&A, **567**, 100 (2014)
[3] Catelan, M., Pritzl, B. J., & Smith, H. A., ApJS, **154**, 633 (2004)
[4] Catelan, M., Minniti, D., Lucas, P. W., et al., in *40 Years of Variable Stars: A Celebration of Contributions by Horace A. Smith*, ed. K. Kinemuchi, C. A. Kuehn, N. De Lee, H. A. Smith, p. 139 (arXiv:1310.1996) (2013)
[5] Catelan, M., & Smith, H. A., *Pulsating Stars* (Wiley-VCH, Weinheim, 2015)
[6] Cioni, M. -R. L., Clementini, G., Girardi, L., et al. A&A, **527**, 116 (2011)
[7] Elorrieta, F., Eyheramendy, S., Jordán, A., et al. A&A, **595**, 82 (2016)
[8] Emerson, J. P., & Sutherland, W. J., in *Ground-based and Airborne Telescopes III.*, SPIE Conf. Ser., Vol. **7733**, ed. Stepp, L. M., Gilmozzi, R., Hall, H. J. (2010)
[9] Feast, M., in *RR Lyrae Stars, Metal-Poor Stars, and the Galaxy*, ed. A. McWilliam, Carnegie Observatory Astrophysics Series, **5**, 170 (2011)
[10] Longmore, A. J., Dixon, R., Skillen, I., et al., MNRAS, **247**, 684 (1990)
[11] Kaluzny, J., Olech, A., Thompson, I. B., et al., A&A, **424**, 1101 (2004)
[12] Majaess, D. J., JAAVSO, **38**, 100
[13] Matsunaga, N., Fukushi, H., Nakada, Y., et al., MNRAS, **370**, 1979 (2006)
[14] Matsunaga, N., preprint (arXiv:1705.02547) (2017)
[15] Minniti, D., Lucas, P. W., Emerson, J. P., et al., New Astron., **15**, 433 (2010)
[16] Navarrete, C., Contreras Ramos, R., Catelan, M., et al., A&A, **577**, 99 (2015)
[17] Navarrete, C., Catelan, M., Contreras Ramos, R., et al., A&A, in press (arXiv:1704.0303) (2017)
[18] Olech, A., Dziembowski, W. A., Pamyatnykh, A. A., et al., MNRAS, **363**, 40 (2005)
[19] Ripepi, V., Marconi, M., Moretti, M. I., et al., MNRAS, **437**, 2307 (2014)
[20] Ripepi, V., Marconi, M., Moretti, M. I., et al., ApJS, **224**, 21 (2016)
[21] Rodríguez, E., & López-González, M. J., A&A, **359**, 597 (2000)
[22] Schechter, P. L., Mateo, M., & Saha, A., PASP, **105**, 1342 (1993)
[23] Sollima, A., Cacciari, C., Arkharov, A. A. H., et al., MNRAS, **384**, 1583 (2008)
[24] Stetson, P. B., PASP, **99**, 197 (1987)
[25] Stetson, P. B., PASP, **106**, 250 (1994)
[26] Weldrake, D. T. F., Sackett, P. D., & Bridges, T. J., AJ, **133**, 1447 (2007)

Properties of Anomalous and Type II Cepheids in the Small and Large Magellanic Clouds

Monika I. Jurkovic[1,2,⋆] and *Martin A. T.* Groenewegen[3,⋆⋆]

[1]*Astronomical Observatory of Belgrade (AOB), Volgina 7, 11 060 Belgrade, Serbia*
[2]*Konkoly Observatory, Research Centre for Astronomy and Earth Sciences, Hungarian Academy of Sciences, H-1121 Budapest, Konkoly Thege Miklós út 15-17, Hungary*
[3]*Koninklijke Sterrenwacht van België (KSB), Ringlaan 3, B-1180 Brussels, Belgium*

Abstract. The Small Magellanic Cloud (SMC) and Large Magellanic Cloud (LMC) give us the possibility to study individual variable star types in a new way. Literature data provide us with photometric information about objects from the ultraviolet to the infrared. Here we would like to show the results of our study of 335 Anomalous and Type II Cepheids in the SMC and LMC detected by OGLE. Using the code More of DUSTY (MoD), a modified version of the DUSTY radiative transfer code, and the assumption that our objects are at a known distance, luminosity and effective temperature were determined. From these data the Hertzsprung-Russell diagram of these objects was compared with the theoretical models. The radius and masses of the examined stars was estimated, too. In the end, we have given the period-luminosity relations for the Anomalous and Type II Cepheids.

1 Introduction

The goal of this work was to these stars using the physical effective temperatures (T_{eff}) and luminosities (L), instead of the previously used color-magnitude diagrams (CMD). This gives a direct comparison with the results of the evolutionary models. In the course of our research we have included the examination of infrared (IR) excess, and the role of binarity, as well. Our detailed results will be published in two recent papers ([3, 4]). We have extended our sample to the semi regular variables, also identified as long period variables (LPVs) in OGLE-III (see [5]).

Type II Cepheids (T2Cs) are low mass ($\sim 0.5-0.6\ M_{\odot}$) pulsating stars, and we follow the classification into subgroups given in the OGLE-III catalog: BL Her (BLH): 1–4 days; W Virginis (including peculiar W Virginis, (p)WVir: 4 – 20 days; RV Tauris (RVT): 20 – 70 days. They are useful distance indicators with L and T_{eff} between the RR Lyrae, Anomalous Cepheids (ACs), and Classical Cepheids (e.g., see [15]).

The ACs are also pulsating stars with pulsating periods ranging from 0.3 to 2.4 days, which overlaps in part with the RR Lyrae and BL Her types. Their average mass is $\sim 1.2\ M_{\odot}$ ([1]). They pulsate in the fundamental mode and first overtone, and these subgroups form different period-luminosity (PL) relations, nevertheless they seem not to reach into the Classical Cepheids region.

⋆mojur@aob.rs
⋆⋆martin.groenewegen@oma.be

Table 1. The distribution of the T2Cs and ACs in the OGLE-III catalog of the SMC and LMC.

	BL Her	(p)W Vir	RV Tau	AC
SMC	17	17	9	6
LMC	64	97	42	83

2 The sample

The 335 stars were taken from the OGLE-III catalog ([11–13]). For the distribution among the different types, see Table 1.

The spectral energy distributions (SEDs) for the 335 stars, as presented in [3], was constructed using the optical, infrared and near-infrared data from the catalogs available at VizieR[1], which included: OGLE, EROS, MACHO, MCPS, DENIS, 2MASS, 2MASS 6X, IRSF, LMC Synoptic Survey, VMC Survey, WISE, Akari, IRAC.

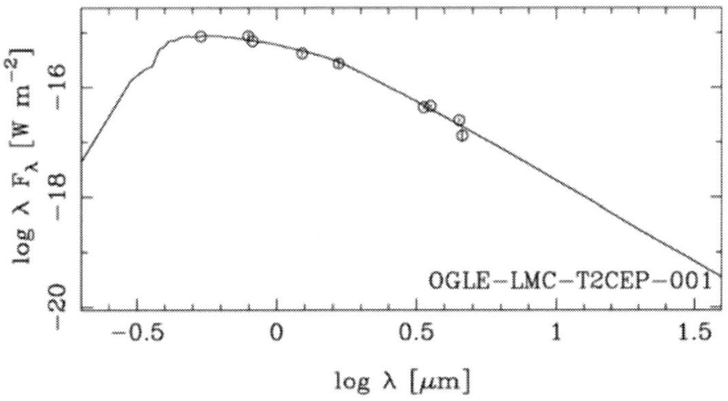

Figure 1. The SED for the OGLE-LMC-T2CEP-001, showing a normal distribution.

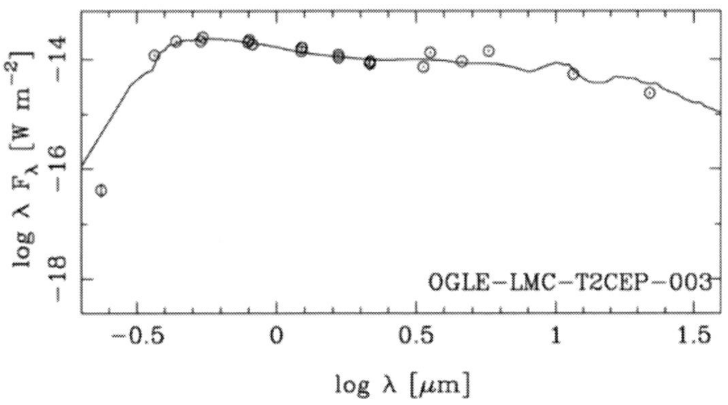

Figure 2. The SED for the OGLE-LMC-T2CEP-003 with a visible IR excess.

[1]http://vizier.u-strasbg.fr/viz-bin/VizieR

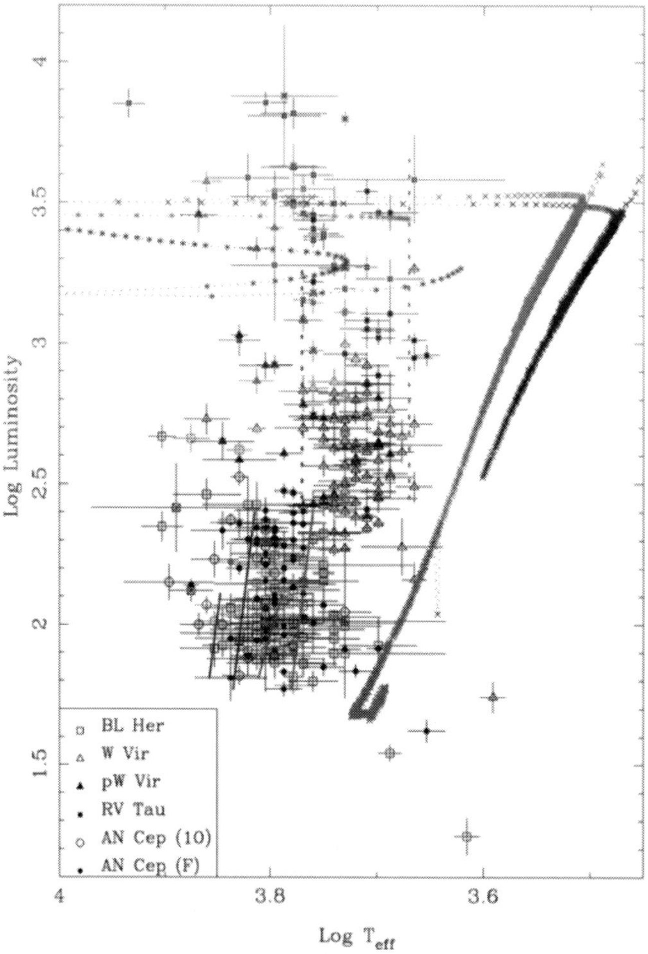

Figure 3. The HRD constructed from the modelled SEDs $T_{\rm eff}$ and luminosity. Stars in the SMC are plotted in red. Stars with an IR excess are marked by a green plus-sign. Stars which show eclipsing or ellipsoidal variations according to OGLE are marked by a blue cross. The tracks plotted with × are the lowest initial mass tracks of [14] for the LMC (0.945 $M_\odot$) and SMC (0.89 $M_\odot$, in red). The final masses are 0.555 and 0.558 $M_\odot$, respectively. The tracks plotted with an asterisk symbol are the lowest initial mass tracks of [10] for a metallicity of Z = 0.01 (1.0 $M_\odot$, current mass 0.534 $M_\odot$) and Z = 0.001 (0.9 $M_\odot$, current mass 0.536 $M_\odot$, in red). In both tracks every tick mark represents 500 years of evolution. The blue and red edges of the fundamental (FU) mode instability strip of BLH (between $\log L \sim 1.81 - 2.1$, for a mass of 0.65 $M_\odot$) and FU ACs (between $\log L \sim 1.77 - 2.4$) are indicated by the solid line (see text). The vertical dotted lines indicate the location of most of the variables at higher luminosity (see text). Taken from [3], with the permission of the authors.

The dust radiative transfer code *More of DUSTY* was used ([2]), which is an extension of DUSTY ([6]). The loop over $T_{\rm eff}$ ran from 2900 to 10000 K. A χ^2 fit was done to search for the model that fitted the data the best, giving us the $T_{\rm eff}$ and L for each object. MARCS model atmospheres were used. The distance for the LMC and SMC were fixed at the following values, $d_{\rm LMC}$ = 50 kpc and $d_{\rm SMC}$ = 61 kpc. A single E(B-V)=0.15 was adopted for all the stars. For the stars where IR excess was detected we fitted a model including dust. Examples of the fitted SEDs are in Figure 2.

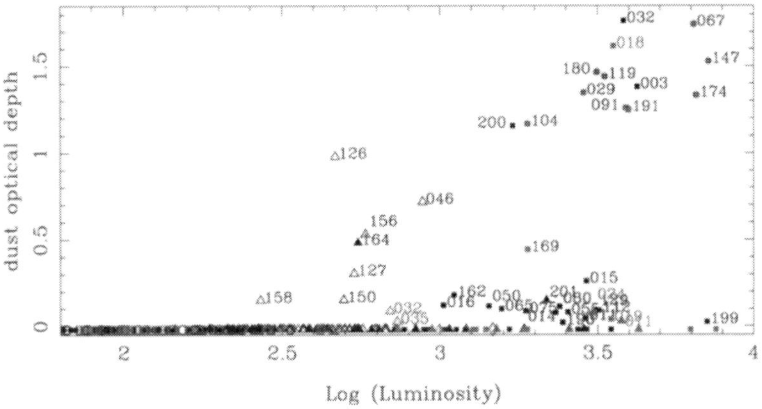

Figure 4. Dependence of the derived dust optical depth against log *L*. Stars in the SMC are plotted in red. Stars with an IRS spectrum are marked by a green plus-sign. Stars with a detectable IR excess are labelled by their identifier. Taken from [3], with the permission of the authors.

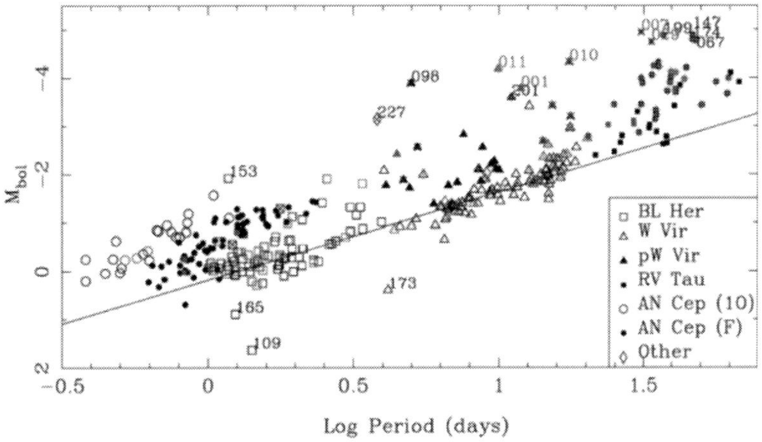

Figure 5. The bolometric magnitude vs. log *P* *PL* relation tor the T2Cs and ACs in the SMC and LMC. Taken from [4] with the permission of the authors.

3 HRD

Figure 3 is the physical Hertzsprung-Russell diagram (HRD) of the T2Cs and ACs in the SMC and LMC. For the detailed description of the symbols, see the caption and [3].

4 The IR excess

Among the examined stars we have found that ~ 60% of RVT, and ~ 10% of the WVir (including the peculiar WVir), show IR excess; see Figure 4. The distributions among subtypes looks like this:

- ACEP: 0/76;
- BLH: 0/81;

- (p)WVir: 8/90 for WVir and 3/24 for pWVir;
- RVT: 30/52.

It was not surprising to find IR excess among the RVT stars, because they are thought to be pAGB objects. What was surprising is the detection of IR excess among the WVir stars below luminosities predicted by single star evolutionary tracks. This IR excess was independently found by [7–9] to be dusty post-RGB stars and possibly related to binary evolution.

5 The PL relation

Using the gathered data we were able to construct the bolometric magnitude vs. log P PL relation for the T2Cs and ACs in the sample; see Figure 5 and [4].

6 Conclusion

We have determined the luminosities and effective temperatures of the sample of 335 T2Cs and ACs in the SMC and LMC. This investigation provided us with data to create the HRD that could be directly compared to the evolutionary model results.

We have derived the PL relation for all the stars in the sample.

We have found infrared excess in 10% of the W Vir stars and 60% of the RV Tau stars. The findings by [9] of the existence of (pulsating) stars that have a lower luminosity value than the luminosity predicted if they would have been evolving as a single star, and that they show IR excess connected to binarity, are also confirmed by us.

For further details, please, look into our articles [3] and [4].

Acknowledgments: M.I.J. acknowledges financial support from the Ministry of Education, Science and Technological Development of the Republic of Serbia through the project 176004, and the Hungarian National Research, Development and Innovation Office through NKFIH K-115709.

References

[1] Fiorentino, G., & Monelli, M., A&A, **540**, A102 (2012)
[2] Groenewegen, M. A. T., A&A, **543**, A36 (2012)
[3] Groenewegen, M. A. T., & Jurkovic, M. I., A&A, in press (arXiv:1705.00886) (2017a)
[4] Groenewegen, M. A. T., & Jurkovic, M. I., A&A, in press (arXiv:1705.04487) (2017b)
[5] Groenewegen, M. A. T., & Jurkovic, M. I., in *these proceedings*
[6] Ivezić, Ž., Nenkova, M., & Elitzur, M., *DUSTY user manual*, (arXiv:astro-ph/9910475) (1999)
[7] Kamath, D., Wood, P. R., & Van Winckel, H., MNRAS, **439**, 2211 (2014)
[8] Kamath, D., Wood, P. R., & Van Winckel, H., MNRAS, **454**, 1468 (2015)
[9] Kamath, D., Wood, P. R., Van Winckel, H., & Nie, J. D., A&A, **586**, L5 (2016)
[10] Miller Bertolami, M. M., A&A, **588**, A25 (2016)
[11] Soszyński, I., Udalski, A., Szymański, M. K., et al., AcA, **58**, 293 (2008)
[12] Soszyński, I., Poleski, R., Udalski, A., et al., AcA, **60**, 17 (2010)
[13] Soszyński, I., Udalski, A., Szymański, M. K., Kubiak, M., Pietrzyński, G., Wyrzykowski, Ł., Ulaczyk, K., & Poleski, R., et al., AcA, **60**, 91 (2010)
[14] Vassiliadis, E & Wood, P. R., ApJ, **413**, 641 (1993)
[15] Wallerstein, G., PASP, **114**, 689 (2002)

Araucaria Project: Pulsating stars in binary systems and as distance indicators

Bogumił Pilecki[1,*], *Wolfgang* Gieren[2,3,**], *Grzegorz* Pietrzyński[1], and *Radosław* Smolec[1]

[1] *Copernicus Astronomical Centre, Polish Academy of Sciences, Bartycka 18, 00-716 Warsaw, Poland*
[2] *Universidad de Concepción, Departamento de Astronomía, Casilla 160-C, Concepción, Chile*
[3] *Millennium Institute of Astrophysics, Santiago, Chile*

Abstract. Pulsating stars, like Cepheids or RR Lyrae stars, are ones of the most important distance indicators. They are also key objects for testing the predictions of stellar evolution and stellar pulsation theory. In the Araucaria Project we have studied these objects since 2002, measuring distances to the galaxies in the Local Group and beyond.
In 2010 we have for the first time confirmed spectroscopically the existence of a classical Cepheid in an eclipsing binary system. This has opened an opportunity to study in great details and with high accuracy (better than 1%) the physical parameters of these very important objects. First dynamical mass determination (M_{cep} = 4.16 ± 0.03 $M_{\odot}$) let us solve the long-standing mass discrepancy problem. Since then we have measured masses for 6 classical Cepheids in binary systems and determined projection factors for three of them. One of the analyzed systems was confirmed to consist of two first-overtone Cepheids.
Type II Cepheids are recently becoming more important as distance indicators and astrophysics laboratory, although our knowledge of these stars is quite limited. Their evolutionary status is also not well understood and observational constraints are needed to confirm the current theories. We are presenting here our first results of the spectroscopic analysis of 4 of these systems. The masses of type II Cepheids seem consistent with the expected 0.5 − 0.6 $M_{\odot}$. We also present first results of the fully modeled pulsator originally classified as peculiar W Vir star. The mass of this star is 1.51 ± 0.09 $M_{\odot}$ and the p-factor 1.3 ± 0.03. It was eventually found not to belong to any typical Cepheid group.

1 Introduction

The Araucaria Project, started in 2000, is a collaboration between astronomers from institutions in Chile, the United States and Europe. Its principal aim is to provide an improved calibration of the local extragalactic distance scale. In the process of setting up the extragalactic distance scale, the greatest difficulty leading to the currently largest contribution on the systematic uncertainty of the Hubble constant lies in the determination of accurate absolute distances to nearby galaxies.

The principal reason for this persisting difficulty is in the ill-known dependencies of stellar standard candles, used to measure the distances of nearby galaxies, on the environmental properties of their host galaxies (metallicity, age of the stellar populations).

*pilecki@camk.edu.pl
**wgieren@astro-udec.cl

Table 1. Distance measurements utilizing P-L relation for classical Cepheids.

Galaxy	Distance modulus	Reference
M33	24.62 ± 0.07	2013, ApJ, 773, 69
NGC 7793	27.68 ± 0.05(stat) ± 0.08(sys)	2010, AJ, 140, 1475
NGC 247	27.644 ± 0.036	2009, ApJ, 700, 1141
NGC 247	27.80 ± 0.09(stat) ± 0.09(sys)	2008, AJ, 136, 1770
WLM	24.924 ± 0.042(stat)	2008, ApJ, 683, 611
NGC 55	26.434 ± 0.037(stat)	2008, ApJ, 672, 266
WLM	25.144 ± 0.03(stat) ± 0.07(sys)	2007, AJ, 134, 594
NGC 55	26.40 ± 0.05(stat) ± 0.09(sys)	2006, AJ, 132, 2556
NGC 3109	25.571 ± 0.024	2006, ApJ, 648, 375
NGC 3109	25.54 ± 0.03	2006, ApJ, 648, 366
NGC 6822	23.312 ± 0.021(stat)	2006, ApJ, 647, 1056
IC1613	24.291 ± 0.035(stat)	2006, ApJ, 642, 216
NGC 300	26.37 ± 0.05(stat) ± 0.03(sys)	2005, ApJ, 628, 695
LMC	18.56 ± 0.04	2005, ApJ, 627, 224
NGC 6822	23.34 ± 0.04(stat) ± 0.05(sys)	2004, AJ, 128, 2815
NGC 300	26.43 ± 0.04(stat) ± 0.05(sys)	2004, AJ, 128, 1167

The Araucaria Project is an effort to remedy this situation for several of the most important stellar candles, including Cepheid variables, RR Lyrae stars, red clump giants, and blue supergiants which all have the potential to provide accurate distance determinations to nearby galaxies once their environmental dependencies are well calibrated.

2 Pulsating stars as distance indicators

Here we present a short summary of the distances to other galaxies measured by the Araucaria Project using the pulsating stars as standard candles. Such measurements are subject to many systematic errors with the reddening being one of the most significant sources thereof. The differences in distance determination may be as high as 30%, as in the case of M33 ([1]). One of the cures for this situation is using a distance measurement method that is as reddening independent as possible. In Araucaria Project we perform deep infrared observations of the galaxies and do a multi-wavelength (expressed as total to selective extinction R) distance moduli fit like the one presented in Figure 3 of [1]. Extrapolating the fit to the value of $R = 0$ we obtain the true distance modulus, and the slope of the fit gives us the average total extinction. Distance moduli determined in this way are also quite independent of the applied reddening law.

In Table 1 we present a list of 16 distances to 10 galaxies obtained using the period-luminosity (P-L) relation for classical Cepheids. In total more than 600 Cepheids were observed in optical and more than 300 in infrared bands. In Table 2 we present a similar list with the distances obtained using the P-L relations for type II Cepheids and RR Lyrae stars. As one can see, classical Cepheids are the most important and most widely used standard candle, as being very luminous they can be seen out to large distances. Nevertheless, lower-mass and older type II Cepheids and RR Lyrae stars offer not only a possibility to study population effects when more than one distance indicator can be observed, but sometimes may be the only way to measure the distance to galaxies where classical Cepheids are not present.

Table 2. Distance measurements utilizing P-L relations for type II Cepheids and RR Lyraes.

Galaxy	Distance modulus	Reference
Type II Cepheids:		
SMC	18.85 ± 0.07(stat) ± 0.07(sys)	2010, AcA, 60, 233
RR Lyraes:		
Carina	20.12 ± 0.02(stat) ± 0.11(sys)	2015, AJ, 150, 3
SMC	18.97 ± 0.03(stat) ± 0.12(sys)	2009, AJ, 138, 1661
LMC	18.58 ± 0.03(stat) ± 0.11(sys)	2008, AJ, 136, 272-279
SDG	19.67 ± 0.02(stat) ± 0.12(sys)	2008, AJ, 135, 1993-1997

3 Pulsating stars in eclipsing binary systems

The results presented above are very important, but such studies utilize the pulsating stars, not telling us anything about the physics behind. To get to know all the parameters of the stars and to study the astrophysics of Cepheids and other pulsators it is very important to analyze this type of stars which are members of eclipsing binary systems. If the system is not eclipsing it can be studied with joint spectroscopic and interferometric observations, but in this case the system has to be located very close to us and have reasonably short orbital period. Up to now no system could be precisely analyzed in this way.

On the other hand, more than ten eclipsing systems with Cepheids were detected so far, and we have analyzed several of them already. The summary of our studies of these systems is presented below.

3.1 Classical Cepheids

Classical Cepheids are radially pulsating giants and supergiants of Population I. Thanks to their period-luminosity relation they are important distance indicators in the local Universe. They form a fundamental part of the cosmic distance ladder, connecting our Milky Way galaxy to galaxies in the Local Group and beyond. They are also key objects for testing the predictions of stellar evolution and stellar pulsation theory. Until recently we did not know their physical properties, however, and there are still many challenges to be faced.

Although Cepheids in binary systems were known to exist, for many years it was hard to find one that would be a member of an eclipsing binary. The breakthrough came with the microlensing surveys like OGLE ([2]) and MACHO ([3]). More than ten candidates in the Small (SMC) and Large (LMC) Magellanic Clouds were proposed after the analysis of their light curves; spectroscopic confirmation was however needed to remove all the doubts. We have then selected the most promising objects and observed them spectroscopically.

OGLE-LMC-CEP-0227 was the first to be confirmed ([4]). It was later reanalyzed by [5] using a more advanced method and more data. Five other Cepheids were confirmed and their parameters measured until now: OGLE-LMC-CEP-1812 ([6]), OGLE-LMC-CEP-1718 A and B ([7]), OGLE-LMC-CEP-2532 ([8]) and OGLE-LMC562.05-9009 ([9]).

For detailed description of the method and the analysis steps we refer the reader to [5] and the papers on other Cepheids that follow. Here we will present just a short summary of the subject.

Radial velocities (RV) were measured using the Broadening Function method implemented in the RaveSpan code ([5, 10]). Orbital solutions were then obtained using the same program and sometimes also with the Wilson-Devinney code ([11]) as well. In RaveSpan the pulsations are included and for the optional analysis with the Wilson-Devinney code they were first subtracted. As a result we

 DOI: 10.1051/epjconf/201715207007

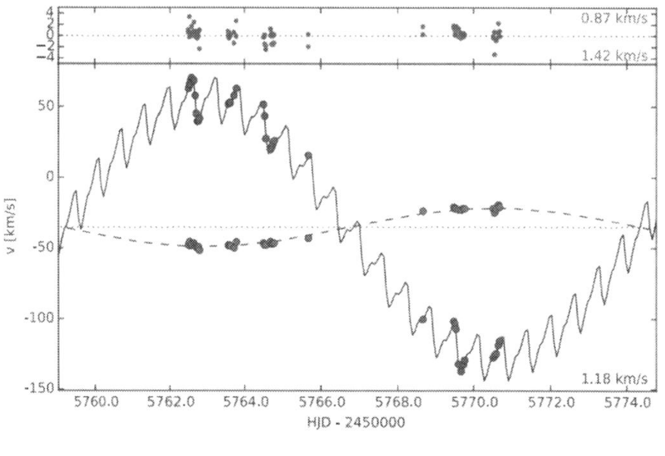

Figure 1. Example radial velocity curve including pulsations and the model obtained with the RaveSpan code.

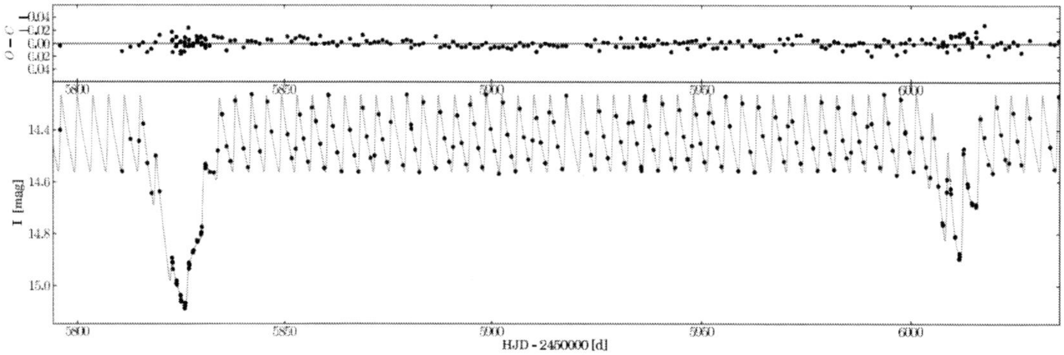

Figure 2. Light curve model of OGLE-LMC-CEP-0227. At the left can be seen the deeper eclipse when the Cepheid is covered by its companion's disk and the amplitude decreases.

have the masses (M) and the system size (scaled by the inclination), the orbital parameters and the pulsational RV curve. The latter is used to measure the radius change after scaling it with the p-factor. Sample radial velocity curve (with the pulsations included) obtained with the RaveSpan is presented in Figure 1.

The photometric data were analyzed using a pulsation-enabled eclipsing modeling tool based on the well-tested JKTEBOP code ([12]) modified to allow the inclusion of pulsation variability. We generate a two-dimensional light curve that consist of purely eclipsing light curves for different pulsating phases. Then a one-dimensional light curve is generated (interpolated from the grid) using a combination of pulsational and orbital phases calculated using a Cepheid and system ephemerides. A model for OGLE-LMC-CEP0227 and its I-band light curve are shown in Figure 2.

From the photometric solution we have the period, the time of the primary minimum, the inclination (i), the fractional radii, the eccentricity, the argument of periastron, the surface brightness ratios, the third light (if present), and the p-factor (p) value. Limb darkening parameters corresponding to

Table 3. Masses and radii of Cepheids

ID	Type	Period [d]	Mass [$M_{\odot}$]	Radius [$R_{\odot}$]	Comment
LMC-CEP-227	FU	3.797086	4.16 (3)	34.9 (2)	p-factor
LMC562.05-9009	FU	2.987846	3.71 (3)	28.6 (2)	p-factor
LMC-CEP-2532	1O	2.035349	3.90 (10)	28.9 (14)	single eclipse
LMC-CEP-1718B	1O	2.480917	4.15 (5)	30.0 (13)	single eclipse
LMC-CEP-1718A	1O	1.963663	4.18 (5)	25.3 (11)	single eclipse
LMC-CEP-1812	FU	1.312903	3.74 (6)	17.4 (9)	prel. p-factor

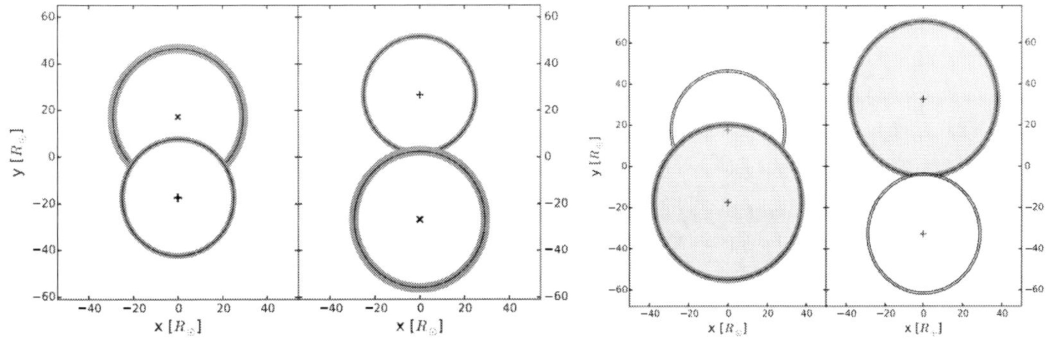

Figure 3. Orbital configuration at the phases of primary and supposed secondary eclipse for OGLE LMC-CEP-1718 and LMC-CEP-2532.

different temperatures are also tested. To obtain an optimal solution and error estimation standard Monte Carlo and Markov Chain Monte Carlo (MCMC) sampling is used.

As for now five systems were analyzed yielding physical parameters for three fundamental (FU) and three first-overtone (1O) Cepheids (see the list in Table 3). Using the obtained mass of the OGLE-LMC-CEP0227 Cepheid (M_{cep} = 4.16 ± 0.03 $M_{\odot}$) we showed that the pulsation theory predicts it correctly and that the theory of evolution has to be adjusted. Expected pulsation mass was 3.98 ± 0.29 $M_{\odot}$ and the evolutionary mass suggested masses between 4.5 and 5 $M_{\odot}$. The parameters of all analyzed Cepheids can be found in the tables in their corresponding papers (see references above).

First direct, distance-independent p-factor measurements were obtained for two of the stars. For OGLE-LMC-CEP0227 we measured p = 1.21 ± 0.05 and for OGLE-LMC562.05-9009 – p = 1.37 ± 0.09. A preliminary p-factor was also obtained for OGLE-LMC-CEP-1812 (paper in prep.), yielding p = 1.27 ± 0.09. As the covered period range is short we cannot exclude the period dependence, but the high scatter of our measurements suggest that other factors may have a bigger impact. For the moment anti-correlation with the mass is the clearest one with the higher mass stars having lower p-factors. For the first two stars mentioned the limb darkening was also estimated and the values resulted to be much lower and the phase dependence significantly different than expected.

The OGLE-LMC-CEP-1718 is especially interesting as both stars are pulsating. This allows us not only to constrain the models with the obtained parameters, but also with the difference or ratio of the parameters (their periods for example). What is also interesting is that the masses of the stars are practically equal.

The first-overtone Cepheids are however particularly unlucky as in both cases only one eclipse is present and it is much harder to precisely determine the parameters. In case of OGLE-LMC-CEP-

DOI: 10.1051/epjconf/201715207007

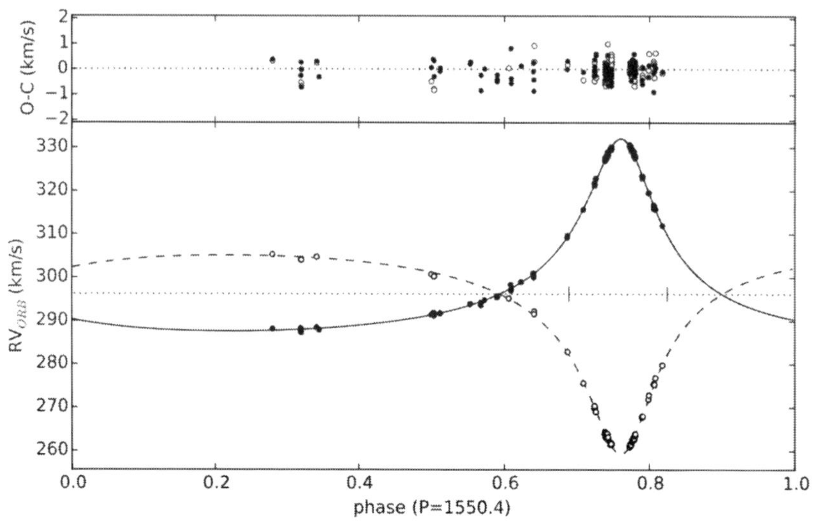

Figure 4. Orbital radial velocity solution for OGLE-LMC562.05-9009. Note the very long period and high eccentricity. The position of eclipses are marked with short vertical lines.

1718 a period-radius relation was used to determine the radii ratio of the stars and only the radii sum was fitted. In case of OGLE-LMC-CEP-2532 we could use the additional information carried by the pulsations, which are directly fitted in our method. The primary eclipse is also much deeper that in case of the former system. The configurations at the times when eclipses should occur for OGLE-LMC-CEP-1718 and OGLE-LMC-CEP-2532 are shown in Figure 3.

The latest analyzed system OGLE-LMC562.05-9009 is very promising, but its long orbital period (1550 days) makes the observational part very challenging. Very high eccentricity ($\sim$ 0.61) makes this task even harder. Nevertheless we could obtain a very precise mass determination ($M_{cep} = 3.71 \pm 0.03\ M_{\odot}$) and another direct measurement of the p-factor value (1.37 ± 0.09). We could also estimate the age of the Cepheid to be about 205 Myr. It is the lowest mass classical Cepheid found so far, and has the highest p-factor among the three stars for which we could measure it. The orbital solution of OGLE-LMC562.05-9009 is presented in Figure 4.We plan to keep observing this object and reanalyze it once enough new data to significantly improve the model is available.

It is interesting to note that with only one exception all analyzed systems with classical Cepheids have mass ratios in the range of 0.97 to 1.0. Only for OGLE-LMC-CEP-1812, the mass ratio is 0.7, making this system a very interesting object to study. A merger scenario was suggested for this system to explain its current properties ([13]).

4 Type II Cepheids

Type II Cepheids are low-mass ($\sim 0.5\ M_{\odot}$) pulsating stars that belong to the disc and halo populations ([14]). They are a much older counterpart of more massive classical Cepheids – they have periods and amplitudes in a similar range, but are about 1.5-2 mag fainter. Similarly they exhibit a tight and well-defined P-L relation and may serve as a good distance indicators, allowing to measure distances both inside and outside our Galaxy ([15–17]).

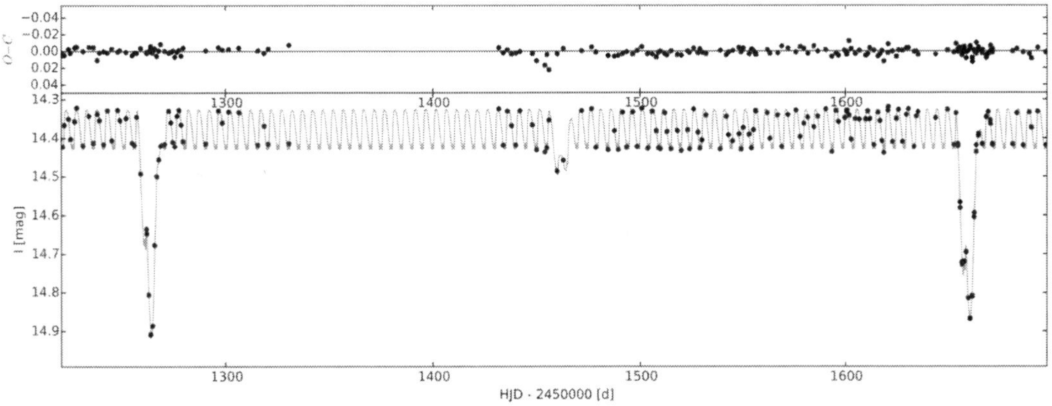

Figure 5. I-band model and OGLE data for OGLE-LMC-T2CEP-098. Note the high difference in the minima depths. The pulsating component is eclipsed at the shallower eclipse.

As compared to classical Cepheids, our knowledge of type II Cepheids is very poor, being more qualitative than quantitative. After the work of Gingold ([18, 19]) they are usually divided into three subgroups (BL Herculis, W Virginis and RV Tauri) depending on the pulsation period, observational properties and their evolutionary status, that nevertheless obey a similar period-luminosity relation. There is however a lack of observational measurements for type II Cepheids that could support the scenario presented by Gingold.

It is now of the highest importance to obtain direct mass measurements for a sample of type II Cepheids to pinpoint their evolutionary status, which is still poorly understood. To measure their masses we need spectroscopic observations of type II Cepheids in eclipsing binary systems. A list of candidates for such systems was given by [16]. We are now collecting observations for 5 systems in LMC and a few more in SMC. Unfortunately only 3 of these systems are clear double-lined spectroscopic binaries.

These systems appear to be very different from the ones that contain classical Cepheids, and much harder to analyze. They have highly unequal minima depths (see Fig. 5) and disk presence is a very common feature. For some of the Cepheids rapid period change was also detected.

We started the analysis with the brightest system that had the best quality light curve, although unfortunately it was of SB1 type. We have obtained the photometric (see Fig. 5) solution and partial orbital solution. The Cepheid companion could not be identified in the spectra, but on the other hand appeared to be many times brighter than the pulsating star. Our method gave us also a tight relation between the p-factor and the mass, i.e. assuming the p-factor we could estimate the mass. This relation limited the mass to the range $0.7 - 2.4\, M_\odot$, with the most probable value close to $1.5\, M_\odot$. Then we have used three mostly independent methods to constrain further the parameters. All of them pointed to a Cepheid mass of about $1.5\, M_\odot$, with the pulsation theory giving the smallest uncertainty ($M_{cep} = 1.51 \pm 0.09\, M_\odot$). To obtain this result we have used the mass-radius relation obtained from the system modeling. Observable properties of this pulsator makes it an outlier, that does not fit into any known variable type. The most probable explanation is that the star has passed through the mass transfer phase during its evolution. More details on the analysis and the derivation of physical parameters for this system can be found in [20].

The light curves of other systems with type II Cepheids are not easy to model, but have the advantage of being of SB2 type. For them we could obtain a preliminary spectroscopic solution.

All three measured $M\sin^3(i)$ are located in the range $0.3 - 0.75\ M_\odot$. As the systems are detached, we do not expect very low inclinations, thus the masses of these Cepheids will not be much higher than the quoted numbers. As the theory expects masses of about $0.5 - 0.6\ M_\odot$, for these cases we see a much better agreement with observations than in case of OGLE-LMC-T2CEP-098. However more detailed studies are necessary to classify correctly these pulsators, as their light curves exhibit features that are not typical for W Virgnis nor BL Herculis stars. In the catalog of the OGLE project ([16]) some of them were classified as peculiar W Virginis, but even among them there are differences.

Contrary to what we found for the systems with classical Cepheids, the measured mass ratios for those with type II Cepheis are very high, being in the range of $M_2/M_{cep} \sim 3 - 10$.

Acknowledgments: We gratefully acknowledge financial support for this work from the Polish National Science Center grant SONATA 2014/15/D/ST9/02248.

References

[1] W. Gieren, M. Górski, G. Pietrzyński, P. Konorski, K. Suchomska, D. Graczyk, B. Pilecki, F. Bresolin, R.P. Kudritzki, J. Storm et al., ApJ, **773**, 69 (2013), 1305.4258
[2] A. Udalski, M.K. Szymański, G. Szymański, AcA, **65**, 1 (2015), 1504.05966
[3] C. Alcock, R.A. Allsman, D.R. Alves, A.C. Becker, D.P. Bennett, K.H. Cook, A.J. Drake, K.C. Freeman, K. Griest, S.L. Hawley et al., ApJ, **573**, 338 (2002), astro-ph/0201481
[4] G. Pietrzyński, I.B. Thompson, W. Gieren, D. Graczyk, G. Bono, A. Udalski, I. Soszyński, D. Minniti, B. Pilecki, Nature, **468**, 542 (2010)
[5] B. Pilecki, D. Graczyk, G. Pietrzyński, W. Gieren, I.B. Thompson, W.L. Freedman, V. Scowcroft, B.F. Madore, MNRAS, **436**, 953 (2013), 1308.5023
[6] G. Pietrzyński, I.B. Thompson, D. Graczyk, W. Gieren, B. Pilecki, A. Udalski, I. Soszynski, G. Bono, P. Konorski, N. Nardetto et al., ApJL, **742**, L20 (2011), 1109.5414
[7] W. Gieren, B. Pilecki, G. Pietrzyński, D. Graczyk, I.B. Thompson, I. Soszyński, P. Konorski, R. Smolec, A. Udalski, N. Nardetto et al., ApJ, **786**, 80 (2014), 1403.3617
[8] B. Pilecki, D. Graczyk, W. Gieren, G. Pietrzyński, I.B. Thompson, R. Smolec, A. Udalski, I. Soszyński, P. Konorski, M. Taormina et al., ApJ, **806**, 29 (2015), 1504.04611
[9] W. Gieren, B. Pilecki, G. Pietrzyński, D. Graczyk, A. Udalski, I. Soszyński, I.B. Thompson, P.G. Prada Moroni, R. Smolec, P. Konorski et al., ApJ, **815**, 28 (2015), 1511.02826
[10] B. Pilecki, P. Konorski, M. Gorski, *RaveSpan - Radial Velocity and Spectrum Analyzer*, in *IAU Symposium*, edited by M.T. Richards, I. Hubeny (2012), Vol. 282, pp. 301–302
[11] R.E. Wilson, E.J. Devinney, ApJ, **166**, 605 (1971)
[12] J. Southworth, P.F.L. Maxted, B. Smalley, MNRAS, **351**, 1277 (2004), astro-ph/0403572
[13] H.R. Neilson, R.G. Izzard, N. Langer, R. Ignace, A&A, **581**, L1 (2015), 1508.02725
[14] G. Wallerstein, PASP, **114**, 689 (2002)
[15] N. Matsunaga, M.W. Feast, I. Soszyński, MNRAS, **413**, 223 (2011), 1012.0098
[16] I. Soszyński, A. Udalski, M.K. Szymański, M. Kubiak, G. Pietrzyński, Ł. Wyrzykowski, O. Szewczyk, K. Ulaczyk, R. Poleski, AcA, **58**, 293 (2008), 0811.3636
[17] D. Majaess, D. Turner, D. Lane, AcA, **59**, 403 (2009), 0909.0181
[18] R.A. Gingold, ApJ, **204**, 116 (1976)
[19] R.A. Gingold, Mem. Soc. Astron. Italiana, **56**, 169 (1985)
[20] B. Pilecki, W. Gieren, R. Smolec, G. Pietrzyński, I.B. Thompson, R.I. Anderson, G. Bono, I. Soszyński, P. Kervella, N. Nardetto et al., ApJ **842**, 110 (2017)

Distance scale calibration based on early-type binaries

Mónica Taormina[1,*], *Grzegorz* Pietrzyński[1], *Bogumił* Pilecki[1], *Ian* Thompson[2], and *Dariusz* Graczyk[1]

[1]*Centrum Astronomiczne im. Mikołaja Kopernika (CAMK), PAN, Bartycka 18, 00-716 Warsaw, Poland*
[2]*Carnegie Observatories, 813 Santa Barbara Street, Pasadena, CA 91101-1292, USA*

Abstract. Our main goal is to establish a firm empirical calibration of the surface brightness vs. $(V-K)$ color relation for early type stars based on high quality spectroscopic and infrared observations of early-type detached eclipsing systems in the Large Magellanic Cloud (LMC). Our calibration of this relation will allow distance determinations accurate to about 2.5% to a single object located well beyond the Magellanic Clouds. This will let us calibrate other distance indicators, including period-luminosity relations for Cepheids. The first step of the project is to determine precise parameters for a sample of B-type systems in the LMC. We have already made a preliminary analysis of light and radial velocity curves for selected objects and measured their masses and radii. Here we present the results for one such system.

1 Introduction

Detached eclipsing double-lined spectroscopic binaries (SB2) offer a unique opportunity to measure directly, and accurately, stellar parameters like mass, luminosity, and radius. For such systems linear dimensions can be measured with a precision of 1% from the analysis of light and radial velocity curves. The distance to an eclipsing binary follows from the linear radii and the angular diameters derived from the absolute surface brightness coming from the surface brightness (SB) vs. $(V-K)$ color calibration. The latter is already well established for late type stars for which accurate stellar angular diameters were determined by interferometry ([2]).

Not much work, however, has been done so far for the early type stars, as there is a lack of such objects in the Solar neighborhood, for which parallaxes or interferometric data could be available (we need independent distance or angular diameter determination). In the latest work on that subject Challouf et al. ([1]) have observed interferometrically a sample of such stars, but could only reduce the uncertainty from about 10% to 7%, which is not enough for precise distance measurements.

Our idea is to observe early-type systems in the Large Magellanic Cloud (LMC) for which we know the distance with a very good accuracy. Using this distance together with the radii from the modeling we can calculate the angular diameters of the stars. With these diameters and observed magnitudes we will be able to calculate the SB and determine the SB-color relation (SBCR) for early type stars with unprecedented precision.

When applied for the early-type stars the eclipsing binary method will serve to accurately measure distances to the galaxies in the Local Group and beyond. Having reliable distances to numerous galaxies with different environmental conditions will be of high importance for the calibration of other standard candles and especially period-luminosity relations for Classical Cepheids. This in turn is one of the most important steps to calibrate the cosmic distance scale.

2 Data

We have selected a sample of 9 objects from the catalog of about 30000 eclipsing binaries in the LMC recently produced by the OGLE collaboration ([3]). The sample consists of detached systems with

*taormina@camk.edu.pl

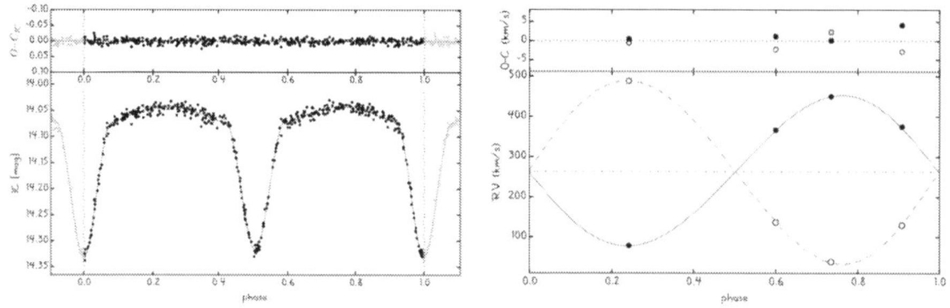

Figure 1. Light and radial velocity curve for one of the system with corresponding preliminary models.

well-covered deep eclipses of similar depths. They are the best-ever discovered objects for the precise parameter determination, which is necessary to calibrate the SBCR for early-type stars.

For the calibration of the SBCR apart from the optical photometry we will need good-quality radial velocity curves and infrared photometry. In the preliminary study we have already collected at least three spectra for all the stars with the MIKE and UVES instruments mounted on the 6.5-meter Magellan Telescope and 8.2-meter VLT, respectively.

3 Analysis and first results

We have performed a simple modeling of all the systems from the sample of 9 eclipsing binaries using both JKTEBOP code and Wilson-Devinney code with the PHOEBE GUI. Using these preliminary models we were able to look for the stars for which the most important parameters can be determined with the lowest possible uncertainty, in particular we made sure that the radii can be measured with a precision better than 1%.

For one object we have obtained 4 observations covering both quadratures and thus letting us check the lower limit of the precision we will get for our objects. Our measured velocities have a scatter of about 2 km/s while the velocity semi-amplitudes are of the order of 200 km/s (see Fig. 1). The radial velocities were measured using the Broadening Function technique ([4]). This technique is superior to the commonly used cross-correlation function as it gives narrower and stronger profiles.

From the preliminary model the masses of the components are 15 and 12 $M_{\odot}$, and their radii 8 and 7 $R_{\odot}$, with the separation between the stars of 30 $R_{\odot}$. The system is a well detached one and will serve for the calibration of the SBCR.

Acknowledgments: Support from the IdP II 2015 0002 64 grant of the Polish Ministry of Science and Higher Education.

References

[1] Challouf, M., Nardetto, N., Mourard, D., et al., A&A, **570**, A104 (2014)

[2] di Benedetto, G. P., MNRAS, **357**, 174 (2005)

[3] Graczyk, D., Soszyński, I., Poleski, R., et al., Acta Astron., **61**, 103 (2011)

[4] Rucinski, S., in *Precise Stellar Radial Velocities*, IAU Colloq. 170, ed. J. B. Hearnshaw & C. D. Scarfe, Astronomical Society of the Pacific Conference Series, **185**, 82 (1999)

The Araucaria Project: Precise distances to Local Group galaxies from near-infrared photometry of RR Lyrae stars

Paulina Karczmarek[1,⋆]

[1] *Warsaw University Observatory, Al. Ujazdowskie 4, 00-478 Warsaw, Poland*

Abstract. We present results of the Araucaria Project's investigation of RR Lyrae stars as distance indicators in nearby galaxies. With an aid of near-infrared period-luminosity-metallicity relations available in the literature we determined distance moduli to five Local Group galaxies with the uncertainty of about 5%.

1 Context

The main goal of the Araucaria Project is to improve the calibration of the cosmic distance scale by studying various stellar distance indicators in nearby galaxies. Special attention is paid to RR Lyrae stars (RRLs), which are found in abundance in many galaxies of the Local Group: Sculptor, Carina, Fornax, LMC, SMC, and more. Near-infrared (NIR) photometry of RRLs, while minimizing the influence of both interstellar extinction and metallicity on the RRL luminosity, reveals a linear relationship between the luminosity and the metallicity and the logarithm of pulsational period, i.e. the period-luminosity-metallicity (PLZ) relation.

2 Methodology and results

Within the Araucaria Project, we adopted a methodology to determine distance moduli from NIR photometry of RRL stars, and used it consistently in order to secure homogeneity and equally high quality of our results. Data collected with world-class infrared cameras SOFI/NTT and HAWK-I/VLT were reduced and calibrated with a pipeline tailored for the purpose of the Araucaria Project ([1, 2]), and yielded random-phase measurements of RRL apparent magnitudes. Using pulsational periods from the literature, we plotted the RRL apparent magnitudes against their log periods (Fig. 1), and performed the linear least squares fitting. The slope of the fit was intentionally fixed using a set of PLZ relations ([3–6]), so the uncertainty of the free parameter – zero point of the calibration – could be substantially reduced. The zero point of the calibration, corrected for the effect of metallicity and interstellar extinction (both from the literature), yielded the true distance modulus.

The PLZ relations, applied to high quality NIR photometry of RRL stars, allowed us to determine distance moduli to five Local Group galaxies: Sculptor ([7]), LMC ([8]), SMC ([9]), Carina ([10]), and Fornax ([11]), all with an uncertainty of about 5% (Table 1). The results agree very well with other determinations available in the literature, but are of considerable better precision.

⋆pkarczmarek@astrouw.edu.pl

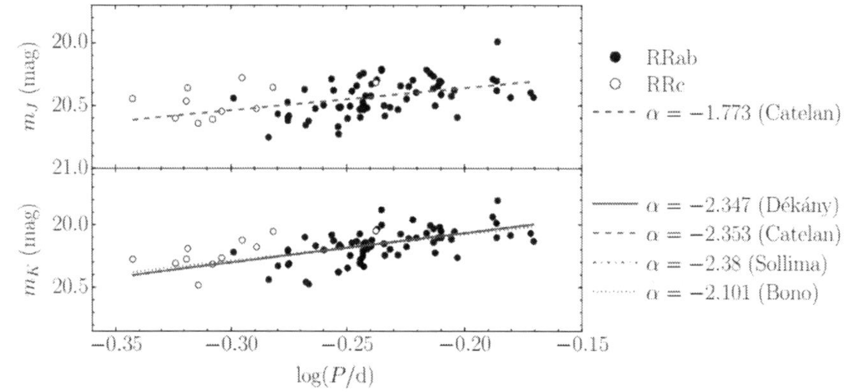

Figure 1. Apparent magnitudes of Fornax RRL stars as a function of their log pulsational period. Lines denote different slopes of PLZ relations, α, extracted from the literature ([3–6]).

Table 1. Distance moduli to five Local Group galaxies determined within the Araucaria Project.

Galaxy	n_{RRL}	$(m-M)_0 \pm \sigma_{syst.} \pm \sigma_{stat.}$	Reference
Sculptor	78	$19.67 \pm 0.02 \pm 0.12$	[7]
LMC	65	$18.58 \pm 0.03 \pm 0.11$	[8]
SMC	34	$18.97 \pm 0.03 \pm 0.12$	[9]
Carina	33	$20.118 \pm 0.017 \pm 0.110$	[10]
Fornax	77	$20.818 \pm 0.015 \pm 0.116$	[11]

The effort put in reducing the statistical error and controlling the systematic error of standard candles, as shown in the case of RRL stars, can significantly improve the accuracy of secondary distance indicators, and ultimately contribute to the refinement of the Hubble constant.

Acknowledgments: I am indebted to all members of the Araucaria Project who contributed to distance determinations in the past 12 years. This project has been supported by the Polish Ministry of Science and Higher Education through the "Ideas Plus" grant IdP II 2015 0002 64.

References

[1] Pietrzyński, G., & Gieren, W., AJ, **124**, 2633 (2002)
[2] Pietrzyński, G., Gieren, W., Fouqué, P., & Pont, F., AJ, **123**, 789 (2002)
[3] Bono, G., Caputo, F., Castellani, V., et al., MNRAS, **344**, 1097 (2003)
[4] Catelan, M., Pritzl, B. J., & Smith, H. A., ApJS, **154**, 633 (2004)
[5] Dékány, I., Minniti, D., Catelan, M., et al., ApJL, **776**, L19 (2013)
[6] Sollima, A., Cacciari, C., Arkharov, A. A. H., et al., MNRAS, **384**, 1583 (2008)
[7] Pietrzyński, G., Gieren, W., & Szewczyk, O., AJ, **135**, 1993 (2008)
[8] Szewczyk, O., Pietrzyński, G., Gieren, W., et al., AJ, **136**, 272 (2008)
[9] Szewczyk, O., Pietrzyński, G., Gieren, W., Ciechanowska, A., Bresolin, F., & Kudritzki, R.-P., AJ, **138**, 1661 (2009)
[10] Karczmarek, P., Pietrzyński, G., Gieren, W., et al., AJ, **150**, 90 (2015)
[11] Karczmarek, P., Pietrzyński, G., Gieren, W., et al., in preparation (2017)

The Araucaria Project. The distance to the Sculptor group galaxy NGC 7793 from near-infrared photometry of Cepheid variables

Bartlomiej Zgirski[1,⋆], *Wolfgang* Gieren[3,4], *Grzegorz* Pietrzyński[1,3], *Paulina* Karczmarek[2], *Marek* Górski[3,4], *Dariusz* Graczyk[1,3,4], *Bogumil* Pilecki[1], *Mónica* Taormina[1], *Ksenia* Suchomska[2], *Piotr* Wielgórski[1], and *Piotr* Konorski[2]

[1]*Nicolaus Copernicus Astronomical Center of the Polish Academy of Sciences, Warsaw, Poland*
[2]*Astronomical Observatory of the University of Warsaw, Warsaw, Poland*
[3]*Universidad de Concepción, Departamento de Astronomia, Concepción, Chile*
[4]*Millennium Institute of Astrophysics, Santiago, Chile*

Abstract. We performed deep near-infrared J and K photometry of a field in the Sculptor Group spiral galaxy NGC 7793 using ESO VLT and HAWK-I instrument. We produced a sample of 14 Cepheids whose locations and periods of pulsation were known from our previous paper on distance determination to this galaxy based on V and I bands. We determined mean J and K magnitudes based on measurements from two nights and produced period-luminosity relations for both filters. Using those near-infrared dependencies together with relations for bands V and I obtained before, we were able to obtain true distance modulus for NGC 7793. We also calculated the mean reddening affecting our Cepheids.

We present first near-infrared period-luminosity relations for Cepheids in NGC 7793. Using these data and optical data published before by Pietrzyński et al. (2010, [2]), we measured accurate distance to this galaxy based on four filters (V, I, J, K).

The near-infrared (NIR) data were collected with HAWK-I and ISAAC cameras installed on VLT telescopes in Cerro Paranal Observatory. From the optical catalogue of 14 Cepheids, we were able to identify 10 on our near-infrared images.

All Cepheids have photometry from 2 HAWK-I nights in 2011 and additionally 2 brightest Cepheids have also 1 data point obtained with ISAAC in 2005.

Assuming well defined slopes of near-infrared relations in the Large Magellanic Cloud by Persson et al. (2004, [1]), we fitted straight lines to period-luminosity relations of Cepheids in NGC 7793 using the least-squares method.

We adopted well determined distance to the Large Magellanic Cloud of Pietrzyński et al. (2013, [3]) as a zero point and fiducial period-luminosity relations for near-infrared bands for the Large Magellanic Cloud established by Persson et al. (2004, [1]).

The observed distance modulus can be expressed in the following way:

$$(m - M)_\lambda = (m - M)_0 + R_\lambda E(B - V), \tag{1}$$

⋆`bzgirski@camk.edu.pl`

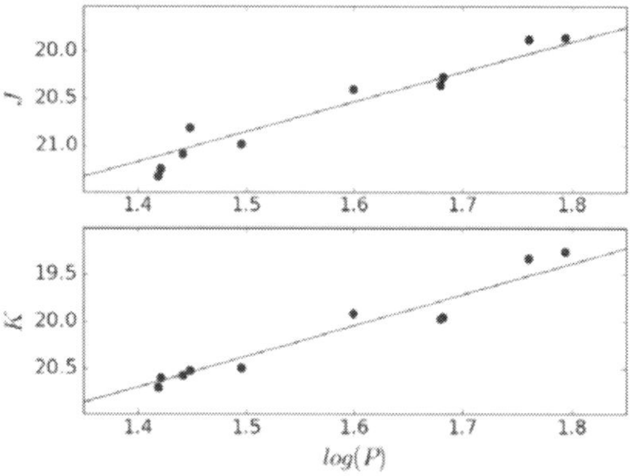

Figure 1. Near-infrared periodluminosity relations for Cepheids in NGC 7793. Corresponding slopes were adopted from Persson et al. (2004, [1]). Periods are in days.

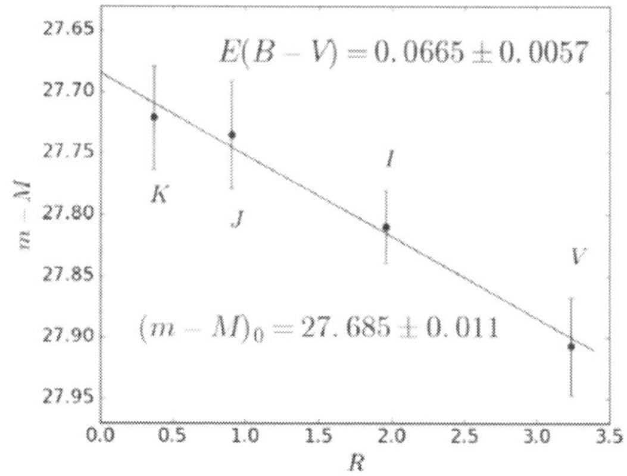

Figure 2. Apparent distance moduli to NGC 7793 as derived in the *VIJK* photometric bands, plotted against the ratio of total to selective extinction as adopted from Schlegel et al. (1998, [4]). The intersection and slope of the best-fitting line give the true distance modulus and the average total reddening, respectively.

where $(m-M)_0$ is the true distance modulus, $E(B-V)$ is color excess and R_λ is a ratio of extinction in given band to the color excess.

Assuming the reddening law of Schlegel et al. (1998, [4]), we calculate the color excess,

$$E(B-V) = 0.0665 \pm 0.0057, \tag{2}$$

and true distance modulus to NGC 7793,

$$(m-M)_0 = 27.685 \pm 0.011 \pm 0.070, \tag{3}$$

which is presently the most accurate distance measurement to NGC 7793.

Acknowledgments: The research leading to these results has received funding from the European Research Council (ERC) under the European Union's Horizon 2020 research and innovation programme (grant agreement No 695099) . Research supported from the IdP II 2015 0002 64 grant of the Polish Ministry of Science and Higher Education.

References

[1] Persson, S. E., Madore, B. F., Krzeminski, W., et al., AJ, **128**, 2239 (2004)
[2] Pietrzyński, G., Gieren, W., Hamuy, M., et al., AJ, **140**, 1475 (2010)
[3] Pietrzyński, G., Graczyk, D., Gieren, W., et al., Nature, **495**, 76 (2013)
[4] Schlegel, D. J., Finkbeiner, D. P., & Davis, M., ApJ, **500**, 525 (1998)

The distance to the Small Magellanic Cloud from multiband period-luminosity relations for classical Cepheids and its dependence on metallicity

Piotr Wielgorski[1,*], *Grzegorz* Pietrzynski[1,2], and *Wolfgang* Gieren[2,3]

[1] *Nicolaus Copernicus Astronomical Center, Polish Academy of Sciences, Bartycka 18, 00-716 Warsaw, Poland*
[2] *Universidad de Concepción, Departamento de Astronomía, Casilla 160-C, Concepción, Chile*
[3] *Millennium Institute of Astrophysics, Santiago, Chile*

Abstract. We present a distance determination to the Small Magellanic Cloud (SMC) using optical and near-infrared photometry from OGLE and IRSF. Applying a multiband method to V, I, J, H and K OGLE and IRSF data, we measured the SMC true distance modulus and mean reddening. Comparing our results with accurate distances to the LMC and SMC obtained from eclipsing binary systems, we conclude that the dependence of Cepheid magnitudes on metallicity is very small (< 0.07 mag/dex) in all bands.

1 Introduction

The Magellanic Clouds, two companions of our Galaxy containing huge numbers of distance indicators, constitute a milestone towards precise and accurate calibration of the cosmic distance scale. While the distance to the Large Magellanic Cloud (LMC) is an excellent zero point for the whole extragalactic distance scale, the Small Magellanic Cloud (SMC) allows us to examine how primary standard candles depend on metallicity. Distances to these galaxies have been measured many times with different methods, and, in particular, using a geometric method based on eclipsing binary systems ([1, 2]). Recent determinations of the SMC distance moduli are very close to 19.0 mag, but the spread around this value, dominated by systematic errors, is larger than 0.6 mag.

2 Data and results

Coordinates of Cepheids, optical V and I photometry and periods of pulsation come from the OGLE III catalogue ([3–5]). Infrared photometry in J, H, K bands is taken from IRSF observations ([6]). From our sample of Cepheids in the SMC, we removed stars with periods shorter than 2.5 d due to the observed nonlinearity of the period-luminosity (P-L) relation in this region.

In order to derive extinction-free P-L relations in the LMC, we used extinction maps published by [7], Galactic extinction towards the LMC from [8], and reddening law by [9]. With the method of least-squares and 3-sigma clipping, we established the Cepheid P-L relations in the LMC. Adopting

*pwielgor@camk.edu.pl

the slopes of the LMC Cepheid P-L relations, we calculated zero points for the P-L relations in the SMC with the least-squares method. The difference in zero points of P-L relations in SMC and LMC in each band is the observed relative distance modulus of these galaxies (i.e., affected by mean extinction towards the SMC Cepheids). Applying a multiband method (e.g., [1]), we found the true relative distance modulus of the SMC and LMC and mean reddening towards the SMC Cepheids. Our results are 0.482 ± 0.005 mag and 0.094 ± 0.004 mag for the relative distance modulus and mean reddening, respectively. Assuming a distance modulus of the LMC of 18.493 mag ([2]), we obtain a distance modulus of the SMC of 18.975 mag.

3 Discussion

Our relative distance determination is in very good agreement with the very accurate [10] determination from late-type eclipsing binaries, which amounts to 0.472 ± 0.026 mag. Assuming that the difference between the relative distance moduli between the LMC and SMC obtained from eclipsing binaries and Cepheids is due to metallicity, we conclude that the metallicity effect on the Cepheid brightness in all bands is very small (< 0.07 mag/dex), and within the errors is consistent with zero. More accurate determination of the metallicity effect require better SMC distance determination from the eclipsing binaries and better determination of Cepheid metallicities in both galaxies. It is worth noticing that our current SMC distance determination from Cepheids is very slightly sensitive to the adopted reddening law, which usually is a very important contribution to the total error budget.

Acknowledgments: The research leading to these results has received funding from the European Research Council (ERC) under the European Union's Horizon 2020 research and innovation programme (grant agreement No. 695099). Support from the IdP II 2015 0002 64 grant of the Polish Ministry of Science and Higher Education is also acknowledged.

References

[1] G. Pietrzyński, W. Gieren, I. Soszyński, F. Bresolin, R.P. Kudritzki, M. Dall'Ora, J. Storm, G. Bono, ApJ **642**, 216 (2006), astro-ph/0601309
[2] G. Pietrzyński, D. Graczyk, W. Gieren, I.B. Thompson, B. Pilecki, A. Udalski, I. Soszyński, S. Kozłowski, P. Konorski, K. Suchomska et al., Nature **495**, 76 (2013), 1303.2063
[3] I. Soszynski, R. Poleski, A. Udalski, M.K. Szymanski, M. Kubiak, G. Pietrzynski, L. Wyrzykowski, O. Szewczyk, K. Ulaczyk, AcA **58**, 163 (2008), 0808.2210
[4] I. Soszyński, R. Poleski, A. Udalski, M.K. Szymański, M. Kubiak, G. Pietrzyński, Ł. Wyrzykowski, O. Szewczyk, K. Ulaczyk, AcA **60**, 17 (2010), 1003.4518
[5] K. Ulaczyk, M.K. Szymański, A. Udalski, M. Kubiak, G. Pietrzyński, I. Soszyński, Ł. Wyrzykowski, R. Poleski, W. Gieren, A.R. Walker et al., AcA **63**, 159 (2013), 1306.4802
[6] D. Kato, C. Nagashima, T. Nagayama, M. Kurita, J.F. Koerwer, T. Kawai, T. Yamamuro, T. Zenno, S. Nishiyama, D. Baba et al., PASJ **59**, 615 (2007)
[7] R. Haschke, E.K. Grebel, S. Duffau, AJ **141**, 158 (2011), 1104.2325
[8] D.J. Schlegel, D.P. Finkbeiner, M. Davis, *Maps of Dust IR Emission for Use in Estimation of Reddening and CMBR Foregrounds*, in *American Astronomical Society Meeting Abstracts* (1997), Vol. 29 of *Bulletin of the American Astronomical Society*, p. 1354
[9] J.A. Cardelli, G.C. Clayton, J.S. Mathis, ApJ **345**, 245 (1989)
[10] D. Graczyk, G. Pietrzyński, I.B. Thompson, W. Gieren, B. Pilecki, P. Konorski, A. Udalski, I. Soszyński, S. Villanova, M. Górski et al., ApJ **780**, 59 (2014), 1311.2340

Applying survival analysis on the Cepheid period-luminosity relation at 70 μm

Chow-Choong Ngeow[1,*]

[1] *Graduate Institute of Astronomy, National Central University, Jhongli 32001, Taiwan*

Abstract.
The survival analysis is a type of statistical technique to deal with data that has upper or lower limits. In this work, I demonstrate an example of applying the survival analysis to re-derive the Cepheid period-luminosity relation at 70 μm, because about 2/3 of the currently available data only contains upper limits in flux.

1 Introduction and motivation

The first observation of Cepheids at 70 μm (with *Spitzer's* MIPS) was presented in [1] (hereafter MAR10) for 29 Galactic Cepheids. However, only 9 Cepheids were detected and possessed the 70 μm photometry. Based on these 9 Cepheids, MAR10 derived the first-ever 70 μm period-luminosity (P-L) relation. For the other 20 Cepheids, MAR10 recorded their 3σ upper limits in flux (or lower limit in magnitude). Measurements with upper limits are not new in astronomy: this kind of data is known as censored data, and statistical techniques that deal with censored data are called survival analysis (e.g., see [2–4]). Therefore, the goal of this work is to re-derive the 70 μm P-L relation using the survival analysis to include Cepheids with upper limit detections.

2 Data and methods

The 70 μm photometry (both detected & lower limits) of these Cepheids were taken from MAR10 (their Table 5). I adopted the "new" distances of the 28 Cepheids from Table 1 of MAR10 (only one Cepheid does not have this "new" distance). The 70 μm magnitudes were then converted to absolute magnitudes with the adopted distances. Extinction is ignored since it is negligible in the mid-infrared (MAR10). To re-derive the P-L relation, I used the code `ASURV` ([5], publicly available at `http://ascl.net/1406.001`), to fit a linear regression to the data.

3 Analysis and (preliminary) results

The EM (expectation-maximization) algorithm and the Buckley-James method available from `ASURV` were applied to fit the 70 μm photometry which included the censored data. Since the P-L relation presented in MAR10 was fitted with weighted linear least squares (LSQ) regression, and current

*`cngeow@astro.ncu.edu.tw`

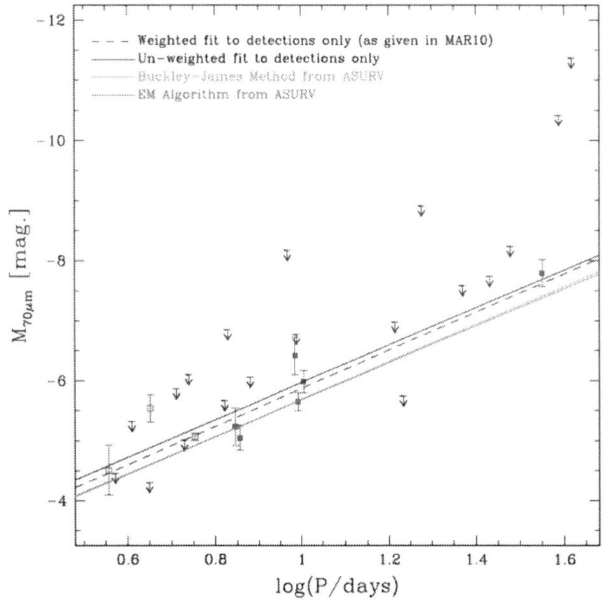

Figure 1. Fitted 70 μm P-L relation with various linear regression methods. Red squares are for fundamental mode (filled symbols) and first overtone (open symbols) Cepheids with detection. Blue symbols are for Cepheids with upper limit in flux.

Table 1. Comparison of the 70 μm P-L relation with different fitting methods.

Method	Slope	Intercept	Dispersion
Unweighted LSQ	-3.12 ± 0.49	-2.85 ± 0.46	0.39
EM Algorithm	-3.08 ± 0.47	-2.60 ± 0.45	0.49
Buckley-James Method	-3.13 ± 0.48	-2.56	0.39

version of the ASURV package does not include the weighted regression, I repeated MAR10 analysis without the weighted fit in order to compare with the results from ASURV. The fitted 70 μm P-L relation from the unweighted LSQ regression, the EM algorithm and Buckley-James method is summarized in Table 1. Figure 1 displays the fitted regressions and the observed data. Including the censored data, the result shows that the slopes of the regression are consistent with the one based on detected data only. The intercepts, on the other hand, show an offset between the two approaches.

Acknowledgments: The author thanks the funding from the Ministry of Science and Technology (Taiwan) under the contract 104-2112-M-008-012-MY3.

References

[1] Marengo, M., Evans, N. R., Barmby, P., Bono, G., Welch, D. L., & Romaniello, M., ApJ, **709**, 120 (2010)

[2] Schmitt, J. H. M. M., ApJ, **293**, 178 (1985)

[3] Feigelson, E. D., & Nelson, P. I., ApJ, **293**, 192 (1985)

[4] Isobe, T., Feigelson, E. D., & Nelson, P. I., ApJ, **306**, 490 (1986)

[5] Lavalley, M., Isobe, T., & Feigelson, E, in *Astronomical Data Analysis Software and Systems I*, ed. D. M. Worrall, C. Biemesderfer, & J. Barnes, ASP Conference Series, **25**, 245 (1992)

San Pedro meeting on Wide Field Variability Surveys: Some concluding comments

Michael W. Feast[1,2,⋆]

[1] *Astronomy Dept., University of Cape Town, South Africa*
[2] *South African Astronomical Observatory, South Africa*

Abstract. This is a written version of the closing talk at the 22nd Los Alamos Stellar pulsation conference on wide field variability surveys. It comments on some of the issues which arise from the meeting. These include the need for attention to photometric standardization (especially in the infrared) and the somewhat controversial problem of statistical bias in the use of parallaxes (and other methods of distance determination). Some major advances in the use of pulsating variables to study Galactic structure are mentioned. The paper includes a clarification of apparently conflicting results from classical Cepheids and RR Lyrae stars in the inner Galaxy and bulge. The importance of understanding non-periodic phenomena in variable stars, particularly asymptotic giant branch variables and R Coronae Borealis stars, is stressed, especially for its relevance to mass-loss in which pulsation may only play a minor role.

1 Introduction

This has been a meeting full of interest. In part, this is due to the range of topics discussed which reflect the overlapping areas in which large scale variability surveys are having a major impact. The traditional pulsators (RR Lyraes, Cepheids, Miras, etc.) remain of major importance for our understanding of stellar structure and evolution. But they are also increasingly important elements in our understanding of the structure and evolution of our own and nearby galaxies as well as being a key to the cosmic distance scale.

The main interest of the founders of these meetings can be correctly described as asteroseismology. The field to which this term now applies has made remarkable advances for our understanding of stars. Currently this new asteroseismology is expanding into the area of Galactic structure by giving us ages, masses and luminosities of large numbers of stars. One sees a bright future for both the traditional pulsators and the new asteroseismology in Galactic studies. But we need to remember that in such work, we are not alone. Major surveys such as Apogee are revealing much about the composition and structure of our Galaxy without any reference to variability and these will need to be considered together with the wide field variability results.

Rather than trying to summarize all the topics discussed at this meeting, I would like simply to mention a few issues which arise, as well as a relevant area which has only been mentioned briefly.

⋆`mwf@ast.uct.ac.za`

2 Some technical issues

2.1 Automated classification

Automated classification and machine learning were discussed in an invited review. Obviously we want to divide variable stars into physically homogeneous groups. But how this is related to observed features such as light curve shapes, remains a learning process and we need to be prepared for surprises. For instance, Pietrzyński et al. ([32]) find that an OGLE star classed as an RR Lyrae variable is in fact an object in a binary with a mass of only 0.26 solar masses. Again, are we correct in placing the metal rich and metal poor RR Lyraes together; for instance, in a single (visual) luminosity sequence (a point raised by Chadid et al., [6])? Published classifications from light curves have often been based on relatively limited material and some types of variable are then rather difficult to distinguish from one another. For instance, only the accurate photometry and densely sampled light curves of the OGLE survey allowed a few classical Cepheids in the extreme outer flared disc of our Galaxy to be correctly identified and studied ([16]).

2.2 Photometric calibration

It is sometimes said that large scale photometric surveys are self-calibrating and do not need to tie themselves to other photometric systems. There is some truth in this. However, in the case of variable star surveys, it is to be hoped that efforts will be made to link the results with other commonly used systems. This is for at least two reasons. First, for at least the brighter variables, other observations are likely to exist. Secondly, in deep surveys the brightest variables will probably be saturated. These stars will need to be included in a variety of analyses. One such, key, place is luminosity calibration. We can expect to have absolute magnitudes in various common systems for various variability classes, e.g. from parallaxes of bright variables, and these will have to be converted into the adopted survey system. The alternative is to make special observations of bright variables in the same system as the deep survey. Linking space-based photometry with ground-based work is a particular problem and is one that is crucial to cosmic distance scale work. The project to observe Milky Way classical Cepheids with good parallaxes directly with the *Hubble Space Telescope* (HST) for a distance ladder calibration is therefore of considerable importance ([33], and the talk by Lucas Macri).

A related problem is the level of consistency between different observers within one photometric system. It is perhaps not always realized how much effort went into the standardization of the $UBVRI$ system. Despite work by various groups there has been, so far as I know, no such massive effort for infrared systems. For instance, in JHK (or K_s) different groups use slightly different filters and different standard stars. In this situation, the transformation between systems (including transformation of reddening laws) is often not well known. Such matters are particularly important in distance scale applications. A warning example is the Local Group dwarf galaxy NGC 6822. Two groups obtained infrared photometry of classical Cepheids in this galaxy. For 11 Cepheids in common the differences are: $\Delta J = 0.126 \pm 0.022$ and $\Delta K_s = 0.061 \pm 0.014$ ([14]). The reason for these differences is not known, but they may be at least partly due to transformation problems. Differences of this size in a nearby galaxy obviously raise warning flags if we are attempting to obtain H_0 to 2.4 percent (± 0.05 mag; see, e.g., Lucas Macri's talk).

2.3 Parallaxes

2.3.1 Gaia

Gisella Clementini gave us a very clear talk on the work she and her colleagues have been doing on the Gaia parallaxes of classical Cepheids and RR Lyrae variables. The Gaia results are very spectacular and will become more so with later data releases. Gisella made some comparison with the pulsation parallaxes (Baade-Wesselink method) of RR Lyrae variables from Muraveva et al. ([28]; the derived parallaxes can be obtained from the data they publish). I also made this comparison. The results are really quite remarkable. For 19 RR Lyraes in common between Gaia and Muraveva, the unweighted mean difference is only 0.02 ± 0.03 mas. The scatter about the mean is only 0.14 mas, whilst the unweighted mean Gaia standard error is twice as much, 0.28 mas. This suggests that the Gaia workers are perhaps being overcautious in their assignments of standard errors. However, when applying the Gaia results to Galactic structure and other problems, well determined standard errors will often be required, if only for assessing any statistical bias. Indeed, Gisella mentioned that different methods of estimating bias gave different results. It will obviously be useful to consider these in detail when the full results are published, together with considerations such as given in Section 2.3.2 below.

2.3.2 Statistical bias

Statistical bias needs to be considered in many problems involving distance estimates whether using parallaxes or other methods. Methods of dealing with distance, or distance related, bias have a long history going back to Eddington ([9, 10]). However, the literature shows that there remain misunderstandings and uncertainties in this area.

An interesting and relevant example is in the papers by Benedict et al. ([2, 3]) which deal with the calibration of the absolute magnitudes of classical Cepheids and RR Lyrae variables using HST parallaxes. The authors of these papers (of which I was one) could not agree on the question of bias corrections and the method to be used in the analyses. The results for two methods were therefore given in the papers. One of these (denoted as A below) involves a Lutz-Kelker (L-K) type correction as recently restated by Benedict et al. ([4]), and this solution was used in a recent H_0 calibration by Reiss et al. ([33]). The other (denoted as B below) takes into account only a (very small) Malmquist-type bias correction.

The problem rests with the manner in which the stars were chosen for observation by the HST and also how the resulting parallaxes are used. In the present case the RR Lyraes and Cepheids were chosen, as the nearest suitable examples of these classes, on the basis of their photometric distance moduli (or relative moduli). In the case of the Cepheids it so happened that a selection according to *Hipparcos* parallaxes gave the same sample. However, these *Hipparcos* parallaxes have significant uncertainties while those of the relative photometric parallaxes are small. Thus it is best to consider the samples as photometrically selected[1]. It is clear that if the photometric parallaxes had no uncertainties (i.e., were the true parallaxes) the HST parallaxes would be distributed about them in an unbiased manner (first equation in the Lutz-Kelker paper, [23]; see, e.g., [11]), and this argument can be generalized. The application of a L-K type bias correction to the HST data is therefore not appropriate. However, samples chosen by photometric parallax will be biased with respect to a volume limited sample and thus, in principle, require a Malmquist-type correction, though in the present cases this is small.

Method A also combines the data in magnitude space which gives rise to biased weighting since this involves π/σ_π and overweights/underweights accidentally too large/ too small parallax measures. Method B avoids this by working in parallax space ([12, 15]). Fortunately in the cases being discussed, the resulting difference between the two analyses was small, the different relative weightings offsetting the difference due to a correction or otherwise for supposed L-K effect.

[1] It should be noted that even if we did consider a *Hipparcos* selected sample it would be the *Hipparcos* uncertainties that would be relevant in any correction for selection bias, not the HST ones. Such corrections would in the present case be large and very uncertain.

As a general comment, it is perhaps worth remarking that, though L-K bias and other statistical corrections may sometimes be appropriate, it is not always recognized that, if these corrections are substantial, their uncertainties can be large and highly asymmetric. Estimates of these uncertainties, in the appropriate situations, can be taken from Koen ([22]) if one takes his N to be the number of stars in the sample, each with a parallax standard error of s.

3 Galactic structure

This meeting has shown how large photometric surveys, extending from the optical to the infrared, are changing our views of our own Galaxy. This is especially true if optical, infrared, spectroscopic and radial velocity data can be combined. New surveys and Gaia data will lead, one can be sure, to further major advances.

Recent variability work has extended from the central nuclear disc to the outer fringes of the halo. The following are brief comments on some of these results, some of which require further extensive follow up work.

3.1 Nuclear stellar disc

In the centre of the Galactic bulge, i.e. in the nuclear stellar disc, Matsunaga et al. ([24, 25], and this meeting) have discovered a few classical Cepheids. The lack of short period Cepheids indicates that star formation rapidly increased there about 25 Myr ago and this activity continued until quite recently. These Cepheids were found using the 1.4m Japanese-South African Infrared Survey Facility (IRSF). We can expect the powerful VISTA telescope to tell us much more about this important region of the Galaxy which is hidden by dust at optical wavelengths.

3.2 The Galactic bulge

Recent years have seen a vast amount of work on the Galactic bulge. This has shown the bulge together with its bar (or bars) to be much more complex than was at one time thought. Some of the variable star work in this area has been covered in talks at this meeting (by, e.g., Soszyński, Minniti and Matsunaga). One importance of variable stars for work of this kind is the possibility of separating populations according to age. An example of this is the recent work of Catchpole et al. ([5]) which examined the distribution of bulge Miras as a function of period. This is an indication of their distribution according to age and shows that the longer period, intermediate age objects populate a bar.

A recent work of particular interest to this meeting is the VISTA survey for RR Lyrae variables towards the Galactic centre ([27], and Dante Minniti, this meeting). The VISTA field covers about 1.5 square degrees though much of this is very heavily obscured. It is suggested that the 14 RR Lyrae stars found are distinct from the general RR Lyrae bulge population and come from globular clusters destroyed in the very high density regions near the centre. This is an interesting possibility. However, even if these stars are at the distance of the centre, distinguishing clearly between the central condensation of the general RR Lyrae population and a separate "globular-derived" group will be difficult. The general bulge RR Lyrae population is expected to be high near the centre since even well out in the bulge the optical OGLE survey finds an average of 171 RR Lyraes per sq. deg. ([34]). The question of whether these RR Lyraes are close to the centre or possibly somewhat nearer is discussed briefly at the end of Section 3.3 below.

3.3 Cepheids and the extreme inner disc

A search for classical Cepheids in the general direction of the Galactic centre has been made by two groups, Matsunaga et al. ([26], and this meeting) using the IRSF, and Dékány et al. ([7, 8], and this meeting) using VISTA. There is reasonable agreement with the Cepheids discovered and, within the limits set by transformations of photometric systems, their photometry. However, the interpretation the two groups give to the data differs, and this difference is of considerable importance for our understanding of the structure and evolution of the inner Galactic disc.

These Cepheids are all highly reddened and much of the difference between the two groups depends on the adopted reddening law ([29] = N6, in the case of Matsunaga et al. [26], and [30] = N9, in the case of Dékány et al. [8]). I can only briefly sketch out my current understanding of this matter which is from the point of view of a member of the Matsunaga group (which includes Nishiyama).

N6 is based on red clump stars, observed with the IRSF, close to the Galactic centre. N9 is based on red clump stars together with red giants from 2MASS transformed by Dékány et al. into the VISTA system. Although the difference between the two reddening laws is not highly significant, it is important in these highly reddened regions. N6 was preferred by Matsunaga et al. partly on the general grounds that the law was derived entirely from IRSF observations as were the Cepheid observations. There are also empirical reasons for adopting N6. Matsunaga et al. ([24, 25]) give good reasons for placing the "nuclear disc Cepheids" (see Sect. 3.1 above) close to the centre which adopting the N6 reddening law does. N9 would place these stars nearer to us. Moreover, there is a 3 kpc region essential devoid of Cepheids around these central stars. This is present independent of the adopted reddening law. It would be off-centred with N9. Such a distribution would be very hard to understand. This "Cepheid-free" region could be due to the sweeping effect of the Galactic bar or perhaps the 3 kpc expanding arm.

In a paper at this meeting, presented on his behalf by G. Hajdu, I. Dékány argued that a $J-H/H-K$ plot favoured N9. This matter will be considered in detail elsewhere. However, it should be noted that neither set of workers use the J band data for distances. Also, it is not clear whether, for instance, the colour-colour relation derived is consistent with Cepheid intrinsic colours or whether the result is statistically significant. It was also argued that the RR Lyraes in the direction of the centre ([27], Sect. 3.2 above) would be closer than the centre if N6 is used. However, these stars would be even nearer to us if N9 had been used. So the argument is not relevant in distinguishing between N6 and N9. Minitti et al. ([27]) place their RR Lyraes near the centre using an estimate of the reddening law by Alonso-García et al. ([1]). The Alonso-García result is an order of magnitude less certain than N6 and the difference from N6 is not highly significant. In fact the large number of RR Lyraes that are expected in the direction of the centre together with the fact that the VISTA survey in this direction is J magnitude limited indicates that the VISTA RR Lyrae stars ([27]) could well be in the foreground. More work on this point would clearly be useful.

3.4 Halo and outer disc

Large scale surveys have played a major role in revealing the structure of the outer regions of our Galaxy and its environment. This work continues, for instance at this meeting Judith Cohen presented velocity data on halo RR Lyraes beyond 50 kpc and the extensive work on RR Lyraes in stellar streams in and beyond the halo is well known. Classical Cepheids in the flared outer disc were mentioned in Section 2.1.

4 Variable stars and mass loss

4.1 AGB variables

Mass loss from stars is important for stellar evolution, the contents of the interstellar medium, Galactic chemical evolution and dust in the Universe. Much of this mass loss comes from variable stars near the top of the asymptotic giant branch (AGB), and this has generally been attributed to the initial ejection of matter by stellar pulsations, followed by the formation of grains and their further ejection by radiation pressure. Yet, perhaps rather strangely, we have heard rather little about mass loss at this meeting. Possibly this is partly due to the embarrassment that adopting this hypothesis has led to little or no success in obtaining a quantitative explanation of the phenomenon. It has been clear for many years (e.g., Whitelock et al., [35]) that for large amplitude AGB variables (Miras) there is a good correlation between period, pulsational amplitude and dust shell mass. More recent studies (e.g., Goldman et al., [18]) get similar results. But it is perhaps time to ask whether these empirical relations are telling us any more than that AGB pulsators can get very large and very cool thus facilitating mass loss and dust formation. Do the correlations tell us anything about the physical causes of mass loss from large, cool, stars? Evidence has been around for some while that though outward velocities associated with pulsations may aid mass loss, other major factors are at work. For instance, C-rich Mira variables in our own and nearby galaxies (e.g., Whitelock et al., [36, 37]) are found to go through protracted "obscuration events" due to dust (soot) formation. In cases where suitable studies have been made (e.g., [13]), the events are found to be due to dust clouds ejected in the line of sight and are not spherically symmetrical ejections.

These ejected clouds are similar to the dust puff events in R Coronae Borealis (RCB) stars and like them have been attributed to ejection from large turbulent convection cells. Support for this conclusion has come, in more recent times, from the very beautiful work by a number of groups who have carried out high resolution work or interferometry on AGB variables. For instance, the work of Ohnaka et al. ([31]) shows the complex structures of gas and dust round the oxygen-rich Mira W Hya. Similar results have been obtained for other mass-losing AGB variables (e.g., [20, 38]). 3-D radiation-hydrodynamic techniques ([17, 21]) are beginning to model these outer structures and asymmetries. Work along these lines may eventually enable AGB mass loss to be understood quantitatively. Nevertheless, there are formidable problems to solve. For instance, what is presumably required is not just some mean understanding of turbulent convection but extreme values. This is analogous to the problem of convective overshooting in stellar interiors. The frequency and size of ejected dust clouds is not understood. Declines can last for many years – e.g., in the case of a recent decline of R For, 18 years ([13]). More observations would be helpful to get proper statistics, and this seems a programme for small automated survey telescopes covering large areas to only moderate depth. These lower cost projects have perhaps the best chance of lasting for the very many years required.

4.2 R Coronae Borealis type variables

One mass loss topic that was discussed at this meeting was that of the RCB variables. A good survey of the RCB problem in general was given by Geoff Clayton. This included a summary of the evidence that pulsation is the cause of mass ejection, which is known to be in puffs or clouds. Considerations such as those just mentioned in the case of AGB stars can be raised regarding pulsation as a physical mechanism for RCB mass loss. Indeed, the evidence of any link of pulsation to mass loss is not, in my view, strong. This is certainly the case for the type star R CrB. Roger Griffin has been monitoring this star for many years with a *Coravel* type radial velocity machine and other instruments. This work together with early work by Herbig and covering more than 50 years (with gaps) is being prepared for publication. One point, relevant to the cause of mass loss, comes from Griffin's observations at the time of the great decline which began in 2007 and still continues. During a period of about 100 days prior to the decline, (part of) the stellar atmosphere was highly disturbed with high velocity outflows ($\sim$ 40 km s^{-1}). It is believed that these outflows eject mass to sufficient distance for dust to form. Dust formation and declines, though spectacular, are simply secondary phenomena of physical processes (probably connected to turbulent convection) on the stellar surface. As with the AGB stars, the frequency of occurrence and lengths of RCB declines is not understood.

5 Summary

Current surveys of pulsating variables are already revealing much new about our own and nearby galaxies, and much more can be expected from forthcoming surveys and their follow-up. Asteroseismology will soon be doing the same, as well as telling us about stellar structure. Surveys of classical pulsators will lead to a fuller physical understanding of them and provide an increasingly firmer base for their use in cosmological and other studies. The reasons for adopting a Galactic disc model with a deficit of classical Cepheids in the inner regions are set out. To make full use of present and future large scale variability surveys attention needs to be given to the problems of photometric standardization. This paper attempts to clarify and settle a problem related to the analysis of HST parallaxes of classical Cepheids and RR Lyrae stars.

Large scale surveys will lead to a better understanding of non-periodic phenomena in, for instance, AGB variables. Understanding these phenomena and their relation to mass loss is of importance both for stellar and galactic evolution.

Acknowledgments: I am very grateful to my colleagues, Noriyuki Matsunaga, Patricia Whitelock and John Menzies, for many helpful discussions and comments. The work was supported by a research grant to MWF from the South African National Research Council (NRF).

References

[1] Alonso-García, J., Dékány, I., Catelan, M., Ramos, R.C., Gran, F., Amigo, P., Leyton, P., & Minniti, D., AJ, **149**, 99 (2015)

[2] Benedict, G. F., MacArthur, B. E., Feast, M. W., Barnes, T. G., Harrison, T. E., Paterson, R. J., Menzies, J. W., Bean, J. L., et al. AJ, **133**, 1810 (2007)

[3] Benedict, G. F., MacArthur, B. E., Feast, M. W., Barnes, T. G., Harrison, T. E., Bean, J. L., Menzies, J. W., Chaboyer, B., et al., AJ, **142**, 187 (2011)

[4] Benedict, G. F., McArthur, B. E., Nelan, E. P., & Harrison, T. E., PASP, **129**, 012001 (2017)

[5] Catchpole, R. M., Whitelock, P. A., Feast, M. W., Hughes, S. M. G., Irwin, I., & Alard, C., MNRAS, **455**, 2216 (2016)

[6] Chadid, M., Sneden, C., & Preston, G. W., ApJ, **835**, 187 (2017)

[7] Dékány, I., Minniti, D., Hajdu, G., Alonso-García, J., Hempel, M., Palma, T., Catelan, M., Gieren, W., & Majaess, D., ApJL, **799**, L1 (2015)

[8] Dékány, I., Minniti, D., Majaess, D., Zoccali, M., Hajdu, G., Alonso-García, J., Catelan, M., Gieren, W., & Borissova, J., ApJL, **812**, L29 (2015)

[9] Eddington, A. S., MNRAS, **73**, 359 (1913)

[10] Eddington, A. S., *Stellar Motions and the Structure of the Universe*, p. 172 (MacMillan, London, 1914)

[11] Feast, M. W., in *Globular Clusters*, ed. C. Martínez Roger, I. Pérez Fournón, & F. Sánchez (Cambridge University Press), p. 251 (1999)

[12] Feast, M. W., MNRAS, **337**, 1035 (2002)
[13] Feast, M. W., Whitelock, P. A., & Marang, F., MNRAS, **346**, 878 (2003)
[14] Feast, M. W., Whitelock, P. A., Menzies, J. W., & Matsunaga, N., MNRAS, **421**, 2998 (2012)
[15] Feast, M. W., in *Planets, Stars and Stellar Systems*, vol. 5, ed. T. D. Oswalt & G. Gilmore (Springer, Dordrecht), p. 829 (2013)
[16] Feast, M. W., Menzies, J. W., Matsunaga, N., & Whitelock, P. A., Nature, **509**, 342 (2014)
[17] Freytag, B., & Höfer, S., A&A, 483, 571 (2008)
[18] Goldman, S. R., van Loon, J. Th., Zijlstra, A. A., Green, J. A., Wood, P. R., Nanni, A., Matsuura, M., Groenewegen, M. A. T., et al., MNRAS, **465**, 403 (2017)
[19] Hanson, R. B., MNRAS, **186**,875 (1979)
[20] Haubois, X., Wittkowski, M., Perrin, G., Kervella, P., Mérand, A., Thébaut, E., Ridgway, S. T., Ireland, M., et al., A&A, **582**, 71 (2015)
[21] Höfner, S., in *The 19th Cambridge Workshop on Cool Stars, Stellar Systems, and the Sun*, ed. G. A. Feiden, p. 1
[22] Koen, C., MNRAS, **256**,65 (1992)
[23] Lutz, T. E., & Kelker, D. H., PASP, **85**, 573 (1973)
[24] Matsunaga, N., Kawadu, T., Nishiyama, S., Nagayama, T., Kobayashi, N., Tamura, M., Bono, G., Feast, M. W., et al., Nature, **477**, 188 (2011)
[25] Matsunaga, N., Fukue, K., Yamamoto R., Kobayashi, N., Inno, L., Genovali, K., Bono, G., Baba, J., et al., ApJ, **799**, 46 (2015)
[26] Matsunaga, N., Feast, M. W., Bono, G., Kobayashi, N., Inno, L., Nagayama, T., Nishoyama, S., Matsuoka, Y., et al., MNRAS, **462**, 414, (2016)
[27] Minniti, D., Ramos, R. C., Zoccali, M., Rejkuba, M., Gonzalez, O. A., Valenti, E., & Gran, F., ApJL, **830**, L14 (2016)
[28] Muraveva, T., Palmer, M., Clementini, G., Luri, X., Cioni, M.-R. L., Moretti, M. I., Marconi, M., Ripepi, V., et al., ApJ, **807**, 127 (2015)
[29] Nishiyama, S., Nagata, T., Kusakabe, N., Matsunaga, N., Naoi, T., Kato, D., Nagashima, C., Sugitani, K., et al., ApJ, **638**, 839 (2006) (=N6)
[30] Nishiyama, S., Tamura, M., Hatano, H., Kato, D., Tanabé, T., Sugitani, K., & Nagata, T., ApJ, **696**, 1407 (2009) (=N9)
[31] Ohnaka, K., Wiegert., G, & Hofmann, K.-H., A&A, **589**, 91 (2016)
[32] Pietrzyński, G., Thompson, I.B., Gieren, W., Graczyk, D., Stepień, K., Bono, G., Moroni, P. G. P., Pilecki, B., et al., Nature, **484**, 76 (2012)
[33] Riess, A. G., Macri, L. M., Hoffmann, S. L., Scolnic, D., Casertano, S., Filippenko, A. V., Tucker, B. E., Reid, M. J., et al., ApJ, **826**, 56 (2016)
[34] Soszyński, I., Dziembowski, W. A., Udalski, A., Poleski, R., Szymański, M. K., Kubiak, M., Pietrzyński, G., Wyrzykowski, Ł., et al., Acta Ast., **61**, 1 (2011)
[35] Whitelock, P. A., Pottasch, S. R., & Feast, M. W., in *Late Stages of Stellar Evolution*, ed. S. Kwok & S. R. Pottasch (Reidel, Dordrecht), p. 269 (1987)
[36] Whitelock, P. A., Feast, M. W., Marang, F., & Groenewegen, M. A. T., MNRAS, **369**, 751 (2006)
[37] Whitelock, P. A., Menzies, J. W., Feast, M. W., Matsunaga, N., Tanabé, T., Ita, Y., MNRAS, **394**, 795 (2009)
[38] Wittkowski, M., Hofmann, K.-H., Höfner, S., Le Bouquin, J. B., Nowotny, W., Paladini, C., Young, J., Berger, J.-K., et al., A&A, **601**, A3 (2017)

AUTHOR INDEX

AUTHOR INDEX

AUTHOR INDEX